Imperial College London
2406696439

AF616379

Economic and Palaeoceanographic Significance of Contourite Deposits

Geological Society books refereeing procedures

The Society makes every effort to ensure that the scientific and production quality of its books matches that of its journals. Since 1997, all book proposals have been refereed by specialist reviewers as well as by the Society's Books Editorial Committee. If the referees identify weaknesses in the proposal, these must be addressed before the proposal is accepted.

Once the book is accepted, the Society Book Editors ensure that the volume editors follow strict guidelines on refereeing and quality control. We insist that individual papers can only be accepted after satisfactory review by two independent referees. The questions on the review forms are similar to those for *Journal of the Geological Society*. The referees' forms and comments must be available to the Society's Book Editors on request.

Although many of the books result from meetings, the editors are expected to commission papers that were not presented at the meeting to ensure that the book provides a balanced coverage of the subject. Being accepted for presentation at the meeting does not guarantee inclusion in the book.

More information about submitting a proposal and producing a book for the Society can be found on its website: www.geolsoc.org.uk.

It is recommended that reference to all or part of this book should be made in one of the following ways:

VIANA, A. R. & REBESCO, M. (eds) 2007. *Economic and Palaeoceanographic Significance of Contourite Deposits*. Geological Society, London, Special Publications, **276**.

VERDICCHIO, G., TRINCARDI, F. & ASIOLI, A. 2007. Mediterranean bottom-current deposits: an example from the Southwestern Adriatic Margin. *In*: VIANA, A. R. & REBESCO, M. (eds) *Economic and Palaeoceanographic Significance of Contourite Deposits*. Geological Society, London, Special Publications, **276**, 199–224.

GEOLOGICAL SOCIETY SPECIAL PUBLICATION NO. 276

Economic and Palaeoceanographic Significance of Contourite Deposits

EDITED BY

A. R. VIANA
Petrobras, Brazil

and

M. REBESCO
Istituto Nazionale di Oceanografia e di Geofisica Sperimentale, Italy

2007
Published by
The Geological Society
London

THE GEOLOGICAL SOCIETY

The Geological Society of London (GSL) was founded in 1807. It is the oldest national geological society in the world and the largest in Europe. It was incorporated under Royal Charter in 1825 and is Registered Charity 210161.

The Society is the UK national learned and professional society for geology with a worldwide Fellowship (FGS) of over 9000. The Society has the power to confer Chartered status on suitably qualified Fellows, and about 2000 of the Fellowship carry the title (CGeol). Chartered Geologists may also obtain the equivalent European title, European Geologist (EurGeol). One fifth of the Society's fellowship resides outside the UK. To find out more about the Society, log on to www.geolsoc.org.uk.

The Geological Society Publishing House (Bath, UK) produces the Society's international journals and books, and acts as European distributor for selected publications of the American Association of Petroleum Geologists (AAPG), the Indonesian Petroleum Association (IPA), the Geological Society of America (GSA), the Society for Sedimentary Geology (SEPM) and the Geologists' Association (GA). Joint marketing agreements ensure that GSL Fellows may purchase these societies' publications at a discount. The Society's online bookshop (accessible from www.geolsoc.org.uk) offers secure book purchasing with your credit or debit card.

To find out about joining the Society and benefiting from substantial discounts on publications of GSL and other societies worldwide, consult www.geolsoc.org.uk, or contact the Fellowship Department at: The Geological Society, Burlington House, Piccadilly, London W1J 0BG: Tel. +44 (0)20 7434 9944; Fax +44 (0)20 7439 8975; E-mail: enquiries@geolsoc.org.uk.

For information about the Society's meetings, consult *Events* on www.geolsoc.org.uk. To find out more about the Society's Corporate Affiliates Scheme, write to enquiries@geolsoc.org.uk.

Published by The Geological Society from:
The Geological Society Publishing House, Unit 7, Brassmill Enterprise Centre, Brassmill Lane, Bath BA1 3JN, UK

(*Orders*: Tel. +44 (0)1225 445046, Fax +44 (0)1225 442836)
Online bookshop: www.geolsoc.org.uk/bookshop

British Library Cataloguing in Publication Data
A catalogue record for this book is available from the British Library.

ISBN: 978-1-86239-226-7

Typeset by Techset Composition Ltd, Salisbury, UK
Printed by The Cromwell Press, Wiltshire, UK

Distributors

North America
For trade and institutional orders:
The Geological Society, c/o AIDC, 82 Winter Sport Lane, Williston, VT 05495, USA
Orders: Tel + 1 800-972-9892
Fax + 1 802-864-7626
E-mail gsl.orders@aidcvt.com
For individual and corporate orders:
AAPG Bookstore, PO Box 979, Tulsa,
OK 74101-0979, USA
Orders: Tel + 1 918-584-2555
Fax + 1 918-560-2652
E-mail bookstore@aapg.org
Website http://bookstore.aapg.org

India
Affiliated East-West Press Private Ltd, Marketing Division, G-1/16 Ansari Road, Darya Ganj,
New Delhi 110 002, India
Orders: Tel. +91 11 2327-9113/2326-4180
Fax +91 11 2326-0538
E-mail affiliat@vsnl.com

Contents

Preface

The sunny summer of 2004 in Florence, Italy, witnessed the meeting of more than 50 people from academia and industry to discuss their ideas about the fascinating but still controversial world of contourites and bottom-current dominated sedimentary environments. Speeches and posters from different stratigraphic, bathymetric and geographical contexts were passionately presented. A panel discussion, carried out after the oral presentations, suggested some future trends in contourites research. Among the most important items suggested were the economic importance of contourite deposits and their stratigraphic–palaeoceanographic relationships.

The growing interest provoked by such themes, previously expressed by the editorial success of the Geological Society Memoir 22 (*Deep-Water Contourites: Modern Drifts and Ancient Series, Seismic and Sedimentary Characteristics*) edited by Stow *et al.* in 2002, was confirmed by the great number of participants in the General Symposium on Contourites held in Florence, 2004, during the 32nd International Geological Congress.

The study of the contourite deposits requires the application of many different theoretical, experimental and empirical resources provided by geophysics, sedimentology, geochemistry, experimental petrology, structural geology, scale modelling and field geology. Following this philosophy, we have edited this volume with the aim of providing an integrated approach for the study of the relevant contourite-related themes highlighted in the Florence meeting: their economic interest and palaeoceanographic implications. Our additional intention in editing this volume is to widen the understanding of the physical mechanisms involved in the sedimentation from contour currents, to better predict and evaluate their role in deposition.

This volume is composed of 16 papers broadly subdivided into two major categories (economic interest and stratigraphic–palaeoceanographic significance), with some of the papers lying between these two research areas. The last paper is dedicated to numerical simulations of contour currents and their impact on sedimentation.

The first five papers have strong economic appeal. **Viana *et al.*** discuss the main aspects of economic interest of contourite deposits, most of them related to the elements of petroleum systems. Modern and ancient cases are retrieved from international literature and presented under this new approach. Some new examples from the SE Brazil margin are presented for the first time, including 3D seismic, core and borehole data.

The contourite sand-rich channels from the North Atlantic described by **Akhmetzhanov *et al.*** constitute very important and well-documented examples of sediment accumulations with large and unexploited potential as reservoir rocks. Similarities to and distinctions from turbidite channels are also addressed.

Llave *et al.* provide us with a discussion on the Quaternary evolution of the contourite depositional system (CDS) in the Gulf of Cadiz based on morphological, structural and stratigraphic analyses using high-resolution seismic lines, borehole data and shallow coring data. Erosion-dominated episodes are contrasted with depositional ones as well as the distribution of coarse-grained versus fine-grained deposits, offering a very detailed temporal and spatial distribution of the various depositional elements that constitute a CDS.

Moraes *et al.* focus on an early Cenozoic case from the Campos Basin. They describe the presence of turbidite beds, reworked by bottom currents and interbedded with sandstones, and discuss their impact in the appraisal of a deep-water oilfield. The authors present a distinction between the classical turbidites, which constitute excellent reservoirs in the study case, and current-reworked sandstones that locally act as reservoir baffles or barriers.

Acting as a link between the first part of this volume, mostly dedicated to the papers that present relevant economic aspects, and the second part, in which discussions on the stratigraphic and palaeoceanographic aspects of contourite systems prevail, **Rebesco *et al.*** distinguish between contourites and turbidites based on swath bathymetry data recently acquired on Drift 7 off the Antarctica Peninsula. The authors discuss the coexistence of different sedimentary processes involving gravity flows and oceanic bottom currents expressed in the resultant sea-floor morphology and sediment accumulation characteristics.

The stratigraphic–palaeoceanographic papers are arranged in two sections. The first section consists of examples of Cenozoic to Quaternary contourites ordered geographically as a 'world ocean tour', beginning on the Pacific margin of the Antarctica Peninsula and continuing across the western and eastern Pacific, then to the south-western Atlantic and finally ending with the Mediterranean and NW Atlantic. The second section includes three fossil cases, ordered from the most recent (Late Cretaceous) to the oldest (Palaeozoic).

The first paper of the stratigraphic–palaeoceanographic part is by **Lucchi & Rebesco** and discusses the palaeoenvironmental and palaeoclimatic conditions for the deposition of glacial contourites along most of the Antarctic margin. Such deposits constitute atypically non-bioturbated, ice-rafted debris rich layers and the authors propose to use them as proxies to define temporal and spatial extension of the Antarctic sea-ice. Such facies coexist with other sediment types and are predominantly derived from sediment-rich gravity flows.

Carter discusses the Canterbury Drifts, SW Pacific Ocean, which were deposited since the Oligocene. The author bases his study on the analysis of data derived from outcrops, marine seismic survey, coring and imaging, and borehole data. The emphasis is on the role of intermediate-depth currents in continental shelf–slope accretion. This approach builds on the hypothesis that slope currents interact with terrigenous derived sediments and the resultant deposit is a slope wedge formed by the welding of slope plastered drifts and the shelf–slope prograding clinoforms. Such a mechanism is probably present worldwide and its importance could be underemphasized as a sedimentary process constructing continental margins.

Robinson *et al.* discuss the impact of glacial–interglacial modifications in the behaviour of the California Counter Current along the California Borderland on the sedimentary record. The authors report a decrease in the intensity of bottom reworking from late Marine Isotope Stage 5 (MIS 5) to the Holocene, expressed in variations in the grain size of associated deposits and in the degree of bioturbation. These observations indicate that sediment transport by bottom currents is not restricted to the western boundary currents but may also be the product of the action of bottom currents on eastern boundary slopes.

Duarte & Viana present a new Cenozoic drift system occurring in the SW Atlantic Ocean. The Santos Drift System is studied using industrial 3D and 2D seismic and borehole data. The authors identify two major drift complexes, a slope plastered drift and a separated drift, and establish their stratigraphic organization related to glacio-eustatic curves and to major climatic–palaeoceanographic events. The study indicates that periods of relative high sea level correspond to phases of increasing drift thickening whereas during predominant lowstands slope drift sedimentation is reduced.

Verdicchio *et al.* deal with the bottom-current deposits along the southwestern Adriatic Margin, Mediterranean Sea. Using high-resolution seafloor imaging and sub-bottom profiling coupled with piston core analysis, the authors study the dramatic palaeogeographical and palaeoceanographic rearrangements that occurred in the Adriatic during the Late Quaternary sea-level oscillations and the depositional response to those modifications.

From the Northern Atlantic, the paper by **Van Rooij *et al.*** discusses the close association of small, mounded contourite drifts and cold-water coral banks, observed along the Porcupine Seabight. The authors propose that the different characteristics of the coral banks development are directly related to climate-driven modifications of the slope current regime and its interaction with tides and slope physiography.

The last paper of this part of the volume presents the Eirik Drift as a long-term barometer of the North Atlantic deepwater flux south of the Greenland margin. **Hunter *et al.***, using seismic stratigraphic techniques, report that the Eirik Drift contains an expanded sedimentary record of bottom and intermediate current intensity variation ranging from the Early Eocene to the Holocene. The authors note that variations in current strength on a decadal to millennial time scale can be related to changes in thermohaline circulation and climate, with a number of internal discontinuities reflecting a variety of palaeoceanographic events.

The next three papers deal with ancient contourite systems ranging from Mesozoic to Palaeozoic ages. **Esmerode *et al.*** propose that the flooding of the NW European craton during the Late Cretaceous trangression created relatively deep epeiric seas into which the oceanographic conditions that prevail on continental margins extended. Such starved basins, instead of presenting flat-lying pelagic successions, are marked by the sedimentary record of the action of strong bottom currents that developed a multitude of imprints on the chalk deposits, such as sediment waves, drifts, moats and extensive unconformities. The authors identified two major episodes of drift deposition in the Danish Basin, one in the Santonian to Campanian and one in the Maastrichtian, developed by the northwestward flow of contour currents.

Georgiev & Botoucharov use borehole data, cores and industry seismic data to present the possibility that a middle Jurassic interbedding of shales and siltstones occurring in the South Moesian platform (Bulgaria) constitutes the sedimentary record of bottom-current processes. The structurally controlled palaeophysiography would have strongly influenced the bottom circulation and hence sediment deposition.

Detailed outcrop studies coupled with palaeogeographical reconstruction lead **Hüneke** to propose that the Devonian calcareous bioclastic successions observed in Germany, Morocco, Austria and

Italy preserve facies characteristics corresponding to contourites. The author observes that the widespread current-induced reworking of calcareous sediments, phosphate formation and major erosion-related hiatuses are associated with major palaeocirculation events that would have occurred as a result of the acceleration of thermohaline currents accompanying the narrowing of the oceanic passageway between the approaching Laurussia and Gondwana continents during the middle and late Devonian.

This volume is completed by a paper by **Lima *et al.*** in which a hydrodynamic numerical model is proposed to study the behaviour of bottom currents flowing along a submarine canyon and adjacent open slope and shelf edge, and their interaction with sediment-rich turbidity currents flowing down-canyon. The model describes the importance of differing current-forcing mechanisms, and estimates their resultant sediment transport under the combined action of turbidity and bottom currents. The editors agree that this approach, as much as physical modelling, is useful to better quantify the impact of bottom currents among the diversity of sedimentary processes occurring in the deep ocean. This may lead to a wider understanding of the role of bottom currents in the geological record and reduce the gap between the different techniques used in earth and oceanic sciences.

No publication can achieve a good scientific standard without the tremendous dedication of the authors and coauthors of the contributions. To all of them we would like to express our deep acknowledgement. Also of huge importance was the role performed by the reviewers, who realized the difficulties of the authors, who may sometimes be too deeply involved with their own data. The reviewers deserve our greatest recognition; they are A. Akhmetzhanov; A. Bouma; A. Carmelenghi; M. Carminatti; S. Ceramicola; E. Cowan; B. de Mol; C. Escutia Dotti; D. Evans; J.-C. Faugères; E. Gonthier; J. Howe; B. Kuvaas; J.-S. Laberg; P. Magalhães; E. Mutti; W. Normark; D. Piper; M. Roveri; I. Soares; G. Stampfli; D. Stow; F. Surlyk; G. Uenzelmann; J. Veevers. Angharad Hills, Commissioning Editor of the Geological Society Publishing House, invited the conveners of the symposium in Florence to edit this Special Publication for the Geological Society of London. Thanks to her and to her continuous and friendly support, we have shared a heavy but pleasant task during this last year and a half.

This list of acknowledgements would not be complete if the editors left out Petrobras and OGS. These institutions provide a clear example of how the association between industry and science may grant the necessary conditions to achieve the supreme objective of fostering the advancement of knowledge. It is thanks to these institutions that we have achieved our major objective with this volume: to provide the readers with the knowledge acquired by the authors.

Adriano R. Viana
Michele Rebesco
April 2006

Petrobras are thanked for their generous contribution to colour printing costs.

The economic importance of contourites

A. R. VIANA, W. ALMEIDA JR, M. C. V. NUNES & E. M. BULHÕES

Petrobras, E&P-Exploration, Rio de Janeiro, RJ 20031-912, Brazil
(e-mail: aviana@petrobras.com)

Abstract: The importance of contour currents in shaping and building continental margins has long been accepted. Their economic implications and the stratigraphic framework in which they are developed remain largely unknown. Data retrieved from sidescan sonar images, seismic profiles and their attribute maps, as well as sea-floor coring, boreholes and the few known outcrops around the world, suggest that bottom currents can locally develop large deposits of relatively coarse-grained sediments. Accumulation of coarse-grained deposits under the influence of bottom currents requires sediment availability, a geologically persistent strong circulation regime and a favourable physiographic setting both for enhancing the currents and for hosting the sediments. The hydrocarbon exploration of oceanic depositional systems demands a better understanding of the role of bottom currents and their implications for petroleum systems such as reservoir and sealing rocks. Such understanding implies additional alternatives for the definition of exploration targets and prospect risk reduction. Correlating seismic anomalies from 3D mapping with core and well logging data reveals the depositional geometry and sedimentological characteristics of coarse-grained contourites. Fine-grained drifts can locally and regionally develop large and thick accumulations, which have an important seal potential for trapping hydrocarbon.

Recent decades have been marked by an increasing number of publications on the recognition and characterization of contourite systems. The great majority of these systems were recognized through high-resolution 2D seismic data locally complemented by sea-floor coring and drilling. Despite the controversy surrounding their lithological characterization, modern contourites have been identified in almost all oceans around the world, and also in confined seas and lakes (see references given by Faugères & Stow 1993; Faugères *et al.* 1993, 1999; Stow & Faugères 1998; Stow *et al.* 2002; and, for lacustrine drifts, Johnson *et al.* 1980; Johnson 1996; Ceramicola *et al.* 2001; Gilli *et al.* 2005).

The difficulties in determining reliable facies characteristics of the interaction between oceanic bottom circulation and sedimentary processes (Viana *et al.* 1998; Stow *et al.* 1998), besides the controversies related to the palaeoceanographic–stratigraphic control on the intensity of thermohaline currents, and to the poorly known economic applications of contourites, relegated these deposits to a secondary rank of interest in the spectrum of deep-sea sediments. The accumulation of oil reserves found in turbidite reservoirs has driven most of the attention of deep-water research, leading to the evolution of facies model concepts (Normark 1970, 1978; Bouma *et al.* 1985; Mutti & Normark 1987, 1991; Normark & Piper 1991; Mutti 1992; Mutti *et al.* 1999, 2003; Piper & Normark 2001), supported by industrial 3D seismic data, well logging, imaging and coring, and corroborated by outcrop field studies. However, contourites not associated with oil-bearing reservoirs have remained a subject of more academic studies, based on traditional shallow penetration high-resolution seismic 2D data and sea-floor piston coring.

The development of swath bathymetry and the widespread utilization of 3D seismic data, which are being gradually transferred to academic centres, provides a wider understanding of the geometry of contourite deposits, related bedforms and possible relationship with coarse-grained accumulations. Nevertheless, these systems are still perceived as a controversial deep-water deposit, where the tractive effects of gravity- or density-driven turbulent flows are the preferred option for explaining the construction of some sedimentary records, which could also be the product of contour currents. Stow *et al.* (1998) presented a discussion on fossil contourites stressing the point that several deep-water deposits described in the literature as contourites could be better considered as fine-grained turbidites. Those workers presented a series of criteria that would help to clarify contourites preserved in the rock record. These criteria have been widely applied, and some outcropping contourites described using such criteria were presented in the Geological Society Memoir on contourites edited by Stow *et al.* (2002).

In the late 1970s to early 1980s, influenced by the suggestion made by Mutti *et al.* (1980) that some rippled, top-truncated, medium- to fine-grained sand-rich oil-bearing horizons of Eocene age in

From: VIANA, A. R. & REBESCO, M. (eds) *Economic and Palaeoceanographic Significance of Contourite Deposits.* Geological Society, London, Special Publications, **276**, 1–23.
0305-8719/07/$15.00

the Campos Basin could constitute contourite sands, the interpretation of other similar deposits, mostly in Oligocene–Miocene age, deep-water sediments, became more frequent in the Campos Basin. Contourite reservoirs are also suggested as occurring in the North Sea (Enjorlas *et al.* 1986). Nevertheless, the recognition of such deposits remains controversial, mostly because of the lack of unequivocal criteria. After that period, concomitantly with the increasing number of fine-grained contourite systems identified in several deep-water settings (see Faugères *et al.* 1999, for references), the attention given to contourite deposits as potential reservoirs remained low.

During the 1990s, papers by Shanmugam *et al.* (1993), Stanley (1993) and Stow *et al.* (1998) tried to redeem the importance of sandy contourites as potential hydrocarbon reservoir rocks. The recognition of palaeoceanographic–climatic changes recorded by contourite rocks was also identified as a crucial application for contourite studies (Faugères & Stow 1993; Robinson & McCave 1994). The continuous expansion of petroleum exploration towards deeper water is responsible for introducing some important new aspects in the study of contourites: the analysis of slope stability areas subjected to strong contour currents and the development of thick fine-grained deposits acting as local to basin-wide sealing rocks.

This paper proposes a broad discussion on the economic implications resulting from the occurrence of contourites in the sedimentary record. We focus on two main aspects: (1) the possibilities of contourites playing an important economic role; (2) contourites acting as reservoirs and seals.

Major implications of contourite studies

The recognition of contourite deposits implies that bottom currents were active and responsible for the accumulation of sediments. The coarser the sediments involved in the deposit, the stronger should be the currents; the thicker the deposit, the longer the current activity. Thus, strong and long-lasting bottom currents ($>10^2$ years) are fundamental requirements for the development of such deposits, as much as the availability of coarse-grained sediments that could be subjected to the impact of such currents. Their occurrence is linked to particular conditions of ocean circulation, which are directly related to climate characteristics, and their interaction with local margin physiography and sediment availability (Faugères & Stow 1993; Hollister 1993; Stow *et al.* 1998, 2002; Viana *et al.* 1998).

Coarse- to fine-grained sediments, along with sea-floor erosion, are common features in contourite accumulations, independently of water depth. Deep-water coarse-grained deposits have long been considered as important hydrocarbon reservoirs, and are generally considered to be the result of various types of gravity flows. Occasionally, coarse-grained contourites can also present the required characteristics for a potential oil-bearing reservoir, such as extent, porosity and permeability. To simplify the contourite concept application, it is herein proposed that the qualifying terms 'shallow-', 'mid-' and 'deep-water' contourites formerly proposed by Viana *et al.* (1998) should be applied only when an unequivocal bathymetric setting is identified and when such a setting is fundamental in characterizing the deposit. In our understanding, contourites are sedimentary deposits resulting from the action of oceanic bottom currents or mostly influenced by them. The qualifying term 'shallow-water' contourite implies that they occur in water settings where the influence of currents derived from storms, tides and trade winds can be recognized in a minor proportion related to the dominant geostrophic surface slope boundary currents. Shallow-water contourites should be considered as part of a large spectrum of rocks that accumulated from the shelf edge to the upper slope setting under the predominant action of surface slope boundary currents. Nevertheless, situations where the downwelling of dense (saline or cold) flows transport and deposit sediments in cross-isobath trends, such as the Cadiz sandy contourite channels (Habgood *et al.* 2003; Akhmetzhanov *et al.* 2007) defy the idea of contour flowing currents and make much more difficult the proper characterization of contourites as being of shallow-, mid- or deep-water origin.

Fine-grained deposits are important constituents of a petroleum system as well. They can develop both sealing facies or permeability barriers and source rock accumulation. This last alternative is more commonly associated with gas hydrate deposits. Large potential gas hydrate accumulations and associated bottom-simulating reflector (BSR) zones are found in contourite deposits all along the Atlantic margin (Blake Outer Ridge (BOR), Dillon & Paull 1983; Kraemer *et al.* 2000; Rio Grande Cone, Silveira & Machado 2004). Sediment permeability is a significant factor in gas hydrate development, and variations in permeability influence zones of hydrate accumulation. Heterogeneities observed in gas hydrate distribution on the BOR (a large drift of clay-rich sediment in the western Atlantic Ocean that contains 30–40 Gt of carbon stored as methane hydrate and underlying free methane gas; Paull *et al.* 2000) are related to modifications in the thermohaline circulation pattern, which induced different styles of drift development, as accumulating terrigenous or biogenic components responded to climatic-induced variations in oceanic circulation (Kraemer *et al.* 2000).

Vigorous slope boundary bottom currents can locally erode the sea floor, causing sediment instability and triggering major mass-flow events, such as those involved in the escape of methane gas from gas hydrate accumulation (Holbrook *et al.* 2002). Identifying relative periods of activity and the factors that induce both local and global-scale current variation through the transport of heat and salt by thermohaline circulation is fundamental in understanding climate changes.

Global climate disequilibria induced by or through the perturbation of ocean circulation can also be evaluated through the study of contourites. Sedimentary current-controlled deposits are directly related to environmental oscillations. Sea-floor erosion can also be responsible for the opening of permeable pathways to gas-free zones in gas hydrate accumulations. Such a mechanism might be associated with sustained, ocean-wide release of methane (Holbrook *et al.* 2002), which could cause large-scale continental slope erosion and also induce major changes in thermohaline circulation, such as during the Palaeocene–Eocene thermal maximum and two intervals in the early Toarcian that were apparently characterized by massive methane release (Dickens *et al.* 1995, 1997; Hesselbo *et al.* 2000; Holbrook *et al.* 2002).

The direct relationship between ocean circulation and climate was recently reassessed by Clark *et al.* (2002) and Rahmstorf (2002). Those workers discussed large-scale ocean circulation as a combination of currents driven directly by winds, currents driven by fluxes of heat and freshwater across the sea surface and subsequent interior mixing of heat and salt (the thermohaline circulation), and tidal action. These driving mechanisms interact in nonlinear ways (as all currents change the distribution of heat and salt) so that no unique relationship exists. Nevertheless, the distinction is useful, particularly when changes in wind or in surface heat and freshwater fluxes are considered in their effects on circulation (Rahmstorf 2002). Clark *et al.* (2002) pointed out that a number of simulations with general circulation models of the coupled ocean–atmosphere system demonstrate the possibility of a reduced Atlantic thermohaline circulation in response to increased greenhouse gas emissions. Both the data and the models suggest that abrupt climate change during the last glaciation originated through changes in the Atlantic thermohaline circulation in response to small changes in the hydrological cycle. Atmospheric and oceanic responses to these changes were then transmitted globally through a number of feedbacks. The palaeoclimate data and the model results indicate that the stability of thermohaline circulation depends also on the mean climate state. Thus, sedimentary proxies such as primary sedimentary structures and bedforms can be used as direct indicators of current regime changes and consequently link them to palaeoclimate oscillations.

Consequently, three main applied aspects must be considered when studying contourites: (1) energy resources; (2) slope stability; (3) palaeoceanographic–palaeoclimatic evolution and mid- to long-term global climate forecasting.

In this paper we focus on the understanding of the direct economic importance of contourites, and on the energy resources aspect, which has major application in petroleum exploration. Recent papers by Llave *et al.* (2001), Reeder *et al.* (2002), Viana *et al.* (2002*b*), Carter *et al.* (2004) and Lu & Fulthorpe (2004), and others presented in this volume (Carter 2007; Duarte & Viana 2007) have dealt with palaeoceanographic aspects of contourites. Despite the occurrence of gas hydrates in some contourite systems around the world (e.g. Bermuda, Southern Brazil) and their direct association in the accumulation of organic-rich sediments and their erosion and release of methane, the incipient exploration of these energy resources leads us to concentrate on wells drilled in hydrocarbon-rich accumulations and their potential analogues.

Petroleum exploration

The presence and the intensity of contour currents may have an impact on several aspects of petroleum systems. Reservoir, sealing and source rock deposition can be directly or indirectly influenced by ocean current activity, allowing or preventing the accumulation of sediments of different types and textures. Also, the modification of sea-floor topography by erosional and depositional processes controlled by bottom currents may be responsible for a readjustment of accommodation space, resulting in the creation of local sub-basins, sediment ponding, or passageways for sediment transfer. Large-scale margin configuration, related to tectonic or depositional features, can cause shelf–slope projections or re-entrants. Such configurations affect the intensity of bottom currents (Fig. 1) with a direct response in the depositional style (Viana & Faugères 1998).

Contourite reservoirs

To be considered as a potential hydrocarbon reservoir, contourite deposits must present some particular characteristics. Coarse-grained sediments, preserving good petrophysical characteristics, such as porosity, permeability, and lateral and vertical

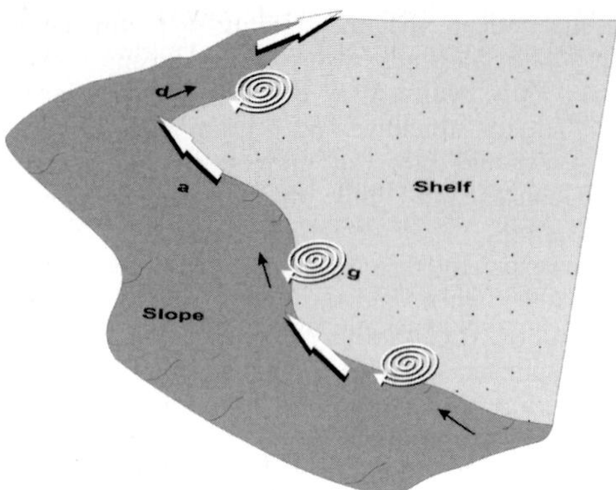

Fig. 1. Schematic representation of current behaviour with respect to margin configuration. Zones of current acceleration (a) are related to seaward margin projections and topographic obstacles that confine the flow; zones of deceleration (d) corresponding to margin re-entrants and flow expansion can induce formation of gyres or eddies (g). Size of arrows indicates current intensity.

transmissibility of fluids, must form part of the deposit. Some reliable criteria for recognizing this, both in seismic profiles and lithology, need to be identified to predict their occurrence.

The main factors controlling geometry and facies characteristics of contourite systems are: (1) intensity and duration of bottom-current regime; (2) the grain-size population available to the current action; (3) sea-floor physiography; (4) margin configuration. These factors regulate where, when and how contourites are deposited into a sedimentary basin.

The persistence and efficiency of a hydrodynamic regime capable of eroding, transporting and redepositing a large volume of sand is fundamental. Strong, semi-permanent bottom currents flowing at $>30 \pm 10$ cm s^{-1} are required for developing sand waves, such as the abyssal dunes of foraminiferal sands on the Carnegie Ridge (Lonsdale & Malfait 1974), and the siliciclastic sand dunes of the Campos Basin upper slope (Viana 1998), the lower Mississippi Fan (Kenyon *et al.* 2002), the Gulf of Cadiz (Habgood *et al.* 2003) and the Faeroe–Shetland Channel (Wynn *et al.* 2002; Masson *et al.* 2004; Akhmetzhanov *et al.* 2007). Two main situations are considered as favourable to the supply of coarse-grained sediments and the action of contour current. The first is a pre-existing sand-rich accumulation whose content is partially transferred to the areas subjected to current action by mechanisms such as shelf currents (currents induced by storms, tides, trade winds, eddies) inducing sand overspill (Viana & Faugères 1998); sediment density flows related to catastrophic floods in a low-gradient, ramp-like margin (hyperpycnal flows in the sense of Mulder *et al.* 2002); and the downwelling of dense saline waters developing sand-rich channels such as in the Gulf of Cadiz (Habgood *et al.* 2003; Mulder *et al.* 2003). The second is a zone subject to isostatic movements (elastic rebound, thermal deformation or halokinesis, intra-plate stresses, etc.) where pre-deposited sand layers are exposed at the sea-floor to sweeping bottom currents (Faugères *et al.* 1999; Davies *et al.* 2004). In both cases, there is no restriction to the water depth.

The largest sand storages in the oceanic realm are on the continental shelves and, to a lesser extent, deep-water sand deposits formerly transferred to the deep sea via any kind of gravity flow, commonly interpreted as turbidites.

Continental margin physiography is extremely important in accelerating or decelerating contour currents. Prominent seaward projections, or regions of relief disturbed by tectonic or adiastrophic structures act as current restrictions enhancing the near-bottom current velocity, whereas re-entrants and canyons reduce current velocity and induce sediment settling (Fig. 1).

The presence of coarse-grained contourites is independent of water depth, although the great majority of sandy contourites have been described from mid-water depth zones (300–2000 m, Viana *et al.* 1998). Studies from different geographical and stratigraphic settings (Mutti *et al.* 1980; Barros *et al.* 1982; Belonin *et al.* 1983; Nelson *et al.* 1993; Shanmugam *et al.* 1993; Famakinwa *et al.* 1997; Castro *et al.* 1998; Parize *et al.* 1998; Viana 1998; Viana & Faugères 1998; Viana *et al.* 1998; Garcia-Moronero & Olmo 2001; Kenyon *et al.* 2002; Cakebread-Brown *et al.* 2003; Isern *et al.* 2004; Rodriguez & Anderson 2004; Akhmetzhanov *et al.* 2007), complemented by a large set of high-resolution 2D and 3D seismic, core, and well log data from the Equatorial and Southwestern Brazilian Atlantic margins, indicate that the occurrence of potentially economic sandy contourites is mostly dependent on the vicinity of a sand-rich area prone to be swept by contour currents.

Processes and main depositional elements. Sands deposited in an outer shelf setting are frequently subjected to hydrographic regimes involving strong storm fronts, tidal currents, and onshelf penetration of superficial slope boundary currents, in the form of meanders and eddies. These can induce an offshelf transport of sand by the development of migrating sand waves. These sands arrive on the upper slope and are there entrained into the slope circulation system. They are redistributed parallel to the isobaths, forming elongated sand-rich deposits (Viana & Faugères 1998; Viana *et al.*

2002*b*). Systems with similar geometry and bathymetric context have been recognized in the Aptian–Albian series of the Vocontian Basin, SE France, by Parize *et al.* (1998; see Viana 1998) and in the modern western Antarctica outer shelf–upper slope by Rodriguez & Anderson (2004), and are now being recognized in the Miocene sequences of the Campos Basin (see below). Several other systems described in the scientific literature with somewhat different interpretation could be reinterpreted as shallow-water contourites (e.g. the upper slope setting of the Brushy Canyon as presented by Mutti 1992; the western Grand Banks of Newfoundland studied by Dalrymple *et al.* 1992; MacNaughton *et al.* 2000).

The upper slope sand deposits of the Campos Basin described by Viana & Faugères (1998), which are indeed contourites deposited in a relatively shallow-water context, involve sediments deposited by both the surface slope boundary current and its underflowing countercurrent. The former are frequently associated with large bedforms comprising coarse-grained sediments, varying from conglomerates to medium- to fine-grained sand, with rare to frequent primary sedimentary structures indicating strong sea-floor currents. The sands deposited in a deeper setting are predominantly fine-grained, with very fine-grained sands and silts deposited after the entrainment of the sediments into the countercurrent by a downslope component of the surface slope boundary current. Rarely, these sands preserve primary sedimentary structures and are moderately to highly bioturbated.

Contourites deposited in relatively shallow water (upper slope setting, controlled by the action of a surface slope boundary current regime) present similar characteristics and are useful as distinguishable architectural elements. The most frequent elements comprise: (1) an adjacent sand-rich shelf with large unidirectional sand waves; (2) an erosional terrace on the top of the slope; (3) overspilled sand from the shelf, induced by slopeward migration of bedforms (grain-by-grain supply) and/or by short-living or short-distance gravity flows (shelf edge instability or distal reach of flood-related flows from neighbouring rivers); (4) mid- to high-amplitude upper slope sand waves (usually, but not always, trending in an opposite direction to the shelf sand waves trend); (5) canyon heads breaching the upper slope terrace and laterally fed by the upper slope sand waves; (6) preceding upper slope fine-grained sedimentation. These elements can be observed from sidescan sonar records and high-resolution seismic data, and sedimentologically calibrated in piston cores (Viana *et al.* 2002).

Slope indentation by submarine canyons creates an alternative path for the migrating sands on the upper slope. These can feed canyon heads with the accumulation of relative unstable sand wedges (e.g. the migrating sand dunes in the SE African continental margin under the action of the Agulhas Current penetrating onto the shelf described by Flemming 1978, 1980; Ramsay 1994; and on the upper slope of the Campos Basin, SE Brazil, described by Viana & Faugères 1998; and Viana *et al.* 2002*a*, *b*), which are easily transferred downslope along the canyon axis. Transferring mechanisms include the progressive increase of the lee-side gradient of the drift-related prograding wedges (Viana 1998), the continuous action of internal waves and tides (Cacchione & Drake 1986; Bogucki *et al.* 1997; Cacchione *et al.* 2002; Lima *et al.* 2007), sediment resuspension by the action of storm currents (Fukushima *et al.* 1985; Baltzer *et al.* 1994), low-amplitude seismic activity, and the connection of the canyon head to a fluvial system during relative sea-level falls.

In deeper waters, thermohaline currents contouring the isobaths or downwelling slopes interact with pre-existing coarse-grained deposits in several ways, such as sweeping the fine-grained population, partially or totally eroding previous deposits, modifying the original depositional geometry, or transferring sediments to a new depocentre, where coarse-grained contourites can be accumulated and preserved. The NE Atlantic is rich in examples, such as those of the Faeroe–Shetland Channel, the Faeroe Bank and the northern Rockall Trough area, where barchan dunes, furrows, contourite sand sheets and channels were identified and analysed by Kenyon (1986), Cochonat *et al.* (1989), Damuth & Olson (2001), Masson (2001), Masson *et al.* (2002, 2004), Wynn *et al.* (2002) and Akhmetzhanov *et al.* (2007). They reported the presence of such deposits as the result of the action of the southward flow of cold deep water from the Norwegian Sea funnelled through narrow topographic passages that extend from Greenland to Scotland, and the incorporation into the flow of coarse-grained sediments derived from gravity processes transferring sediments from shallower settings. In the eastern Gulf of Mexico sand dunes develop after the reworking of the sand deposits of the Mississipi deep-sea fan by bottom currents (Kenyon *et al.* 2002). In the western Gulf of Mexico, Niedoroda *et al.* (2003) and Bryant & Slowey (2004) observed vigorous bottom flow of the westward branch of the Loop Current reaching speeds of the order of 100 cm s^{-1}. Such strong velocities probably result from constriction of the Loop Current against the Sigsbee Escarpment by the Coriolis effect, resulting in a giant field of mega-furrows, and the sweeping and erosion of flanks and summits of salt knolls as illustrated in Figure 2, modified from Bryant & Slowey (2004). The incorporation of dissolved

Fig. 2. A 3D rendered surface showing the action of the deep westward branch of the Loop Current at the foot of the Sigsbee Escarpment, northwestern Gulf of Mexico. The presence of a large field of mega-furrows should be noted. The depth and alignment of the furrows indicate the intensity of the currents. The currents pass over and erode the summit of the Green Knoll (GN, *c.* 15.5 km of its NW–SE-trending long axis a–a′). The incorporation of salt from the knoll affects the density and the current velocity (1). Current intensity is also augmented at the foot of the topographic obstacles (2) such as knolls and escarpments. Equally, lateral restrictions of flows can concentrate currents towards the converging axis (3), where currents accelerate. Figure modified from Bryant & Slowey (2004).

salt locally increases the flow density, and this effect is accompanied by an increase in the current's erosion and transport capacity. Density-enhanced bottom currents are expected to occur in other salt-deformed basins where evaporitic structures crop out at the sea floor, such as those occurring on both sides of the South Atlantic, or in areas with a high exchange of heat or salt, as exemplified by the Mediterranean Outflow Water penetrating into the Gulf of Cadiz (Mulder *et al.* 2002; Habgood *et al.* 2003; Llave *et al.* 2007).

In all studied cases, the presence of coarse-grained contourites in deep water is commonly associated with a set of medium- to large-scale bedforms. An order of increasing current velocity as proposed by Masson *et al.* (2004) includes fine-grained sediment waves and mounded contourite deposits, contourite sand sheets, coarse-grained sediment waves, barchan-like dunes, sand ribbons, channels, comet or obstacle marks, furrows and erosional scours.

Deep-sea unconformities. Diachronous, regional unconformities found at the base of contourite systems (the basal erosional surface of Faugères *et al.* 1999) are overlaid by onlapping upslope-migrating layers. These erosional features can be easily misinterpreted as sequence boundary unconformities, in the sense of Vail *et al.* (1977) and Posamentier & Vail (1988). However, their regional extent and abrasive aspect, with along-slope trending rather than downslope scouring characteristics, suggest a contour current-related erosion. The South Atlantic Brazilian margin is prone to such unconformities, such as the examples described by Viana (2001), Gomes & Viana (2002), Lima

(2003) and Duarte & Viana (2007). The physical coincidence between some of these unconformities and sequence boundaries is also probable, and has been demonstrated for the Santos Drift example (Duarte & Viana 2007), where deep-water drift boundaries were correlated to regional sequence boundaries tied with exploratory wells. The upslope-migrating package, a frequent feature on separated or plastered drifts (Faugères *et al.* 1999), is occasionally misinterpreted as lowstand prograding wedges in the sense of their sequence stratigraphic position. These two different interpretations lead to differing interpretations of sedimentary facies and either penalize or favour any exploratory prospects. The stratigraphic–palaeoceanographic reconstruction of drift evolution along the SE Brazil margin, from shallow to deep water, based on isotopic and biostratigraphic data, some corroborated by ^{14}C dating (Viana 1998; Viana & Faugères 1998), permits us to propose the profile presented in Figure 3, where we identify the period from the latest lowstand to the early transgressive phase (20–12 ka BP) as that of greater bottom-current activity, inducing effective sea-floor erosion and accumulation of coarse-grained deposits. In such a scheme, sands are preferentially accumulated in the upper slope setting and at the foot of topographic obstacles, where pre-existing sand is available for resedimentation under the action of currents. This scenario, conceived for the Brazilian margin, must be checked for other margins, where timing of sediment delivery to the slope and contour current acceleration may diverge.

The impact of sea-floor topography. The continental slopes and rises often present complex sea-floor topography such as salt walls, fault planes, knolls, mud or salt diapirs, submarine volcanoes and escarpments, among other features. The size of the relief and its trend in relation to the current flow, associated with the Coriolis effect, play an important role in the interaction of accelerated currents with unconfined, exposed sand supply. In a strong bottom-current regime coarse-grained material present on the sea floor may be incorporated into flows as bed-load supply, developing different bedforms and depositional geometries as a function of the intensity and duration of the flow regime, and accommodation space.

The intensity, width and position of the current core change with time. Flow regime modifications, responding to high- or low-frequency climatic oscillations or to margin physiographic modifications caused by local tectonic or erosional activity, affect the width and locus of the current core. Dynamic changes in the sea-floor topography related to such processes may also constrain (and accelerate) or relax (and decelerate) the flow, influencing the related depositional style.

Along-slope current re-accelerations are geographically frequent, and often related to physiographic changes such as margin projections (Figs 1 and 3), topographic restrictions (straits, passages, sills), and changes in gradient (which, along with their coincidence with water masses boundaries, were advocated by Habgood *et al.* (2003) to explain the abrupt termination of the sand deposits in the Gil Eanes contourite channel in the Gulf of Cadiz). Margin projections may induce the formation of meanders and gyres, which locally play an important role as sedimentary agents by sweeping the sea bottom, and resuspending and incorporating sediments along their path ('the sea-floor polishing effect', Viana *et al.* 1998; Viana & Stow 2000). The gyres can develop independently of the water masses involved (Fig. 4) and at any water depth, as demonstrated by the results of the HEBBLE project (McCave 1976; Nowell *et al.* 1982, 1985; McCave *et al.* 2002), and also recently exemplified by the break-up of both the Atlantic Deep Western Boundary Current and the North Brazil Current into several eddies along their path in the eastern and equatorial South America margins (Dengler *et al.* 2004). Thus, alternation between current acceleration and deceleration along its path is expected, as well as short-term reversals in its sense of flow. These modifications result in different types of sediment record.

Some examples from the SE Brazilian margin are presented here to illustrate the importance of the sea floor in controlling the flow and development of sedimentary accumulations involving coarse-grained sediments. One case, in a lower slope–continental rise setting, shows the presence of Cenozoic sand layers exposed at the sea floor as a result of salt diapir uplift. The seismic lines of Figure 5a–c show high-amplitude, hummocky packages corresponding to upper Oligocene–lower Miocene turbiditic sands exposed at the sea floor along the flank of an escarpment developed by salt halokinesis. A broad channel has been carved as a result of the restriction of the current flow imposed by the escarpment; the sand layers have been eroded and the sediments redistributed downstream. A map of the sea-floor seismic amplitudes obtained in that area was calibrated with the sediment facies using several piston cores (Viana 2001; Viana *et al.* 2001). A wide regional sand sheet, of more than 500 km^2, laterally passing into a sand ribbon zone, was observed downstream of the zone of current erosion of the Oligo-Miocene sands (Fig. 5b). Seismic horizon slices at different depths indicate that the sand sheet passes downslope into a contourite mud-rich drift, in an unconfined setting, with the development of a sediment wave field and a furrow

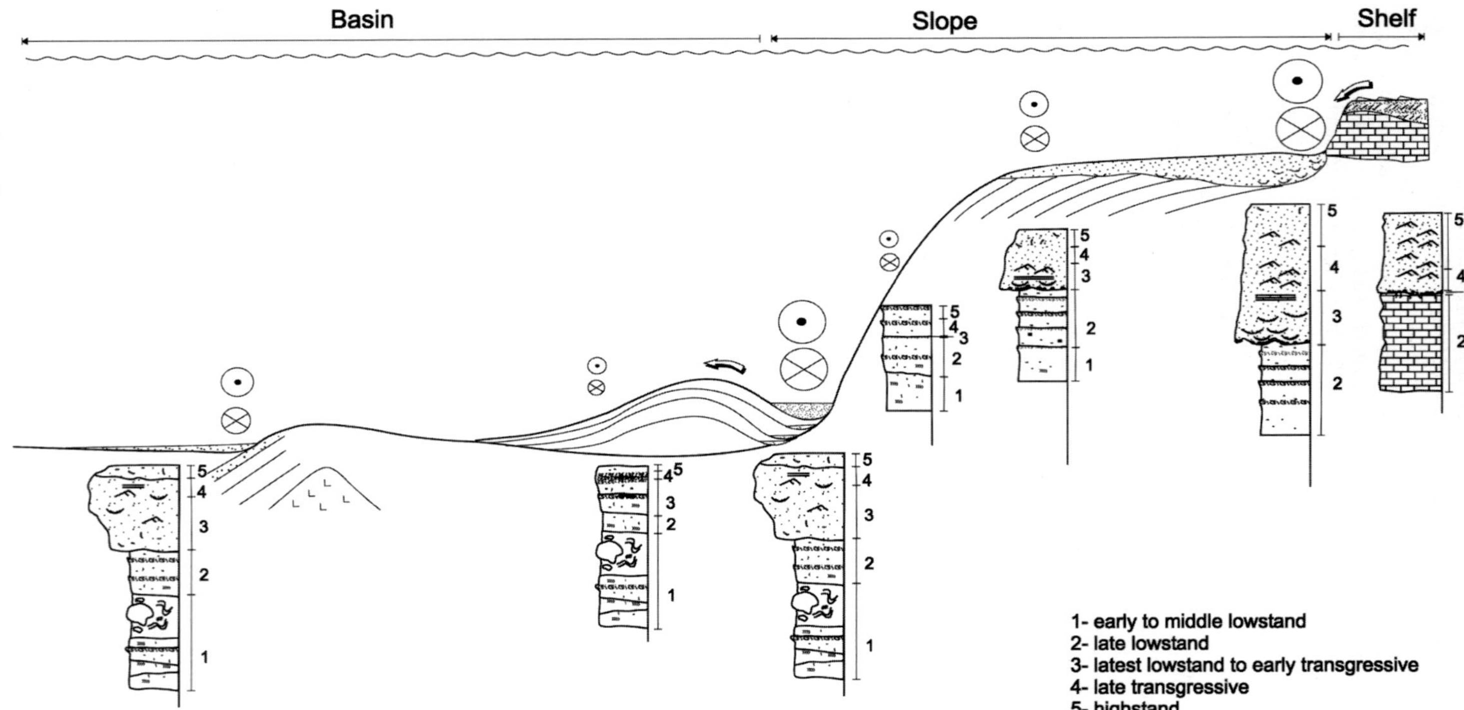

Fig. 3. Schematic profile suggesting the location of preferred sites for coarse-grained contourite deposits, and their relationship with relative sea level. From this graph, based on Viana & Faugères (1998) and Viana *et al.* (2002*b*), the period from the maximum lowstand to the late transgressive phase corresponds to the period of stronger currents and to the deposition of coarser-grained material.

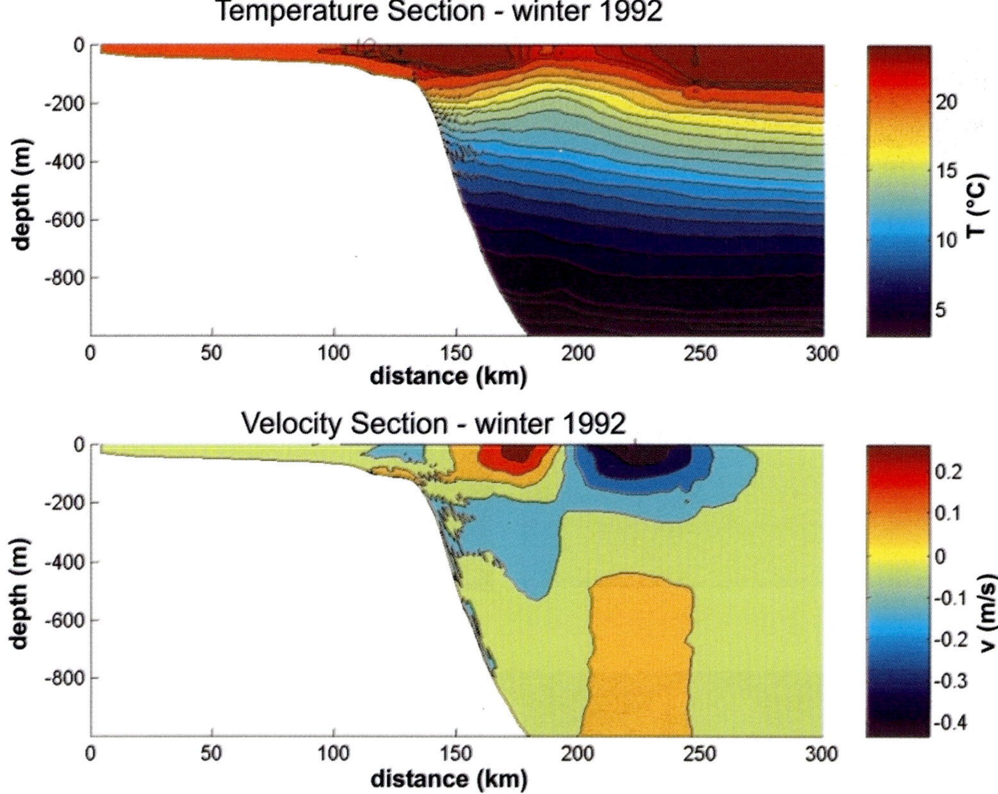

Fig. 4. Temperature and velocity sections through the Campos Basin, SE Brazil margin, showing the presence of a meso-scale eddy and its impact in a deep-water setting where a water column of >400 m depth is affected.

system (Fig. 6). Cores retrieved in the sand ribbon zone confirm this depositional transition, and show an alternating deposition of decimetre-thick fine- to very fine-grained sands and mud layers (Fig. 6).

The impact on sedimentation of a vigorous current, flowing above a large region of a margin strongly marked by topographic modifications, is illustrated in a lower Eocene section from the Santos Basin on the SE margin of Brazil, imaged by 3D seismic data (Fig. 7). The general trend of the slope in this section is east–west and salt withdrawal resulted in the development of elongated mini-basins with a NNE–SSW trend. The mini-basins acted as current corridors in which the intensity of the bottom flow is increased because of the lateral topographic constriction. In a deep-water setting, a large field of current-carved furrows was identified in amplitude coherence maps. The furrows are aligned parallel to the axis of the salt mini-basins (Fig. 7). Some of the mini-basins are depocentres for sands transferred from shallower settings by gravity currents. Independently of the sediment conduit (mini-basin axis, fill and spill, or salt crestal troughs; Fig. 7), the sands emplaced by gravity currents are exposed to current action and, consequently, their original depositional geometry is greatly modified. In this example, the sands are delivered to the core of the mini-basin via avulsion from a salt crestal trough used as sediment fairway, developing an elongated lobe. This geometry was subsequently modified by the currents, and by the sands incorporated into the furrows (Fig. 7).

These currents, following the NNE trend of the mini-basin's axis, encountered the east–west-trending continental slope obliquely. The Eocene slope in that area possesses a narrow and steep upper slope escarpment (Moreira *et al.* 2001) and a wide terrace at its base, where shallow water-derived coarse- and fine-grained sediments accumulated, producing a chaotic seismic pattern. In our analysis the terrace was the initial depocentre of sands exported from the shelf by various mechanisms

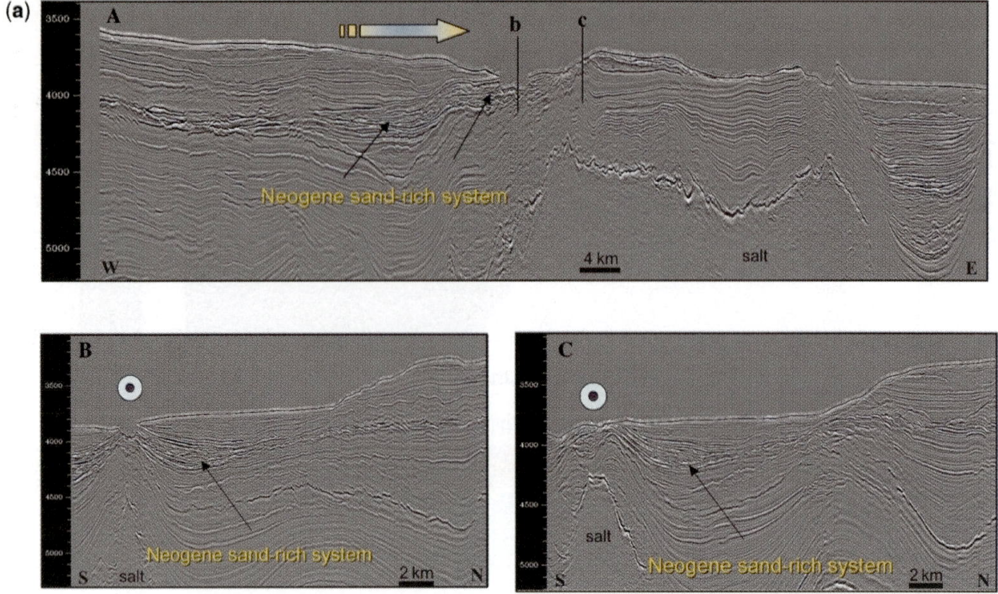

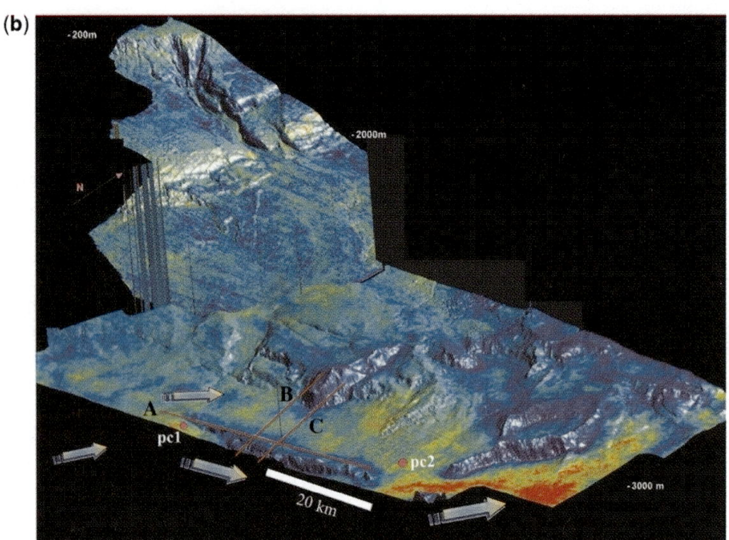

Fig. 5. (**a**) Seismic lines indicating a high-amplitude sand-rich package truncated at the sea floor by bottom currents, after uplift by salt ascent. (**b**) Oblique 3D view of the sea floor indicating the position of the seismic lines of (**a**). The topographic obstacle created by the salt wall is fundamental in accelerating bottom currents. Pink dots indicate the position of piston cores that retrieved the Neogene sands cropping out at the sea floor (pc1) and the sands redeposited as a contourite sheet (pc2).

such as delta front flows and/or erosion of the shelf edge by oceanic processes inducing water sapping at the face of the slope setting, resulting in shelf break–upper slope instability. The impingement of the strong NE-flowing bottom currents against the east–west slope escarpment reworked the shelf-derived sediments at the escarpment base and started development of ENE-migrating barchan sand dunes above the terrace, as seen in the seismic amplitude map of Figure 8. Extensive fields of

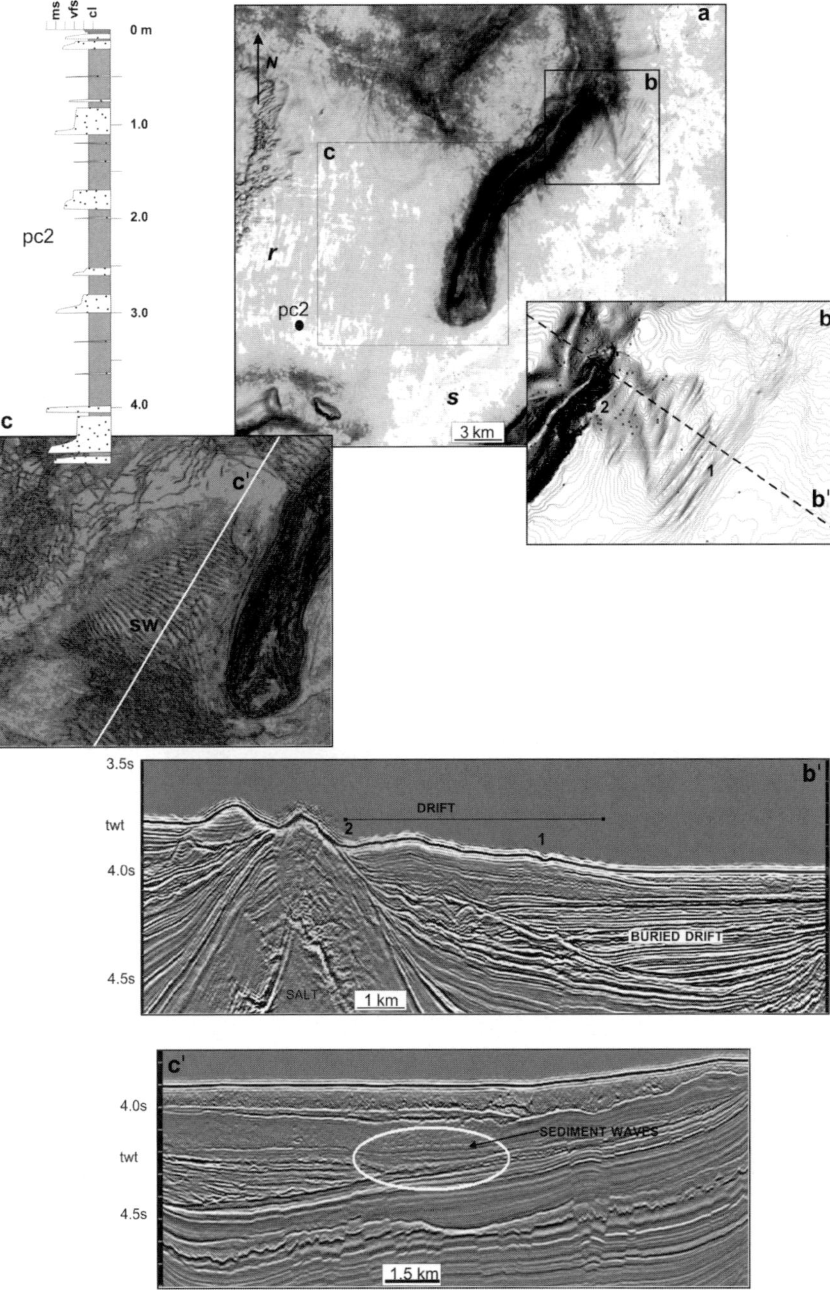

Fig. 6. (**a**) Amplitude map of sea bottom of the area adjacent to that shown in Figure 5, where the sand-rich Neogene section is eroded. Amplitude anomalies are presented as ribbon-like features (*r*) and as sheets (*s*). Piston cores retrieved from these anomalies (pc2) indicated the presence of contourite sands. (**b**) Detail of a moat-drift system and furrows influenced by the presence of the salt wall. *B*′ is a seismic line indicating the construction of moat-drift deposits related to the topographic obstacle caused by the salt wall. (**c**) Horizon slice from a seismic reflector with a crenulated pattern identified below and beyond a huge mass-flow deposit. The seismic image of this reflector, obtained from a coherence map, suggests a field of sediment waves migrating northeastward, following the flow pattern of the ribbon-like features and the furrows observed on the modern sea floor, and in agreement with the suggested current direction based on the construction of moat-drift deposits in the area. The seismic line c′ indicates the position of the horizon slice and seismic pattern associated with the sediment waves. twt, two-way travel time.

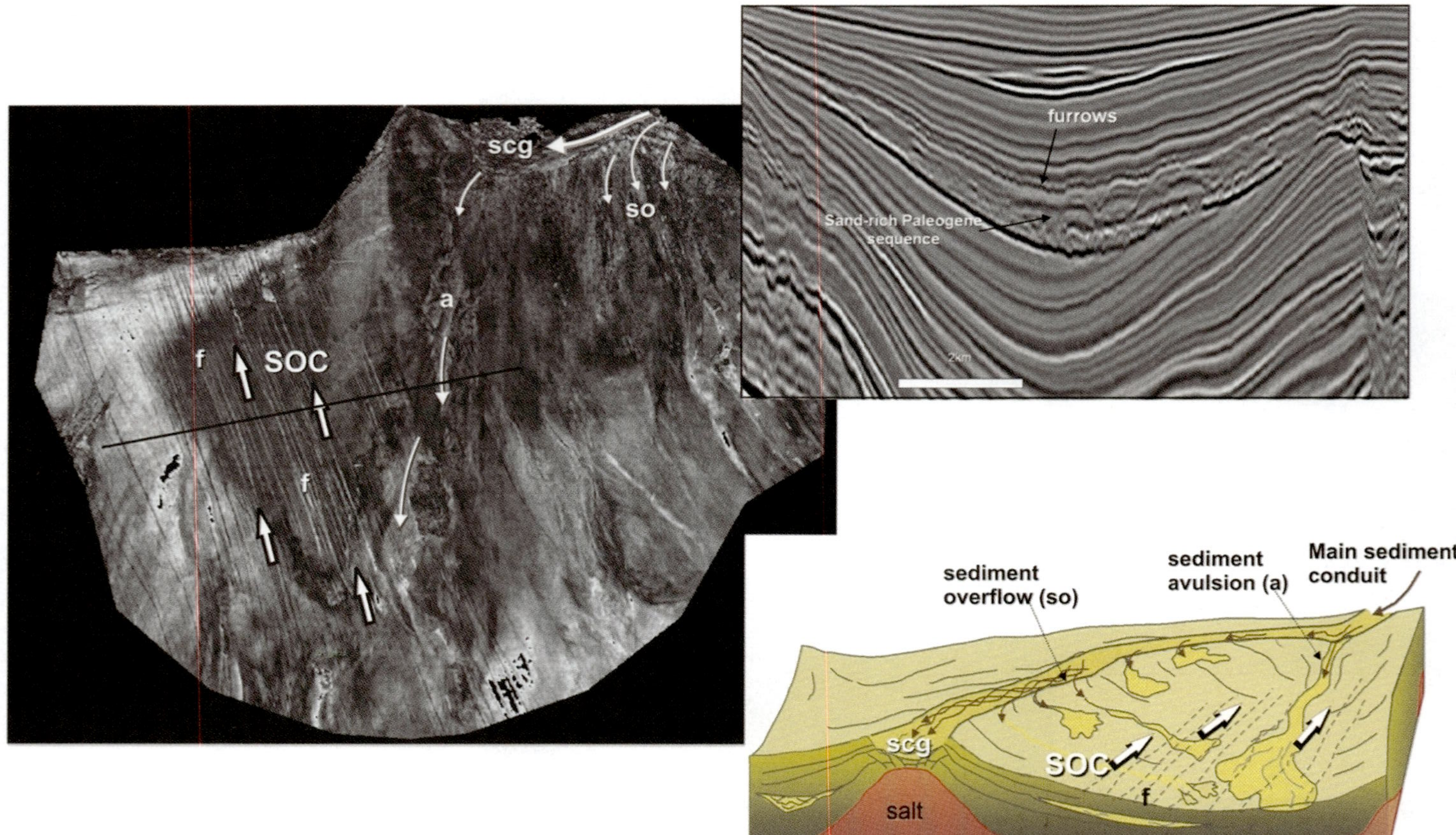

Fig. 7. Linear features (furrows; f) scouring the latest Palaeocene–earliest Eocene sea floor of the Santos Basin deep-water area. Horizon slice of the furrow surface, indicating the seismic line, shows the furrows interacting with the high-amplitude patches (sands?) delivered to the core of the mini-basin by avulsion or spillover from the channel developed within a salt crestal graben (scg). A schematic representation of these coupled processes is shown. SOC, Southern Ocean Current.

barchan dunes, sand waves and mega-furrows have developed on the modern sea floor of the Gulf of Mexico, in water depths ranging from 2000 to 3000 m, as a result of bottom currents flowing at about 100 cm s^{-1} (Bryant *et al.* 2000; Kenyon *et al.* 2002; Niedoroda *et al.* 2003; Bryant & Slowey 2004). Similar strong bottom currents probably caused the formation of the barchan dunes identified in the uppermost Palaeocene–lower Eocene section of the Santos Basin, indicating bottom-current activity in a period when coarse-grained sediments were delivered to the basin to form potential reservoirs.

Large-scale analysis of 3D seismic data for the Santos Basin suggests that slope instability could be related to the action of strong slope currents. In such a case, vigorous SW-flowing slope surface currents, in a pattern similar to that of the present-day Brazil Current (see discussion by Duarte & Viana 2007), would erode and induce instability at the tip of the shelf edge clinoforms prograding from the shelf edge, where the gradient increases. Such erosion induces the introduction of sediments into the basin via successive mass flows, whose heads are aligned for several tens of kilometres along the isobaths. Figure 9 presents seismic imaging of this kind of erosive feature, more than 50 km long, which is highly linear and conspicuous. We consider that such erosion exposed aquifers and induced water sapping, a continuous low-rate flow of water-charged sediments. Such flows are probably the explanation for the sinuous channels crossing the slope and the mass-flow deposits accumulated at the foot of the upper slope escarpment. They developed a series of short, narrow and parallel slope aprons that evolved to a chaotic relief of mass-flow deposits of intermingled sand and mud that developed above the mid-slope terrace where the barchan dunes were found (Fig. 8). Above this terrace dominated by the chaotic deposits, NE-flowing deep-water currents are accelerated because of the constriction of the upper slope escarpment and the complex local relief. Here the bottom currents have eroded a sinuous, more than 30 km long, slope-parallel channel, with drifted contouritic sediments developing an asymmetric channel–levee system (Figs 8 and 9). These examples illustrate the close interaction between abrupt slope gradient changes and strong currents, frequently inducing destabilization of the adjacent slope in any bathymetric setting, with the development of small- to mid-scale mass movement deposits.

Seismic and well log characteristics of sandy contourites. Given the credible evidence for the existence of coarse-grained deposits associated with contour currents, the recognition and characterization of their depositional elements from seismic and well log data is an essential element. The diagnostic criteria for recognition of contourites in seismic lines were presented by Faugères *et al.* (1999), who focused on the distinction of contourites from turbidites, as well as on the identification of seismic features that correspond to coarse-grained contourites, to provide predictive tools for reservoir identification. Linear, and largely continuous strong seismic amplitudes were observed in the lower Pliocene contourite gas-bearing sands with more than 30% porosity in the Spanish Gulf of Cadiz (Cakebread-Brown *et al.* 2003). These biogenic gas accumulations exhibit seismic amplitude anomalies (amplitude v. offset), and their presence in the Gulf of Cadiz region, the characterization of their seismic geomorphology and the regional occurrence of a potential source rock open the possibility of a new giant play in the Algarve basin. Also in the Gulf of Cadiz, Garcia-Mojonero & Olmo (2001), on the basis of seismic lines and one exploratory well, identified upper Pliocene–Quaternary contourite deposits consisting of more than 800 m thick sand–shale intercalations, with an average net-to-gross (total sand content) of 75%, with the sand beds reaching 40 m thickness with good lateral continuity and porosities >30%. Below are some examples of contourites deposited both in shallow and deep water based on industrial 3D seismic data, shot along the Brazilian SE margin, coupled with exploratory and appraisal well data and piston cores.

In the first example, concerning upper slope contourites, a seismic amplitude map close to the Serravallian unconformity (middle Miocene) was obtained from the shelf edge down to the basin (Fig. 10). Strong negative amplitudes, calibrated by well data, correspond to coarse-grained deposits. They occur at the top of the slope as a >10 km long and 3–5 km wide field of linear ridges that are parallel to the slope. The ridges grade downslope and downstream to weak positive amplitude reflections, corresponding to finer-grained sediments. The seismic lines of Figure 10 suggest the presence of a small upper slope escarpment with an adjacent terrace above, bearing high-amplitude reflectors corresponding to the location of the ridges. Locally, these reflections develop a hummocky clinoform seismic pattern, suggesting a migration of the bedforms above the terrace. Recurrent, stacked similar features are observed, suggesting that the depositional system was maintained during a relatively long period (>1 Ma), alternating with continuous low-amplitude, high-frequency reflectors, interpreted as periods of fine-grained deposition, related to the weakening of the current regime.

Linear seismic amplitude anomalies are also observed on the modern deep sea floor, as discussed

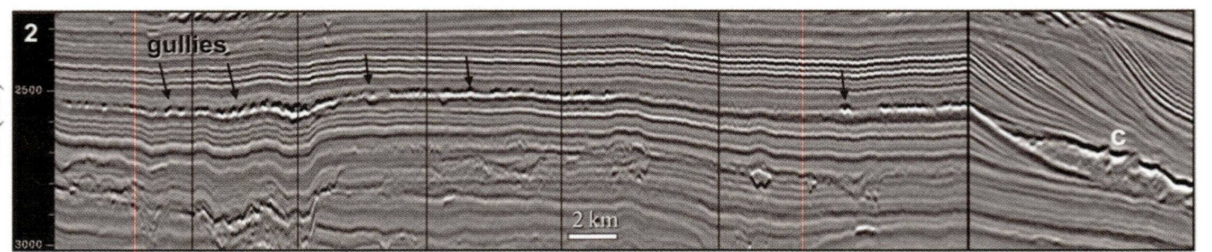

Fig. 9. Seismic amplitude image showing detail of slope erosion by boundary currents inducing a conspicuous, linear water-sapping front (wsf) and related slope apron deposits, evolving downslope to chaotic features carved by a sinuous, parallel-to-the-slope channel (c). The drifted material from the channel path develops an asymmetric levee system (l). The white bar indicates the position of the seismic line 1, showing a narrow extension of the upper slope escarpment where porous, water-saturated beds (strong white reflections truncated against the escarpment face indicated by the arrow) cropped out against the lower–middle Eocene sea floor. Composite yellow seismic line 2 crosses the several gullies and sediment aprons formed by the water sapping (small black arrows) and its termination indicates their genetic relationship with the chaotic, sand-rich mound carved by the channel. A huge mass-flow feature (mf) is observed, indicating local unsteadiness of the chaotic deposits, probably because of movement of adjacent salt walls. White arrows indicate the sense of flow of surface currents (sc) and deep currents (dc).

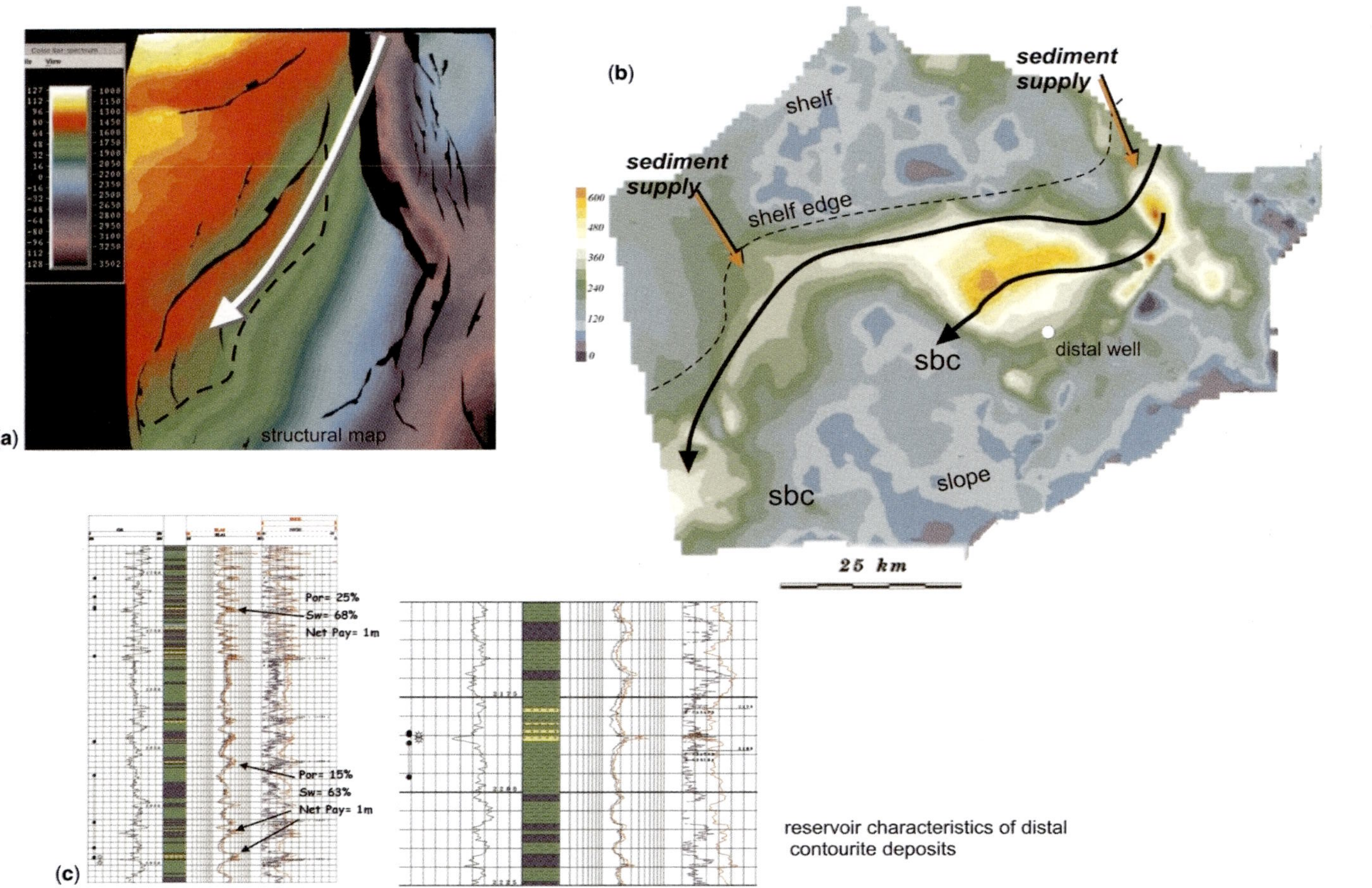

Fig. 11. (**a**) Structural map of middle Miocene interval where contourite sands predominate. The closure against the fault (dashed line) provides an interesting structural trap with higher sand content close to the fault, which may also have served as migration pathway. (**b**) Isopach map of the lower to middle Miocene sand-rich interval indicating thicker accumulation on a trend parallel to the isobaths; the sediments were introduced to the slope via point sources (incised valleys?) and distributed by the slope boundary current (sbc). (**c**) Gamma-ray, resistivity and density logs of a distal well with measured porosity, water saturation and average net pay in oil-bearing sand beds.

changes, and another, downslope trending, related to the width of the current core and to the spacing from that core, both marked by a general decrease in current speed and thus, in the transported or accumulated grain size.

The pattern and distribution of bedforms are indicative of the bottom-current intensity. The examples from the Faeroe–Shetland Channel and the Gulf of Cadiz channels (Habgood *et al.* 2003; Masson *et al.* 2004; Akhmetzhanov *et al.* 2007) and Campos Basin upper slope terrace (Viana & Faugères 1998; Viana *et al.* 2002*b*) are good illustrations of such flow oscillations and their related erosional–depositional features. The erosional pavements, gravel patches, furrows–grooves–striations, sand ribbons, sand dunes, current ripples and fine-grained sediment waves are subsequently deposited with a decrease in the current intensity.

A large spectrum of sedimentary facies is attributed to contourites. Most of these are similar, being deposited under other current regimes where bed-load and suspended-load transport can coexist with different intensities as a function of the basin morphology and environmental constraints. Structureless or faintly to well-defined cross-stratified, very fine-grained to coarse-grained siliciclastic or bioclastic sands develop individual sets a few millimetres to several decimetres thick. The high-frequency alternation between coarser and finer sediments is indicative of high-frequency current changes and is marked by truncations, abrupt changes in grain-size stacking or in millimetre to centimetre thick passages from normal to inverse grading. Slope intraclasts are frequently preserved at the location of major current acceleration and, because traction prevails in a coarse-grained contourite system, they probably do not travel very far from the point of their plucking. Traction structures are dominant, even in very thin sand streaks where millimetre thick layers of very fine to fine-grained sand occur, probably accumulated by rolling or saltation under the action of bottom currents transporting sediments delivered to the basin by different mechanisms or from the reworking of previously deposited sands. Nevertheless, suspension may occur where current velocity surpasses the threshold of bed-load transport or when the suspended sediments of gravity flows or in benthic storms are entrained by any kind of contour currents (Adams & Weatherly 1981; Nowell *et al.* 1982; Nowell & Hollister 1985; McCave 1986). Bioturbation is common, occurring in several degrees of intensity and suggesting oxygenation of the bottom water. Sharp erosive contacts are also a frequent feature. They indicate an abrupt increase of bottom current intensity and separate any textural class susceptible to current action.

The 'normal' behaviour of an oceanic flow is marked by current intensity changes along its path as well as over time. Bottom-current reversals, tidal imprints and extreme high- to low-frequency oscillations in the current intensity are commonly observed (McCave 1976; Nowell & Hollister 1985; De Madron & Weatherly 1994; Huthnance 1995; Habgood *et al.* 2003), and induce the development of sedimentary structures similar to tidal or flood-related hyperpycnal flow-derived deposits (in the sense of Mulder *et al.* 2003). The distinction of contourites from sediments deposited under such conditions requires good sampling, palaeogeographical reconstructions, and vertical and lateral facies association analysis. The activity of bedforms and bioturbation respond also to current variability. A schematic representation of the sedimentary facies distributed along a coarse-grained contourite system is proposed in Figure 12. Sedimentary facies from various sites on the SE Brazil margin were chosen to illustrate this conceptual representation. The general model is a representation of the facies changes along the path of a slope contour current, which crosses a sand-rich area (sediment source) and distributes the sediment load downstream. The lateral confinement of the flow is provided by any topographic feature, either pre-existing or constructed by the current itself. The current intensity decreases downstream, inducing a diversity of sedimentary facies and bedforms as shown in Figure 12. A longitudinal trend of coarser-grained sediments accompanies the current core. If the current is strong enough to erode the substrate to form channels, the development of levees and oblique sediment waves is also expected, as shown in Figure 7, where the funnelling of the currents against the foot of the upper slope escarpment has resulted in a channel–levee-like system.

Final remarks

The recognition of deep-water coarse-grained deposits as contourites still remains controversial. A gravity-flow culture prevails upon the idea of contour-current action mostly as a result of: (1) the early establishment of conceptual models proposing facies characterization and the geometry of the gravity-flow accumulations; (2) gravity-flow deposits of sand are volumetrically predominant at least in the modern ocean; (3) most sandy contourites involve sediment transport on the upper slope, a zone that was poorly investigated in the 1960s to 1980s. The continuous scientific and industrial research in deep-water realms is providing new data that have been used in some attempts to characterize the coarse-grained contourites and indicate their economic potential. Controversy is an important triggering mechanism to stimulate further efforts in identifying and describing contourites. We hope that this paper will be useful in providing impetus to propose further ideas concerning the characterization and the economic importance of contourites as well as the continuity of the effort in determining the role of palaeoceanographic changes in bottom-current regimes and their geological record, and thus provide a basis for a better understanding of the wide spectrum of depositional systems in the marine environment.

A large number of Petrobras colleagues were fundamental in development of the ideas presented in this paper. Their contribution occurred in several ways, through discussions, seismic mapping, and mostly following their criticisms and enthusiasm in pushing forward the efforts to broaden the understanding of the role of bottom currents in shaping and building sedimentary deposits of economic interest along the Brazilian margin. Among them, R. Kowsmann was involved in the 'industrial marine geology' approach to contourites. Discussions with M. Carminatti and C. E. Souza Cruz provided extremely fruitful intellectual exchange. J.-C. Faugères was involved since the start of the evolution of these thoughts, and motivated the pursuit of the contourite task. The briefer than desired moments of conversation with E. Mutti and his geological wisdom greatly influenced how the ideas were adjusted and helped to point out the doubts that remain unsolved after this overview. We sincerely thank J. Howe and D. Piper for their precise scientific suggestions and help in adjusting the manuscript into a readable format. M. Rebesco, in the double role of first author's friend and co-editor of this volume, was fundamental in pushing forward during the periods when the production of this paper was menaced by the endless and absorbing tasks of petroleum exploration.

References

ADAMS, C. E. & WEATHERLY, G. L. 1981. Suspended-sediment transport and benthic boundary-layer dynamics. *In*: NITTROUER, C. A. (ed.) *Sedimentary Dynamics of Continental Shelves*. Elsevier, New York, 1–18.

AKHMETZHANOV, A., KENYON, N. H., HABGOOD, E., VAN DER MOLLEN, A. S., NIELSEN, T., IVANOV, M. & SHASHKIN, P. 2007. North Atlantic contourite sand channels. *In*: VIANA, A. R. & REBESCO, M. (eds) *Economic and Palaeoceanographic Significance of Contourite Deposits*. Geological Society, London, Special Publications, **276**, 25–47.

BALTZER, A., COCHONAT, P. & PIPER, D. J. W. 1994. *In situ* geotechnical characterization of sediments on the Nova Scotian Slope, eastern Canadian continental margin. *Marine Geology*, **120**(3–4), 291–308.

BARROS, M. C., POSSATO, S. & GUARDADO, L. R. 1982. Carapebus Member (Eocene), Campos Basin, Brazilian offshore; deepsea fan turbidites winnowed by bottom currents. *AAPG Bulletin*, **66**, 545–546.

BELONIN, M. D., SHIMANSKY, V. V., BREHUNCOV, A. M., DESCHENIA, N., BORODKIN, V. & HAFIZOV, S. 2003. Application of paleofacial reconstructions for forecasting oil traps in turbidite systems, Lower Cretaceous, Western Siberia. *In*: *AAPG International Conference, Barcelona, September 2003*.

BOGUCKI, D., DICKEY, T. & REDEKOPP, L. 1997. Sediment resuspension and mixing by resonantly generated internal solitary waves. *Journal of Physical Oceanography*, **27**, 1181–1196.

BOUMA, A. H., NORMARK, W. R. & BARNES, N. E. 1985. *Submarine Fans and Related Turbidite Systems*. Springer, New York.

BRYANT, W., DELLAPENNA, T., SILVA, A., BEAN, D. & DUNLAP, W. 2000. Massive bed-forms, mega-furrows, on the continental rise at the base of the Sigsbee Escarpment, northwest Gulf of Mexico. AAPG Annual Meeting, New Orleans, LA, Abstracts.

BRYANT, W. R. & SLOWEY, N. C. 2004. Deep-sea furrows: physical characteristics, mechanisms of formation and associated environmental processes, a joint industry project. Texas A&M University, Department of Oceanography and Offshore Technology Research Center, internal report. World Wide Web address: http://deeptow.tamu.edu/jip/index.html.

CACCHIONE, D. A. & DRAKE, D. E. 1986. Nepheloid layers and internal waves over continental shelves and slopes. *Geo-Marine Letters*, **6**, 147–152.

CACCHIONE, D. A., PRATSON, L. F. & OGSTON, A. S. 2002. The shaping of continental slopes by internal tides. *Science*, **296**, 724–727.

CAKEBREAD-BROWN, J., GARCIA-MOJONERO, C., BORTZ, R., SCHWARZHANS, W., CORTES, L. & OLMO, W. M. 2003. Offshore Algarve, Portugal: a prospective extension of the Spanish Gulf of Cadiz Miocene Play. *In*: *AAPG International Conference, Barcelona, September 2003*.

CARTER, R. M. 2007. The role of intermediate-depth currents in continental shelf–slope accretion: Canterbury Drifts, SW Pacific Ocean. *In*: VIANA, A. R. & REBESCO, M. (eds) *Economic and Palaeoceanographic Significance of Contourite Deposits*. Geological Society, London, Special Publications, **276**, 129–154.

CARTER, R. M., FULTHORPE, C. S. & LU, H. 2004. Canterbury Drifts at Ocean Drilling Program Site 1119, New Zealand: climatic modulation of southwest Pacific intermediate water flows since 3.9 Ma. *Geology*, **32**(11), 1005–1008.

CASTRO, R. D., MAGALHÃES, P. M., MOLITERNO, A. M. C., SANTOS, V. S. S., BLAUTH, M. & CADDAH, L. F. G. 1998. Feeder canyon evolution model deduced from the stacking pattern of a Tertiary turbidite system in Campos Basin, Brasil. *AAPG Bulletin*, **82**(10), 1883–1884.

CERAMICOLA, S., REBESCO, M., DE BATIST, M. & KHLYSTOV, O. 2001. Seismic evidence of small-scale lacustrine drifts in Lake Baikal

(Russia). *Marine Geophysical Researches*, **22**, 445–464.

Clark, P. U., Pisias, N. G., Stocker, T. F. & Weaver, A. J. 2002. The role of the thermohaline circulation in abrupt climate change. *Nature*, **415**, 863–869.

Cochonat, P., Ollier, G. & Michel, J. L. 1989. Evidence for slope instability and current-induced sediment transport, the RMS *Titanic* wreck search area, Newfoundland Rise. *Geo-Marine Letters*, **9**, 145–152.

Dalrymple, R. W., LeGresley, E. M., Fader, G. B. J. & Petrie, B. D. 1992. The western Grand Banks of Newfoundland: transgressive Holocene sedimentation under the combined influence of waves and currents. *Marine Geology*, **105**, 95–118.

Damuth, J. E. & Olson, H. C. 2001. Neogene–Quaternary contourite and related deposition on the West Shetland Slope and Faeroe–Shetland Channel revealed by high-resolution seismic studies. *Marine Geophysical Researches*, **22**(5–6), 369–398.

Davies, R., Cloke, I., Cartwright, J., Robinson, A. & Ferrero, C. 2004. Post-breakup compression of a passive margin and its impact on hydrocarbon prospectivity: an example from the Tertiary of the Faeroe–Shetland Basin, United Kingdom. *AAPG Bulletin*, **88**(1), 1–20.

De Madron, X. D. & Weatherly, G. 1994. Circulation, transport and bottom boundary layers of deep currents in the Brazil Basin. *Journal of Marine Research*, **52**, 583–638.

Dengler, M., Schott, F. A., Eden, C., Brandt, P., Fischer, J. & Zantopp, R. J. 2004. Break-up of the Atlantic western boundary current into eddies at 8°S. *Nature*, **432**, 1018–1020.

Dickens, G. R., O'Neil, J. R., Rea, D. K. & Owen, R. M. 1995. Dissociation of oceanic methane hydrate as a cause of the carbon isotope excursion at the end of the Paleocene. *Paleoceanography*, **10**, 965–971.

Dickens, G. R., Castillo, M. M. & Walker, J. C. G. 1997. A blast of gas in the latest Paleocene: simulating first-order effects of massive dissociation of oceanic methane hydrate. *Geology*, **25**, 259–262.

Dillon, W. P. & Paull, C. K. 1983. Marine gas hydrates, II. Geophysical evidence. *In*: Cox, J. L. (ed.) *Natural Gas Hydrates: Properties, Occurrences, and Recovery*. Butterworth, Woburn, MA, 73–90.

Duarte, C. S. L. & Viana, A. R. 2007. Santos Drift System: stratigraphic organization and implications for late Cenozoic palaeocirculation in the Santos Basin, SW Atlantic Ocean. *In*: Viana, A. R. & Rebesco, M. (eds) *Economic and Palaeoceanographic Significance of Contourite Deposits*. Geological Society, London, Special Publications, **276**, 171–198.

Enjorlas, J. M., Gouadain, J., Mutti, E. & Pizon, J. 1986. New turbiditic model for the Lower Tertiary sands in the South Viking Graben. *In*: Spencer, A. M. (ed.) *Habitat of Hydrocarbons on the Norwegian Continental Shelf*. Norwegian Petroleum Society, Oslo, 171–178.

Famakinwa, S. B., Shanmugam, G. & Nieto, J. 1997. Bottom-current reworked sands—a 'new' latent reservoir type in deep-water sedimentation, Equatorial Guinea and Nigeria. *AAPG Bulletin*, **81**(13), 34.

Faugères, J.-C. & Stow, D. A. V. 1993. Bottom-current-controlled sedimentation: a synthesis of the contourite problem. *Sedimentary Geology*, **82**(1–4), 287–297.

Faugères, J.-C., Mezerais, M. L. & Stow, D. A. V. 1993. Contourite drift types and their distribution in the North and South Atlantic Ocean basins. *Sedimentary Geology*, **82**(1–4), 189–203.

Faugères, J.-C., Stow, D. A. V., Imbert, P. & Viana, A. 1999. Seismic features diagnostic of contourite drifts. *Marine Geology*, **162**, 1–38.

Flemming, B. W. 1978. Underwater sand dunes along the southeast African continental margin—observations and implications. *Marine Geology*, **26**, 177–198.

Flemming, B. W. 1980. Sand transport and bedform patterns on the Continental Shelf between Durban and Port Elizabeth (southeast Africa continental margin). *Sedimentary Geology*, **26**, 179–205.

Fukushima, Y., Parker, G. & Pantin, H. M. 1985. Prediction of ignitive turbidity currents in Scripps Submarine Canyon. *Marine Geology*, **67**(1–2), 55–81.

Garcia-Mojonero, C. & Olmo, W. M. 2001. One sea level fall and four different gas plays: the Gulf of Cadiz Basin, SW Spain. *In*: *Petroleum Systems of Deep Water Basins: Global and Gulf of Mexico Experience*. GCSSEPM, Bob. F. Perkins Research Conference, December 2001.

Gilli, A., Anselmetti, F. S., Ariztegui, D., Beres, M., McKenzie, J. A. & Markgraf, V. 2005. Seismic stratigraphy, buried beach ridges and contourite drifts: the Late Quaternary history of the closed Lago Cardiel basin, Argentina (49°S). *Sedimentology*, **52**, 1–23.

Gomes, P. O. & Viana, A. R. 2002. Contour currents, sediment drifts and abyssal erosion on the northeastern continental margin off NE Brazil. *In*: Stow, D. A. V., Pudsey, C. J., Howe, J. A., Faugères, J.-C. & Viana, A. R. (eds) *Deep-Water Contourites: Modern Drifts and Ancient Series, Seismic and Sedimentary Characteristics*. Geological Society, London, Memoirs, **22**, 239–248.

Habgood, E. L., Kenyon, N. H., Masson, D. G., Akhmetzhanov, A., Weaver, P. P. E., Gardner, J. & Mulder, T. 2003. Deep-water sediment wave fields, bottom current sand channels and gravity flow channel-lobe systems: Gulf of Cadiz, NE Atlantic. *Sedimentology*, **50**, 483–510.

Hesselbo, S. P., Grocke, D. R., Jenkyns, H. C., Bjerrum, C. J., Farrimond, P., Morgans-Bell, H. S. & Green, O. R. 2000. Massive dissociation of gas hydrate during a Jurassic oceanic anoxic event. *Nature*, **406**, 392–395.

Holbrook, W. S., Lizarralde, D., Pecher, I. A., Gorman, A. R., Hackwith, K. L., Hornbach, M. & Saffer, D. 2002. Escape of methane gas through sediment waves in a large methane hydrate province. *Geology*, **30**(5), 467–470.

HOLLISTER, C. D. 1993. The concept of deep-sea contourites. *Sedimentary Geology*, **82**, 5–11.

HUTHNANCE, J. M. 1995. Circulation, exchange and water masses at the ocean margin: the role of physical processes at the shelf edge. *Progress in Oceanography*, **35**, 353–431.

ISERN, A. R., ANSELMETTI, F. S. & BLUM, P. 2004. A Neogene carbonate platform, slope, and shelf edifice shaped by sea level and ocean currents, Marion Plateau (Northeast Australia). *In*: EBERLI, G. P., MASAFERRO, J. L. & RICK SARG, J. F. (eds) *Seismic Imaging of Carbonate Reservoirs and Systems*. AAPG Memoirs, **81**, 291–307.

JOHNSON, T. C. 1996. Sedimentary processes and signals of past climate change in the large lakes of the East African Rift Valley. *In*: JOHNSON, T. C. & ODADA, E. O. (eds) *The Limnology, Climatology and Paleoclimatology of the East African Lakes*. Gordon & Breach, Amsterdam.

JOHNSON, T. C., CARLSON, T. W. & EVANS, J. E. 1980. Contourites in Lake Superior. *Geology*, **8**, 437–441.

KENYON, N. H. 1986. Evidence from bedforms for a strong poleward current along the upper continental slope of Northwest Europe. *Marine Geology*, **72**, 187–198.

KENYON, N. H., AKHMETZHANOV, A. M. & TWICHELL, D. C. 2002. Sand wave fields beneath the Loop Current, Gulf of Mexico: reworking of fan sands. *Marine Geology*, **192**, 297–307.

KRAEMER, L. M., OWEN, R. M. & DICKENS, G. R. 2000. Lithology of the upper gas hydrate zone, Blake Outer Ridge: a link between diatoms, porosity, and gas hydrate. *In*: PAULL, C. K., MATSUMOTO, R., WALLACE, P. J. & DILLON, W. P. (eds) *Proceedings of the Ocean Drilling Program, Scientific Results*, **164**. Ocean Drilling Program, College Station, TX, 229–236.

LLAVE, E., HERNANDEZ-MOLINA, F. J., SOMOZA, L., DIAZ-DEL-RIO, V., STOW, D. A. V., MAESTRO, A. & ALVEIRINHO DIAS, J. M. 2001. Seismic stacking pattern of the Faro–Albufeira contourite system (Gulf of Cadiz): a Quaternary record of paleoceanographic and tectonic influences. *Marine Geophysical Researches*, **22**, 487–508.

LLAVE, E., HERNÁNDEZ-MOLINA, F. J., SOMOZA, L., STOW, D. A. V. & DÍAZ DEL RÍO, V. 2007. Quaternary evolution of the contourite depositional system in the Gulf of Cadiz. *In*: VIANA, A. R. & REBESCO, M. (eds) *Economic and Palaeoceanographic Significances of Contourite Deposits*. Geological Society, London, Special Publications, **276**, 49–79.

LIMA, A. F. 2003. *Comparação dos sistemas sedimentares profundos da bacia sudeste-sul do Brasil com ênfase no sistema misto Colúmbia*. PhD thesis, University of São Paulo.

LIMA, J. A. M., MOLLER, O. O., VIANA, A. R. & PIOVESAN, R. 2007. Hydrodynamic modelling of bottom currents and sediment transport in the Canyon São Tomé (Brazil). *In*: VIANA, A. R. & REBESCO, M. (eds) *Economic and Palaeoceanographic Significance of Contourite Deposits*. Geological Society, London, Special Publications, **276**, 329–342.

LONSDALE, P. & MALFAIT, P. I. 1974. Abyssal dunes of foraminiferal sand in Carnegie Ridge. *Geological Society of America Bulletin*, **85**, 1679–1712.

LU, H. & FULTHORPE, C. S. 2004. Controls on sequence stratigraphy of a middle Miocene–Holocene, current-swept, passive margin: offshore Canterbury Basin, New Zealand. *Geological Society of America Bulletin*, **116**(11–12), 1345–1366.

MACNAUGHTON, R. B., NARBONNE, G. M. & DALRYMPLE, R. W. 2000. Neoproterozoic slope deposits, Mackenzie Mountains, northwestern Canada: implications for passive-margin development and Ediacaran faunal ecology. *Canadian Journal of Earth Sciences*, **37**, 997–1020.

MASSON, D. G. 2001. Sedimentary processes shaping the eastern slope of the Faeroe–Shetland Channel. *Continental Shelf Research*, **21**, 825–857.

MASSON, D. G., HOWE, J. A. & STOKER, M. S. 2002. Bottom current sediment waves, sediment drifts and contourites in the northern Rockall Trough. *Marine Geology*, **192**, 215–237.

MASSON, D. G., WYNN, R. B. & BETT, B. J. 2004. Sedimentary environment of the Faroe–Shetland and Faroe Bank Channels, north-east Atlantic, and the use of bedforms as indicators of bottom current velocity in the deep ocean. *Sedimentology*, **51**(6), 1207–1241.

MCCAVE, I. N. 1976. *The Benthic Boundary Layer*. Plenum, New York.

MCCAVE, I. N. 1986. Local and global aspects of the bottom nepheloid layers in the world ocean. *Netherland Journal of Deep-Sea Research*, **20**(2–3), 167–181.

MCCAVE, I. N., CHANDLER, R. C., SWIFT, S. A. & TUCOLKE, B. E. 2002. Contourites of the Nova Scotian continental rise and the HEBBLE area. *In*: STOW, D. A. V., PUDSEY, C. J., HOWE, J. A., FAUGÈRES, J.-C. & VIANA, A. R. (eds) *Deep-Water Contourite Systems: Modern Drifts and Ancient Series, Seismic and Sedimentary Characteristics*. Geological Society, London, Memoirs, **22**, 21–38.

MORAES, M. A. S., MACIEL, W. B., BRAGA, M. S. S. & VIANA, A. R. 2007. Bottom-current reworked Palaeocene–Eocene deep-water reservoirs of the Campos Basin, Brazil. *In*: VIANA, A. R. & REBESCO, M. (eds) *Economic and Palaeoceanographic Significance of Contourite Deposits*. Geological Society, London, Special Publications, **276**, 81–94.

MOREIRA, J. L. P., NALPAS, T., JOSEPH, P. & GUILLOCHEAU, F. 2001. Stratigraphie sismique de la marge Éocène du Nord du bassin de Santos (Brésil); relations plateforme/systèmes turbiditiques; distortion des séquences de dépôt. *Comptes Rendus de l'Académie des Sciences, Earth and Planetary Sciences*, **332**, 491–498.

MULDER, T., LECROART, P., VOISSET, M., ET AL. 2002. The Gulf of Cadiz. A key area for understanding paleoclimate record and oceanic circulation. *EOS Transactions, American Geophysical Union*, **83**(43), 481–488.

MULDER, T., SYVITSKI, J. P. M., MIGEON, S., FAUGÈRES, J.-C. & SAVOYE, B. 2003. Marine hyperpycnal flows: initiation, behavior and related deposits. A review. *Marine and Petroleum Geology*, **20**, 861–882.

MUTTI, E. 1992. *Turbidite Sandstones*. San Donato Milanese: AGIP—Istituto di Geologia, Università di Parma, Parma.

MUTTI, E. & NORMARK, W. R. 1987. Comparing examples of modern and ancient turbidite systems: problems and concepts. *In*: LEGGETT, J. K. & ZUFA, G. G. (eds) *Marine Clastic Sedimentology: Concepts and Case Studies*. Graham and Trotman, London, 1–38.

MUTTI, E. & NORMARK, W. R. 1991. An integrated approach to the study of turbidite systems. *In*: WEIMER, P. & LINK, H. (ed.) *Seismic Facies and Sedimentary Processes of Submarine Fans and Turbidite Systems*. Springer, New York, 75–106.

MUTTI, E., BARROS, M., POSSATO, S. & RUMENOS, L. 1980. Deepsea fan turbidite sediments winnowed by bottom currents in the Eocene of the Campos Basin, Brazilian offshore. *International Association of Sedimentologists, First European Regional Meeting, Abstracts*, 114.

MUTTI, E., TINTERRI, R., REMACHA, E., MAVILLA, N., ANGELLA, S. & FAVA, L. 1999. *An introduction to the analysis of ancient turbidite basins from an outcrop perspective*. AAPG Course Notes, **39**.

MUTTI, E., TINTERRI, R., BENEVELLI, G., DI BIASE, D. & CAVANNA, G. 2003. Deltaic, mixed and turbidite sedimentation of ancient foreland basins. *Marine and Petroleum Geology*, **20**, 733–755.

NELSON, C. H., BARAZA, J. & MALDONADO, A. 1993. Mediterranean undercurrent sandy contourites, Gulf of Cadiz, Spain. *Sedimentary Geology*, **82** (1–4), 103–131.

NIEDORODA, A. W., REED, C. W., HATCHETT, L. *ET AL.* 2003. Bottom currents, deep sea furrows, erosion rates, and dating slope failure-induced debris flows along the Sigsbee Escarpment in the Deep Water Gulf of Mexico. *OTC 2003 Proceedings*.

NORMARK, W. R. 1978. Growth patterns of deep-sea fans. *AAPG Bulletin*, **54**, 2170–2195.

NORMARK, W. R. 1991. Fan valleys, channels and depositional lobes on modern submarine fans: characters for recognition of sandy turbidite environments. *AAPG Bulletin*, **62**, 912–931.

NORMARK, W. R. & PIPER, D. J. W. 1985. Initiation processes and flow evolution of turbidity currents: implications for the depositional record. *In*: OSBOURNE, R. H. (ed.) *From Shoreline to abyss*. SEPM, Special Publications, **46**, 207–230.

NOWELL, A. R. M. & HOLLISTER, C. D. 1982. *Deep Ocean Sediment Transport*. Elsevier, Amsterdam.

NOWELL, A. R. M., HOLLISTER, C. D. & JUMARS, P. A. 1985. High Energy Benthic Boundary Layer Experiment: HEBBLE. *EOS Transactions, American Geophysical Union*, **63**, 594–595.

NOWELL, A. R. M., MCCAVE, I. N. & HOLLISTER, C. D. 1985. Contributions of HEBBLE to understanding marine sedimentation. *Marine Geology*, **66**, 397–409.

PARIZE, O., VIANA, A. R., FAUGÈRES, J. C., IMBERT, P. & RUBINO, J.-L. 1998. Stratigraphical organization of upper slope deposits of passive margin: comparison between outcrop (Apto-Albian clastic deposits of south-east France) and modern sedimentation (Campos Basin, Brazil). *SEPM Research Conference STRATACON—Strata and Sequences on Shelves and Slopes—Sicily, Italy, Abstracts, Volume*.

PAULL, C. K., MATSUMOTO, R., WALLACE, P. J. & DILLON, W. P. (eds) 2000. *Proceedings of the Ocean Drilling Program, Scientific Results, 164*. Ocean Drilling Program, College Station, TX.

PIPER, D. J. W. & NORMARK, W. R. 2001. Sandy fans—from Amazon to Hueneme and beyond. *AAPG Bulletin*, **85**, 1407–1438.

POSAMENTIER, H. W. & VAIL, P. R. 1988. Eustatic control on clastic deposition II—sequence and systems tracts models. *In*: WILGUS, C. K., HASTINGS, B. S., KENDALL, C. G. ST. C., POSAMENTIER, H. W., ROSS, C. A. & VAN WAGONER, J. C. (eds) *Sea Level Changes: an Integrated Approach*. SEPM, Special Publications, **42**, 125–154.

RAHMSTORF, S. 2002. Ocean circulation and climate during the past 120,000 years. *Nature*, **419**(12 Sept), 207–214.

RAMSAY, P. J. 1994. Marine geology of the Sodwana Bay shelf, Southeast Africa. *Marine Geology*, **120**, 225–247.

REEDER, M. S., ROTHWELL, G. & STOW, D. A. V. 2002. The Sicilian gateway: anatomy of the deep-water connection between East and West Mediterranean basins. *In*: STOW, D. A. V., PUDSEY, C. J., HOWE, J. A., FAUGÈRES, J.-C. & VIANA, A. R. (eds) *Deep-Water Contourite Systems: Modern Drifts and Ancient Series, Seismic and Sedimentary Characteristics*. Geological Society, London, Memoirs, **22**, 171–190.

ROBINSON, S. G. & MCCAVE, I. N. 1994. Orbital forcing of bottom currents enhanced sedimentation on Feni drift, NE Atlantic, during the Mid-Pleistocene. *Paleoceanography*, **9**(6), 943–972.

RODRÍGUEZ, A. B. & ANDERSON, J. B. 2004. Contourite origin for shelf and upper slope sand sheet, offshore Antarctica. *Sedimentology*, **51**, 699–711.

SHANMUGAM, G., SPALDING, T. D. & ROFHEART, D. H. 1993. Process sedimentology and reservoir quality of deep-marine bottom-current reworked sands (sandy contourites): an example from the Gulf of Mexico. *AAPG Bulletin*, **77**, 1241–1259.

SILVEIRA, D. P. & MACHADO, M. A. P. 2004. *Bacias sedimentares brasileiras, Bacia de Pelotas*. Phoenix Bulletin, Fundação Paleontológica Phoenix, Aracaju, Brazil, **63**.

SOUZA CRUZ, C. E. 1995. Estratigrafia e sedimentação de águas profundas do Neogeno da Bacia de Campos, estado do Rio de Janeiro, Brasil. PhD thesis, Instituto de Geociências—Universidade Federal do Rio Grande do Sul, Porto Alegre.

SOUZA CRUZ, C. E. 1998. South Atlantic paleoceanographic events recorded in the Neogene deep water section of the Campos Basin, Brazil. *AAPG Bulletin*, **82**(10), 1883–1984.

STANLEY, D. J. 1993. Model for turbidite-to-contourite continuum and multiple process transport in deep marine settings: examples in the rock record. *Sedimentary Geology*, **82**(1–4), 241–255.

STOW, D. A. V. & FAUGÈRES, J.-C. (eds) 1998. Contourites, Turbidites and Processes Interaction. *Sedimentary Geology Special Issue*, **115**(1–4).

STOW, D. A. V., FAUGÈRES, J. C., VIANA, A. R. & GONTHIER, E. 1998. Fossil contourites, a critical review. *Sedimentary Geology*, **115**(1–4), 3–32.

STOW, D. A. V., PUDSEY, C. J., HOWE, J. A., FAUGÈRES, J.-C. & VIANA, A. R. (eds) 2002. *Deep-Water Contourites: Modern Drifts and Ancient Series, Seismic and Sedimentary Characteristics*. Geological Society, London, Memoirs, **22**.

VAIL, P. R., TODD, R. G. & SANGREE, J. B. 1977. Seismic stratigraphy and global changes of sea level, Part five: chronostratigraphic significance of seismic reflections. *In*: PAYTON, C. E. (ed.) *Seismic Stratigraphy—Applications to Hydrocarbon Exploration*. AAPG Memoirs, **26**, 99–116.

VAIL, P. R., AUDEMARD, F., BOWMAN, S. A., EISNER, P. N. & PEREZ-CRUZ, C. 1991. The stratigraphic signature of tectonics, eustasy and sedimentology—an overview. *In*: EINSELE, G., RICKEN, W. & SEILACHER, A. (eds) *Cycles and Events in Stratigraphy*. Springer, Berlin, 617–659.

VIANA, A. R. 1998. Le rôle et l'enregistrement des courants océaniques dans les dépôts de marges continentales: la marge du bassin sud-est Brésilien. PhD thesis, Bordeaux I University.

VIANA, A. R. 2001. Seismic expression of shallow- to deep-water contourites along the south-eastern Brazilian margin. *Marine Geophysical Researches*, **22**(5–6), 509–521.

VIANA, A. R. & FAUGÈRES, J. C. 1998. Upper slope sand deposits; the example of Campos Basin, a latest Pleistocene–Holocene record of the interaction between alongslope and downslope currents. *In*: STOKER, M. S., EVANS, D. & CRAMP, A. (eds) *Geological Processes on Continental Margins; Sedimentation, Mass-Wasting and Stability*, Geological Society, London, Special Publications, **129**, 287–316.

VIANA, A. R. & STOW, D. A. V. 2000. Seafloor polishing and sand spillover at the outer shelf to upper slope boundary. Spotlight, Contourite Watch, Issue 4, Dec. 2000, Southampton, *IGCP Newsletter*, **432**, 8–9.

VIANA, A. R., FAUGÈRES, J.-C., KOWSMANN, R. O., LIMA, J. A. M., CADDAH, L. F. G. & RIZZO, J. G. 1998. Hydrology, morphology and sedimentology of the Campos Continental Margin, Offshore Brazil. *Sedimentary Geology*, **115**(1–4), 133–158.

VIANA, A. R., ALMEIDA, C. W., SCHREINER, S., *ET AL.* 2001. Improving interpretation of seafloor geology from the integration of conventional marine geology tools and 3D seismic. 7th International Congress, Brazilian Geophysical Society, Salvador, Expanded Abstracts, CD-ROM.

VIANA, A. R., ALMEIDA, W. JR. & ALMEIDA, C. F. W., 2002*a*. Upper slope sands—the late Quaternary shallow-water sandy contourites of Campos Basin SW Atlantic margin. *In*: STOW, D. A. V., PUDSEY, C. J., HOWE, J. A., FAUGÈRES, J.-C. & VIANA, A. R. (eds) *Deep-Water Contourite Systems: Modern Drifts and Ancient Series, Seismic and Sedimentary Characteristics*. Geological Society, London, Memoirs, **22**, 261–270.

VIANA, A. R., HERCOS, C. M., ALMEIDA, W. JR, MAGALHÃES, J. L. C. & ANDRADE, S. B. 2002*b*. Evidence of bottom current influence on the Neogene to Quaternary sedimentation along the northern campos slope, SW Atlantic Margin. *In*: STOW, D. A. V., PUDSEY, C. J., HOWE, J. A., FAUGÈRES, J.-C. & VIANA, A. R. (eds) *Deep-Water Contourite Systems: Modern Drifts and Ancient Series, Seismic and Sedimentary Characteristics*. Geological Society, London, Memoirs, **22**, 249–259.

WYNN, R. B., MASSON, D. G. & BETT, B. J. 2002. Hydrodynamic significance of variable ripple morphology across deep-water barchan dunes in the Faroe–Shetland Channel. *Marine Geology*, **192**, 309–319.

North Atlantic contourite sand channels

A. AKHMETZHANOV[1], N. H. KENYON[1], E. HABGOOD[1],
A. S. VAN DER MOLLEN[2], T. NIELSEN[3], M. IVANOV[4] & P. SHASHKIN[4]

[1]*National Oceanography Centre, Southampton, Empress Dock, Southampton SO14 3ZH, UK (e-mail: Andrey.Akhmetzhanov@noc.soton.ac.uk)*

[2]*Netherlands Institute of Applied Geoscience TNO–National Geological Survey, Kriekenpitplein 18, P.O. Box 80015, 3508 TA, Utrecht, The Netherlands*

[3]*Geological Survey of Denmark and Greenland, Thoravej 8, 2400 Copenhagen NV, Denmark*

[4]*UNESCO–MSU Centre for Marine Geosciences, Geology Faculty, Lomonosov Moscow State University, GSP-2, Leninskie Gory, 119899, Moscow, Russia*

Abstract: Two sand-rich channelized depositional systems, formed by strong contour currents, were studied west of the Faeroe Bank Channel and in the Gulf of Cadiz. Both are areas beyond the exit of constrictions where water overflows from the Norwegian Sea and the Mediterranean Sea, respectively. West of the Faeroe Bank, newly mapped channels are developed mainly under the influence of a geostrophic current and are characterized by significant lateral migration, which determines the marked cross-sectional asymmetry and the architecture of the deposits. The pathways of the Mediterranean Undercurrent in the Gulf of Cadiz are complex, with the greater proportion flowing under geostrophic conditions along a terrace but with some of the denser water becoming ageostrophic and descending downslope owing to gravity. A series of 'peel-off' channels is formed, with the largest one, Gil Eanes, being about 40 km long. Most of the channel fills consist of medium–coarse sand. Levees are mainly silts with a higher sand content in the vicinity of the channel. Both depositional systems have a variety of contourite sand channels, which in most respects are remarkably similar. In both cases there are stretches where the flow is ageostrophic, with water descending downslope for as much as 400 m before resuming geostrophic flow at deeper levels. In each case the main pathway of the densest water is the shallowest and several branches turn off to the left of this main pathway before bending to the right under the influence of Coriolis forces. In both cases there are channel fills of medium–coarse sand, probably cross-bedded, and up to 200 ms thick. Sheets of sand with a thickness of a few metres to a few tens of metres are common. Similarities to turbidite channels are the aggradational nature of some channel floors and the flanking muddy or silty sediment waves. Contourite channel depositional complexes are distinguished from turbiditic ones by their coarsening-up rather than fining-up sand units, the asymmetry in channel architecture, the presence of regional unconformities, and the distribution pattern with well-marked boundaries of current-derived deposits.

Geostrophic contour currents are known to be important agents of sediment transport in the deep sea. Giant bodies of muddy sediment formed by contour currents are recognized in many parts of the slope and rise (e.g. McCave & Tucholke 1986; Faugères *et al.* 1993). Normally sand makes up a rather insignificant portion of the net sediment volume (Viana *et al.* 1998). However, there are several sedimentary systems where sand does play an important role, forming significant accumulations that can be considered as potential hydrocarbon reservoirs. Two such systems were studied during several cruises conducted by the UNESCO–IOC Training-through-Research (TTR) Programme and by the Challenger Division of the National Oceanography Centre, Southampton, in the areas lying west of the Faeroe Bank Channel and in the Gulf of Cadiz (Fig. 1).

Methods

Datasets used for this study were collected with a wide range of echosounders, sidescan sonars, seismic systems, bottom sampling systems and cameras. In the Faeroe Bank area the survey included 10 kHz OKEAN long-range sidescan sonar, a single-channel seismic system with 1.8 l air-gun source and 30 kHz OREtech deep-towed

From: VIANA, A. R. & REBESCO, M. (eds) *Economic and Palaeoceanographic Significance of Contourite Deposits*. Geological Society, London, Special Publications, **276**, 25–47.
0305-8719/07/$15.00

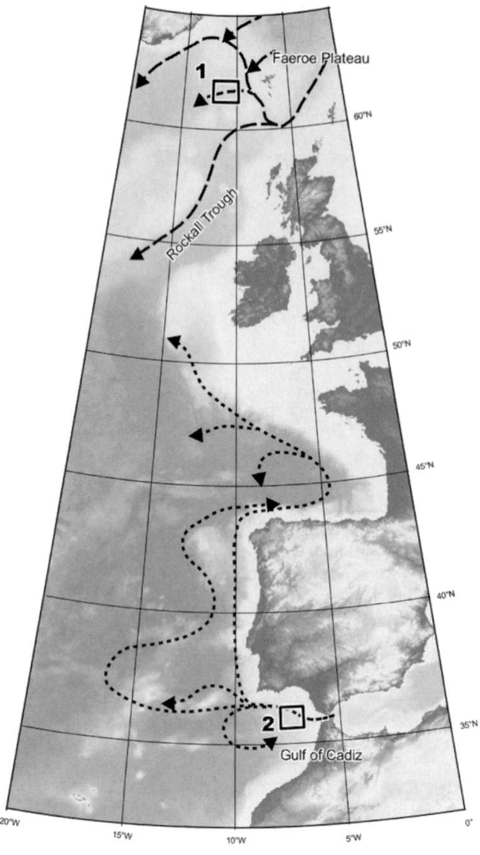

Fig. 1. General map showing study areas. Norwegian Sea Outflow pathways adapted from Hansen & Østerhus (2000), Mediterranean Outflow pathways adapted from Iorga & Lozier (1999).

sidescan sonar with built-in bottom 7 kHz profiler. Acoustic and seismic data from the Gulf of Cadiz area were collected using 11–12 kHz SEAMAP long-range sidescan sonar, operated by the US Naval Oceanographic Office and the Naval Research Laboratory, 30 kHz TOBI deep-towed sidescan sonar, operated by the NOC, Southampton, and a single-channel seismic system. Underwater television and bottom sampling with conventional box and large-diameter gravity corers and a giant piston corer were used for groundtruthing.

Acoustic data were digitally enhanced and geographically registered. Seismic processing included predictive deconvolution and band-pass filtering. Cores were visually logged with subsequent measurement of magnetic susceptibility using a Bartington MS2 system with a MS2C core logging sensor. A grain-size analysis was undertaken on selected samples from the Faeroe Bank Channel area with a Fritsch Analysette A22 laser particle sizer.

Area west of Faeroe Bank Channel

The cold Norwegian Sea Outflow Water (NSOW) flows with considerable speed through the Faeroe Bank Channel, which is the deepest passageway for bottom water out of the Norwegian Sea into the North Atlantic (Fig. 2). To the south of the Faeroe Islands the outflow is confined between the Faeroe platform and Faeroe Bank where, as a result of its topographic restriction, its speed reaches 1 m s^{-1} (Crease 1965; Saunders 1990). To the NE of the Faeroe Bank the channel broadens rapidly and as a result the outflow speed can decelerate to about 20 cm s^{-1} on average (Crease 1965). Part of the flow carries on along the slope of the Faeroe–Iceland Ridge with its core at a depth of between about 900 and 1100 m, where its pathway is marked by a zone of sandy bedforms (Dorn & Werner 1993). To the NW of the Faeroe Bank, published bathymetric maps (Crease 1965; Fleisher *et al.* 1974; Hansen & Meincke 1979; British Oceanographic Data Centre 1997) show the presence of channels in the depth range of 1200–1400 m. Bathymetry derived from satellite-altimetry with shipborne sounding control (Smith & Sandwell 1994) also shows a pattern of anastomosing channels (Fig. 2a). Several researchers have reported the presence of dense, cold (3°) Norwegian Sea Outflow Water associated with these channels (Crease 1965; Kuijpers *et al.* 1998). The channels are believed to be formed by a fast-flowing bottom current representing an outlet of NSOW that diverts to the left from the main pathway along the Faeroe–Iceland Ridge after it leaves the Faeroe Bank Channel mouth. The current is known to have been operating in this area since the Early Miocene (Boldreel *et al.* 1998; Kuijpers *et al.* 1998) or possibly even since earlier Oligocene as suggested by Davies *et al.* (2001), which indicates that the channels have had a relatively long evolution. Additional sidescan sonar, seismic and groundtruthing data have resolved the channel pattern and their morphology, and the processes of sediment transport and deposition. They were obtained during one of the ECs ENAM-II cruises onboard R.V. *Pelagia* in 1995 and on the seventh TTR cruise onboard R.V. *Professor Logachev* in 1997.

Channel pattern and geometry

Sea-floor sonographs. Long-range OKEAN sidescan sonar and single-channel seismic surveys conducted west of the mouth of the Faeroe Bank

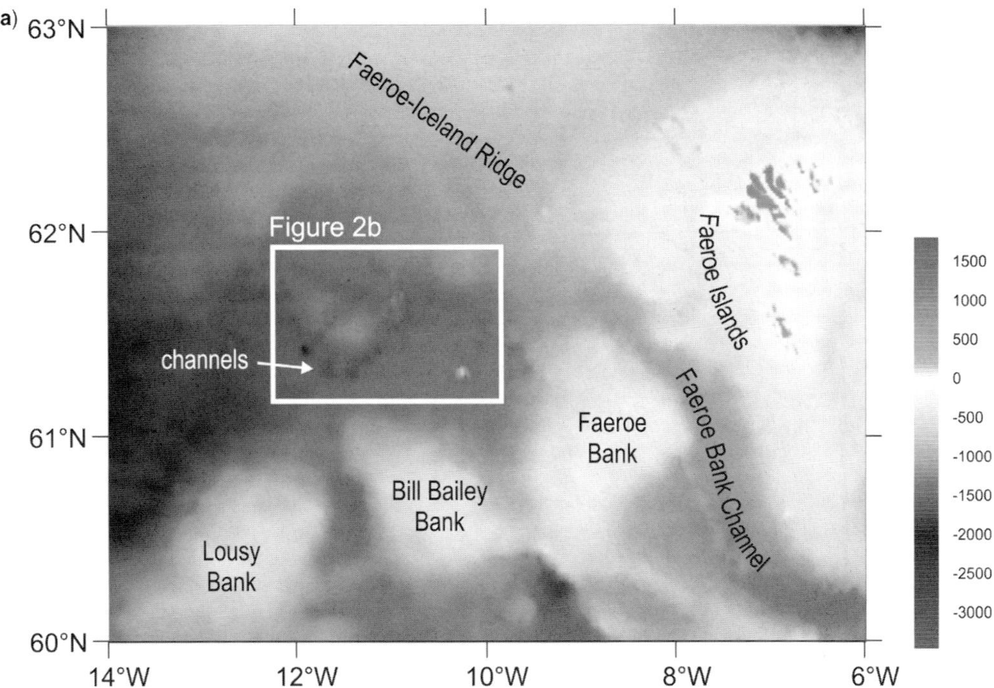

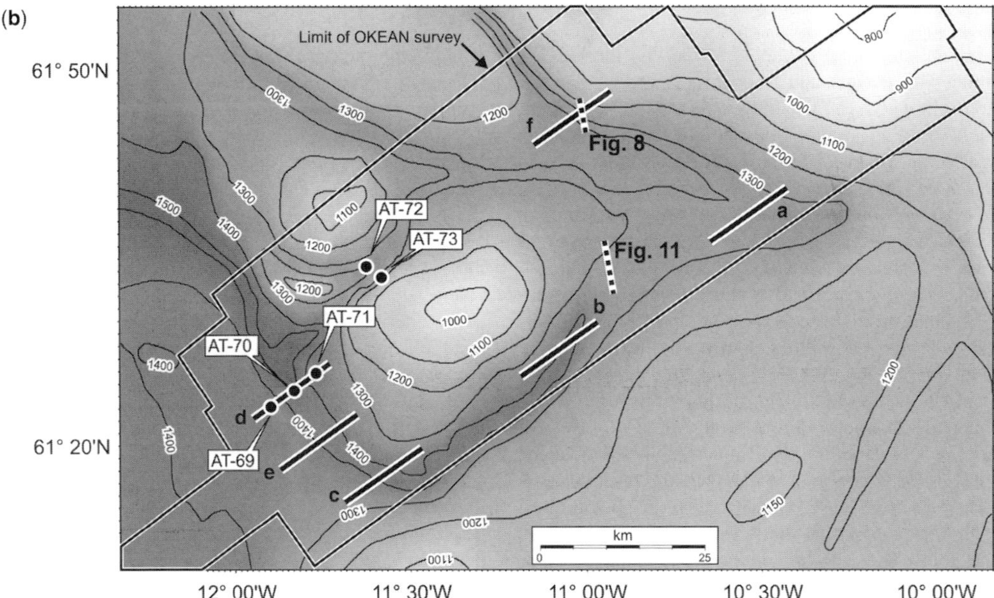

Fig. 2. (**a**) Satellite-derived bathymetry image showing channels west of the Faeroe Bank; (**b**) location map of the study area (lower-case letters correspond to seismic sections shown in Fig. 7).

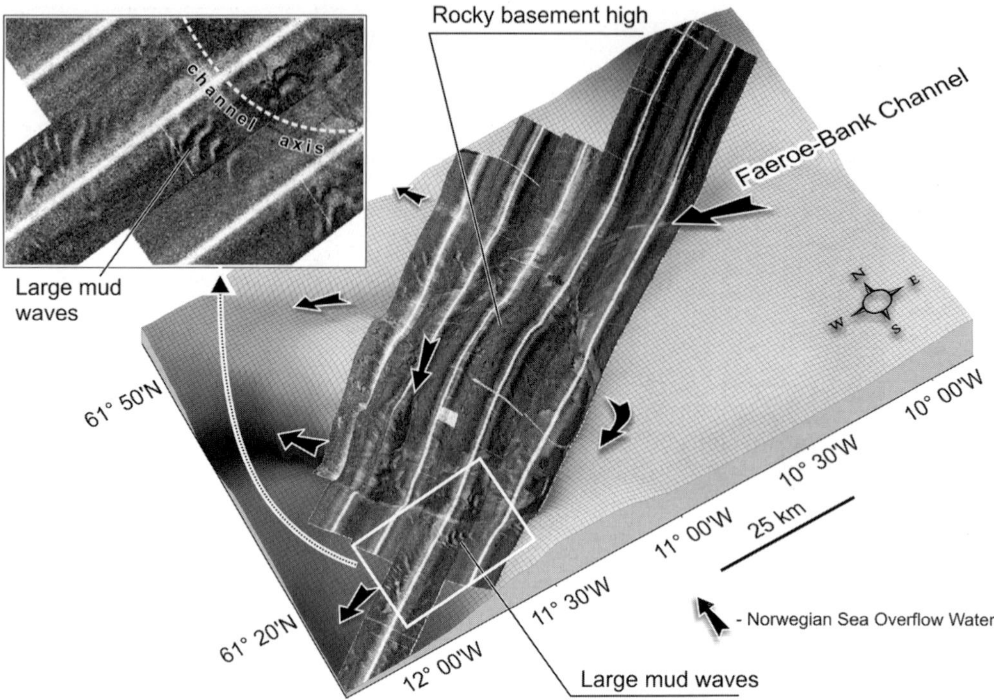

Fig. 3. Perspective view of the OKEAN mosaic draped over bathymetry. Major NSOW pathways are indicated by arrows. High backscatter is shown by darker shades.

Channel show that channels formed by the NSOW outlet have an anastomosing pattern that is controlled by a flat-topped acoustic basement high standing about 200 m above the surrounding sea floor (Fig. 2b). The outflow water enters the area from the east, where the fast-flowing bottom current sweeps the sea floor, forming an extensive area of erosion and non-deposition clearly seen on seismic records (see below). OKEAN sonographs also show a large-scale sea-bed lineation of a higher acoustic backscatter, indicating the presence of a strong bottom current (Fig. 3).

Most of the bottom water forming the newly mapped channels is the main flow of NSOW coming out of the Faeroe Bank Channel. However, there is evidence on the TOBI 30 kHz sonographs (Kuijpers *et al.* 1998) that there is a contribution from a proportion of the NSOW approaching from the south (Fig. 4). This is Norwegian Sea water that overflows the Wyville-Thomson Ridge and follows the bottom contours at a depth of about 1000 m. It flows clockwise around the Faeroe Bank and Bill Bailey Bank (Boldreel *et al.* 1998; Kuijpers *et al.* 1998) as shown by areas of sea-bed erosion and coarse sediment transport, observed on TOBI sonographs (Kuijpers *et al.* 1998; Fig. 4). It is postulated here that this branch of the NSOW rejoins the main flow and that the water masses combine before flowing into the North Atlantic. The shallowest filament flows along the Faeroe–Iceland Ridge at a depth of about 1000 m (Dorn & Werner 1993). At the basement high the flow bifurcates and forms several distinct channels bending around either side of the plateau. The more prominent channel runs around the southern side of the plateau, becoming progressively deeper as it does so (Fig. 5). It makes an almost 90° change of direction around the southwestern corner of the plateau and then runs to the NW at a depth of about 1400–1500 m. Another filament flows north of the plateau, bifurcating twice before one branch eventually rejoins the clockwise flow at a depth of 1400–1500 m. This is deeper than used to be supposed for fast-flowing NSOW in this region (e.g. Hansen & Meincke 1979) but in keeping with there being several cores (e.g. Hansen & Østerhus 2000).

An OKEAN 10 kHz sonograph shows that this channel has a generally weakly backscattering floor and left-hand bank, whereas the right-hand bank has stronger backscatter. Such a backscatter pattern was also observed on a 30 kHz TOBI sonograph that partially overlaps with the OKEAN

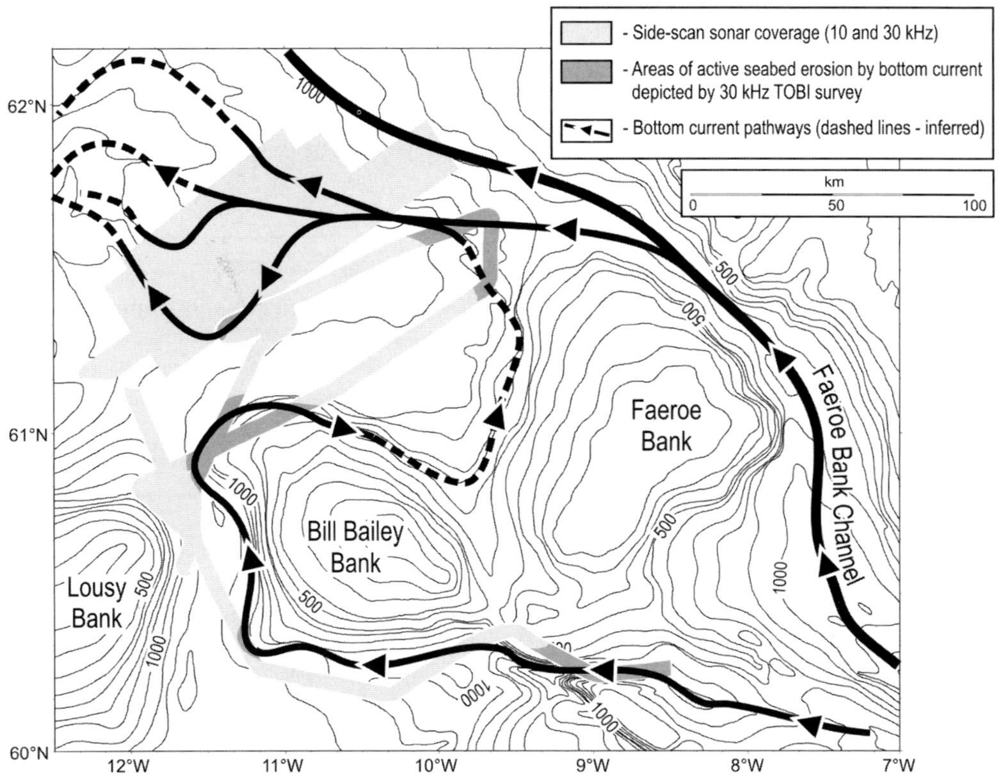

Fig. 4. Main bottom-current pathways in the Faeroe Bank area. The pathway around the Bill Bailey Bank is from Kuijpers *et al.* (1998).

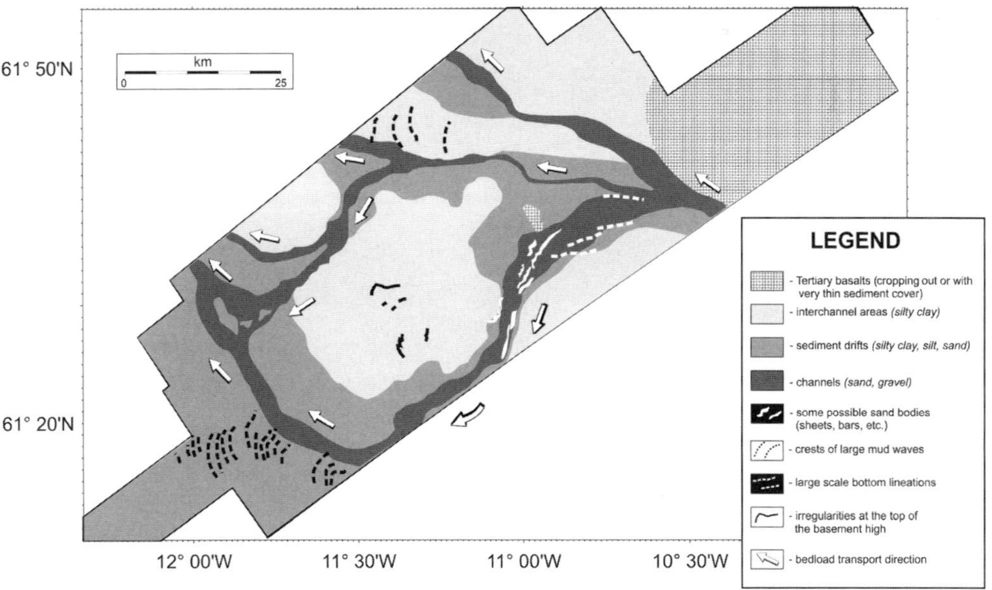

Fig. 5. Map of the main features and bedload transport pathways west of the Faeroe Bank Channel based on integrated interpretation of seismic, sidescan sonar and bottom sampling data.

sonograph and provides a more detailed picture of this fragment of the channel (Fig. 6). Apparently a transition from the channel floor and the right-hand wall is marked by the sharp boundary between low and high backscatter areas, where low backscatter corresponds to the channel floor and high to the channel wall. Next to this boundary there is a 300 m wide zone of transverse bedforms. These are believed to be sand waves, with a wavelength of about 100 m (Kuijpers *et al.* 1998), based on their shape and on the low backscattering characteristics, as low backscattering is usually associated with sand on sonographs (e.g. Belderson *et al.* 1972). A similar backscatter pattern was observed

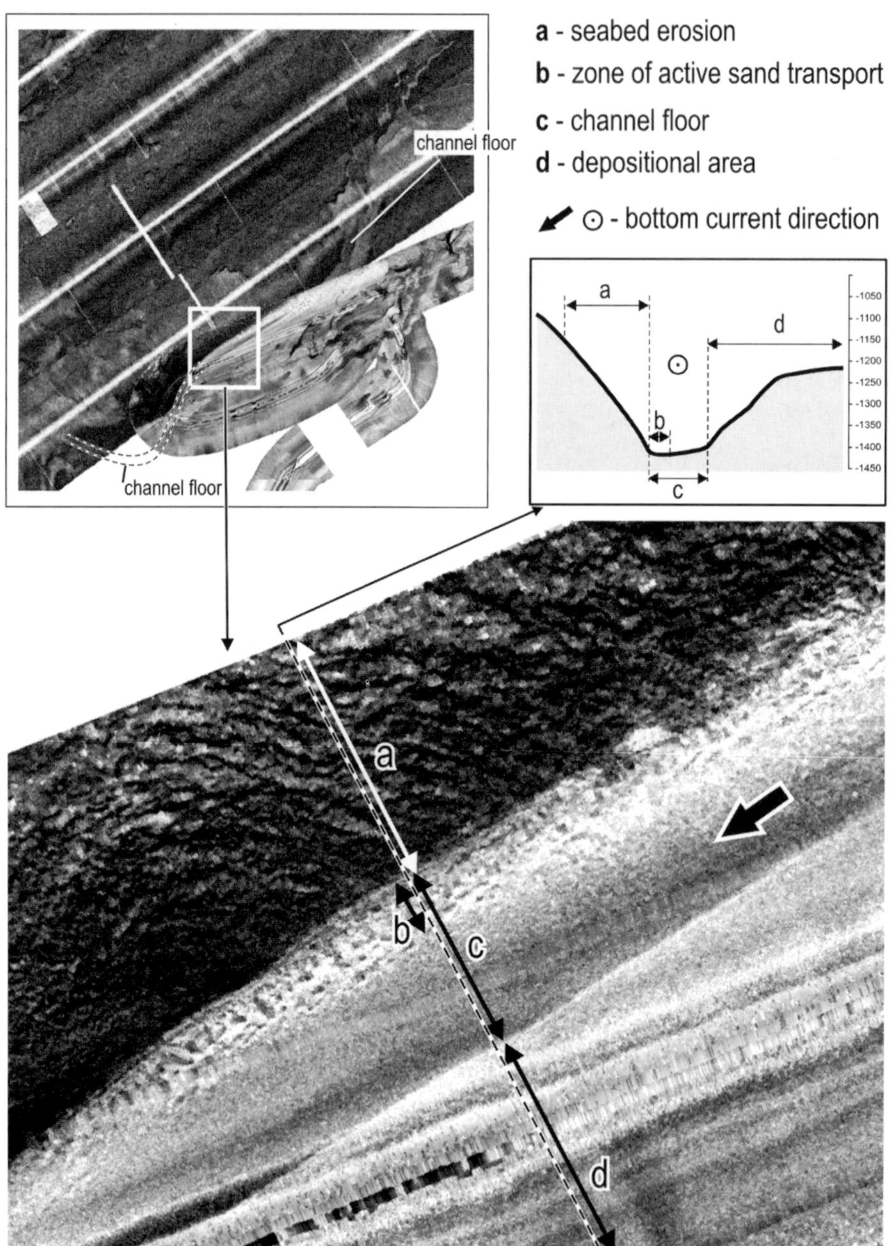

Fig. 6. Details of a contourite channel on the OKEAN and TOBI images. High backscatter is shown by darker shades.

on the OREtech 30 kHz sonograph that runs across one of the northwesterly directed channels. The slightly higher resolution sonograph also shows a characteristic sea-bed current lineation on the channel floor, similar to that reported by Kuijpers *et al.* (1998) on the TOBI sonograph to the north of the Faeroe Bank. A series of large mud waves with a wavelength of about 1.5 km appears on the OKEAN sonograph that covers the channel running south of the plateau. They are on the outside of the major channel bend. Waveforms of similar size were also noticed on the TOBI sonograph further to the SW, although they are more probably associated with the NSOW branch that flows around the Bill Bailey Bank.

Seismic and sub-bottom profiler expression. NSOW pathways in the study region are identified on the seismic records as areas of non-deposition or erosion or as well-marked channels. The former are encountered to the east where the Faeroe Bank Channel outlet enters the region and a strong bottom current prevents deposition of fine sediments.

The lower part of the seismic succession represents an acoustic basement that is topped by a pronounced high-amplitude reflector (Fig. 7). Its internal structure is mostly chaotic, although in the upper part prolonged parallel reflectors are common. The upper part of the succession comprises a pattern of lower-amplitude continuous parallel, often sinuous, reflectors. The surficial succession becomes more complex within the banks to the left of the channels, with three units being identified. The lower unit (Unit I) forms a sedimentary drape over the acoustic basement and has a set of parallel continuous reflectors. Near the channels the upper boundary of the unit is marked by an erosional unconformity and the overlying middle unit (Unit II) has a pattern of sigmoidal reflectors. The thickness of the unit is about 100 ms and it pinches out away from the channels. Toward the channel the erosive character of its base becomes more obvious and high-amplitude events associated with channel fills can be recognized. The seismic pattern suggests that the channels have migrated to the NE; that is, to the right of the flow direction. The upper boundary of the middle unit is also an erosional unconformity and most of the upper unit (Unit III) comprises a set of parallel continuous reflectors. The erosion at the base of the upper unit becomes more pronounced towards the channel, where the unit lies directly on the basement and forms most of the bank of the channel. The bank is a broad elevation about 75 m high and the thickness of the unit here reaches 450 ms. In the lower part of the unit sigmoidal reflectors are observed whereas the uppermost part reveals a characteristic pattern of parallel continuous reflectors typical for contourite drifts (Faugères *et al.* 1999). The seismic pattern also suggests that the banks or walls to the left of the channels are formed mainly as a result of lateral accretion. The right-hand banks or walls are either erosional or have a reduced sedimentary succession suggesting either very low sedimentation rates or erosion. A high-resolution 3.5 kHz profile running across one of the channels also shows that the right-hand bank is more gentle and erosive whereas the left-hand one is steeper because of accumulation of presumed fine-grained sediment (Fig. 8). According to the thickness of individual layers resolved on the record the sedimentation rate is five times higher on the left-hand bank compared with the surrounding area.

Seismic records show that channels have flat floors and are 2–3 km wide and about 180 m deep. High-amplitude reflectors beneath some of the channels suggest the presence of a possible coarse aggradational deposit that can be up to 200 ms thick. High-resolution data also suggest the presence of the coarse sediment on the channel floor, where an acoustically transparent lens-like sediment body about 10 m thick is observed on the record. This agrees with observations from 3.5 kHz TOBI profiler records obtained from the main channel, where a sand body with a thickness <10 m was reported (Kuijpers *et al.* 1998).

Sediment cores

Five cores were collected from the banks and channel floors to groundtruth sidescan sonographs and to calibrate different morphological elements of contourite channels. Although no age determinations were performed on these cores it was possible to perform partial correlation with existing cores from the area (Kuijpers *et al.* 1998) on the basis of the presence of an ash layer and available magnetic susceptibility (MS) data.

Channel banks. Three cores from banks recovered mainly a silty–clayey sequence topped by a Holocene sand layer (Fig. 9). Grain-size analysis revealed typical contourite successions (Stow 1979) with a coarsening-upward trend. Correlation with cores collected to the south of the area (Kuijpers *et al.* 1998) showed that cores from left-hand banks contain an extended sedimentary succession as a result of higher sedimentation rates compared with cores from right-hand banks. Marine Isotope Stages (MIS) are allocated based on attribution of the ash layer to the boundary of MIS 2–3 or to the top of MIS 3 and on the comparison with the MS curve of the core ENAM-33 by Kuijpers *et al.* (1998). Clays are taken to represent extreme glacial periods; for instance, MIS 2 clayey intervals recognized in

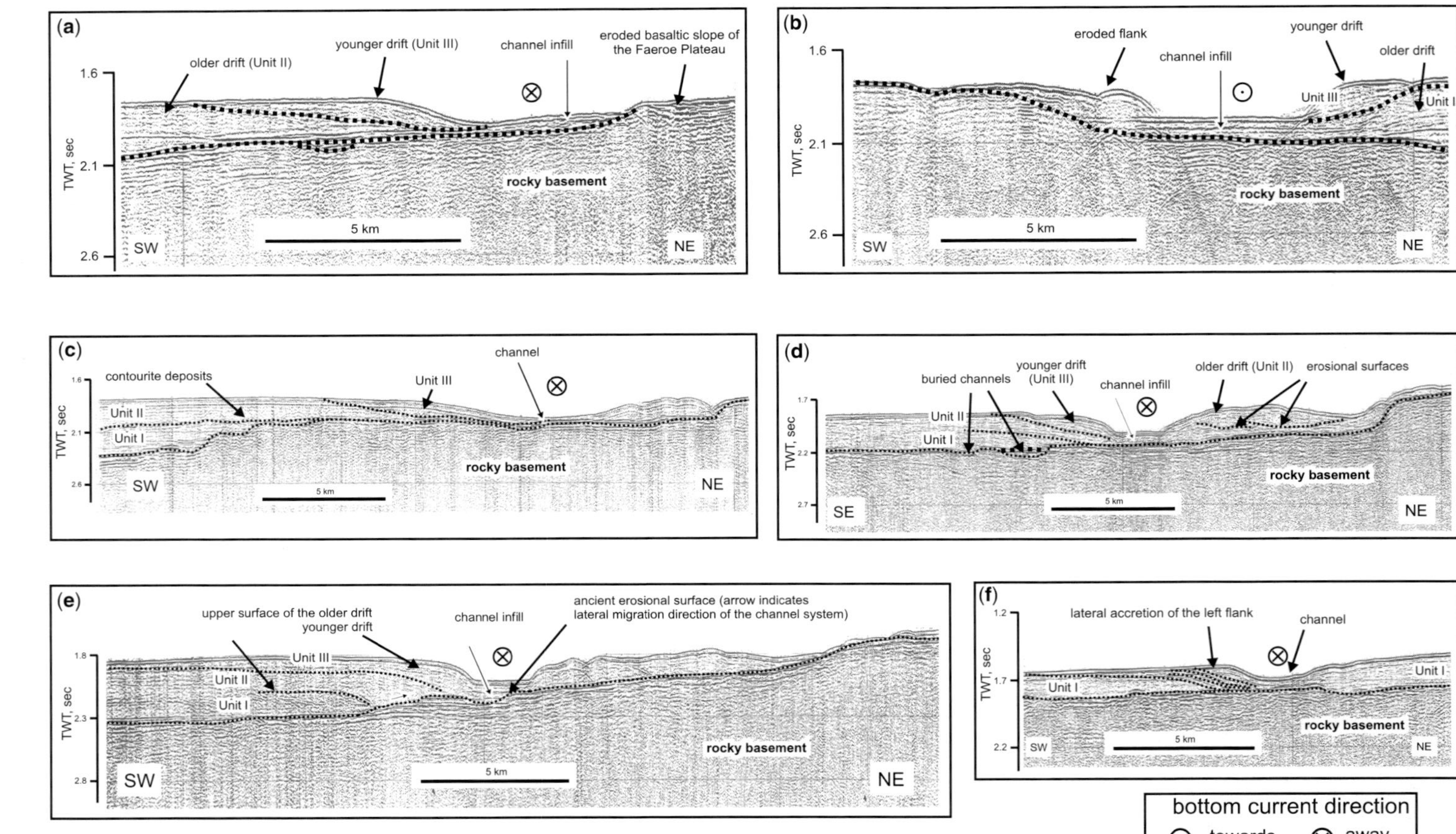

Fig. 7. Seismic profiles across channels showing asymmetry of deposition or erosion and migration of channels to the right. TWT, two-way travel time.

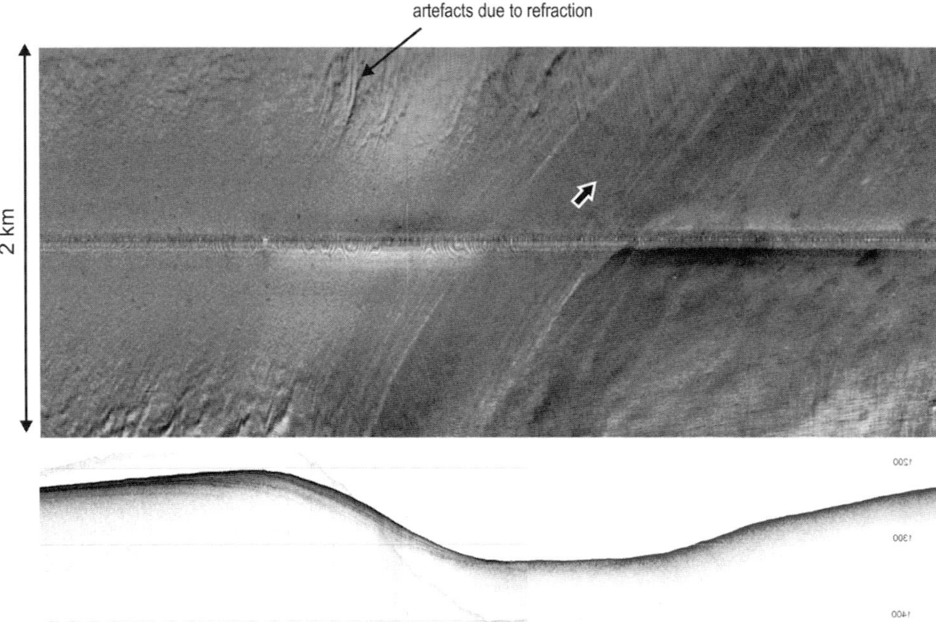

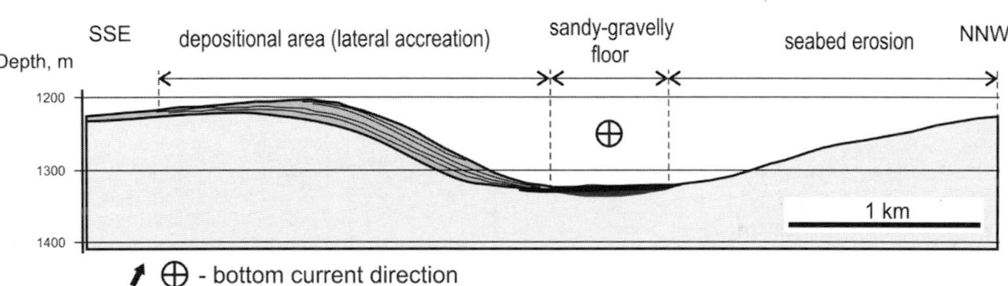

Fig. 8. High-resolution sidescan sonar and bottom profile across a contourite channel showing migration of channel to the right.

cores AT-69G and AT-73G (left-hand banks) are characterized by 2.1–2.5 times higher sedimentation rates than MIS 2 clay in the core AT-71G, taken from a right-hand bank. The sedimentary section in core AT-71G also has a higher content of sandy intervals, which indicates more active bottom dynamics. These observations are in a good agreement with seismic and 3.5 kHz data indicating active deposition of fine-grained material on the left-hand banks of channels and erosion or suppressed sedimentation on the right-hand ones.

Channel floor. Two cores taken from channel thalwegs recovered mainly sandy deposits (Fig. 10). Core AT-70G collected from the main channel contained a coarsening-upward, 50 cm long succession of fine–medium well-sorted terrigenous sand with an interbedded layer of foraminiferal sand. Clay content varies from 15 to 40%. Core AT-72G was taken from a distributary channel to the north of the plateau and the succession is also composed of interbedded layers of terrigenous and foraminiferal sands. Some sandy intervals are coarsening upwards. Both cores have coarse sand–gravel top layers, which are thought to have formed as a result of increased activity of NSOW in the Holocene (Stage 1).

Geometry of sand bodies

Seismic, sidescan sonar and coring data indicate that most of the sand is transported along the channels. High-amplitude reflectors observed on the seismic records beneath channels suggest that

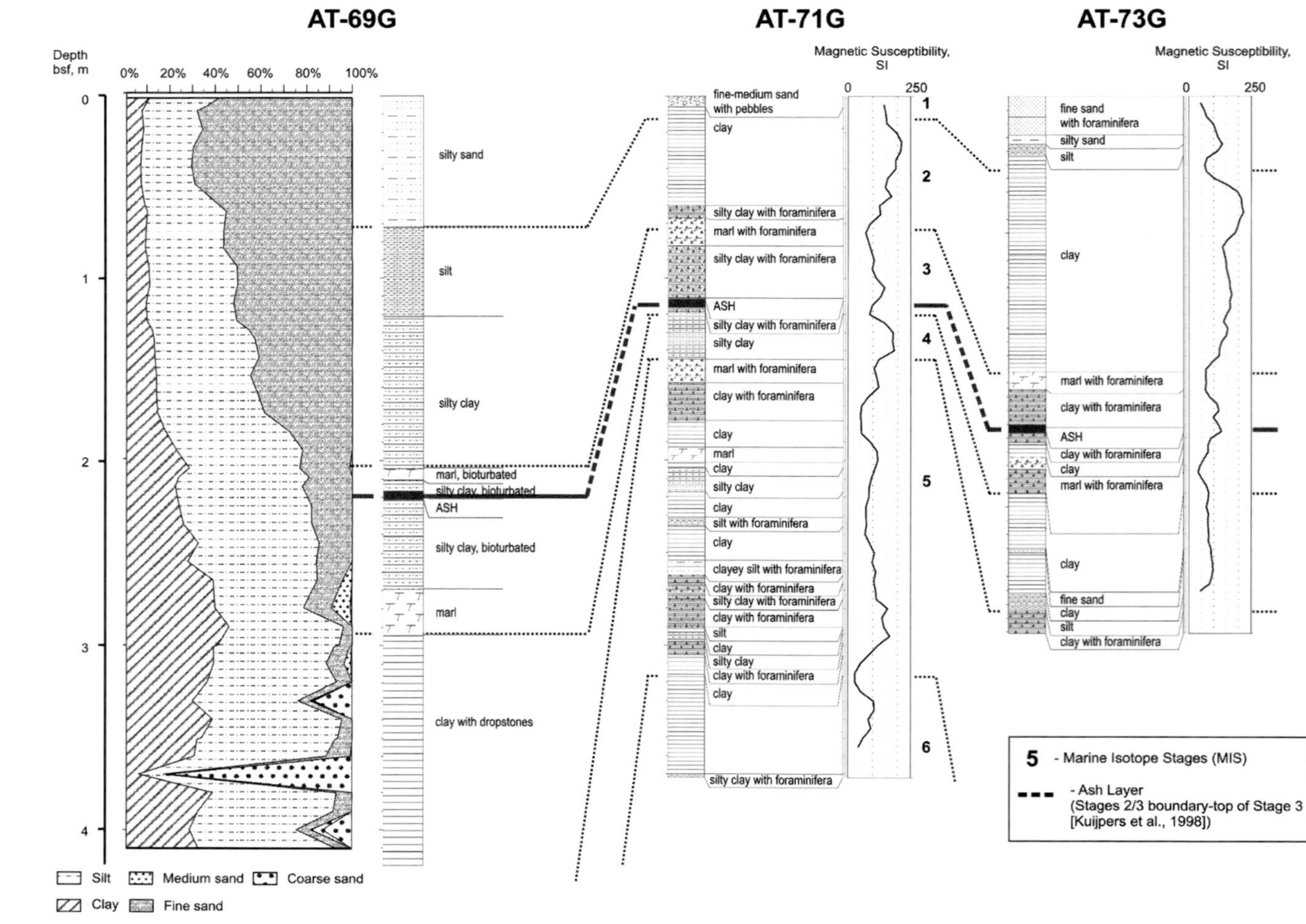

Fig. 9. Logs of cores from contourite channel walls. bsf, below sea floor.

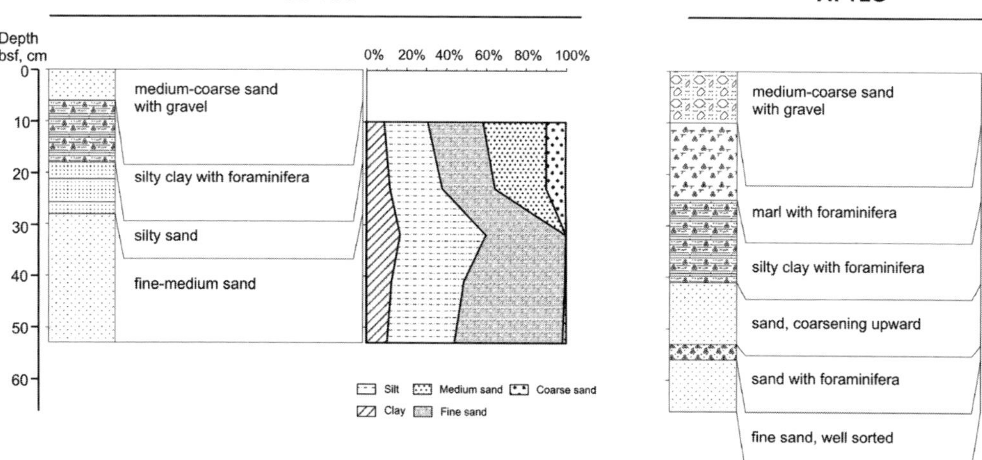

Fig. 10. Logs of cores from contourite channel thalwegs.

sand forms a significant part of channel infills. High-resolution sidescan sonar data obtained across the southern channel show fields of sand about 2 km wide covered in places with trains of sand waves with 50 m wavelength (Fig. 11). These sand fields in turn are fragments of linear low backscattering features about 10–20 km long, also seen on the 10 kHz OKEAN sonograph. Deep-towed 3.5 kHz profiles across one of the sand fields reveal the presence of an acoustically transparent lens-like sedimentary body, about 7 m thick, which is interpreted as a homogeneous sandy deposit (Fig. 11). Similar lens-like sand bodies were observed in the eastern Porcupine Seabight (Akhmetzhanov 2003), in another area of strong bottom current where a large number of such bodies were stacked to form a sheet of sand up to 10 m thick. The presence of similar sheets can also be expected in the study area as shown by seismic records. Unit II, with its sigmoidal high-amplitude reflectors, is interpreted as a sequence of such buried sand sheets formed during channel migration.

Cores collected from the flanks of nearby banks also contain sand-rich intervals, some of which are about 2 m thick (Kuijpers *et al.* 1998) and mainly consist of winnowed Foraminifera shells (50–75% $CaCO_3$). Sand-rich intervals seem to be more common on the right-hand side of these channels.

Gulf of Cadiz

One of the largest known expanses of sand laid down in the deep sea by a contour current is in the Gulf of Cadiz, where the Mediterranean Undercurrent decreases in speed as it comes out of the Strait of Gibraltar (Fig. 12a). The bedform distribution and the geometry of the sandy deposits are best known in their proximal setting (Heezen & Johnson 1969; Kenyon & Belderson 1973; Nelson *et al.* 1993; Mulder *et al.* 2003) where there is a sequence of bedform zones of gravel and sand similar to those mapped on current-swept continental shelves. They are spread across a terrace on the continental slope at depths between about 700 and 900 m. Several filaments of the Undercurrent plunge downslope along channels (Fig. 12b), as shown by Madelain (1970) and Kenyon & Belderson (1973). An extensive dataset has been obtained over the three most proximal channels with a 12 kHz SEAMAP long-range sidescan sonar, a deep towed, 30 kHz TOBI sidescan sonar, single-channel seismic and underwater video systems (Kenyon *et al.* 2000, 2001; Habgood *et al.* 2003). The widest of the channels (Channel 2 of Kenyon & Belderson 1973) is controlled by NE–SW-trending diapiric ridges but the two most proximal channels, a newly discovered channel and Channel 1 of Kenyon & Belderson (1973), named the Gil Eanes Channel by Kenyon *et al.* 2000), are not controlled by underlying topography, at least at present. Their location is thought to be associated with the unconstrained bottom current peeling off downslope through gaps in a longitudinal sandy contourite ridge (Kenyon & Belderson 1973), which acts as a barrier to restrict the main flow of the MOW to the broad current-swept terrace.

Channel pattern and geometry

Sea-floor sonographs and 3.5 kHz profiler records. At least five downslope running channels are identified from the available bathymetric data

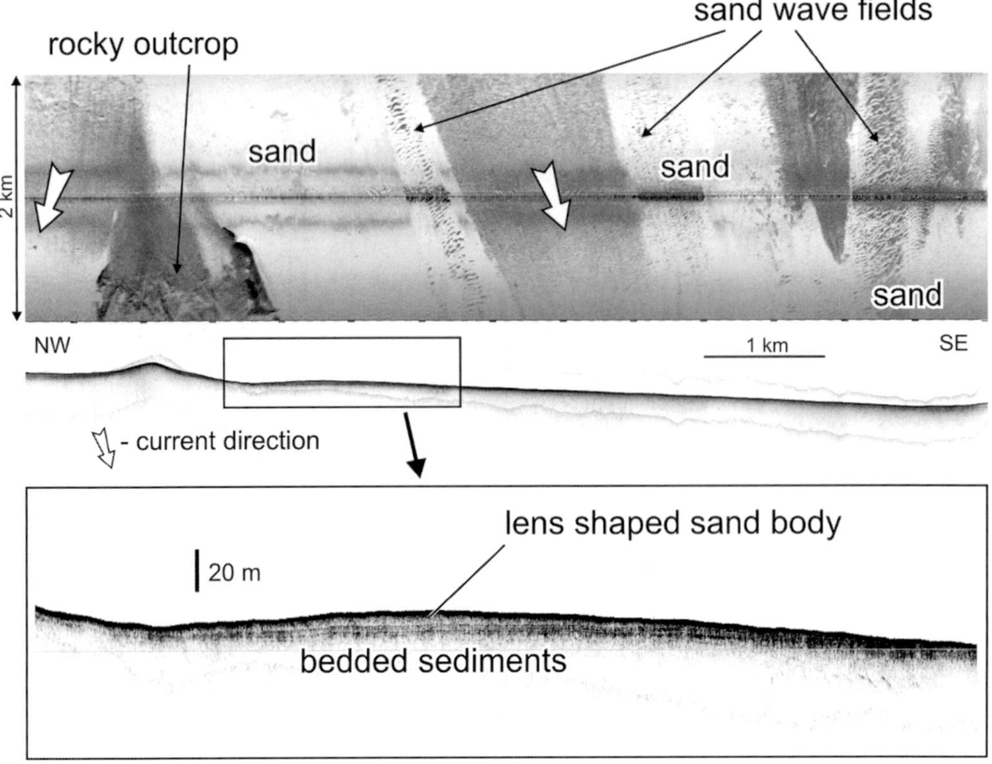

Fig. 11. High-resolution sidescan sonar and sub-bottom profiler records showing sand sheets (lower backscattering shown by lighter shades) on a contourite channel floor.

(Fig. 12a) and are best seen on the 11 kHz SeaMap sonograph mosaic (Fig. 13). The largest channel, Channel 2 of Kenyon & Belderson (1973), is controlled by NE–SW-trending diapiric ridges, which act as a partial barrier across the MOW pathway. The floor of this channel is characterized by relatively high backscatter resulting from scoured rock outcrop and basal conglomerate, across which sand is transported as longitudinal and transverse bedforms.

There are also non-topographically controlled channels (the 'free-standing' channels of Habgood *et al.* (2003)), of which the largest is the Gil Eanes Channel. All of the channels run downslope and trend perpendicular to the main pathway of the MOW (Fig. 13). These channels and their setting have been described by Habgood *et al.* (2003) and so will not be described in detail here. The head of the Gil Eanes Channel is *c.* 6 km wide, and is at a depth of 900 m, and the channel extends downslope for 40 km, terminating at 1250 m water depth. The flanks of the channel are covered with a dense pattern of sediment waves. The morphology and lithology of these flanking waves changes with depth as the influence of the main flow of the MOW lessens. There seems to be a continuum of sediment waves related to this change in current speed, divided here into sand waves, muddy sand waves and mud waves. Their appearance in plan and profile has been shown by Habgood *et al.* (2003).

Sand waves. The sand waves are confined to a part of the terrace, and to the floors and levees of the downslope channel (Fig. 12). With the higher resolution of the TOBI sidescan system several sandy bedforms are differentiated. Most of the sand waves on the terrace are small asymmetrical sand waves (wavelength about 40 m) on the back of large asymmetrical sand waves (wavelength about 300 m). The 3.5 kHz profiles show small, overlapping hyperbolae, with vertices that are tangential to the sea floor. Shorter straight-crested transverse waves with wavelengths of 200–260 m are also common.

Muddy sand waves. This type of wave is mostly found on the flanks of the Gil Eanes Channel. In

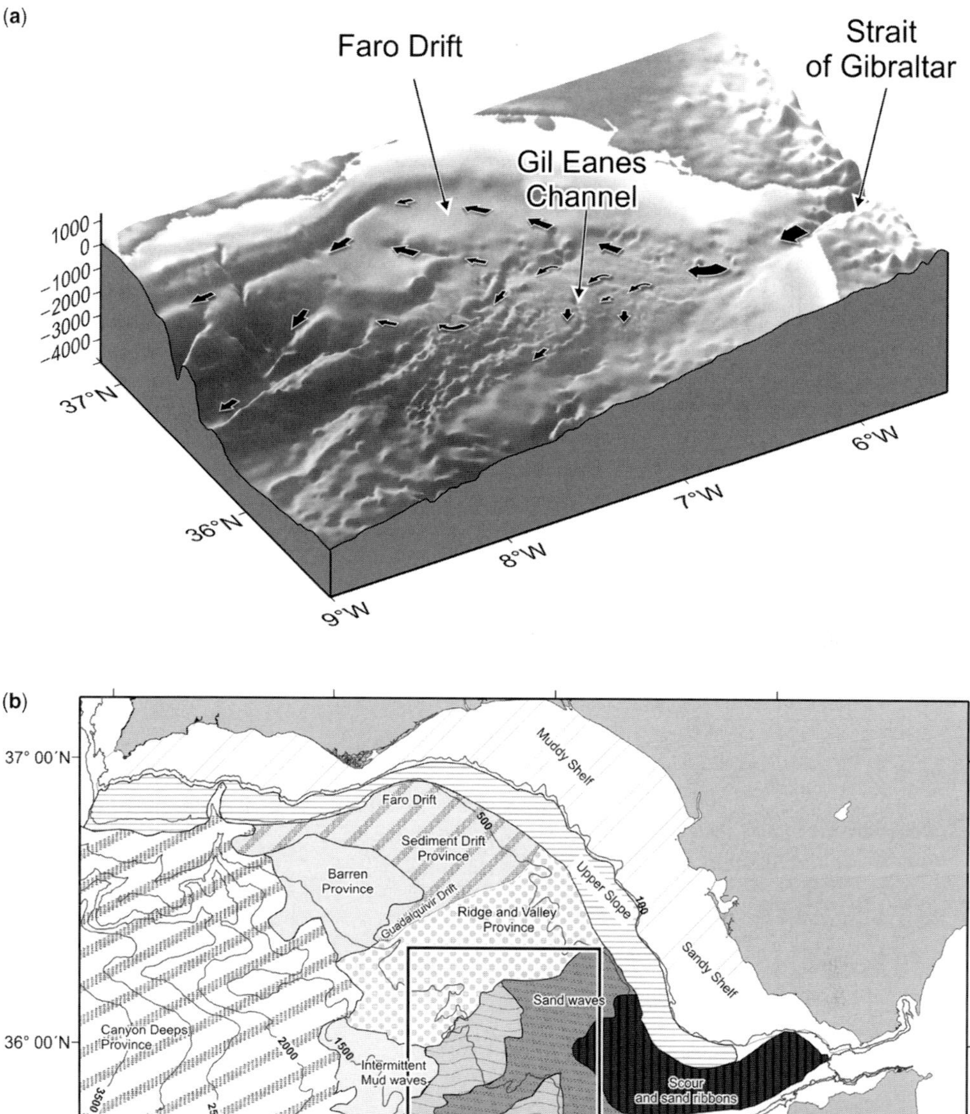

Fig. 12. (**a**) Perspective view of the Gulf of Cadiz sea-floor topography. Arrows show major Mediterranean Outflow Water pathways. (**b**) Physical provinces of the Gulf of Cadiz (modified from Heezen & Johnson 1969; Kenyon & Belderson 1973; Nelson *et al.* 1993, 1999; Baraza *et al.* 1999; Maldonado *et al.* 1999; Habgood *et al.* 2003).

plan view these waves are mainly sinuous to crescentic with short crestlines of *c.* 600 m. Wavelengths range from 300 m to over 600 m but with an average of 350 m. The echo response from 3.5 kHz profiles is of steep-sided, overlapping hyperbolae, thus the waves are interpreted as having relatively steep sides and narrow troughs. This hyperbolic response is similar to the profile type IIB-2 from the western equatorial Atlantic (Damuth 1975) and has been associated with both turbidity-current processes and bottom-current processes.

Mud waves. These waves occur in deeper water beyond the fields of muddy sand waves on the

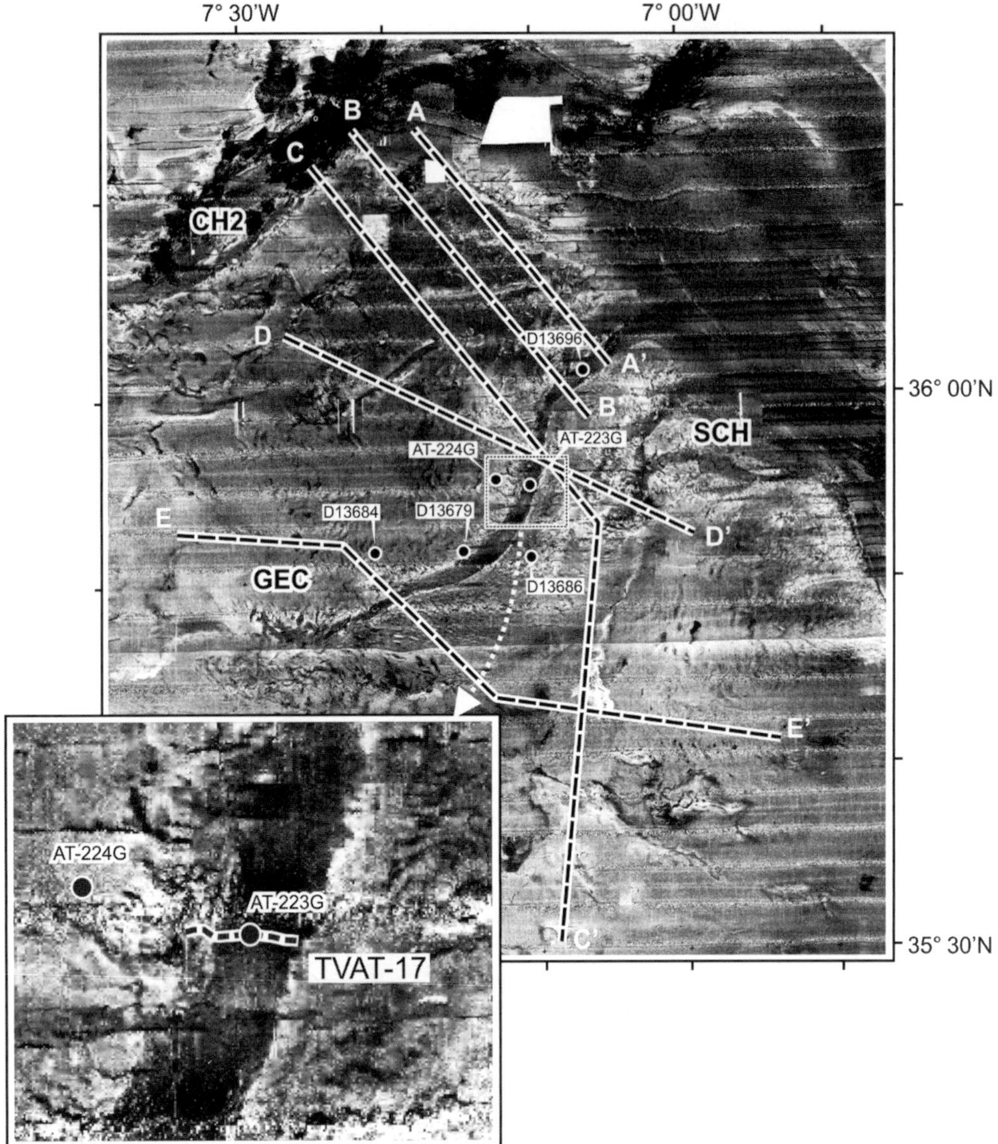

Fig. 13. A portion of SeaMap sea-floor imagery showing the Gil Eanes Channel (GEC) and smaller (SCH) peel-off channels. Seismic lines, sampling stations and TV run discussed in the text are also shown. CH2 is Channel 2 of Kenyon & Belderson (1973), which is topographically steered by diapiric ridges. High backscatter is shown by darker shades.

flanks of the channels. The currents that form them are weaker than those that form the sandier waves. The wave orientation on both flanks is at *c.* 40° to the channel, making a V shape that points down-channel. The mud waves are generally straight crested, with wavelengths between 600 and 1000 m and amplitudes of 25–35 m. On our seismic profiles they have broad, low-amplitude, non-overlapping hyperbolae, often with parallel sub-bottom reflectors equivalent to the type II-B4 of Damuth (1975).

Seismic expression. A set of seismic lines was obtained across the Gil Eanes Channel and adjacent areas of the contourite drift (Fig. 13). Four major seismic units are recognized (Fig. 14). The lowermost unit, Unit I, has a chaotic seismic signature and its upper surface shows a number of diapir-like features protruding into the overlying sedimentary sequence. This unit is attributed to the Olistostrome Unit and Middle Miocene deposits described by Maldonado *et al.* (1999). Unit II is about 0.8 s thick and forms a sedimentary drape over the Olistostrome Unit. It has parallel continuous reflectors and is correlated with the Lower Miocene–Pliocene strata of Maldonado *et al.* (1999). The upper unit, Unit III, is found where the peel-off channels occur and is an Upper Pliocene–Quaternary sequence according to Maldonado *et al.* (1999). The unit tends to fill the sub-basins formed by undulations of the surface of the middle unit and has a lens like-geometry, thinning out at its flanks, which is a characteristic of contourite deposits. The thickness of the unit increases downslope from about 200 to 600 ms. At its base there is an erosional unconformity, otherwise it onlaps or downlaps on the underlying strata. The internal structure of Unit III is complex and three seismic facies have been recognized.

Facies 1. The facies has a chaotic character and yet is acoustically transparent and is mainly found in the uppermost part of the sequence in the areas where sand waves and muddy sand waves are seen on the sonographs. The facies is found within channel levees, and is interpreted as a mainly sandy sequence formed by aggradation of sand waves and muddy sand waves outside the peel-off channels.

Facies 2. The facies has prolonged or short parallel reflectors onlapping or downlapping on the underlying Unit II. The facies is more common in the lower part of Unit III and is interpreted as a primarily silty sequence with some sand beds.

Facies 3. The facies is characterized by short, high-amplitude coherent reflectors and is found mainly underneath channels. It is interpreted as coarse-grained channel fills.

The Gil Eanes Channel, the largest of the channels associated with Unit III, has a U-shaped profile and is about 120 m deep. At the upper part of the slope where Unit III is only about 100 ms thick the channel is incised into Unit II. The channel has an asymmetric profile with a higher right-hand 'levee', which shows an erosional character with truncated beds of Unit II cropping out at the sea floor. The left levee crest is about 75 m lower and has an aggradational nature. From the vertical distribution of Facies 3 it is inferred that the position of the channels through time was relatively stable. In the deeper part of the slope to the west of the Gil Eanes Channel, where small channels are rare, the sequence is dominated by Facies 2.

Sediment cores

Channel flanks. Four cores were taken from different locations along the Gil Eanes Channel flanks (Fig. 13) and their composition is shown in Figure 15.

Core D13684, from the outer part of the western flank of the Channel, has an almost 8 m long sequence of fine-grained sediments with the upper 1.6 m being homogeneous silty mud underlain by 6.4 m of bioturbated mud that contains a 55 cm thick fine silty sand interval. The sand coarsens up in its lower part and fines up in its upper part. Core AT-224G, which was taken upslope and closer to the channel, has 2.5 m of homogeneous, well-sorted clayey silt with foraminfer casts and shell fragments. Core D13679, from the top of the western flank of the Gil Eanes Channel, consists of 373 cm of muddy sands and silts. From visual examination there is a major cycle composed of three main units with gradational contacts between each. However, a closer examination of the continuous wet bulk density profiles identifies a series of smaller cycles. A bioturbated muddy fine sand unit, 116 cm thick, has three coarsening-up–fining-up sequences, each between 30 and 40 cm thick. This is overlain by a similar coarsening-up–fining-up muddy sand fining up into a muddier silt, 60 cm thick. A 14 cm thick muddy unit grades up into muddy sand of similar composition to the base unit; the sand appear to coarsen up throughout and is 142 cm thick.

Core D13686, from the eastern levee of the Gil Eanes Channel (Fig. 15), consists of a 292 cm long cyclical series of coarsening-up muddy fine sands, with sharp erosive tops directly overlain by bioturbated mottled muds. Several phases of bioturbation can be determined within the muds and *Chondrites* is often identifiable. Lenses of muddy silt are common within the sand and silty lenses within the muddy bases. Muddy intervals make up between 4 and 8 cm of a cycle with the muddy sands between 20 and 25 cm of a cycle.

The study of core logs confirmed observations from seismic and sub-bottom profiler records suggesting that there is a more sandy character for the eastern flank of the channel.

Channel floor. Deposits from the channel floor were tested by two cores and an underwater TV line. Core D13696 was collected from the mouth of the Gil Eanes Channel at a water depth of 854 m and recovered 363 cm of biogenic sand (Fig. 15). The core can be subdivided into four

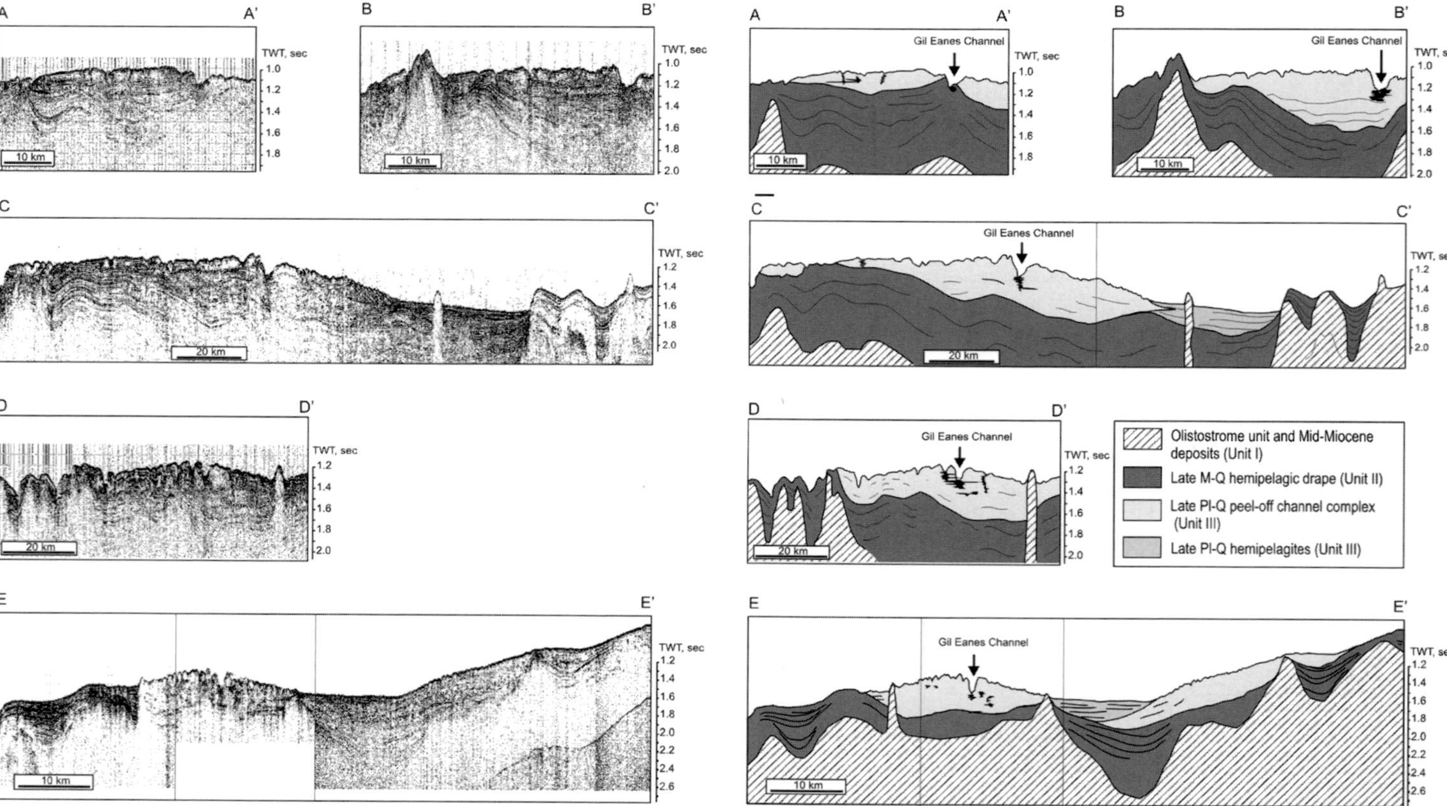

Fig. 14. Interpretation of seismic profiles across the Gil Eanes Channel showing downslope evolution of the channel complex. Profiles are located on Figure 13.

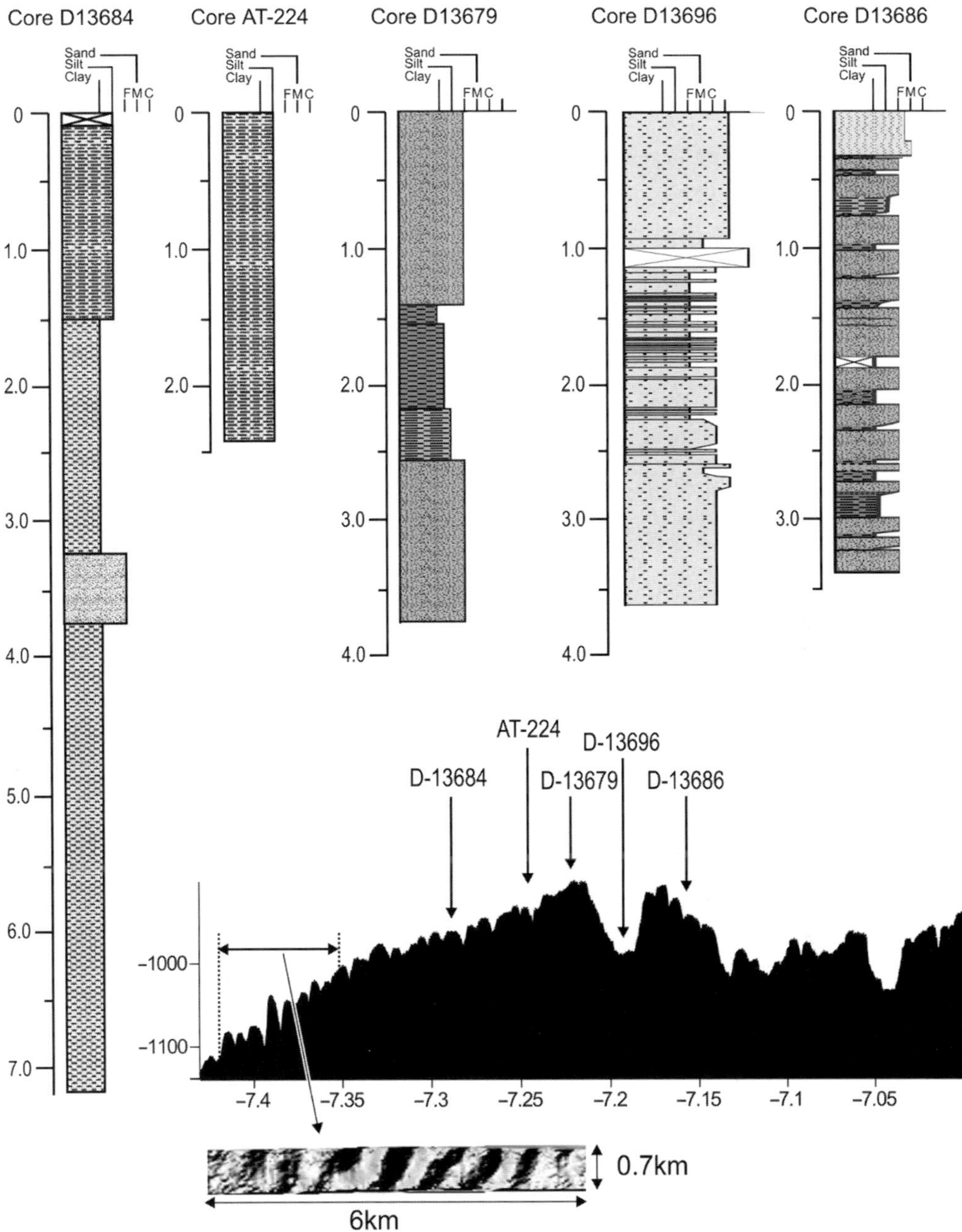

Fig. 15. Core transect and profile across the Gil Eanes Channel and its flanking levee-like features. Available multibeam data and cores show that there are mud waves on the deeper part of the 'levees'.

intervals: a poorly sorted, coarse to very coarse, massive biogenic sand (363–279 cm); a unit with thick coarse sands interbedded with thin fine biogenic sand (279–269 cm); an interbedded sequence of fine–coarse biogenic sands with sharp tops and bottoms (259–114 cm); and a poorly sorted, very coarse biogenic sand, similar to the base unit but having more shelly fragments. Core AT-223G, taken 12 km down the channel, recovered only a small amount of well-sorted medium–coarse sand.

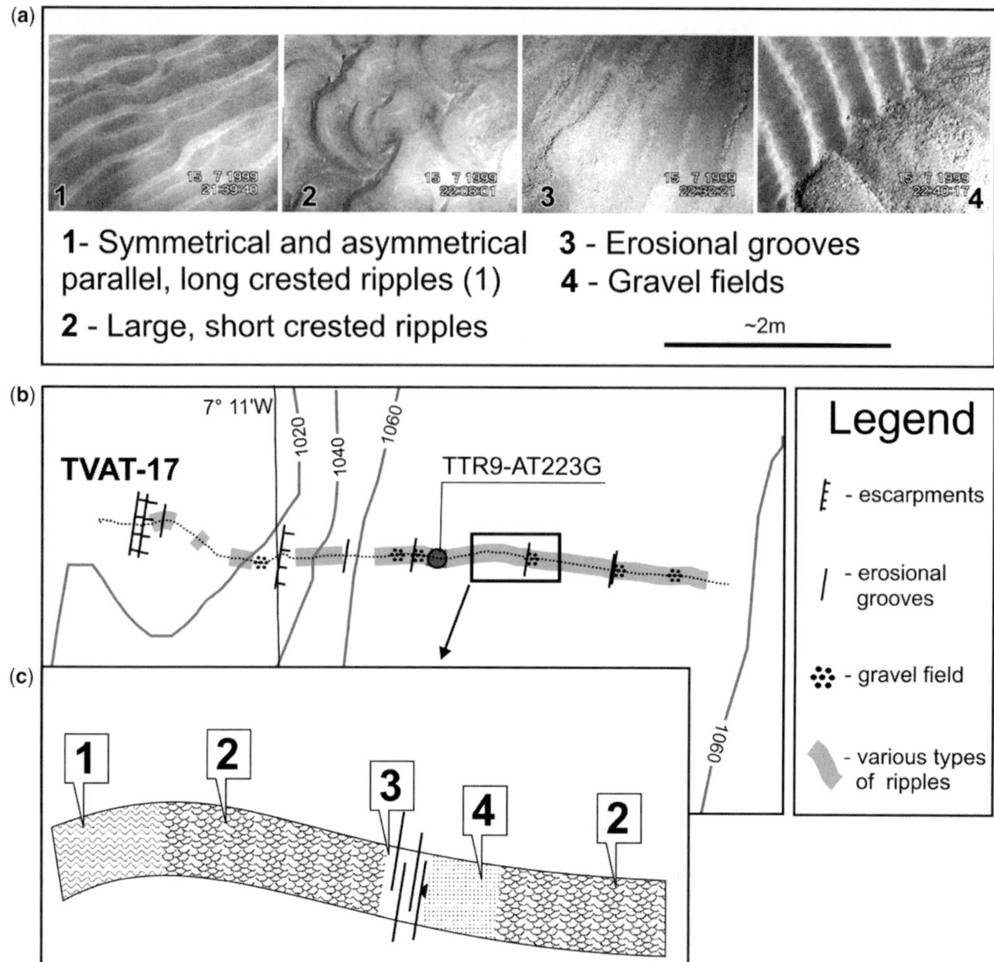

Fig. 16. (**a**) Video stills showing main types of small bedforms observed across the Gil Eanes Channel thalweg; (**b**) interpretation of an underwater video transect (TVAT-17); (**c**) sequence of bedforms reflecting variations of current velocity across a secondary flow filament.

The character of the sea bed in the vicinity of the core was observed during a deep-towed camera run across the channel, the location of which is shown in Figure 13, that was briefly discussed by Habgood *et al.* (2003). The channel wall has a smoothed silty–clay surface that is intensively bioturbated. Escarpments (Fig. 16), a few metres high, are believed to be formed by sediment failures where the channel wall is undercut by current erosion near the base. The erosion has caused striations. Near the base of the wall ripples become larger and have more irregular shapes. Below 1060 m the profile crosses the relatively flat channel floor on which there are ubiquitous fields of ripples of various shapes and abundant gravels (Fig. 16). The most common are large linguoid ripples, which have a denser distribution towards the centre of the channel. Other frequently encountered bedforms include fields of gravel with parallel ripples, sand streaks and erosional grooves.

Bedforms on the channel floor form the following sequence, which is repeated across the channel floor (Fig. 16): parallel symmetrical ripples, parallel asymmetric ripples, large linguoid ripples and gravel fields with longitudinal sand ribbons and erosional grooves. This sequence is very similar to the one observed in shallow environments, where it was related to changes of bed shear stress (Allen 1968). In the Gil Eanes Channel it is thought to reflect the current velocity distribution across longitudinal flow filaments. The parallel symmetrical ripples correspond to lower velocities in the outer part of the filaments and the erosional grooves to higher velocities in the inner part of the filaments. Using ripple morphology as a proxy for bottom-current velocity (e.g. Lonsdale & Spiess 1977) the peak current velocities within flow filaments are estimated at 17–22 cm s^{-1} for parallel ripples, 30–50 cm s^{-1} for large linguoid ripples and above 50 cm s^{-1} for gravel fields and erosional grooves.

Discussion

Styles of contourite channels

In the studied areas there are both geostrophic, contour-following channels, which seem to descend gently in the direction of flow, and ageostrophic channels that run down relatively sleep slopes. Well-marked channels are developed where the bottom current is constricted between a topographic high, such as the basement plateau in the Faeroe Bank area, and an actively growing sediment drift. Such a constriction will cause flow acceleration, which may in turn be followed by stronger erosion on the right-hand side of the channel. Channels in the Faeroe Bank area resemble the moats known from many contourite drift systems worldwide (Faugères *et al.* 1993).

The Mediterranean Undercurrent forms a complex channel pattern in the Gulf of Cadiz, which also comprises both geostrophic and ageostrophic modes. Here too, a significant portion of the flow has geostrophic behaviour, following contours, and often is cited as an upper core and a lower core (Ambar & Howe 1979; Baringer & Price 1997). Eventually, muddy sediment drifts such as the upper Faro Drift (Gonthier *et al.* 1984; Stow *et al.* 1986) are built in the area where the current speed has fallen. SW–NE-trending diapiric ridges topographically steer several filaments of the Undercurrent to the left of its main course but they reassert themselves on the alongslope course once the ridges are passed (Kenyon & Belderson 1973). As a result, the channel distribution has an anastomosing pattern, separated by lozenge-shaped drift bodies, similar to the pattern formed by the channels found west of the Faeroe Bank. A portion of the Undercurrent, known as the lower core, descends downslope without topographic steering and forms peel-off channels including the Gil Eanes Channel and several smaller channels, and thus shows ageostrophic behaviour, being driven downslope primarily by a density contrast. It is not clear from the present dataset whether sediment failure is involved in the initiation of these contourite channels. There are several small-scale, slightly sinuous channels with associated sand lobes found beyond the contourite channels (Habgood *et al.* 2003), which indicate that sandy accumulations at the ends of these channels fail and are redeposited downslope by gravity flows. At depths of about 1200–1300 m the SeaMap image shows that channels terminate abruptly. According to oceanographic observations at these depths the MOW reaches its density equilibrium level and by mixing with colder North Atlantic Central Water (Baringer & Price 1997) becomes intermediate in character and lifts off the sea floor (Thorpe 1976). This explains the build-up of sand that has then repeatedly failed and produced the gravity-driven channel and lobe systems.

Factors controlling channel development and sand distribution

Both studied channel systems are formed by persistent bottom currents, which were operating over a long period. Recent drilling in the Faeroe–Shetland Channel showed that the onset of deep-water exchange between Arctic and North Atlantic Oceans was established in the early Oligocene (Davies *et al.* 2001). As the Faeroe Bank Channel is a part of this gateway, the thermohaline bottom current has operated in the area for the last 35 Ma. Contourite channels are primarily formed by filaments of fast-flowing current but their geometry can be affected by a growth of sediment drift predominantly on the outer side (left-hand side in the northern hemisphere) of the current pathway. This process seems to be compensated by erosion on the opposite flank, where the current is piled up against the slope by the Coriolis force. As a result, channels migrate laterally. They can be steered by large-scale, resistant sea-bed features, such as the basaltic plateau in the Faeroe Bank Channel area and the diapiric ridges in the Gulf of Cadiz, which can split the flow into several filaments and thus produce the anastomosing pattern. Channels may be narrower in the vicinity of such obstacles, as a result of the resistance of the materials.

A geostrophic bottom current restricted within a channel is often strong enough to be able to transport and deposit sand. The source for this sand is to be sought in the vast amounts of terrigenous material introduced into the area during Late Pliocene–Pleistocene glacial cycles and reworked by the strong bottom current. Sandy sediment is mainly concentrated along the channels and represents a lag deposit, whereas fines are washed out by the fast bottom current and are deposited on the flanks, forming levee-like features. As a channel migrates it forms a continuous sheet of sand overlain by a muddy levee sequence.

The Mediterranean Undercurrent has been operating in the Gulf of Cadiz for almost 5 Ma, since the Early Pleistocene, and has formed several large, fine-grained drifts such as the Faro Drift in the north (e.g. Stow *et al.* 1986). Most of the

Gil Eanes sequence has a seismic signature typical for a Quaternary section (Rodero *et al.* 1999). The formation of this and other peel-off channels can be closely related to the southeastward progradation of shelf prodeltaic deposits during late Quaternary time. Such an abundant supply into the area affected by strong bottom currents resulted in further reworking and formation of significant accumulations of coarse sediments in the upper part of the slope. The latter in turn are redeposited further downslope, primarily by a descending portion of the Undercurrent, although gravity flows may be important at all stages of channel development as there is the potential for rapid accumulation of sand on what are, in some places, steep gradients. Study of the Gil Eanes Channel revealed that its infill is a sand-rich sequence. Seismic records show the presence of an aggradational channel–levee complex with little lateral migration of the channel in time. Levees are composed of silts with increased sand content in the upper part of the sequence where currents are stronger.

Comparison with other channel systems

Channels in the deep sea are formed by two main processes: gravity flows and bottom currents. Turbidity currents are the most common type of gravity flows and are responsible for the formation of the majority of the channels. Bottom currents, and geostrophic contour currents in particular, are also widespread in the deep sea and often form large accumulations called drifts, consisting primarily of fine-grained sediments. Channels are not a typical element of bottom-current systems but may be common in the ocean gateway areas such as the Vema Channel (Johnson 1984) and the areas described here. Previous studies revealed that both contourite and turbidite systems have similar features and can be confused with each other. For instance, similar depositional lobes were reported from the Vema contourite channel and from the Cap Ferret turbidite system (Faugères *et al.* 1998).

Distinguishing between sandy turbidites and sandy bottom-current deposits on seismic data or in the rock record is important because of the economic potential of sands as hydrocarbon reservoirs. The examples studied here show that significant sand accumulations can be formed in contour-current systems and that these present-day examples can be considered as analogues for hydrocarbon reservoirs.

Channel–levee complexes are known to be an important element of turbidite systems but they are also found in areas swept by both the NSOW and the MOW. Channels in the area west of the Faeroe Bank have a remarkable asymmetry, with one flank being erosional and the other aggradational, and their architecture can be compared with fluvial systems on land. Seismic records from the area, showing prolonged discontinuities and buried palaeo-channels, have a clear resemblance to seismic records across the Vema Channel, suggesting a similar origin for both channel systems. The presence of sandy infill in the Faeroe Bank Channel indicates that peak current speed is about 30–50 cm s^{-1} (e.g. Johnson *et al.* 1982), which is higher than the current speed reported from the Vema Channel (about 20 cm s^{-1}) (Johnson 1984). Mud waves in the Faeroe Bank Channel region are found only on the outside of channel bands, as they are for turbidite channel systems such as the Monterey (Normark *et al.* 1980; McHugh & Ryan 2000), Toyama (Nakajima *et al.* 1998), Var (Migeon *et al.* 2001) and other systems. This suggests that flow stripping, which is believed to cause sediment wave formation on the flanks of channelized turbidite systems (Piper & Normark 1983), happens mostly at the bends and otherwise most of the flow is concentrated within the channels.

The Gil Eanes Channel forms an aggradational sequence with a stacking pattern of palaeo-channels that is very similar to that for turbidite channel sequences. A variety of sediment waves covers the levees and outer part of the channel (Fig. 15), indicating that overflowing of the descending MOW plume is a frequent, if not near-continuous, process. Fields of sediment waves are not usually as extensive on the levees of turbidite channels, where they mainly occur at bends. However, waves do occur beyond the mouths of turbidite channels, where the turbidity currents become unconfined (e.g. Wynn *et al.* 2000).

The sediments deposited by the Gil Eanes and other peel-off channels have a limited distribution as the traction process that moves them terminates relatively abruptly where the current lifts off the sea bed. Such abrupt terminations of flow and the resultant sand accumulations are not typical for turbidity currents, which usually travel a long way beyond the channel mouth, forming varied and extensive sheet-like deposits as on the Mississipi Fan, Gulf of Mexico (Garrison *et al.* 1982), Indus Fan, Indian Ocean (Kolla & Coumes 1987), Var sedimentary system, Mediterranean Sea (Mulder *et al.* 1998), Pochnoi Fan, Bering Sea (Kenyon & Millington 1995), and in many other turbidite systems.

Conclusions

The study of two channelized depositional systems controlled by contour currents or bottom currents revealed that they can form significant sand accumulations in the deep sea. Channel–levee complexes can be formed by both geostrophic and ageostrophic bottom currents. The channels formed by the NSOW, west of the Faeroe Bank Channel, are developed under the influence of both geostrophic currents and ageostrophic currents and are characterized by significant lateral migration, usually to the right, which determines the channel architecture with its cross-sectional asymmetry. Prolonged unconformities marking the onset of the bottom-current dominated environment and also resulting from channel migration are typical of this system. Sand bodies are mostly associated with channel infills, whereas outside the channels, sequences are typically fine grained.

The bottom-current system in the Gulf of Cadiz formed by the Mediterranean Undercurrent has a complex pattern with a large portion of the water mass flowing under geostrophic conditions, but some of the densest water becomes ageostrophic and descends downslope as a result of gravity. As a result, a series of channels is formed with the largest one, Gil Eanes, being about 40 km long. Medium–coarse sand and fine gravels are predicted to form most of the channel infills, and sand is also abundant in the upstream areas, where fields of sand waves form on the broad slope terrace developed by the Undercurrent. Flanking the channels are features resembling levees. These consist mainly of silty sequences that have a significantly higher sand content in the shallower water.

Both of the studied depositional systems have a variety of contourite sand channels but in most respects they are remarkably similar. The main pathway of the densest water is the shallowest and several branches turn off to the left of this main pathway, and descend to a lower level before bending to the right, parallel to the main pathway, under the influence of Coriolis forces. In both cases there are proximal deposits of sand deposited as the flow becomes unconfined and the current speed slackens. In the Gulf of Cadiz these consist of extensive sheets of cross-bedded sand of the order of several tens of kilometres across and up to several tens of metres thick. The speed may pick up again as the current flows down a gentle gradient and further coarse deposits, both sand and gravel, accumulate within channels, reaching several hundreds of metres thick. At the ends of some channels in the Gulf of Cadiz there are concentrations of sand formed as the current lifts off the sea floor. Similar accumulations occur where a canyon intersects the pathway of the contour current. These deposits can fail and form sandy lobes on the floor of the nearest basin or, in the case of a canyon, on the canyon floor (Habgood *et al.* 2003).

Similarities between bottom-current–contourite sand channels and turbidite sand channels are the aggradational nature of some channel floors and the flanking waves of finer-grained sediment. Contourite channels are distinguished from turbidite channels by their more marked asymmetry in channel profile and sand distribution, the presence of regional unconformities and their shape in plan view.

Two factors that are found in both the studied areas and may be of particular importance for the formation of contourite channels with reservoir potential are the high sand supply (including 'pure' sands or large volumes of sand-enriched sediments), and the long history of strong bottom-current existence.

This is a contribution of the UNESCO–IOC Training-through-Research (TTR) programme and of the EUROSTRATAFORM project, EC contract EVK3-CT-2202-00079, funded by the European Commission's Fifth Framework Programme under 'Energy, Environment and Sustainable Development'. The EC is not liable for any use that may be made of information contained herein. We are grateful for the thoughtful comments of D. Piper, which helped to improve the overall look of the manuscript.

References

AKHMETZHANOV, A. M. 2003. Sandy sediments of deep-water bottom current depositional systems. *Vestnik Moskovskogo Universiteta, Geology Series*, **5**, 39–47 [in Russian].

ALLEN, J. R. L. 1968. *Current Ripples; Their Relation to Patterns of Water and Sediment Motion.* North-Holland, Amsterdam.

AMBAR, I. & HOWE, M. R. 1979. Observations of the Mediterranean outflow—I. Mixing in the Mediterranean outflow. *Deep-Sea Research*, **26A**, 535–554.

BARAZA, J., ERCILLA, G. & NELSON, C. H. 1999. Potential geologic hazards on the eastern Gulf of Cadiz slope (SW Spain). *Marine Geology*, **155**, 191–215.

BARINGER, M. O. & PRICE, J. F. 1997. Mixing and spreading of the Mediterranean outflow. *Journal of Physical Oceanography*, **27**, 1654–1677.

BELDERSON, R. H., KENYON, N. H., STRIDE, A. H. & STUBBS, A. R. 1972. *Sonographs of the Seafloor: a Picture Atlas.* Elsevier, Amsterdam.

BOLDREEL, L. O., ANDERSEN, M. S. & KUIJPERS, A. 1998. Neogene seismic facies and deep-water gateways in the Faeroe Bank area, NE Atlantic. *Marine and Petroleum Geology*, **152**, 129–140

BRITISH OCEANOGRAPHIC DATA CENTRE 1997. *GEBCO Digital Atlas.* British Oceanographic Data Centre on behalf of IOC and IHO.

CREASE, J. 1965. The flow of Norwegian Sea water through the Faroe Bank Channel. *Deep-Sea Research*, **12**, 143–150.

DAMUTH, J. 1975. Echo character of the Western Equatorial Atlantic floor and its relationship to the dispersal and distribution of terrigenous sediments. *Marine Geology*, **18**, 17–45.

DAVIES, R., CARTWRIGHT, J., PIKE, J. & LINE, C. 2001. Early Oligocene initiation of North Atlantic deep water formation. *Nature*, **410**, 917–920.

DORN, W. U. & WERNER, F. 1993. The contour-current flow along the southern Iceland–Faeroe Ridge as documented by its bedforms and asymmetrical channel fillings. *Sedimentary Geology*, **82**, 47–59.

FAUGÈRES, J. C., MEZERAIS, M. L. & STOW, D. A. V. 1993. Contourite drift types and their distribution in the North and South Atlantic Ocean basins. *Sedimentary Geology*, **82**, 189–203.

FAUGÈRES, J. C., IMBERT, P., MEZERAIS, M. L. & CREMER, M. 1998. Seismic patterns of a muddy contourite fan (Vema Channel, South Brazilian Basin) and a sandy distal turbidite deep-sea fan (Cap Ferret system, Bay of Biscay); a comparison. *Sedimentary Geology*, **115**, 81–110.

FAUGÈRES, J. C., STOW, D. A. V., IMBERT, P. & VIANA, A. 1999. Seismic features diagnostic of contourite drifts. *Marine Geology*, **162**, 1–38.

FLEISCHER, U., HOLZKAMM, F., VOLLBRECHT, K. & VOPPEL, D. 1974. Die Struktur des Island-Faroer-Rückens aus geophysikalischen Messungen. *Deutsche Hydrographische Zeitschrift*, **27**, 97–113.

GARRISON, L. E., KENYON, N. H. & BOUMA, A. H. 1982. Channel systems and lobe construction in the Mississippi Fan. *Geo-Marine Letters*, **2**, 31–39.

GONTHIER, E. G., FAUGÈRES, J. C. & STOW, D. A. V. 1984. Contourite facies of the Faro Drift, Gulf of Cadiz. *In*: STOW, D. A. V. & PIPER, D. J. W. (eds) *Fine-Grained Sediments; Deep-Water Processes and Facies.* Geological Society, London, Special Publications, **15**, 275–292.

HABGOOD, E. L., KENYON, N. H., AKHMETZHANOV, A. M., WEAVER, P. P. E., MASSON, D. G., GARDNER, J. & MULDER, T. 2003. Deep water sediment wave fields, contourite channels and associated depositional sand lobes in the Gulf of Cadiz, NE Atlantic. *Sedimentology*, **50**, 483–510.

HANSEN, B. & MEINCKE, J. 1979. Eddies and meanders in the Iceland–Faroe Ridge area. *Deep-Sea Research*, **26**, 1067–1082.

HANSEN, B. & ØSTERHUS, S. 2000. North Atlantic–Nordic Seas exchanges. *Progress in Oceanography*, **45**, 109–208.

HEEZEN, B. C. & JOHNSON, G. L. 1969. Mediterranean undercurrent and microphysiography west of Gibraltar. *Bulletin de l'Institut Océanographique, Monaco*, **67**, 1–28.

IORGA, M. C. & LOZIER, M. S. 1999. Signatures of the Mediterranean outflow from a North Atlantic climatology, 1, Salinity and density fields. *Journal of Geophysical Research*, **104**(C11), 25985–26009.

JOHNSON, D. A. 1984. The Vema Channel; physiography, structure, and sediment–current interactions. *Marine Geology*, **58**, 1–34.

JOHNSON, M. A., KENYON, N. H. & BELDERSON, R. H. 1982. Sand transport. *In*: STRIDE, A. H. (ed.) *Offshore Tidal Sands: Processes and Deposits.* Chapman and Hall, London, 58–94.

KENYON, N. H. & BELDERSON, R. H. 1973. Bedforms of the Mediterranean undercurrent observed with side-scan sonar. *Sedimentary Geology*, **9**, 77–99.

KENYON, N. H. & MILLINGTON, J. 1995. Contrasting deep-sea depositional systems in the Bering Sea. *In*: PICKERING, K. T., HISCOTT, R. N., KENYON, N. H., RICCI LUCCHI, F. & SMITH, R. D. A. (eds) *Atlas of Deep Water Environments: Architectural Style in Turbidite Systems.* Chapman & Hall, London, 196–202.

KENYON, N. H., IVANOV, M. K., AKHMETZHANOV, A. M. & AKHMANOV, G. G. (eds) 2000. *Multidisciplinary Study of Geological Processes on the North East Atlantic and Western Mediterranean Margins. Preliminary results of geological and geophysical investigations during the TTR-9 cruise of R/V* Professor Logachev, *June–July, 1999.* IOC Technical Series, **56**.

KENYON, N. H., IVANOV, M. K., AKHMETZHANOV, A. M. & AKHMANOV, G. G. (eds) 2001. *Interdisciplinary Approaches to Geoscience on the North East Atlantic Margin and Mid-Atlantic Ridge. Preliminary results of investigations during the TTR-10 cruise of R/V* Professor Logachev, *July–August, 2000.* IOC Technical Series, **60**.

KOLLA, V. & COUMES, F. 1987. Morphology, internal structure, seismic stratigraphy and sedimentation of Indus Fan. *AAPG Bulletin*, **71**, 650–677.

KUIJPERS, A. H., ANDERSEN, M. S., KENYON, N. H., KUNZENDORF, H. & VAN WEERING, T. C. E. 1998. Quaternary sedimentation and Norwegian Sea Overflow pathways around Bill Bailey Bank, northeastern Atlantic. *Marine Geology*, **152**, 101–127.

LONSDALE, P. & SPIESS, F. N. 1977. Abyssal bedforms explored with a deeply towed instrument package. *Marine Geology*, **23**, 57–75.

MADELAIN, F. 1970. Influence de la topographie du fond sur l'écoulement Méditerranéen entre le Détroit de Gibraltar et le Cap Saint-Vincent. *Cahiers Océanographique*, **22**, 63–72.

MALDONADO, A., SOMOZA, L. & PALLARES, L. 1999. The Betic orogen and the Iberian–African boundary in the Gulf of Cadiz; geological evolution (central North Atlantic). *Marine Geology*, **155**, 9–43.

MCCAVE, I. N. & TUCHOLKE, B. E. 1986. Deep current-controlled sedimentation in the western North Atlantic. *In*: VOGT, P. R. & TUCHOLKE, B. E. (eds) *The Western North Atlantic Region.* Geological Society of America, Boulder, CO, 451–468.

MCHUGH, C. M. G. & RYAN, W. B. F. 2000. Sedimentary features associated with channel overbank flow: example from the Monterey Fan. *Marine Geology*, **163**, 199–215.

MIGEON, S., SAVOYE, B., ZANELLA, E., MULDER, T., FAUGÈRES, J. C. & WEBER, O. 2001. Detailed seismic-reflection and sedimentary study of turbidite sediment waves on the Var Sedimentary Ridge (SE France): significance for sediment transport and deposition and for the mechanisms of sediment-wave construction. *Marine and Petroleum Geology*, **18**, 179–208.

MULDER, T., SAVOYE, B., PIPER, D. J. W. & SYVITSKI, J. P. M. 1998. The Var submarine sedimentary system: understanding Holocene sediment delivery processes and their importance to the geological record. *In*: STOKER, M. S., EVANS, D. & CRAMP, A. (eds) *Geological Processes on Continental Margins: Sedimentation, Mass-wasting and Stability*. Geological Society, London, Special Publications, **129**, 145–166.

MULDER, T., VOISSET, M., LECROART, P., *ET AL.* 2003. The Gulf of Cadiz: an unstable giant contouritic levee. *Geo-Marine Letters*, **23**, 7–18.

NAKAJIMA, T., SATOH, M. & OKAMURA, Y. 1998. Channel–levee complexes, terminal deep-sea fan and sediment wave fields associated with the Toyama Deep-Sea Channel system in the Japan Sea. *Marine Geology*, **147**, 25–41.

NELSON, C. H., BARAZA, J. & MALDONADO, A. 1993. Mediterranean undercurrent sandy contourites, Gulf of Cadiz, Spain. *Sedimentary Geology*, **82**, 103–131.

NELSON, C. H., BARAZA, J., MALDONADO, A., RODERO, J., ESCUTIA, C. & BARBER, J. H. JR. 1999. Influence of the Atlantic inflow and Mediterranean outflow currents on late Quaternary sedimentary facies of the Gulf of Cadiz continental margin. *Marine Geology*, **155**, 99–129.

NORMARK, W. R., HESS, G. R., STOW, D. A. V. & BOWEN, A. J. 1980. Sediment waves on the Monterey Fan levee: a preliminary physical interpretation. *Marine Geology*, **37**, 1–18.

PIPER, D. J. W. & NORMARK, W. R. 1983. Turbidite depositional patterns and flow characteristics, Navy submarine fan, California Borderland. *Sedimentology*, **30**, 681–694.

RODERO, J., PALLARES, L. & MALDONADO, A. 1999. Late Quaternary seismic facies of the Gulf of Cadiz Spanish margin; depositional processes influenced by sea-level change and tectonic controls. *Marine Geology*, **155**, 131–156.

SAUNDERS, P. M. 1990. Cold outflow from the Faroe Bank Channel. *Journal of Physical Oceanography*, **20**, 29–43.

SMITH, W. H. F. & SANDWELL, D. T. 1994. Bathymetric prediction from dense satellite altimetry and sparse shipboard bathymetry. *Journal of Geophysical Research*, **99**, 21803–21824.

STOW, D. A. V. 1979. Distinguishing between fine-grained turbidites and contourites on the Nova Scotian deep water margin. *Sedimentology*, **26**, 371–384.

STOW, D. A. V., FAUGÈRES, J. C. & GONTHIER, E. 1986. Facies distribution and textural variation in Faro Drift contourites; velocity fluctuation and drift growth. *Marine Geology*, **72**, 71–100.

THORPE, S. A. 1976. Variability of the Mediterranean undercurrent in the Gulf of Cadiz. *Deep-Sea Research*, **23**, 711–727.

VIANA, A. R., FAUGÈRES, J. C. & STOW, D. A. V. 1998. Bottom-current-controlled sand deposits; a review of modern shallow- to deep-water environments. *Sedimentary Geology*, **115**, 53–80.

WYNN, R. B., WEAVER, P. P. E., ERCILLA, G., STOW, D. A. V. & MASSON, D. G. 2000. Sedimentary processes in the Selvage sediment-wave field, NE Atlantic; new insights into the formation of sediment waves by turbidity currents. *Sedimentology*, **47**, 1181–1197.

Quaternary evolution of the contourite depositional system in the Gulf of Cadiz

E. LLAVE[1], F. J. HERNÁNDEZ-MOLINA[2], L. SOMOZA[1], D. A. V. STOW[3] & V. DÍAZ DEL RÍO[4]

[1]*Instituto Geológico y Minero de España, Ríos Rosas, 23, 28003 Madrid, Spain (e-mail: e.llave@igme.es)*

[2]*Facultad de Ciencias del Mar, Universidad de Vigo, 36200 Vigo, Spain*

[3]*Southampton Oceanography Centre, University of Southampton, Waterfront Campus, Southampton SO14 3ZH, UK*

[4]*Instituto Español de Oceanografía, C/Puerto Pesquero s/n, 29640 Fuengirola, Spain*

Abstract: This paper provides for the first time a detailed vertical and spatial representation of Quaternary evolution of the contourite depositional system (CDS) in the Gulf of Cadiz, based on the results of careful morphological, structural and stratigraphic analyses using high-resolution seismic reflection profiles as well as oil company borehole data, and piston and gravity cores. Different drifts observed on the stratigraphic architecture allow us to propose a regional Quaternary evolution for the whole system, in which three major stages can be identified. (1) In the Early Pleistocene to Mid-Pleistocene, the CDS was mainly dominated by depositional processes, where the upper and lower cores of the Mediterranean Outflow Water (MOW) generated the mounded elongated Cadiz–Faro–Albufeira drift in the transition between the middle and upper slope, and the equivalent Huelva–Guadalquivir drift on the middle slope. During this stage the main erosive features were established close to the Strait of Gibraltar. (2) In the Mid-Pleistocene to Late Pleistocene, two important changes in the CDS took place. One occurred at the transition between the middle and upper slope, related to a change in the upper branch of the MOW, when a mixed drift began to develop, burying the eastern part of the Cadiz–Faro–Albufeira mounded elongated and separated drift. The second change is observed on the central area of the middle slope, related to the lower branch of the MOW, where a large contourite channel (the Guadalquivir channel) progressively eroded the western part of the mounded Huelva–Guadalquivir drift. Laterally an extensive sheeted drift buried the previous mounded deposits. (3) In the Late Pleistocene to Holocene, in the northern area of the CDS, a plastered drift started to be developed in the transitional zone between the upper and middle slope. On the middle slope, the mounded elongated Huelva–Guadalquivir drift was not developed and more erosive processes became dominant as the lower core of the MOW intensified. In the sector close to the Straits of Gibraltar, a field of broad seabed forms was generated. These three evolutionary stages have been controlled by tectonics, including recent diapiric movement, Guadalquivir Bank uplift, and reactivation along several fault systems and anticline–syncline structures. Tectonics has been a key factor in the sea-floor morphological changes, which has caused new pathways for the core and branches of the MOW, and consequently has produced the contourite stratigraphic and architectural changes. Superimposed on these tectonic changes, both climatic and eustatic changes during the Quaternary (but especially from the Mid-Pleistocene) have controlled the development of vertical contourite stratigraphy. The general conclusion of this study is that the contourite depositional system of the Gulf of Cadiz has changed from a dominantly depositional system to a dominantly erosive one during the Quaternary.

It is possible to deduce, from the characteristics of contourite deposits (their morphology, internal seismic facies and position in the oceanic basin), the pathway of the water mass that was responsible for their development. This is particularly relevant when buried contourite drifts are found in the sedimentary record of a basin, because it then becomes possible to reconstruct the original palaeoceanographic conditions (Hernández-Molina *et al.* 2006). In addition, the stratigraphic architecture of contourite deposits provides essential indicators for determining the evolution of marine environments and palaeoceanographic conditions of the water masses with time. Long-term records may provide information about the distribution and timing of large-scale sedimentary changes related to bottom-current circulation changes (or other factors that control those changes, such as

From: VIANA, A. R. & REBESCO, M. (eds) *Economic and Palaeoceanographic Significance of Contourite Deposits*. Geological Society, London, Special Publications, **276**, 49–79.
0305-8719/07/$15.00

tectonics), as expressed by unconformities and changing styles of deposition.

The contourite depositional system (CDS) along the northern margin of the Gulf of Cadiz has been defined and characterized very recently as a unique system that results from Mediterranean Outflow Water (MOW) interaction with the sea floor (Fig. 1) (Hernández-Molina *et al.* 2003; Llave 2003). There have been numerous studies focused on the influence of the MOW on the sedimentary stacking pattern of some specific parts of the CDS (e.g. Madelain 1970; Kenyon & Belderson 1973; Mélières 1974; Gonthier *et al.* 1984; Faugères *et al.* 1985; Stow *et al.* 1986, 2002; Nelson *et al.* 1993, 1999; Llave *et al.* 2001, 2004*a*, 2006; García 2002; Stow *et al.* 2002; Habgood *et al.* 2003; Mulder *et al.* 2003). Special attention has been paid to the establishment of depositional facies models (Stow 1979, 1982; Faugères *et al.* 1984, 1993, 1999; Gonthier *et al.* 1984) and their mechanism of formation (Stow & Faugères 1993, 1998; Stow *et al.* 1996; Shanmugam 2000; Stow & Mayall 2000) to achieve the possibility of reconstructing the depositional history of drift development, of demonstrating the changing seismic character of contourites and geometry of the drifts, and of placing specific phases of deposition within a palaeocirculation framework. A cyclic pattern of deposition and erosion as a consequence of changes in the intensity of MOW through time has been recognized (Nelson *et al.* 1993). Cyclicity has also been described from the CDS seismic and sediment facies, which are interpreted as resulting from fluctuations in grain size as a consequence of varying MOW intensity, which in turn can be related to palaeoclimatic changes (Llave *et al.* 2001; Stow *et al.* 2002; Llave 2003).

In spite of this large volume of previous work, there has been no overall synthesis dealing with the evolution of the CDS as a whole. The main goal of this paper, therefore, is to present for the first time an attempt at reconstruction of the Quaternary evolution of the CDS in the Gulf of Cadiz, through an integration of tectonic, stratigraphic and palaeoceanographic changes. We find six major types of contourite drift deposits within the CDS and emphasize differences in the stratigraphic stacking patterns, and we then elucidate their evolution within a general tectonostratigraphic framework. Emphasis is placed on the varied depositional and erosive features of Quaternary contourite deposits, as identified from regional seismic data, and how these relate to changes in basin geometry and palaeocirculation.

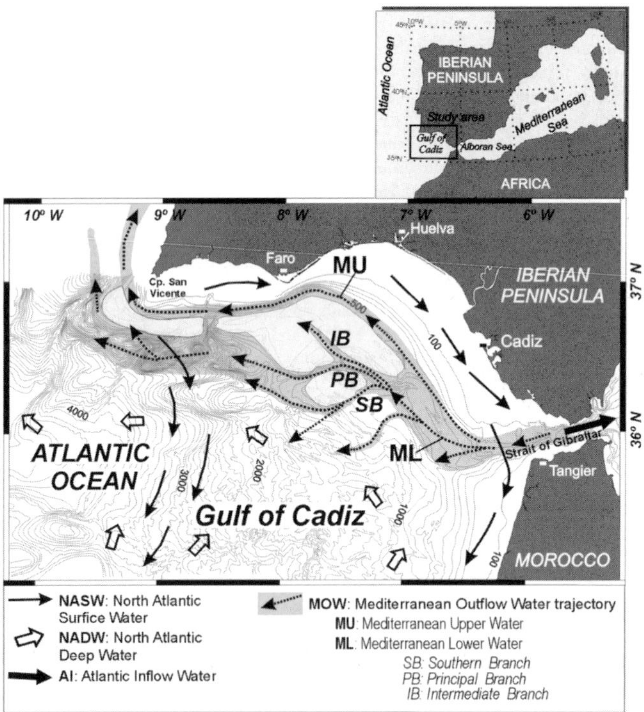

Fig. 1. Location of the Gulf of Cadiz in a regional bathymetric map (Heezen & Johnson 1969) indicating the general circulation patterns of MOW. (Modified from Hernández-Molina *et al.* 2003.)

Gulf of Cadiz: geological and oceanographic setting

Geological framework

The southwestern margin of the Iberian Peninsula, at the eastern end of the Azores–Gibraltar zone, is the location of the diffuse plate boundary between Europe and Africa. Distinct periods of crustal deformation, fault reactivation and halokinesis related to the movement between Eurasia and Africa (Malod & Mauffret 1990; Srivastava *et al.* 1990; Maldonado *et al.* 1999; Gutscher *et al.* 2002; Alves *et al.* 2003; Gutscher 2004; Medialdea *et al.* 2004) are known to have controlled the tectonostratigraphic evolution of this part of the Iberian Peninsula. The tectonic structure of this area (Fig. 2) is a consequence of the distinct phases of rifting from the Late Triassic to the Early Cretaceous (Murillas *et al.* 1990; Pinheiro *et al.* 1996; Wilson *et al.* 1996; Borges *et al.* 2001) and its later deformation during the Cenozoic, especially in the Miocene (Ribeiro *et al.* 1990; Alves *et al.* 2003). Since the late Miocene, the NW–SE compressional regime has developed simultaneously with the extensional collapse of the Betic–Rif orogenic front, by westward emplacement of a giant 'olistostrome', the Cadiz Allochthonous Unit, and by very high rates of basin subsidence coupled with strong diapiric activity (Perconig 1960–1962; Torelli *et al.* 1997; Flinch & Vail 1998; Maldonado *et al.* 1999; Gracia *et al.* 2003; Medialdea *et al.* 2004). By the end of the Early Pliocene the connection between the Atlantic and the Mediterranean through the Strait of Gibraltar opened, leading, after the end of the Messinian salinity crisis, to the establishment of the circulation pattern as it is known nowadays (Thunell *et al.* 1991; Nelson *et al.* 1993). The margin evolved towards more stable conditions during the Late Pliocene–Quaternary (Maldonado *et al.* 1999; Somoza *et al.* 1999; Maestro *et al.* 2003; Fernández-Puga 2004; Medialdea *et al.* 2004). Some neotectonic reactivation is also evident, as expressed by the occurrence of mud volcanoes and diapiric ridges (Díaz del Río *et al.* 2003; Llave 2003; Somoza *et al.* 2003;

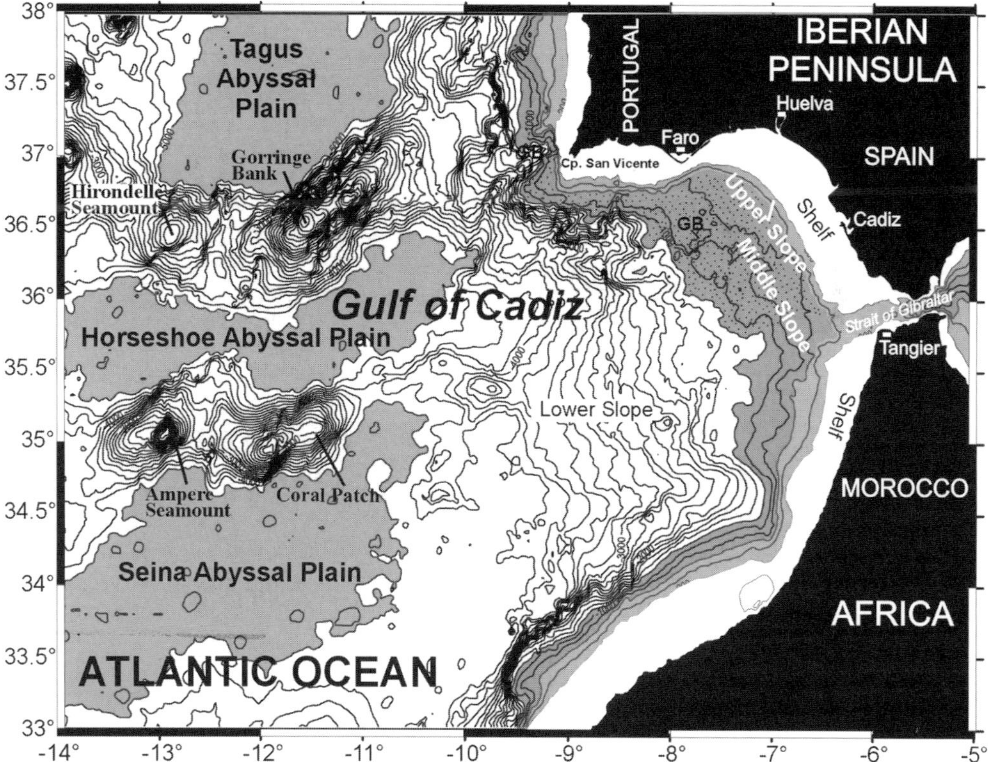

Fig. 2. Regional bathymetric map (in metres, from satellite data of Smith & Sandwell 1997) of the continental margin of the Gulf of Cadiz (modified from Hernández-Molina *et al.* 2006). The contourite depositional system (CDS) of the Gulf of Cadiz is shown by the dotted area.

Fernández-Puga 2004), and fault reactivation (Maestro *et al.* 1998; Lobo *et al.* 2003).

Sea-bed topography

The major part of the Gulf of Cadiz comprises a giant outward bulge sloping to the west, with irregular surface relief (Fig. 2). Based on the slope gradient and morphological features, the principal physiographic features of this broad slope are (Malod & Mougenot 1979; Baraza & Nelson 1992; Nelson *et al.* 1993): a shelf-break located between 100 and 140 m depth; a steeper (2–3°) upper slope between 150 and 400 m depth and 19 km wide; two gently dipping (>1°) wide terraces located at water depths of 500–750 and 800–1200 m on the middle slope, crossed by channels and ridges that trend NE; a smooth lower slope (0.5–1°), between 800 and 4000 m depth; and the abyssal plains, at water depths greater than 4300 m, separated by submarine banks (or seamounts) trending ENE. For the most part, this continental slope lacks submarine canyons, except in the western area of the Algarve margin and close to the Strait of Gibraltar (Hernández-Molina *et al.* 2006).

Oceanographic setting

The oceanographic setting in the Gulf of Cadiz is characterized by intense hydrographic dynamics controlled by the exchange of water masses through the Strait of Gibraltar. This is determined by the overflow at a depth of 40–200 m of dense, highly saline and warm Mediterranean Outflow Water (MOW) near the bottom, and the turbulent, less saline, cool-water mass of Atlantic Inflow Water (AI) on the surface (Fig. 1; e.g. Madelain 1970; Mélières 1974; Zenk 1975; Ambar *et al.* 1976; Ambar & Howe 1979; Baringer & Price 1999; Nelson *et al.* 1999). Because of its higher density compared with the ambient Atlantic waters, the Mediterranean overflow forms a turbulent flux ranging from 150 to 200 m wide, which moves along a straight channel in a WSW direction at a speed of more than 200 cm s^{-1} (Ambar & Howe 1979). It then progressively sinks northwestwards, descending from the 300 m deep strait down the continental slope of the eastern Gulf of Cadiz (e.g. Madelain 1970; Mélières 1974; Thorpe 1975; Zenk 1975; Gardner & Kidd 1983; Ochoa & Bray 1991; Baringer & Price 1999; Nelson *et al.* 1999).

Above the MOW, the water masses in the Gulf of Cadiz comprise the North Atlantic Surface Water (NASW), which flows at the surface to a water depth around 100 m, and the North Atlantic Central Water (NACW), which flows between 100 and 700 m depth. Parts of these water masses together constitute the Atlantic Inflow (AI) towards the Alboran Sea through the Strait of Gibraltar, characterized by high temperatures and moderate salinities (12–16 °C, 34.7–36.25‰).

In the Gulf of Cadiz, part of the deep, warm and more saline MOW joins the NADW. This mixture flows very slowly southwards along the western part of the Atlantic Ocean (Zenk 1975; Knauss 1978). The NADW component is characterized by low temperatures (3–8°) and 34.95–35.2‰ salinity (Caralp 1988, 1992), flowing southwards from its source in the Greenland–Norwegian Sea region (e.g. Reid 1994).

MOW itself is composed mainly (90%) of Levantine Intermediate Water (LIW) and, to a much lesser extent, of Western Mediterranean Deep Water (WMDW) (Bryden & Stommel 1984), and is characterized as warm and saline (13 °C, 36.5‰) with an oxygen content of 4‰ (Madelain 1970; Ambar & Howe 1979). As it moves westwards, MOW shows a decrease in temperature, salinity and velocity. It is influenced by both Coriolis force and topography, being divided into two main cores (Fig. 1): (1) Mediterranean Upper Water (MU), which corresponds to the small shallow core described by Ambar *et al.* (1999), which moves as a warm, moderately saline flux (3.7 °C, 37.07‰) between depths of 400 and 600 m at the base of the upper slope as far west as Cape San Vicente, with an average velocity of about 46 cm s^{-1} (Ambar & Howe 1979); (2) Mediterranean Lower Water (ML), which constitutes the more saline (37.42‰), lower core and the MOW's principal nucleus, at a depth of 600–1200 m and with an average velocity of *c.* 20–30 cm s^{-1} (Zenk & Armi 1990; Baringer 1993; Bower *et al.* 1997).

The ML described here corresponds to the sum of the two branches (upper at 800 m and lower at 1200 m depth) described previously by Madelain (1970), Zenk (1970, 1975), Ambar & Howe (1979) and Ambar (1983). The study area is characterized by an irregular slope morphology, as shown by bathymetry data and acoustic imagery (Mulder *et al.* 2002, 2003; Hernández-Molina *et al.* 2003), which diverts the ML flow, such that it subdivides into three minor branches between the Cadiz and Huelva meridians (Fig. 1; Madelain 1970; Kenyon & Belderson 1973; Mélières 1974; Zenk 1975; Ambar 1983; Nelson *et al.* 1993, 1999; Borenäs *et al.* 2002). These include: (1) the Intermediate Branch (IB), which moves northwestwards through the Diego Cao channel; (2) the Principal Branch (PB), which is believed to transport, at present, the MOW's major flow (Madelain 1970), through the Guadalquivir channel south of the Guadalquivir Bank, which modifies the flux toward the SE; (3) the Southern Branch (SB), which follows a steep valley towards the SW through the Cadiz channel (Fig. 1).

As far west as 8°W, the MOW constitutes a bottom layer along the base of the upper slope and middle slope, but further west the MOW density range reaches a neutral buoyancy that characterizes an intermediate oceanic level (Madelain 1970; Gardner & Kidd 1983; Ochoa & Bray 1991; Ambar *et al.* 2002). It thereby loses contact with the sea floor at 1000 m depth in the eastern sector and at 1400 m depth in the western sector (Gardner & Kidd 1983; Baringer & Price 1999).

Palaeoceanography

The palaeocirculation of the MOW, from the last maximum glacial stage to the present, has been studied in detail based on benthic and planktonic Foraminifera, and sedimentological, mineralogical and geochemical analyses. However, there is still considerable controversy on definition of the age of the warmer and colder periods, on limits in the time scale and on sample resolution to study the variation of flow (and the effect on the precision of faunal data). It is generally agreed that during the last glacial stage (20–18 ka BP) there was significant vertical exchange between water masses (Caralp 1988, 1992). The direction of exchange through the Gibraltar gateway was similar to the present one, with MOW flow to the west (Caralp 1988, 1992; Grousset *et al.* 1988; Vernaud-Grazzini *et al.* 1989), but the relative intensity of flow is disputed.

Some researchers (Faugères *et al.* 1984, 1985; Stow *et al.* 1986; Nelson *et al.* 1993; Sierro *et al.* 1999; Pailler & Bard 2002) have suggested that during the first and second stages of the last deglaciation (15–13 ka BP), and especially during the Bolling Allerod (14–11 ka BP), there was a marked increase in MOW intensity, and a decrease in the MOW flow between 11 and 10 ka BP (the Younger Dryas stage). During the Early Holocene (10–7 ka BP), the intensity was once again somewhat reduced and a quasi-permanent thermocline was established. Finally, an increase in flow of MOW took place to its present level during the Late Holocene (since 7 ka BP). Other workers have considered that MOW intensification did not occur during the deglaciation stages, but was most marked during the cooler episodes (Caralp 1988, 1992; Vernaud-Grazzini *et al.* 1989; Cacho *et al.* 2000; Schönfeld & Zahn 2000). Recent studies have supported this view and suggested that the MOW played a stronger role during cold intervals and Heinrich events in deeper waters than during warmer intervals (Cacho *et al.* 2000; Llave *et al.* 2004*a*, 2006). During these cool stages, a smaller and denser MOW was developed and mixed vigorously with North Atlantic waters (Baringer & Price 1999). However, this view is still controversial, as the MOW volume would have decreased with the reduced cross-section of Gibraltar Strait during the glacial sea-level lowstands (Gardner & Kidd 1983; Bryden & Stommel 1984; Zahn 1997; Matthiesen & Haines 1998), thereby diminishing the exchange between the Mediterranean Sea and the Atlantic Ocean (Bethoux 1984; Bryden & Stommel 1984; Duplessy *et al.* 1988). On the other hand, owing to this diminished exchange coupled with lowered temperatures (Paterne *et al.* 1986; Rohling *et al.* 1998; Ambar *et al.* 1999) and a drier Mediterranean, the MOW formed during these cool conditions would have had a significantly higher salinity and hence greater density (Zahn *et al.* 1987; Thunell & Williams 1989; Schönfeld 1997; Zahn 1997; Cacho *et al.* 2000; Ambar *et al.* 2002). This would have led to an intense and deeper MOW (Thomson *et al.* 1999; Schönfeld & Zahn 2000; Rogerson 2002), creating a stronger interaction with the sea floor at greater depths, and hence facilitating the transport and deposition of coarser material, and so also leading to higher sand contents in contourites (Llave *et al.* 2004*a*, 2006).

Dataset and methods

This study uses a broad database comprising both seismic and sediment data. Seismic data have been obtained during several cruises supported by Spanish Research Council projects under a Hispano-Portuguese collaboration during the last 7 years including high-resolution (sparker 3000, 4000 and 7500 J, airgun) and very high-resolution seismic data (3.75 kHz and TOPAS). Further details on the seismic network have been given by Hernández-Molina *et al.* (2003). The seismic data were collected on board R.V. *Francisco de Paula Navarro*, R.V. *Cornide de Saavedra* and R.V. *Hesperides* during the oceanographic research cruises FADO 97/11, ANASTASYA 99/09, TASYO 2000, ANASTASYA 00/09 and ANASTASYA 01/09. The seismic emission frequency used was between 100 and 1500 Hz, giving an average vertical resolution of 1.5 m. The emission shot interval was 3 s, giving a horizontal resolution of about 7 m. A differential global positioning system (GPS) navigation system was used. Borehole data from the north Cadiz margin have been made available by Portuguese and Spanish oil companies (Fig. 3). Several Calypso piston cores, up to 20 m long, were recovered during the IMAGES V cruise on board R.V. *Marion Dufresne*, and a number of shorter gravity cores (<3 m long) were

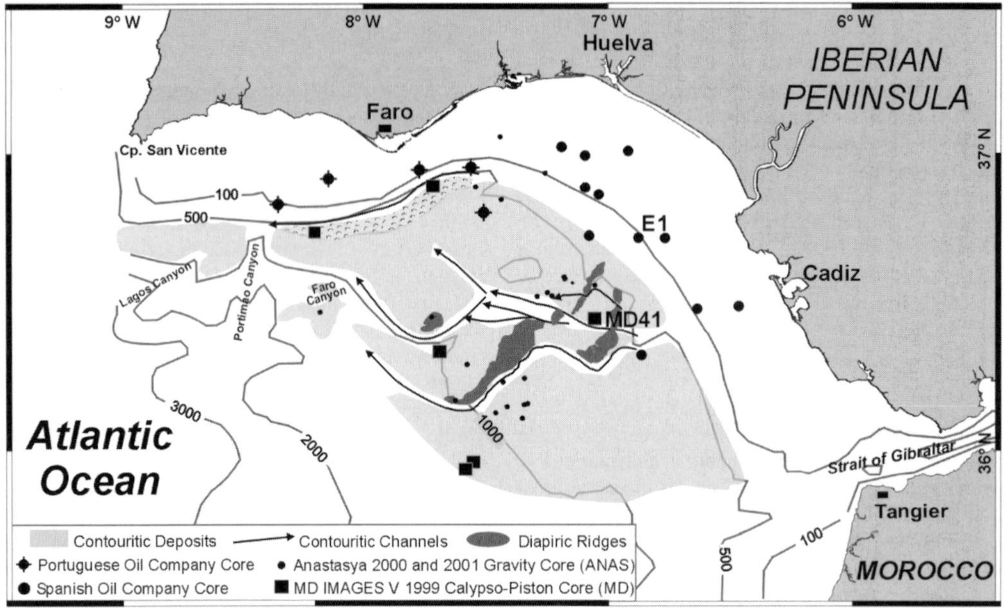

Fig. 3. Location of cores within the main sedimentary features of the contourite depositional system. The labels correspond to the dated examples shown in Figures 6 and 10 (E1 and MD41). The bathymetry is in metres.

obtained during ANASTASYA 2000 and 2001 cruises on board R.V. *Cornide Saavedra* (Fig. 3). All cores are from water depths between 500 and 1300 m.

For this study, the distribution, seismic stratigraphy and sedimentary stacking of the depositional contourite features are considered. Regional mapping of the distribution and boundaries of the seismic units was carried out, but only in the central, north and west sector of the middle slope of the Gulf of Cadiz. In the sector close to the Strait of Gibraltar it was not possible because of the presence of many recent tectonic features (mainly the diapiric ridges), which make it difficult to correlate the regional stratigraphy with the same degree of detail. Major results have been published where a more detailed explanation of the seismic stratigraphy of the contourite deposits along the middle slope has been given (see Llave *et al.* 2001, 2006; Hernández-Molina *et al.* 2002, 2006; Stow *et al.* 2002; Llave 2003). However, in this paper we focus on certain features such as the following.

(1) The development of a general chronostratigraphy of the depositional sequences for the Quaternary sedimentary record. For this we have used the following information. (a) Correlation of the dense network of low-resolution to very high-resolution single-channel and multi-channel seismic reflection lines with the results of core and borehole data from various surveys. We used cores and borehole data obtained by oil company drilling (Fig. 3) to establish the chronological framework for the Pliocene and Quaternary, and Calypso giant piston and standard gravity cores were used for the Late Pleistocene and Holocene chronostratigraphy (Llave *et al.* 2001, 2004*a*, *b*, 2006; Llave 2003). (b) Correlation of our sequences and units with the stratigraphic seismic results obtained previously by other workers on the continental margin made at different scales: (i) Mesozoic and Cenozoic sedimentary record (Baldy *et al.* 1977; Malod 1982; Mougenot & Vanney 1982; Mougenot 1988; Riaza & Martínez del Olmo 1996; Tortella *et al.* 1996; Maldonado *et al.* 1999; Terrinha *et al.* 2002; Fernández-Puga 2004; Medialdea *et al.* 2004); (ii) Pliocene and Quaternary (Rodero 1999; Rodero *et al.* 1999; Hernández-Molina *et al.* 2002); (iii) Quaternary (Llave *et al.* 2001, 2004*a*; Llave 2003); (iv) Late Pleistocene–Holocene (Hernández-Molina *et al.* 1994, 2000; Lebreiro 1995; Lobo 1995, 2000; Lebreiro *et al.* 1997; Somoza *et al.* 1997; Lobo *et al.* 2001, 2002, 2005*a*; Llave *et al.* 2004*b*, 2006); (v) Late Holocene (Lobo *et al.* 2003, 2005*b*; Llave *et al.* 2004*b*).

(2) The development of a detailed chronostratigraphy for the Upper Quaternary sedimentary record based on correlation of two Calypso piston cores (isotope-dated) with very high-resolution seismic profiles, using a sound velocity in sediments of 1600 m s^{-1} (Llave *et al.* 2004*b*).

(3) A sequence stratigraphic analysis of the Quaternary and Late Pleistocene–Holocene contourite deposits.

(4) The interpretation of climatic changes in relation to palaeoceanographic evidence in the CDS.

(5) The consideration of the tectonic influence on palaeoceanography and the evolution of contourite depositional architecture.

The contourite depositional system: present-day morphosedimentary features

Five morphosedimentary sectors within the CDS in the middle slope of the Gulf of Cadiz have been distinguished by Hernández-Molina *et al.* (2003) and Llave (2003): (1) proximal scour and sand ribbons sector; (2) overflow–sedimentary lobe sector; (3) channels and ridges sector; (4) active contourite depositional sector; (5) submarine canyons sector (Fig. 4).

Sector 1: proximal scour and sand ribbons sector

This is characterized by a smooth platform located alongslope in the SE zone close to the Strait of Gibraltar, between 500 and 800 m water depth. It is an extensive area dominated by an abrasion surface and several erosive scour features oriented NE–SW, with a smooth 'V' shaped expression and truncated reflectors in the seismic profiles. There are also some depositional features, including a SE–NW oriented sequence of bedforms, comprising ripple marks, sand ribbons and sand waves (first described by Kenyon & Belderson 1973) showing regular and symmetrical morphologies and high acoustic reflectivity.

Sector 2: overflow–sedimentary lobe sector

This is adjacent to and seawards of Sector 1, between 800 and 1600 m water depth, and has a fan shape 65 km long and 60 km wide. It constitutes a large sedimentary lobe with surface bed-forms, including small sandy and muddy lobate bodies and wave-fields, which in the seismic profiles show asymmetrical morphologies, smooth topographies and medium to low reflectivity. Erosive features are also important, including several large furrows with a NE–SW orientation, which display gravitational features

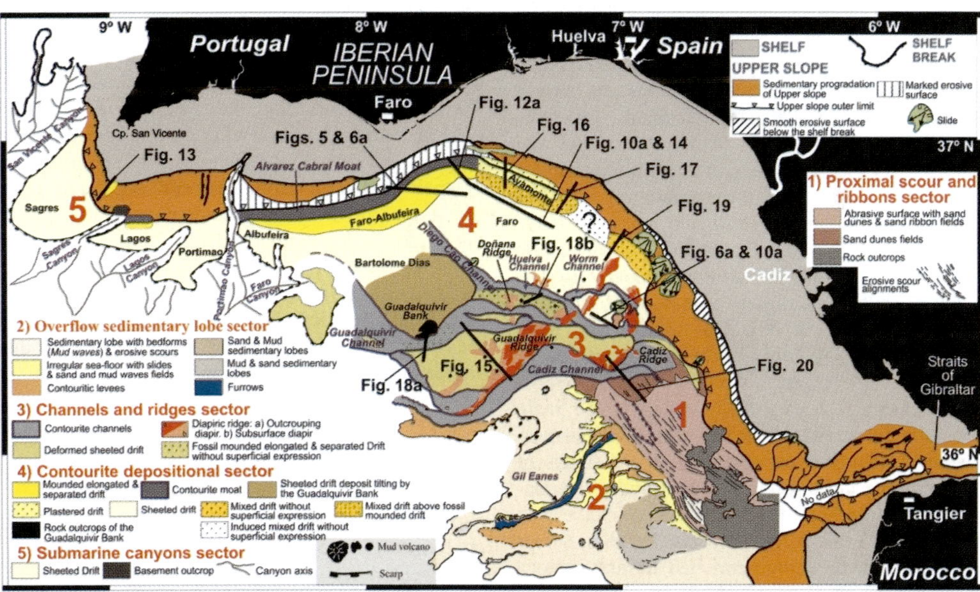

Fig. 4. Morphosedimentary map of the contourite depositional system on the middle slope of the Gulf of Cadiz with the location of the seismic examples (sparker seismic profiles). Morphosedimentary sectors: (1) proximal scour and sand ribbons sector; (2) overflow sedimentary lobe sector; (3) channels and ridges sector; (4) contourite depositional sector; (5) submarine canyon sector. (Modified from Hernández-Molina *et al.* 2003.)

on their margins. Although Hernández-Molina *et al.* (2003) interpreted this sector as an overflow–sedimentary lobe, and Mulder *et al.* (2003) as an unstable giant contourite levee, these hypotheses are not contradictory, as we can see gravitational features, and these are by numerous sea-floor emissions of hydrocarbon-enriched fluids, many fields of mud volcanoes, hydrocarbon seepage and hydrocarbon-derived carbonate chimneys (Somoza *et al.* 2002, 2003; Díaz del Río *et al.* 2003).

Sector 3: channels and ridges sector

This is located in the central area of the middle slope, between 800 and 1600 m water depth. It is dominated by the presence of significant erosive and tectonic elements. Five main contouritic channels known as the Cadiz, Guadalquivir, Huelva, Diego Cao and Gusano channels represent the main erosive features. These channels are asymmetrical, with lengths of 10 km to over 100 km, widths of 1.5–10 km, and depths of *c.* 10–350 m, generally presenting a deeper and more abrupt northern flank. They are 'S'-shaped, with alongslope zones oriented NW–SE, which change to downslope zones oriented NE–SW as a result of interaction between the bottom current pathway and the irregular slope morphology. These irregularities in the sea floor are marked by the occurrence of some structural relief features: the Guadalquivir, Cadiz and Doñana diapiric ridges and the Guadalquivir Bank uplift. The Guadalquivir and Cadiz diapiric ridges are NE–SW-oriented elongate outcropping ridges, from 300 to 1100 m depth. The Doñana ridge crops out only in restricted areas. The Guadalquivir Bank is a structural basement high made up of Palaeozoic and Mesozoic rocks of the Iberian margin (Medialdea *et al.* 2004). It is located in the southern part of the Bartolomeu Dias sheeted drift and has a NE–SW orientation. The principal contourite channels are located along the SE flank of the diapiric ridges; however, several marginal valleys with irregular morphologies and a NE–SW trend have been detected on the NW flanks of the diapiric ridge. All of these erosive features were established over a broad deformed sheeted drift, which is the main depositional morphology in this sector. Some mass movement elements are observed on the flanks of the diapiric structures and on the margins of the contourite channels. Finally, several mud volcanoes can be observed in this sector associated with sea-floor emissions of hydrocarbon-enriched fluids (Somoza *et al.* 2002, 2003).

Sector 4: active contourite depositional sector

This is developed in the central and NW areas of the middle slope. It is characterized by the dominance of depositional features represented by the following. (1) The mounded elongated and separated Faro–Albufeira drift, located at 500 m depth on the South Portuguese margin. (2) A sheeted drift complex, forming the basinward prolongation of the mounded drift and characterized by a planar and horizontal morphology. Three main sheeted drift segments have been differentiated: the Faro–Cadiz, at a water depth of 600 m, which is 20 km wide and 30 km long; the Bartolomeu Dias, at 750 m depth, which is 30 km wide and 45 km long; and the Albufeira, at 850 m depth, which is 10 km wide and 24 km long. (3) A plastered drift, located to the east of the mounded drift, between 300 and 600 m depth, about 35 km long and 12 km wide. The smooth contourite terrace is scoured by an important erosive contouritic channel parallel to the slope named the Alvarez Cabral Moat, which is 80 km long and 4–11 km wide, with a 'U'-shaped cross-section. There are also gravitational features defined on the mounded drift and Faro–Cadiz sheeted drift.

Section 5: submarine canyons sector

This sector is located in the western area of the middle slope and is characterized by the occurrence of a number of channels cutting across the slope from NE to SW, including the Portimao, Lagos, Sagres and San Vicente submarine canyons, with steep margins and erosive surfaces. These erosive features cut through and delineate several sheeted drifts at around 1000 m water depth, including the Portimao (16 km long and 14 km wide), Lagos (24 km long and 12 km wide), and Sagres (26 km long and 30 km wide) sheeted drifts. There is also a small mounded elongated and separated drift (the Lagos drift) at a water depth of 950 m. It is 8 km long, 6 km wide, and has 75 m relief. It is developed along the left margin of a small erosive moat, parallel to the south coast of San Vicente, and is about 8 km long and 4 km wide. Some mass-wasting elements are observed on the sheeted drifts near the canyons.

Seismic stratigraphic and chronostratigraphic framework

Within the Quaternary sedimentary record of the contourite deposits, two main depositional sequences have been identified regionally: Q-I and Q-II. These two sequences are separated by a marked continuous reflector of high reflectivity that shows distinct erosion in parts (Fig. 5); this is defined as the MPR discontinuity. This discontinuity marks a change in seismic facies, separating deposits with a more aggrading pattern (Q-I) from those above with a more progradational

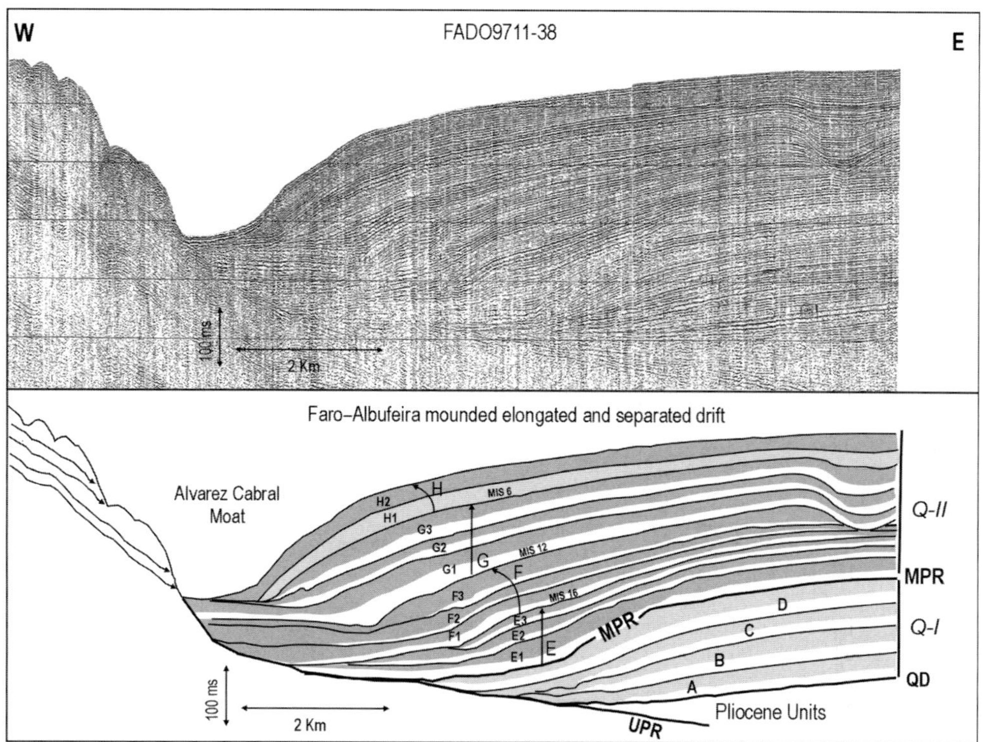

Fig. 5. Sparker seismic profile and line drawing through the western zone of the Faro–Albufeira mounded drift located in Sector 4, indicating the main seismic units within the Quaternary contourite sedimentary record. The reflective seismic facies are shown in grey, and the more transparent seismic facies in white (see Fig. 4 for location).

stacking pattern (Q-II). Both depositional sequences show a generally progradational stacking pattern in the mounded drift (Fig. 5) that tends to become aggradational laterally in the sheeted drift (Fig. 5).

The same facies trend is observed within the CDS: (1) a transparent zone at the base; (2) smooth, parallel reflectors of moderate to high amplitude in the upper part; (3) a high-amplitude erosive continuous surface at the top (Fig. 5). This cyclicity has helped to differentiate the superimposition of four minor depositional units in Q-I and Q-II (from A to D and from E to H) bounded by minor erosive and reflective surfaces, which are interpreted as discontinuities (Fig. 5). A further cyclic repetition of seismic facies is seen in Q-II, allowing the differentiation of minor seismic units (E_1, E_2, E_3, F_1, F_2, F_3, G_1, G_2, G_3, H_1 and H_2). The uppermost sequence, H is composed of four minor units: *a*, *b*, *c* and *d*, bounded by minor discontinuity surfaces. These minor units show vertical and gradual changes from transparent facies at the bottom to reflective facies near the top, which has allowed the differentiation of 10 subunits: a_1, a_2 and a_3; b_1, b_2, b_3 and b_4; c; d_1 and d_2 (Fig. 6a and b). The detailed description of these units is beyond the scope of the present paper, and further details have been given by Llave *et al.* (2001, 2006) and Llave (2003).

The general spatial distribution of depositional sequences Q-I and Q-II has been mapped from the central sector (Sector 3) towards the north and west (Sectors 4 and 5). The main structural features in the region, including the diapiric mounds and ridges and the Guadalquivir Bank uplift, have had a marked effect on the distribution of these depositional sequences. The most important depocentres in depositional sequence Q-I are located in the following areas (Fig. 7a): (1) the Faro–Albufeira drift, where the maximum thickness is about 200 ms; (2) south of the Faro–Cadiz sheeted drift, with a maximum thickness of 250 ms; (3) the central part of the Bartolomeu Dias sheeted drift, where a depocentre up to 300 ms (two-way travel time; twtt) thick is observed. The main depocentres of depositional sequence Q-II are located in similar areas and with similar orientations to those of sequence Q-I, as follows (Fig. 7b): (1) the Faro–Albufeira drift, where the maximum thickness is about 375 ms; (2) south of the Faro–Cadiz sheeted drift, up to 250 ms (twtt) thick; (3) the central part of the Bartolomeu Dias sheeted drift, where the depocentre is up to 300 ms (twtt) thick.

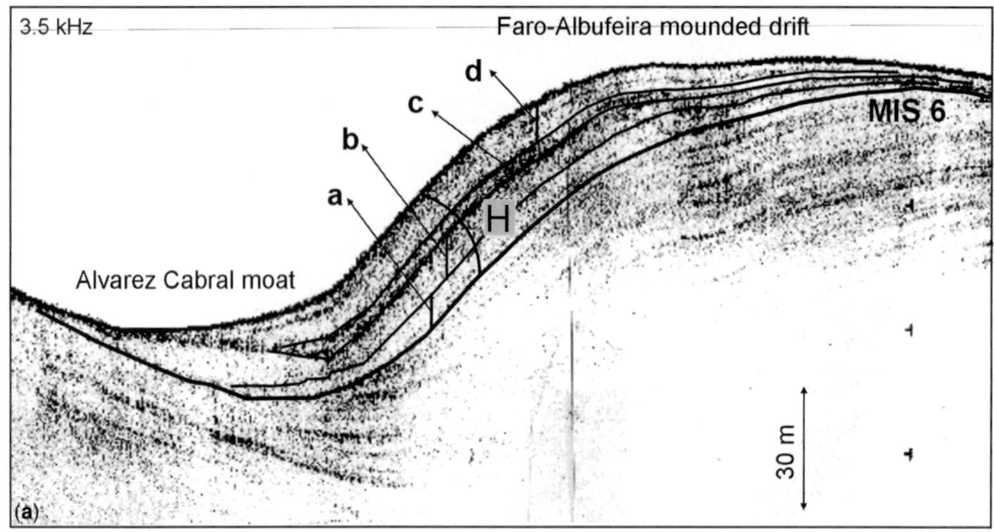

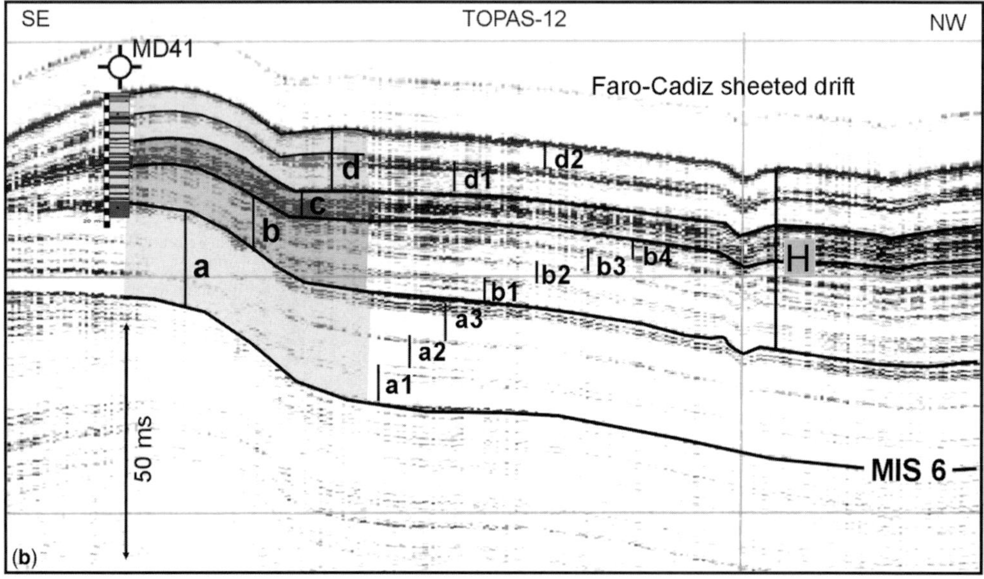

Fig. 6. (**a**) A 3.75 kHz seismic profile and line drawing through the mounded Faro–Albufeira drift indicating the main Late Pleistocene seismic units; (**b**) Topas seismic profile and line drawing through the Faro–Cadiz sheeted drift indicating the main seismic units within the main Late Pleistocene seismic units (modified from Llave *et al.* 2006) (see Fig. 4 for location).

A regional Quaternary chronostratigraphic framework is shown in Figure 8, where it can be determined that depositional sequences Q-I and Q-II are Quaternary in age, and the main discontinuities QD, MPR and MIS12 are dated at around 1.8 Ma, 900 ka and 400 ka, respectively (Llave 2003; Llave *et al.* 2004*a*, *b*, 2006), corresponding in age to the base of the Quaternary, the mid-Pleistocene and the late part of the mid-Pleistocene, respectively (Fig. 10a). In more detail, the youngest depositional sequence (H) is Late Pleistocene–Holocene in age (Fig. 10a). The oldest seismic unit *a* within sequence H was deposited between MIS 6 (135 ka) and Heinrich event H6 (57 ka), seismic unit *b* was deposited between H6 (57 ka) and H3 (32 ka), seismic unit *c* between H3 (32 ka) and H2 (24 ka), and the youngest seismic unit *d* was deposited between H2 (24 ka) and the present (Llave 2003; Llave *et al.* 2004*a*, *b*, 2006).

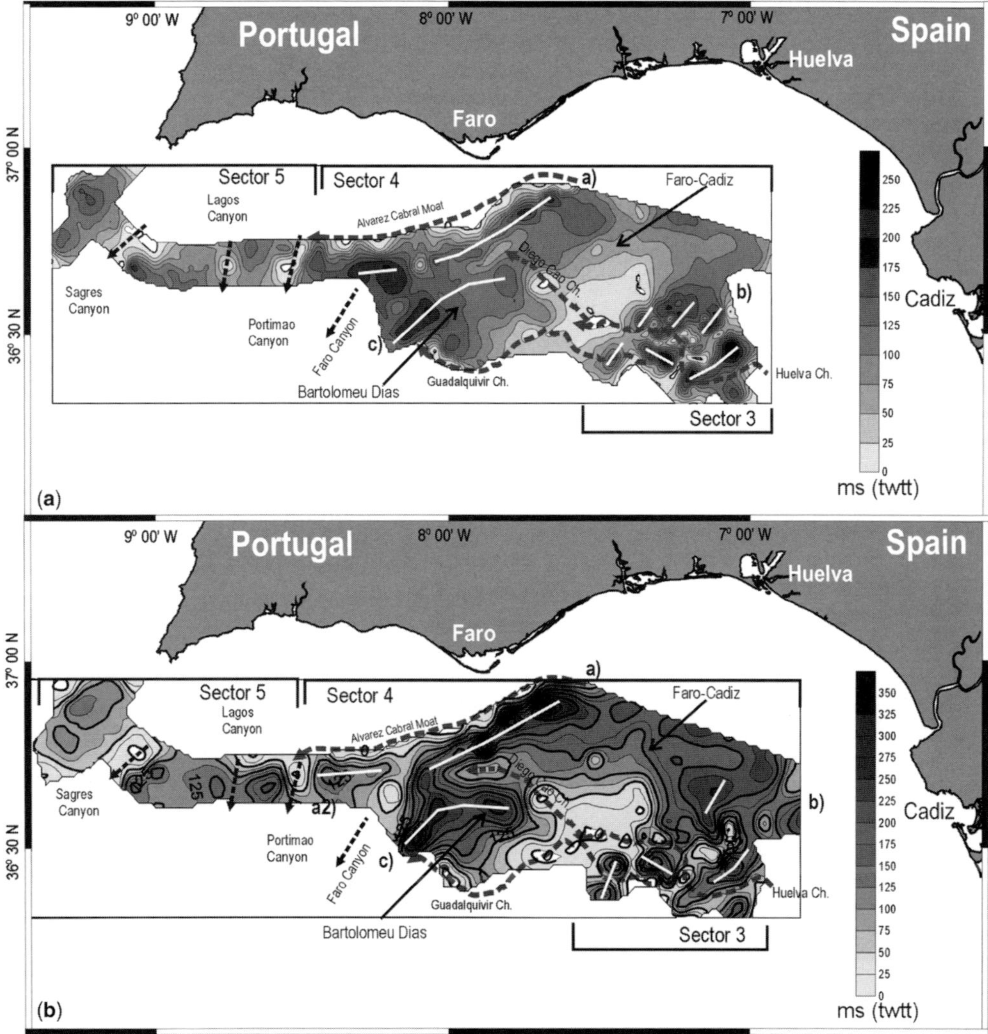

Fig. 7. Distribution and main depocentres of seismic units Q-I (**a**) and Q-II (**b**). The main depocentres are shown by white lines and lower-case letters.

Stratigraphic architecture of the major drift deposits

Within the CDS, significant differences in the Quaternary stratigraphic stacking pattern of the same major and minor depositional sequences are identified in the active and fossil (buried) drifts. Defining these changes is essential to understanding the evolution of the depositional system and for identifying major palaeoceanographic events.

Six contourite drifts have been identified (Fig. 11). Four of them follow the nomenclature of Faugères *et al.* (1999): (1) mounded elongated and separated; (2) sheeted; (3) plastered; and (4) wave fields, on the present sea floor. The other two are (5) fossil mounded drifts and (6) mixed drifts, so named because they are composed of inactive mounded drifts.

Active drifts

Mounded elongated and separated drifts. These drifts are located in Sectors 4 and 5 of the CDS. Three mounded elongated and separated drifts (hereinafter referred to as mounded drifts) are identified: the Faro–Albufeira, Lagos and Sagres drifts.

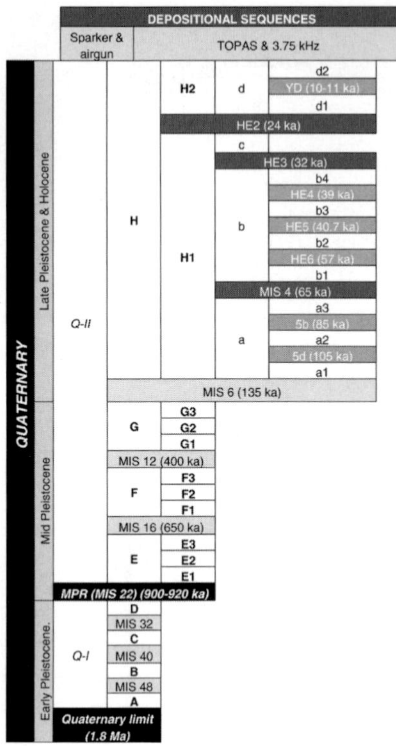

Fig. 8. Chronostratigraphic table of the Quaternary depositional sequences described from different resolution seismic profiles analyses.

The mounded Faro–Albufeira drift (Fig. 4) is located on the middle slope of Sector 4 at about 500–600 m depth. Two zones can be differentiated: an eastern proximal zone where the drift has a general SW trend, and a more distal western zone where the drift has a westward trend. These two zones follow the curved margin morphology, so that the drift is parallel to the upper slope, but separated from it by the Alvarez Cabral moat. The drift displays an asymmetrical mound shape with a steep, slump-prone northern flank and a gentler, smooth southern flank. It is around 80 km long, 12–20 km wide and about 75–100 m high.

Llave *et al.* (2001) considered that the Faro–Albufeira mounded drift is part of a system composed of five morphological elements; these are, from the upper slope to the middle slope: erosive surface on the upper slope; Alvarez Cabral moat; mounded elongated and separated drift; sheeted drift; erosive features over the drift. These elements can be recognized on the present sea floor, but also have been identified in each of the major and minor discontinuities defined previously within the Quaternary sedimentary record of the drift.

Major (Q-I and Q-II) and minor (A–H) depositional sequences described previously are vertically and spatially related, forming a 600 ms (twtt) thick Quaternary contourite deposit (Figs 4 and 5). Sequences exhibit a layered reflection pattern, the reflectors ranging from parallel and continuous to divergent or convergent, resulting in non-uniform thickness of the component layers with an evident

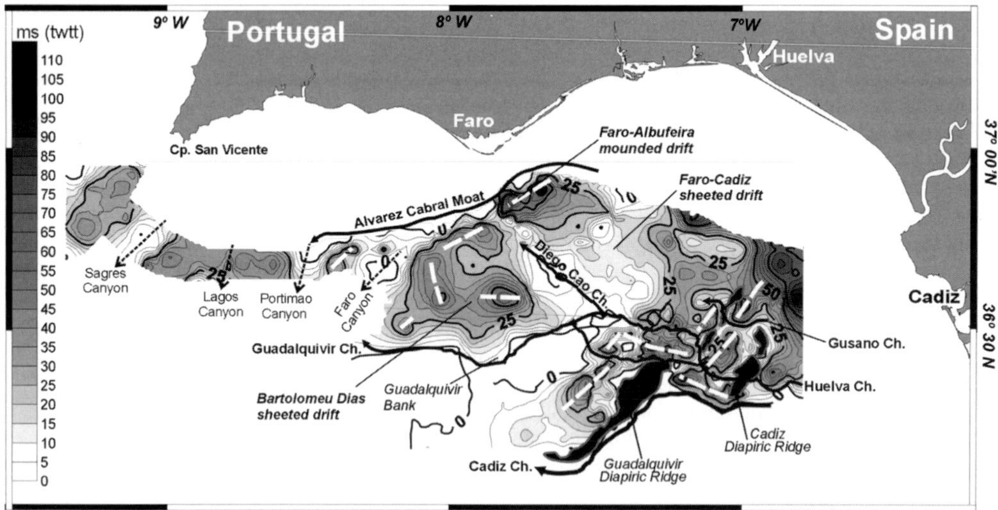

Fig. 9. Distribution and main depocentres of seismic unit H. The main depocentres are shown by white lines and lower-case letters.

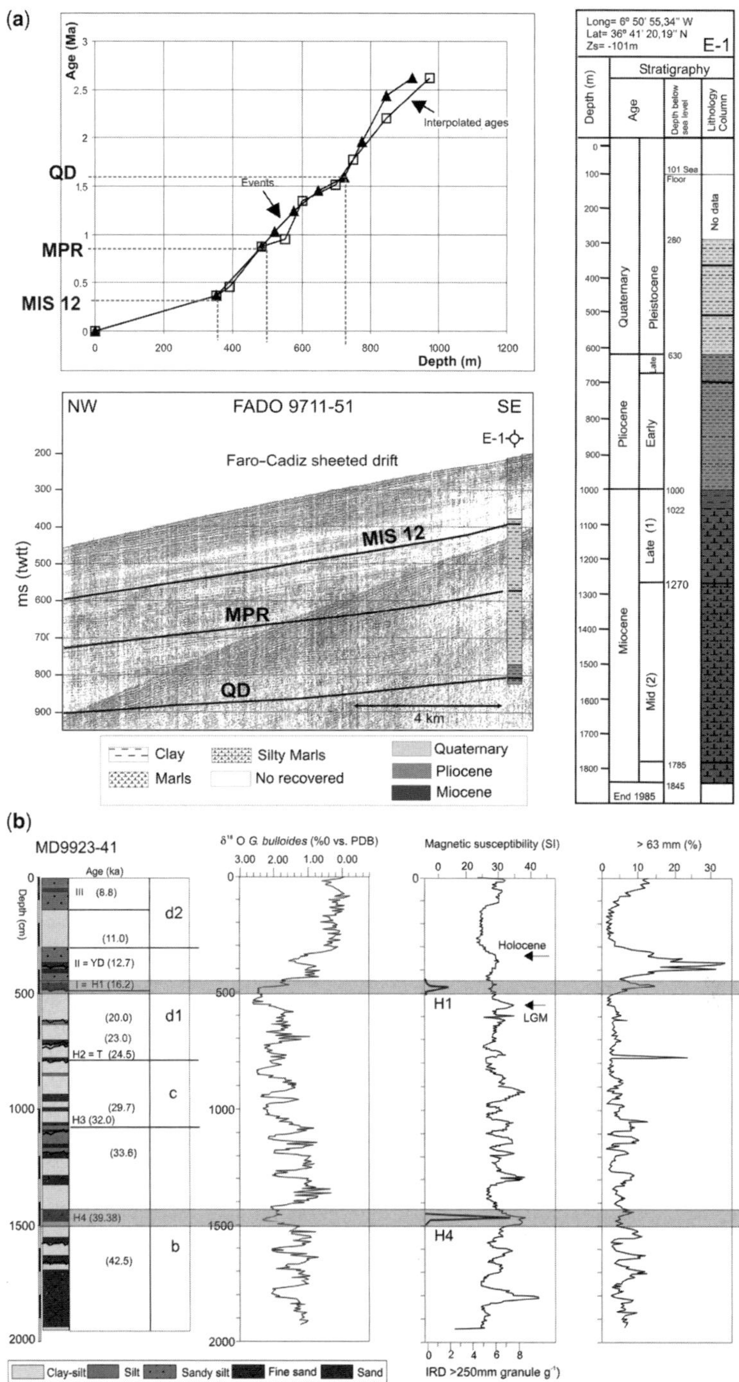

Fig. 10. (**a**) Example of the correlation between borehole E1 (see Fig. 3 for location) and a sparker seismic profile. The chronology of the discontinuities is based on calcareous nannofossil analyses, with three events that correspond to QD, MPR and MIS 12 (for detail see Llave *et al.* 2004*b*). (**b**) Example of the correlation between TOPAS seismic profile and Calypso piston core MD-41 (see Fig. 3 for location) and its $\delta^{18}O$, magnetic susceptibility and grain-size curves (after Mulder *et al.* 2002; Llave *et al.* 2006).

TYPES OF CONTOURITES					
	Deposits	Location		Stacking pattern	Sketch
ACTIVE DRIFTS	Elongate mounded & separated	Distal part of upper slope-Middle	Faro–Albufeira	Vertical and lateral prograding stacking pattern with downlap terminations	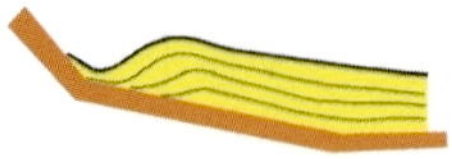
		Upper slope	Sagres		
	Plastered	Upper slope	Ayamonte	Parallel stacking pattern with onlap terminations	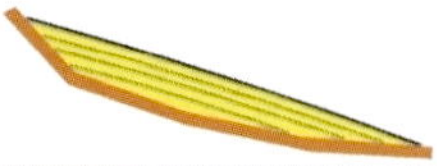
	Sheeted drift	Middle slope	- Faro-Cadiz - Bartolomeu Dias - Albufeira - Portimao - Lagos - Sagres	Vertical parallel or subparallel stacking pattern	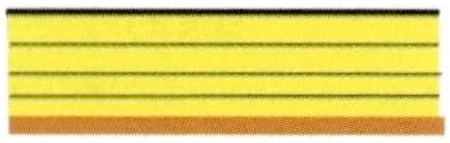
FOSSIL DRIFTS	Elongate mounded & separated	East of Faro–Albufeira active mounded drift	Cadiz	General prograding stacking pattern and change to aggrading stacking pattern in recent units	
		South of Guadalquivir Bank	Guadalquivir		
		Between Huelva and Guadalquivir channels	Huelva		
MIXED DRIFT		Middle slope	Ayamonte	Alternating aggrading stacking pattern units connected with the upper slope and prograding stacking pattern units separated from the upper slope	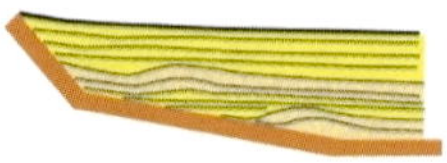
BED FORMS		Middle slope: Sectors 1 and 2	Sectors 1 and 2 sedimentation	Bedforms in Unit H fossilizing the sheeted drift	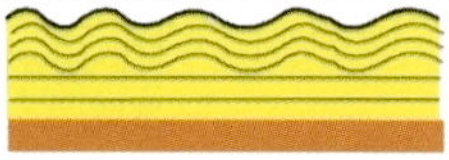

Fig. 11. Summary of the different types of contourite drifts described in the middle slope of the Gulf of Cadiz.

overall upslope direction of lateral sediment accretion. The most important change in the stacking pattern of the Faro–Albufeira mounded drift is associated with the MPR discontinuity, marking the change from an oblique progradational pattern to a more sigmoid progradational pattern (Figs 5 and 12a). This more progradational stacking pattern is observed in the western zone of the Faro–Albufeira drift, where depositional sequences are downlapping against the upper boundary of the lower sequences (Fig. 5), whereas in the eastern zone a change is observed from a sigmoid progradational configuration of the oldest sequences ($A–G_1$) (for the nomenclature see Llave *et al.* 2001; Llave 2003) to a more aggradational stacking pattern of the two youngest sequences (G_2–H) (Fig. 12a). Nevertheless, in the western zone a coeval change in the stacking pattern can be determined. Regarding the changes observed after MPR, depocentre locations of the minor sequences A–D display a NE migration, and a similar migration of depocentres has been determined for each of the sequences E–H (Fig. 12b). Nevertheless, a clear depocentre migration towards the SW is identified by comparing sequences D–E, in relation to the MPR discontinuity (Fig. 12b).

The mounded Lagos drift is located on the middle slope of Sector 5 with an east–west trend at around 950 m water depth, parallel to the upper slope, and separated from it by the Lagos moat (Fig. 4). It has an asymmetrical mound shape, and is around 8 km long, 6 km wide and about 75 m high. The Quaternary deposits reach about 240 ms (twtt) thick. Sequences display a similar stratigraphic stacking pattern to that of the Faro–Albufeira mounded drift in the western zone.

The mounded Sagres drift is located on the upper slope of Sector 5 south of Cape San Vicente, close to the shelf break between 300 and 500 m depth (Fig. 4). It has an asymmetrical shape, and is about 3 km long and 50 m high. The present sea floor represents a significant erosive surface, more prominent than any equivalent surface within the sedimentary record (Fig. 13). Sequences have a progradational stacking pattern towards the shelf break through the upper slope. They display a sigmoid to oblique–parallel configuration along with a significant discontinuity at the base.

Sheeted drifts. Within the CDS, seven major sheeted drifts have been recognized (Fig. 4), which occupy a large part of the basin floor of the

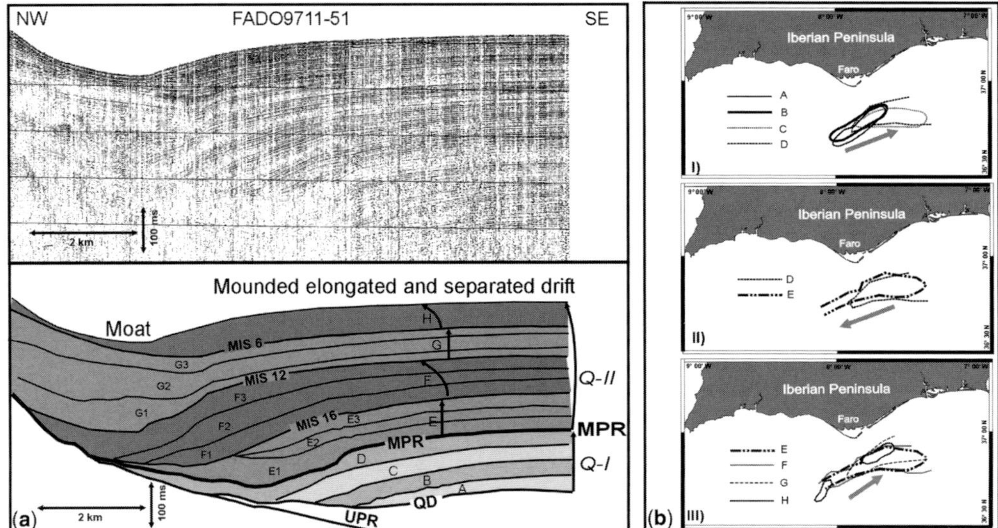

Fig. 12. (**a**) Sparker seismic profile and line drawing through the eastern zone of the Faro–Albufeira mounded drift located in Sector 4, indicating the main seismic units within the Quaternary contourite sedimentary record. The change related to the MPR discontinuity should be noted. Also, a change from a sigmoid progradational configuration of the oldest sequences (A–G) to a more progradational stacking pattern of the two youngest sequences (G and H) can be identified (see Fig. 4 for location). (**b**) Sketches of the changes observed in the western zone of the Faro–Albufeira mounded drift depositional sequences depocentres.

middle slope. They are located in Sector 4 (Faro–Cadiz; Bartolomeu Dias and Albufeira drifts, which form the basinward prolongation of the Albufeira–Faro mounded drift) and Sector 5 (Portimao, Lagos and Sagres drifts). In Sector 3, contourite deposits are mainly sheeted drifts but are widely deformed by diapirirism activity. Sheeted drifts are around 400 ms (twtt) thick on average in Sector 4, around 300 ms (twtt) on average in Sector 5, and up to 600 ms (twtt) in the deformed sheeted drifts of Sector 3. The *Faro–Cadiz drift* is the largest sheeted drift, with a maximum thickness up to

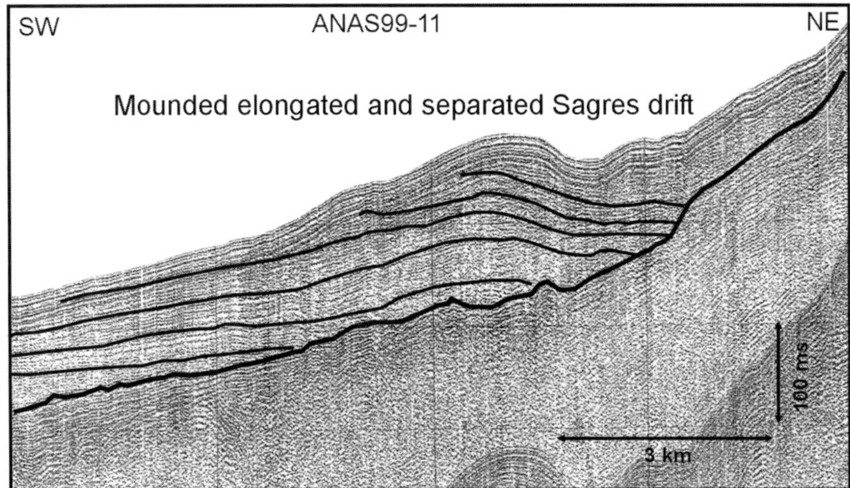

Fig. 13. Sparker seismic profile of the mounded Sagres drift located in Sector 5, which displays a sigmoid to oblique–parallel configuration along a discontinuity at its base (see Fig. 4 for location).

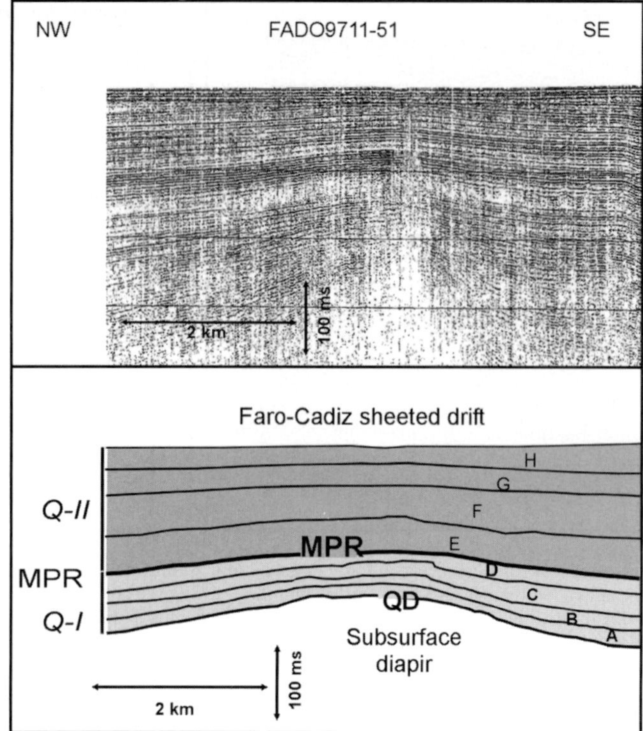

Fig. 14. Sparker seismic profile and line drawing through part of the Faro–Cadiz sheeted drift indicating the main seismic units within the Quaternary contourite sedimentary record. The stratigraphic stacking pattern of the sequences that form the sheeted drifts is mainly aggradational (see Fig. 4 for location).

500 ms and a width of several tens of kilometres. Its lateral extent is controlled by the upper slope to the east, several channels and diapiric ridges to the south, and the Diego Cao channel to the west (Fig. 4). The stratigraphic stacking pattern of the sequences that make up the sheeted drifts is mainly aggradational (Fig. 5b), with stratified, parallel (or subparallel) seismic facies, showing high lateral continuity (Fig. 14). Additionally, in Sector 3, the sequences are affected by low-frequency and low-amplitude anticline–syncline structures (Fig. 15).

Plastered drift. The plastered drift is located between the middle slope and the distal part of the upper slope, between 300 and 600 m water depth (Fig. 4), and develops eastward from the Faro–Albufeira mounded drift. It has a convex shape on the upper slope changing to concave on the middle slope. It is 35 km long and 12 km wide, with a thickness around 100 ms (twtt), forming depositional sequence H, and shows an aggradational stacking pattern and lens shape (Fig. 16). One intriguing feature of this drift is that it started

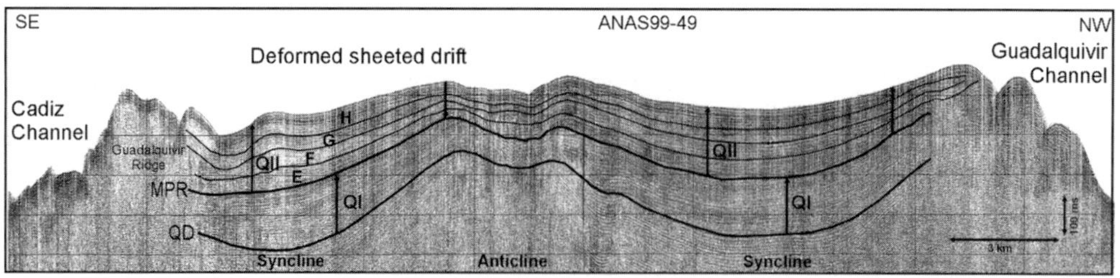

Fig. 15. Sparker seismic profile and line drawing through part of the deformed sheeted drift located in Sector 3 affected by low-frequency, low-amplitude anticline–syncline structures (see Fig. 4 for location).

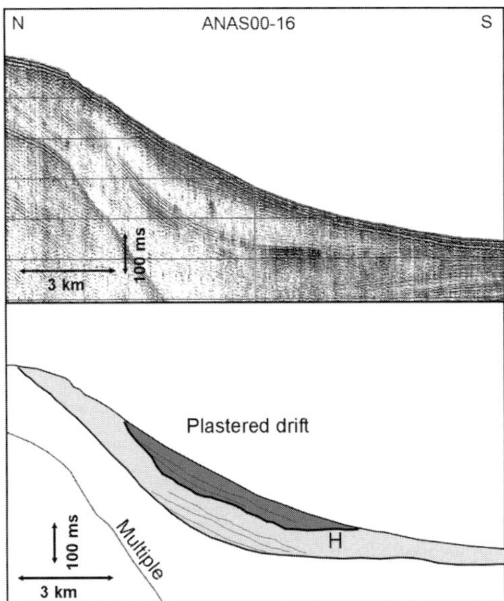

Fig. 16. Sparker seismic profile and line drawing through part of the plastered drift and the Ayamonte mixed drift located in Sector 4 (see Fig. 4 for location).

to be progressively developed some time within the deposition between depositional sequences G and H. This change in the stacking pattern of the upper slope could be coeval with the change in sedimentary architecture identified on the mounded Faro–Albufeira drift. On the other hand, within depositional sequence H a remarkable erosive surface is identified. Above this surface, the reflector configuration of the youngest deposits is one of onlap and downlap (Fig. 16).

Non-active mounded or fossil mounded drifts

In Sector 3 and part of Sector 4 of the middle slope, the main Late Quaternary deposits are sheeted drifts, deformed by local tectonics in Sector 3 and the alternation of mounded or sheeted drifts in Sector 4. However, beneath these deposits, three main mounded elongated and separated drifts have been identified, referred to here as fossil mounded drifts. The criterion for use of this nomenclature is based on the cessation of the mounded drift developing and their burial by a different contourite architectural style, generally sheeted drifts (Hernández-Molina *et al.* 2003; Llave *et al.* 2003*a*, *b*, 2006).

These fossil mounded deposits are: (1) the Cadiz fossil mounded drift (CFD); (2) the Guadalquivir fossil mounded drift (GFD); (3) the Huelva fossil mounded drift (HFD).

The Cadiz fossil mounded drift (CFD) is located in the NE zone of Sector 4, between 500 and 600 m depth, in the transition between the upper and middle slopes. It is 35 km long and 12–20 km wide. The CFD is composed only of sequence Q-I, and the stratigraphic stacking pattern of its minor sequences (A–D) displays a progradational configuration towards the upper slope (Fig. 17). Discontinuities bounding these minor sequences are locally depressed at the transition between the middle slope and the upper slope, which represents a system of fossilized moats laterally connected towards the west with the Alvarez Cabral moat (Fig. 17). These moats have no morphological expression on the present sea floor because they are completely filled and buried by sequence Q-II. Consequently, this mounded drift is an abandoned eastern part of the Faro–Albufeira mounded drift, which ended its development as a mounded elongated from and separated after the MPR discontinuity (Fig. 17).

The Guadalquivir fossil mounded drift (GFD) and the Huelva fossil mounded drift (HFD) are located in the central sector of the middle slope. They are characterized by depositional sequences with a progradational sigmoidal-to-oblique landward configuration prograding over a palaeoslope (Fig. 18a). Fossil mounded drifts are separated by a fossil moat from their palaeoslope, which will hereinafter be designated as the Guadalquivir fossil moat. The stratigraphic stacking pattern of the depositional sequences is different in both the GFD and HFD, showing a differential time interval for their activity as mounded elongated and separated drift, especially during the deposition of Q-II.

The Guadalquivir fossil mounded drift (GFD), is located south of the Guadalquivir Bank, at a depth of 1100 m. It is 15 km long and 6 km wide. In the GFD, despite strong erosion caused by the present Guadalquivir channel activity in this area, it can be observed that the fossil mounded drift deposits, *c.* 300 ms (twtt) thick, exhibit a prograding upslope stacking pattern (Q-I) (Fig. 18a). This depositional sequence has been buried and fossilized by a depositional sequence (Q-II) with an aggrading stratified stacking pattern, *c.* 175 ms (twtt) thick (Fig. 18a). Therefore, the MPR discontinuity constitutes in this sector an important erosive truncated surface, which separates the prominent northeasterly prograding body (Q-I) from a more aggrading one (Q-II) (Fig. 18a).

The Huelva fossil mounded drift (HFD) is located on the southern margin of the present-day Huelva Channel, at a depth of 650 m (Fig. 4). It is 32 km long and 6 km wide (Fig. 4). The HFD is characterized by a prominent northeasterly progradation of the internal reflectors, whereas the main axis of the body trends northwestwards. It is more than 400 ms (twtt) thick and comprises several

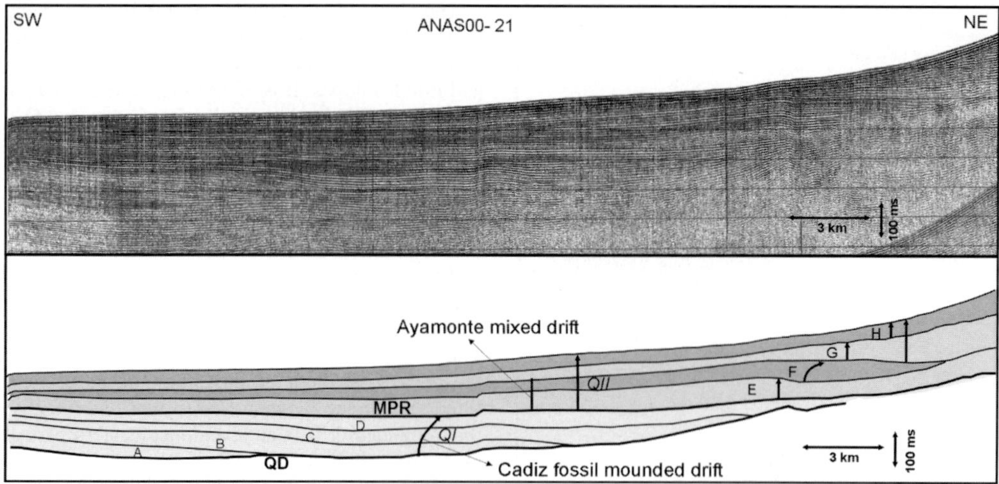

Fig. 17. Sparker seismic profile and line drawing through part of the Cadiz fossil mounded drift buried by the Ayamonte mixed drift located in Sector 4. It should be noted that the discontinuities bounded by the minor sequences of Q-I (A–D). Form depressions in the transition between the middle slope and the upper slope, which represent fossilized moats laterally connected towards the west with the Alvarez Cabral moat. Within Q-II, an aggradational stacking pattern is observed in E, and in connection with the upper slope units, followed by a mounded drift (F), and then by aggrading units connected with the upper slope depositional system (G and H) (see Fig. 4 for location).

sequences (A–H). This drift is buried by the most recent seismic unit (H), showing an aggrading stacking pattern, characterized by stratified, subparallel, high- and low-amplitude reflectors, 50–70 ms (twtt) thick (Fig. 18b). Consequently, the change in the depositional style of the mounded drift is related to the base of the seismic unit H (Fig. 18b).

The basinward prolongation of the three fossil mounded drifts (CFD, GFD and HFD) has an aggrading stacking pattern with alternating transparent and reflective units, constituting a sheeted–deformed contourite drift in Sector 3 and sheeted contourite drift in Sector 4.

Mixed drift

The mixed drift is located in the northeastern and eastern zone of Sector 4 (Fig. 4), and was developed during sequence Q-II above the Cadiz fossil mounded drift described above. It comprises an alternation of sequences with sheeted drift seismic facies characteristics and sequences with mounded and separated drift characteristics (Figs 17 and 19). This drift, named the Ayamonte mixed drift, has no morphological expression on the present sea floor, but is around 200 ms (twtt) thick within the Quaternary deposits (Q-II) (Figs 17 and 19).

The stratigraphic stacking pattern of the mixed drift is complex and two zones have been determined. In the northeastern zone the stratigraphic stacking pattern displays an alternation of sheeted and mounded deposits for sequences E and F, but a general aggrading stacking pattern for sequences G and H (Fig. 17). Sequences with sheeted drift facies are connected with the sequences of the upper slope but mounded drift facies at the top are disconnected from the upper slope (Fig. 17). The main change in the depositional style is related to the base of sequence G. Discontinuities bounding sequences E and F have depressions at the transition between the middle slope and the upper slope, which represent fossilized moats laterally connected towards the west with the Alvarez Cabral moat. These moats have no morphological expression on the present sea floor (Figs 4 and 17). In the southeastern zone, across the diapiric ridges, a different stratigraphic stacking pattern is evident. An alternation in the seismic facies has been identified within sequences EH. Transparent and massive facies in the lower part of the sequence have sheeted facies, but the reflective upper part of the sequence is characterized by a mounded and progradational seismic facies, which is visible in the youngest sequence (H; Fig. 19). At the top of the mounded and reflective facies an erosive surface is developed as the present sea floor in this area (Fig. 19).

Sediment wave fields

Sedimentary wave fields are widespread on the present sea floor in Sector 2, and especially in

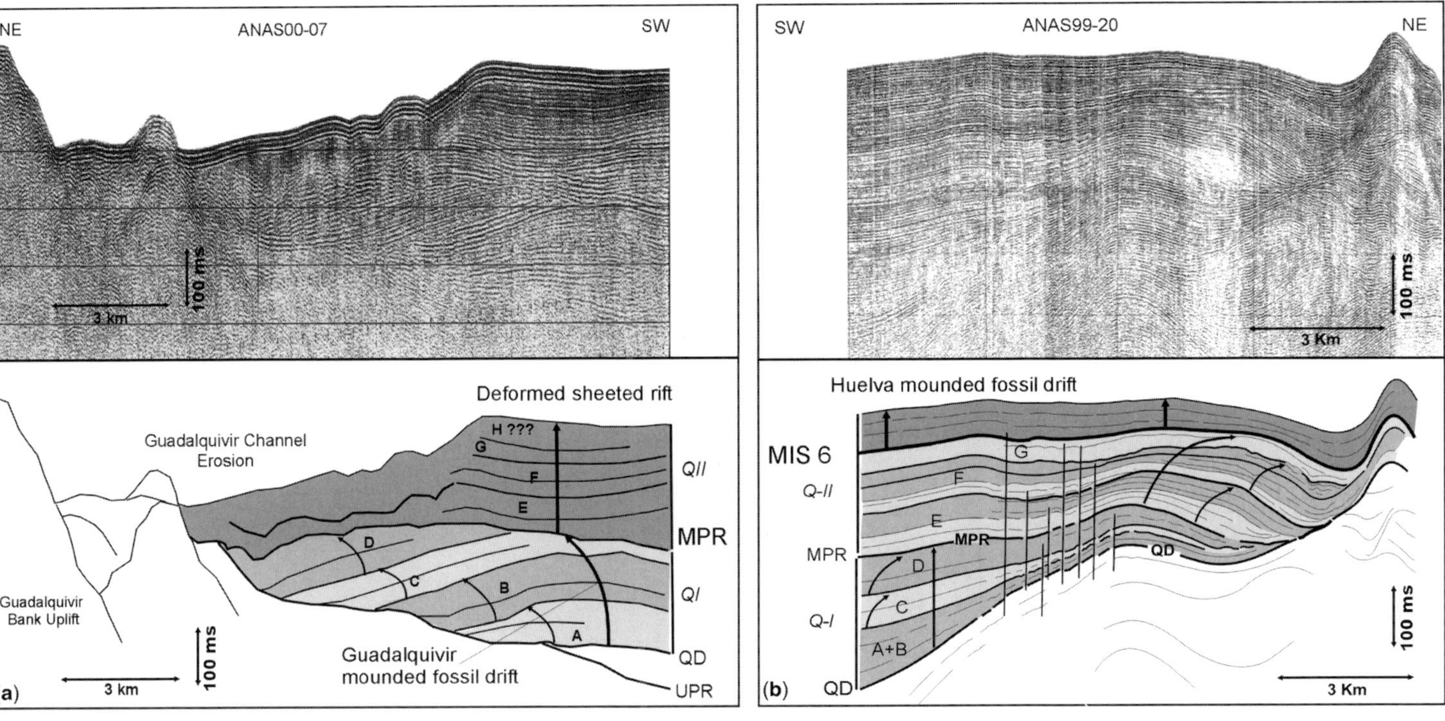

Fig. 18. (**a**) Sparker seismic profile and line drawing through part of the Guadalquivir fossil mounded drift. This drift is fossilized by the most recent aggradational seismic unit (H) (see Fig. 4 for location). (**b**) Sparker seismic profile and line drawing through part of the Huelva fossil mounded drift. The MPR discontinuity constitutes in this sector an important erosive truncated surface, which separates the prominent northeastwards prograding body (Q-I) from a more aggrading one (Q-II).

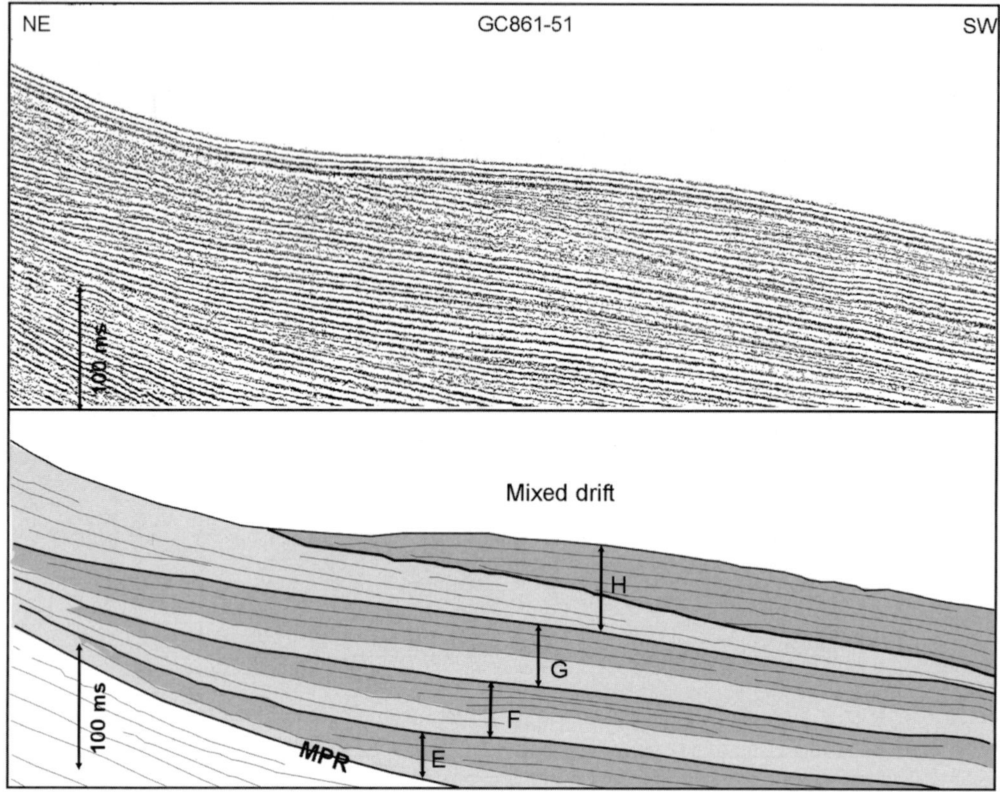

Fig. 19. Airgun seismic profile and line drawing through part of theAyamonte mixed drift located in the western zone of Sector 4. An alternation in the seismic facies has been identified within sequences E–H (see text for further details). It should be noted that the base of each depositional sequence is connected with the sedimentation of the upper slope; however, the top of the units is separated from the upper slope by erosive surfaces (see Fig. 4 for location).

Sector 1 of the CDS (Fig. 4). They represent secondary sedimentary features superimposed on the most recent sequence (H). This sequence displays a low-angle progradation towards the NW to aggradational configuration with high-amplitude and lateral continuity reflectors. Sedimentary wave fields have undulating seismic facies composed of bedforms of various wavelengths and low amplitude (Fig. 20). The stratigraphic stacking pattern of the Quaternary deposits of Sector 1 shows that the sedimentary wave field occurs only in the youngest sequence H. Older sequences (G, F, etc.) display sheeted contourite facies with less acoustic response and without the undulating wave facies (Fig. 20).

Major morphostructural features

The development and distribution of contourite deposits have been strongly influenced by the major morphostructural features in the Gulf of Cadiz, including the movement of diapirs, leading to both outcropping and buried diapiric ridges, the uplift of the Guadalquivir Bank, the displacement along several fault systems, and the evolution of anticline–syncline structures (Figs 4 and 21).

The areas where the deposits are more clearly deformed by recent tectonics are Sector 3 and the southern zone of Sector 4. The main structures that affect the contourite deposits in these sectors are (Figs 4 and 21): (1) the Cadiz and Guadalquivir diapiric ridges (GDR and CDR); (2) the Doñana buried diapiric ridge (DBDR); (3) the Guadalquivir Bank (GB). Diapiric ridges (GDR, CDR and DBDR) constitute elongated structures oriented NE–SW to NNE–SSW (Figs 4 and 21), which have undergone several episodes of activity from the Middle Miocene to the present (Maldonado *et al.* 1999; Maestro *et al.* 2003; Fernández-Puga 2004; Llave *et al.* 2006). In general, it has been

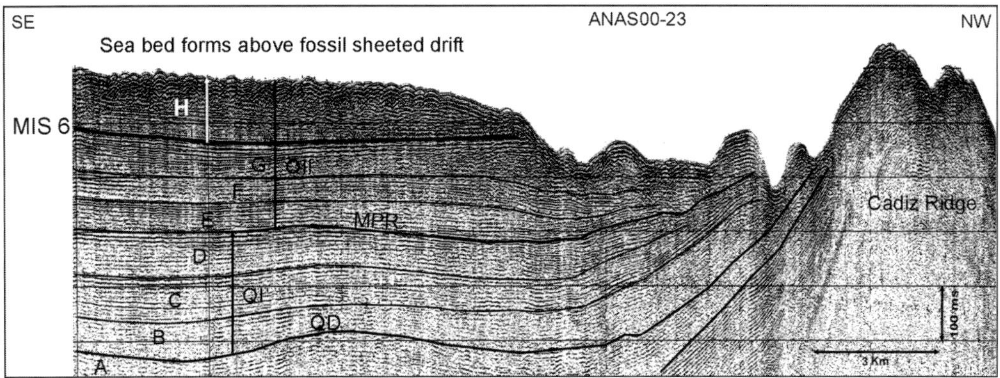

Fig. 20. Stratigraphic stacking pattern of the Quaternary deposits of Sector 1, showing that the sedimentary wave facies and seismic facies of sequence H are not identified in the oldest sequences, which display sheeted contourite facies with less acoustic response and without undulated facies (see Fig. 4 for location).

observed that the contourite sequences are synsedimentary with the emplacement of these tectonic structures during the Quaternary, especially observed within the depositional sequences A–G (Fig. 21b), the most recent sedimentation (depositional sequence H) being post-sedimentary (Fig. 21c). Several isolated diapiric outcrops and buried diapirs affect contourite sedimentation locally (Díaz del Río *et al.* 2003; Somoza *et al.* 2003). The contourite seismic units deposited during diapiric dome growths are characterized by thinning toward the axis of the diapiric uplift and only minor thickening into relatively distant peripheral sinks (Fig. 21b). These structures have determined not only the deformation of the sheeted drift and their extension in Sector 3, but have also conditioned the irregularities of the sea floor and the location of the Guadalquivir and Cadiz channels. The Guadalquivir Bank represents a structural high made up of Palaeozoic and Mesozoic basement rocks of the Iberian margin (Fernández-Puga 2004; Medialdea *et al.* 2004). This structure, located in the southern part of the Bartolomeu Dias sheeted drift (Figs 4 and 21), has played an important role in this area, affecting the hydrodynamic system and the accommodation space for Pliocene–Quaternary sedimentation, as we will see in the Quaternary evolution of the CDS. Its uplift from Pliocene to recent times (Maestro *et al.* 2003; Fernández-Puga 2004; Llave *et al.* 2006) has produced changes in the depositional geometry. Several discontinuities within the Bartolomeu Dias contourite sheeted drift have developed, as well as the generation of a monocline in the northern Bartolomeu Dias zone affecting especially the Q-I unit (Fig. 21d and e). The contourite sedimentation and MOW pathways have been conditioned not only by these morphostructural features, but also by the submarine canyons' trajectories (Fig. 21a and f).

Quaternary evolution of the contourite depositional system: a discussion

The contourite depositional system (CDS) of the Gulf of Cadiz began its development after the opening of the Strait of Gibraltar at the end of the Messinian (e.g. Madelain 1970; Mélières 1974; Gonthier *et al.* 1984; Faugères *et al.* 1985; Nelson *et al.* 1993, 1999; Llave *et al.* 2001, 2004*a*, 2006; García 2002; Mulder *et al.* 2002, 2003; Stow *et al.* 2002; Habgood *et al.* 2003; Hernández-Molina *et al.* 2003, 2006; Llave 2003). Since then, it has been influenced principally by tectonics, global climate and sea-level changes. The influence of global climate and sea-level changes is observed in the cyclic nature of the depositional sequences, not only the seismic facies but also the distribution of depocentres, especially observed in the northern part of the CDS, in the Faro–Albufeira mounded elongated and separated drift (Llave *et al.* 2001; Llave 2003). Based on the stratigraphic stacking patterns identified in the six types of contourite drifts described, three main stages have been recognized in the evolution of the CDS influenced by tectonics (Fig. 22): (1) Early Pleistocene to Mid-Pleistocene; (2) Mid-Pleistocene to Late Pleistocene; and (3) Late Pleistocene to Holocene.

Early Pleistocene to Mid-Pleistocene CDS: depositional stage (from QD to MIS 22)

The regional paleogeography during this first evolutionary stage of the Quaternary, which covers the time interval of sequence Q-I (depositional

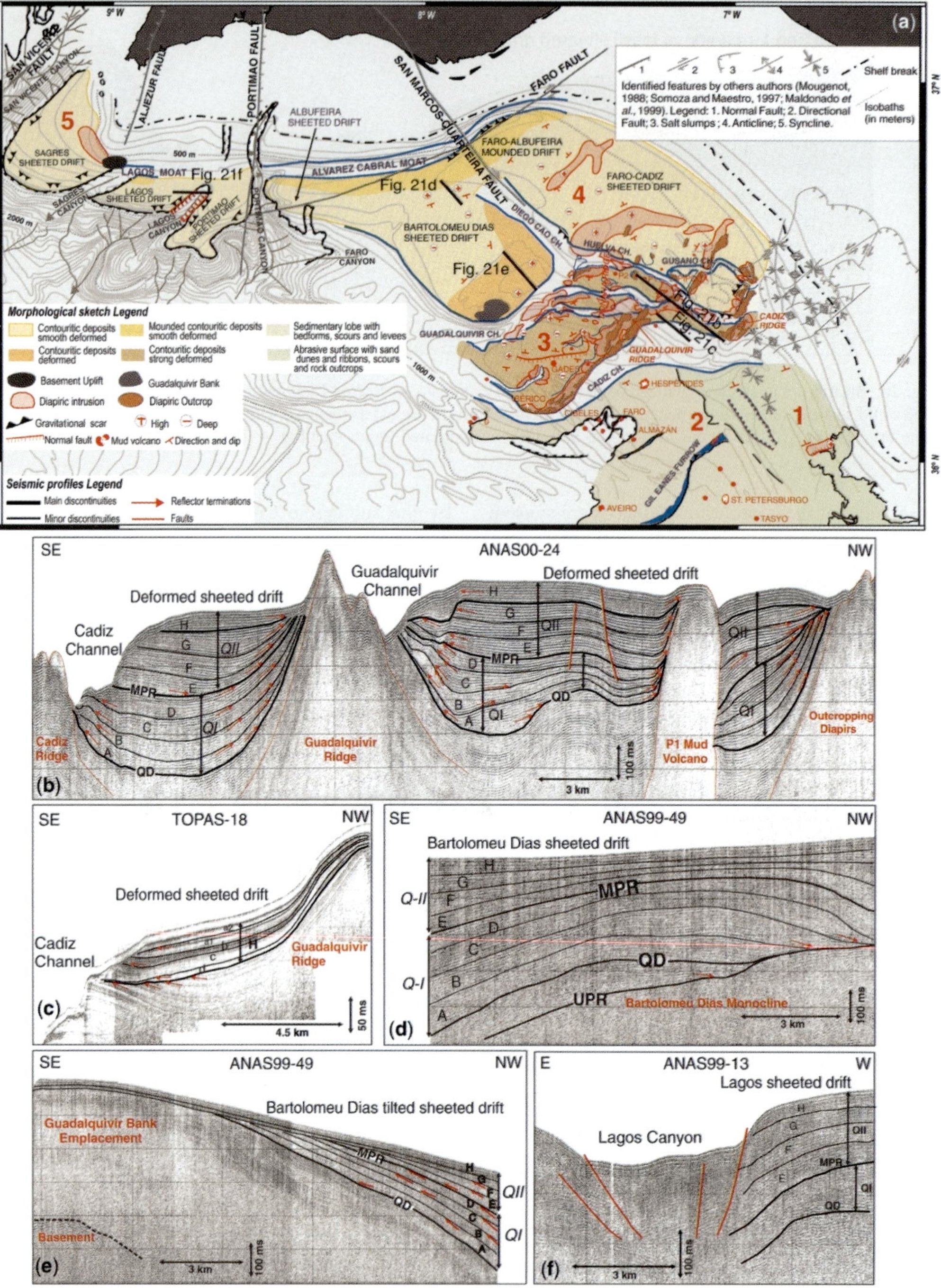

Fig. 21. Morphostructural sketch of the main geological features that affect the contourite depositional system on the middle slope of the Gulf of Cadiz, with some examples of these structural characteristics shown on sparker seismic profiles.

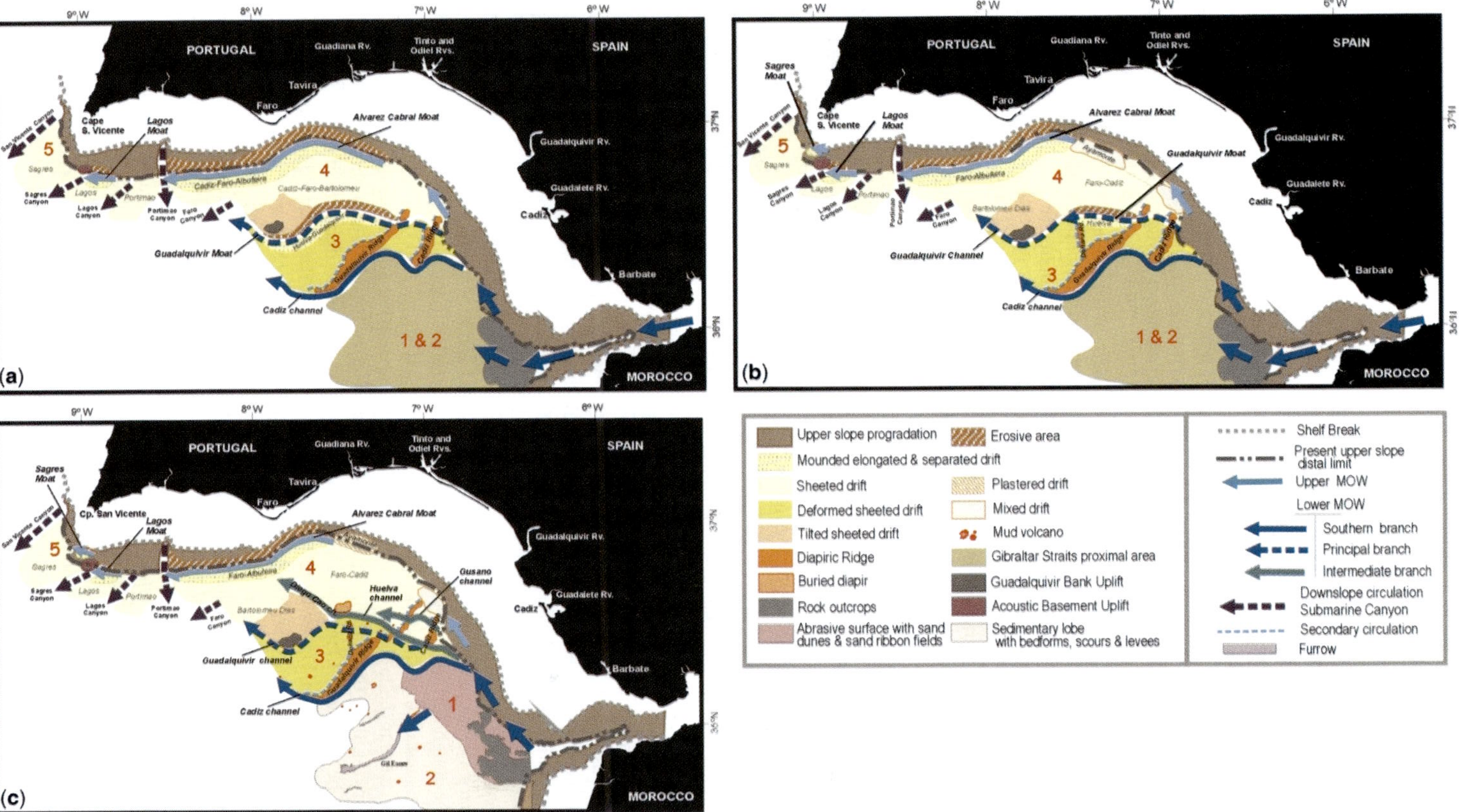

Fig. 22. Interpretative sketch of palaeogeography and palaeoceanography with a reconstruction of the evolution of the contourite depositional system during: (**a**) the Early to Mid-Pleistocene, when the depositional sequences A–D developed; (**b**) the Mid- to Late Pleistocene, when depositional sequences E–G developed; (**c**) the Late Pleistocene and Holocene, when depositional sequence H developed. (See text for details.)

sequences A–D), is shown in Figure 22a. Sectors 1 and 2 of the CDS were characterized by the development of broad sheeted drifts bounded by extensive abrasion surfaces. In the northern part of Sector 3, the wide mounded elongated and separated Huelva–Guadalquivir drift (Fig. 18) was developed along the southern side of the Guadalquivir moat (Fig. 22a). The basinward prolongation of this mounded drift is composed of a large sheeted drift (Fig. 22a), whereas in the southern part of Sector 3, the Cadiz Channel was active between the Cadiz and Guadalquivir diapiric ridges as a large contourite channel (Fig. 22a).

During this period, the Alvarez Cabral moat and the mounded elongated and separated Cadiz–Faro–Albufeira drift were already active in the northern part of Sector 4 (Figs 12 and 17), being located close to the diapiric structures of Sector 3 and extending as far west as Portimao Canyon. As can be observed, the eastern lateral extension with respect to the present Faro–Albufeira mounded drift is consequently greater (Fig. 4), this being zthe reason why it has been named the Cadiz–Faro–Albufeira drift (Fig. 22a). The basinward prolongation of this mounded drift was represented by broad sheeted drifts (the Faro–Cadiz–Bartolomeu Dias and Albufeira drifts). In Sector 5, the main contourite sheeted drifts (the Portimao, Lagos and Sagres drifts) were deposited. Only the mounded elongated and separated Lagos drift was already active in the northern part of the Lagos shelf (Fig. 22a).

Based on the aforementioned regional palaeogeographical reconstruction, the regional MOW palaeoceanographic conditions are considered during this time. From Early Pleistocene to Mid-Pleistocene times, contourite depositional processes dominated in Sectors 2–5. Erosive processes were dominant in Sector 1 close to the Strait of Gibraltar, giving rise to the rough surface.

Sheeted drifts identified in Sectors 1 and 2 are indicative of a tabular behaviour of the MOW (Fig. 22a). Following this, three main branches of the MOW can be defined. One branch can be related to the Mediterranean Upper Core (MU) of the MOW, and the other two to the Mediterranean Lower Core (ML).

The MU flowed along the base of the upper slope. This core became turbulent after the diapiric structures located in the northern part of Sector 3, favouring enhancement of the MU and development of the Alvarez Cabral moat (Fig. 22a). This moat extended to the west as far as Portimao Canyon and along the MU trajectory, and the flow generated both erosion in the northern margin (upper slope) and the mounded Cadiz–Faro–Albufeira drift in the south (Fig. 22a). Westward from the Portimao Canyon, the MU evolved into a more tabular flow (Fig. 22a), once more becoming locally turbulent close to the Sagres Canyon as a result of its interaction with local topography. This led to the development of the short Lagos moat and the mounded elongated and separated Lagos drift (Fig. 22a).

The ML flowed along the middle slope and its interaction with the Cadiz and Guadalquivir diapiric ridges produced two branches (Fig. 22a), the southern and northern branches. The southern branch interacted with the Cadiz diapiric and Guadalquivir ridges, where it was intensified and generated a turbulent flow through the Cadiz channel (Fig. 22a). The northern branch flowed through the Cadiz and Guadalquivir ridges and became a turbulent and intensified flow, which generated the Guadalquivir moat (Fig. 22a). The northern branch through this moat formed an erosive surface to the north and developed the Huelva–Guadalquivir mounded drift to the south (Fig. 22a). A wide sheeted drift constituted the basinward prolongation of these mounded drifts, where the ML became tabular (Fig. 22a). The circulation of the ML through the north of the diapiric structures also generated marginal valleys that eroded the adjacent sheeted drifts (Fig. 22a).

Mid-Pleistocene to Late Pleistocene CDS: transitional stage (from MIS 22 to MIS 6)

Most of sequence Q-II (depositional sequences E–G) was deposited during this second major evolutionary stage. The characteristics of Sectors 1 and 2 were nearly the same as for the first stage, and the most important change produced in the stacking pattern of the contourite deposits occurred in Sector 3. The westernmost part of the Guadalquivir–Huelva mounded elongated and separated drift located south of the Guadalquivir Bank, in the area of Sector 3, became inactive (fossil Guadalquivir mounded drift, FGD) (Figs 18 and 22b). In its place, extensive sheeted drift sequences were generated after the MPR discontinuity, fossilizing the previous mounded deposits. It can also be observed in this zone that the Guadalquivir moat passes into the Guadalquivir channel (Fig. 22b). On the other hand, the eastern part of this Guadalquivir–Huelva mounded drift (from now on named the Huelva mounded drift), between the Guadalquivir diapiric ridge on the right and the Doñana diapiric ridge on the left, was still active (Figs 18 and 22b).

Another change observed during this period took place in the eastern and northern parts of Sector 4, where the Ayamonte mixed drift developed (Figs 16, 17, 19 and 22b), fossilizing part of the Faro–Cadiz sheeted drift and the eastern part of the Cadiz–Faro–Albufeira mounded drift (this part is named the fossil Cadiz mounded drift,

FCD). Therefore it can be observed that the active mounded drift deposits during this stage migrated northward, as did the Alvarez Cabral moat, being located at almost the same place as at present (Figs 4 and 22b). However, the palaeogeographical situation of the rest of Sectors 4 and 5, from the Mid-Pleistocene to the Late Pleistocene, was similar to that of the previous stage.

Changes in the architecture style of the contourite deposits and the evolution between the first and the second stages are probably related to tectonic activity in the Guadalquivir Bank and diapiric ridge area (Figs 21 and 22b). Diapiric ridge movements and Guadalquivir Bank uplift were most probably related to the compressive regime that took place at 740–450 and 295–225 ka as described by Rodero (1999) and Rodero *et al.* (1999).

Based on the regional palaeogeographical reconstruction for this second stage, the MOW palaeoceanographic conditions can be considered. From Mid- to Late Pleistocene, contourite depositional processes were dominant in Sectors 2–5. Erosive processes were more dominant in Sector 1, close to the Straits of Gibraltar (Fig. 22b).

New local oceanographic conditions in the Upper and Lower Cores (MU and ML) of the MOW were established.

The focus of the MU fluctuated northwards during this stage after passing the diapiric ridge in the areas where the Ayamonte mixed drifts are developed. In the westernmost part of these mixed drifts, the MU was sufficiently enhanced to excavate the Alvarez Cabral moat (Fig. 22b). The flow of this MU through the moat extends to the west as far as Portimao Canyon as a turbulent regime that generated erosion along the northern margin of the moat and the mounded Faro–Albufeira drift growth along the southern flank (Fig. 22b). This MU evolved into a tabular flow towards the west of Portimao Canyon, in Sector 5, but became turbulent again following interaction with local irregularities of the middle slope, generating the short Lagos moat and the mounded elongated and separated Lagos drift (Fig. 22b).

During this second stage, the ML circulated along the middle slope and was also divided into two branches (southern and northern). The southern branch flowed through the Cadiz contourite channel as in the first evolutionary stage. On the other hand, the northern branch showed more turbulent behaviour after crossing the diapiric ridges and was focused along the Guadalquivir moat, which developed an erosive surface to the north and the Huelva mounded drift to the south (Fig. 22b). Laterally, the Guadalquivir moat continues to the Guadalquivir channel; mounded deposits did not develop, but the deposits changed laterally into a sheeted drift (Fig. 22b). This change in the type of the contourite development is puzzling. We hypothesize that: (1) along the Guadalquivir channel, the condition of the northern branch of the ML could have been too energetic because of its interaction with the Guadalquivir Bank, so erosive processes dominated along the southern boundary of that bank (Fig. 22b); (2) the mounded drift was developed after the MPR but because the high intensity of the MOW it was later eroded; (3) the Guadalquivir Bank Uplift of ancient highly consolidated material provided insufficient material to allow construction of a mounded drift on the opposite flank. Clearly, this situation was different from that of the first stage. On the other hand, in the central area of Sector 3, the tabular ML favoured development of a broad sheeted drift (Fig. 22b).

Late Pleistocene to Holocene CDS: the present configuration (from MIS 6 to present)

During this last evolutionary stage, the upper part of sequence Q-II was deposited (depositional sequence H). Most of the present contourite features were developed during this stage; in particular, most of the erosive characters observed in Sector 3 (Fig. 22c).

In Sector 1, more erosive features began to be developed, as an abrasion surface with erosive scours alignment and furrows. Numerous sedimentary waves were also developed, not only in Sector 1 but also in Sector 2. This change in the system is determined by the high acoustic response of depositional sequence H in comparison with the previous sequences (Fig. 20).

In Sector 3, new contourite channels were formed, including the Gusano and Huelva channels and other minor branches related to the Guadalquivir channel (Fig. 22c). In the northern part of Sector 3, the previous mounded elongated and separated Huelva drift disappeared completely (fossil Huelva mounded drift, FHD), being buried by the youngest sheeted drift and eroded by the Huelva contourite channel (Fig. 22c). This change in the depositional style of the margin took place just before depositional sequence H (Figs 18 and 21).

Another change during this third stage is observed on the Ayamonte mixed drift, in the eastern and northern parts of Sector 4. Drift deposits again migrated northward with an aggradational stacking pattern (Fig. 17). Laterally to the NW, a plastered drift started to develop in the transition zone between the upper and middle slopes (Fig. 16), located to the east of the Faro–Albufeira mounded drift and Alvarez Cabral moat, which reached its present location (Fig. 22c). The palaeogeographical situation of Sector 5 during this stage was similar to that of the previous one.

These changes from the second to the third evolutionary stages can be related to coeval tectonic movements in the Strait of Gibraltar described by Zazo *et al.* (1999), and in the reactivation of the diapiric ridges and isolated diapiric bodies of Sector 3 (Pérez-Fernández 1997; Llave *et al.* 2006). Our hypothesis is that recent tectonic activity conditioned a new sea-bottom morphology, especially regarding the Guadalquivir and Cadiz diapiric ridge configuration on the sea floor (Fig. 22c), and this controlled new flow pathways, with the strongest flowing branches of the MOW over the middle slope. Consequently, a general circulation of the MOW similar to the present one was established during this stage.

Immediately outside the Strait of Gibraltar, the MOW generated erosive features in those areas close to the Strait and, where the velocity decreased, numerous depositional bedforms and sedimentary lobes were developed (Fig. 22c).

The Mediterranean Upper Branch (MU) flowed along the transition between the middle and upper slopes. Tectonic changes in the diapiric ridges (Pérez-Fernández 1997) helped condition the lateral migration of the MU towards the north, in shallower areas than during the second stage, and with a more tabular flow character. New circulation of the MU along the upper slope generated the plastered drift. The main change in the morphology of the Algarve margin from a NW to a SW bathymetric trend was important enough to generate the MU path again in the transition between the upper and middle slopes as far west as the Portimao canyon. Therefore, in the northern part of Sector 4, the MU was enhanced and excavated the *Alvarez Cabral moat*, developing an erosive surface to the north and the *mounded Faro–Albufeira drift* to the south (Fig. 22c). When the MU passed the Portimao Canyon it evolved into a tabular flow in Sector 5, although it became turbulent locally over the Lagos shelf, leading to development of the *Lagos moat* and the *mounded elongated and separated Lagos drift* (Fig. 22c).

The Mediterranean Lower Branch (ML) had a complex interaction with the sea-floor topography generated by the diapiric ridges, which led to progressive erosion through several contourite channels. The three main branches identified at present (Southern, Principal and Intermediate branches) were active during this third stage (Fig. 22c). Part of the ML flow (the Southern branch) was deflected by the linear NE–SW-trending Cadiz and Guadalquivir diapiric ridges, and focused through the Cadiz channel. The other two branches of the ML followed and sculpted different channels: the Principal branch followed the main Guadalquivir channel; the Intermediate branch first followed the Huelva channel and then linked up with the Diego Cao channel. A fourth minor, unnamed branch probably followed the Gusano channel before joining the Huelva channel (Fig. 22c).

Conclusions

According to stratigraphic stacking pattern studies of the CDS, four major contourite drift types have been identified: mounded elongated and separated drifts; sheeted drifts; plastered drifts; deformed sheeted drifts. All these drifts have a morphological expression on the present sea bottom. In this study, we recognize a further two buried drift types, mixed drifts and fossil mounded drifts, which occur in the Quaternary sedimentary record buried by other younger non-mounded elongated and separated deposits.

The nature and changes in the CDS facies have been correlated with climatic cycles during the Quaternary, especially those from the Mid-Pleistocene. Nevertheless, superimposed on these climatic changes, tectonics during the Quaternary has determined, in the short term, the local thickness, geometry and development of different types of contourite drifts owing to its contribution to new oceanographic conditions. Tectonics has represented a key long-term factor in sea-floor morphology changes, which has controlled at every stage new pathways in the cores and branches of the MOW. Consequently, the behaviour of these cores and branches controlled the contourite stratigraphic architecture. The understanding of this stratigraphic architecture has helped us to recognize three main stages during the evolution of the CDS affected by tectonics. (1) In the Early Pleistocene to Mid-Pleistocene, the CDS was mainly dominated by depositional processes, where the Upper and Lower cores of the Mediterranean Outflow Water (MOW) generated the Cadiz–Faro–Albufeira mounded elongated and separated drift in the transition between the middle and upper slope, and the equivalent Huelva–Guadalquivir drift on the middle slope. During this stage the main erosive features were established close to the Strait of Gibraltar. (2) In the Mid-Pleistocene to Late Pleistocene, two important changes in the CDS took place. One occurred in the transition between the middle and upper slope, related to a change in the upper branch of the MOW, where a mixed drift began to develop, burying the eastern part of the Cadiz–Faro–Albufeira mounded elongated and separated drift. The second change is observed on the central area of the middle slope, related to the lower branch of the MOW, where a large contourite channel (the Guadalquivir channel) progressively eroded the western part of the mounded Huelva–Guadalquivir drift. Laterally an extensive sheeted drift buried the previous mounded deposits. (3) In the Late Pleistocene to Holocene, in the northern area of the CDS, a plastered drift started to be developed in the transitional zone between the upper and middle slope, close to the eastern part of Faro–Albufeira mounded drift. On the middle slope, the mounded elongated Huelva–Guadalquivir drift was not developed and more erosive processes became dominant as the lower core of the MOW intensified. In the sector close to the Straits of Gibraltar, broad sea-bed forms were generated.

These evolutionary stages were most probably a function of synsedimentary tectonic activity, including diapiric ridge movement, bank uplift and compressive structures. The general conclusion of this study is that the contourite depositional system of the Gulf of Cadiz changed from a predominantly depositional system during the Early to Mid Pleistocene, to a predominantly erosive system during the Late Pleistocene.

This work was supported by the CICYT PB94-1090-C03-03 (FADO), MAR-98-02-0209 (TASYO), REN2002-04117-C03-01 (GADES), REN2002-11668-E/MAR (Eurocore-Euromargins 01-LEC-EMA06F), REN2002-11669-E/MAR (Eurocore-Euromargins 01-LEC-EMA24F) and is part of the IGCP-432 'Contourites, Bottom Currents and Palaeocirculations' project.

References

ALVES, T. M., GAWTHORPE, R. L., HUNT, D. W. & MONTEIRO, J. H. 2003. Cenozoic tectono-sedimentary evolution of the western Iberian margin. *Marine Geology*, **195**, 75–108.

AMBAR, I. 1983. A shallow core of Mediterranean Water off western Portugal. *Deep-Sea Research*, **30**(6A), 677–680.

AMBAR, I. & HOWE, M. R. 1979. Observations of the Mediterranean Outflow I. Mixing in the Mediterranean Outflow. *Deep-Sea Research*, **26A**, 535–554.

AMBAR, I., HOWE, M. R. & ABDULLAH, M. I. 1976. A physical and chemical description of the Mediterranean outflow in the Gulf of Cadiz. *Deutsche Hydrographische Zeitschrift*, **29**, 58–68.

AMBAR, I., ARMI, L., BOWER, A. & FERREIRA, T. 1999. Some aspects of time variability of the Mediterranean water off south Portugal. *Deep-Sea Research I*, **46**, 1109–1136.

AMBAR, I., SERRA, N., BROGUEIRA, M. J., *ET AL.* 2002. Physical, chemical and sedimentological aspects of the Mediterranean outflow off Iberia. *Deep-Sea Research II*, **49**, 4163–4177.

BALDY, P., BILLOT, G., DUPEUBLE, P. A., MALOD, J., MOITA, I. & MOUGENOT, D. 1977. Carte géologique du plateau continental sud-portugais et sud-espagnol (Golfe de Cadix). *Bulletin de la Société Géologique de France*, **7**(XIX),703–724.

BARAZA, J. & NELSON, C. H. 1992. Clasificación y dinámica de formas de fondo en el Golfo de Cádiz: Implicaciones de la corriente profunda mediterránea en los procesos sedimentarios durante el Pliocuaternario. Sociedad Geológica de España. *III Congresso Geología de España y VIII Congresso Latinoamericano de Geología Salamanca, 1992, Spain, Simposios tomo* **2**, 477–486.

BARINGER, M. O. 1993. *Mixing and dynamics of the Mediterranean outflow*. PhD thesis, Woods Hole Oceanographic Institution, Woods Hole, MA.

BARINGER, M. O. & PRICE, J. F. 1999. A review of the physical oceanography of the Mediterranean Outflow. *Marine Geology*, **155**, 63–82.

BETHOUX, J. P. 1984. Paléo-hydrologie de la Méditerranée au cours des derniers 20000 ans. *Oceanologica Acta*, **7**(1), 43–48.

BORENÄS, K. M., WAHLIN, A. K., AMBAR, I. & SERRA, N. 2002. The Mediterranean outflow splitting—a comparison between theoretical models and CANIGO data. *Deep-Sea Research II*, **49**, 4195–4205.

BORGES, J. F., FITAS, A. J. S., BEZZEGHOUD, M. & TEVES-COSTA, P. 2001. Seismotectonics of Portugal and its adjacent area. *Tectonophysics*, **337**, 373–387.

BOWER, A., ARMI, L. & AMBAR, I. 1997. Lagrangian observations of meddy formation during a Mediterranean undercurrent seeding experiment. *Journal of Physical Oceanograph*, **27**(11), 2545–2575.

BRYDEN, H. L. & STOMMEL, H. M. 1984. Limiting processes that determine basic features of the circulation in the Mediterranean Sea. *Oceanologica Acta*, **7**, 289–296.

CACHO, I., GRIMALT, J. O., SIERRO, F. J., SHACKLETON, N. & CANALS, M. 2000. Evidence for enhanced Mediterranean thermohaline circulation during rapid climatic coolings. *Earth and Planetary Science Letters*, **183**, 417–429.

CARALP, M. H. 1988. Late Glacial to Recent deep-sea benthic Foraminifera from the northeastern Atlantic (Cadiz Gulf) and western Mediterranean (Alboran Sea): paleoceanographic results. *Marine Micropaleontology*, **13**, 265–289.

CARALP, M. H. 1992. Paléohydrologie des bassins profonds nord-marocain (Est et Ouest Gibraltar) au Quaternaire terminal: apport des foraminifères benthiques. *Bulletin de la Société Géologique de France*, **163**(2), 169–178.

DÍAZ DEL RÍO, V., SOMOZA, L., MARTÍNEZ-FRÍAS, J., *ET AL.* 2003. Vast fields of hydrocarbon-derived carbonate chimneys related to the accretionary wedge/olistostrome of the Gulf of Cadiz. *Marine Geology*, **195**, 177–200.

DUPLESSY, J. C., SHACKLETON, N. J., FAIRBANKS, R. G., LABEYRIE, L., OPPO, D. & KALLEL, N. 1988. Deepwater source variations during the last climatic cycle and their impact on the global deepwater circulation. *Paleoceanography*, **3**, 343–360.

FAUGÈRES, J.-C., GONTHIER, E. & STOW, D. A. V. 1984. Contourite drift molded by deep Mediterranean outflow. *Geology*, **12**, 296–300.

FAUGÈRES, J. C., FRAPPA, M., GONTHIER, E., DE RESSEGUIER, A. & STOW, D. A. V. 1985. Modèle et facies de type contourite a la surface d'une ride sédimentaire édifiée par des courants issus de la veine d'eau méditerranéenne (ride du Faro, Golle de Cadix). *Bulletin de la Société Geologique de France*, **I**(1), 35–47.

FAUGÈRES, J. C., MÉZERAIS, M. L. & STOW, D. A. V. 1993. Contourite drift types and their distribution in the North and South Atlantic Ocean basins. *Sedimentary Geology*, **82**(1–4), 189–203.

FAUGÈRES, J. C., STOW, D. A. V., IMBERT, P. & VIANA, A. 1999. Seismic features diagnostic of contourite drifts. *Marine Geology*, **162**, 1–38.

FERNÁNDEZ-PUGA, M. C. 2004. *Diapirismo y estructuras de expulsión de gases hidrocarburos en el talud continental del Golfo de Cádiz*. PhD thesis, University of Cádiz.

FLINCH, J. F. & VAIL, P. R. 1998. Plio-Pleistocene sequence stratigraphy and tectonics of the Gibraltar Arc. *In*: DE GRACIANSKY, P. C., HARDENBOL, J., THIERRY, J. & VAIL, P. R. (eds) *Mesozoic and Cenozoic Sequence Stratigraphy of European Basins*. SEPM, Special Publications, **60**, 199–208.

GARCÍA, M. 2002. *Caracterización morfológica del sistema de canales y valles submarinos del talud medio del Golfo de Cádiz (SO de la Península Ibérica): implicaciones oceanográficas*. Tesis de Licenciatura, University of Cadiz.

GARDNER, J. V. & KIDD, R. B. 1983. Sedimentary processes on the Iberian continental margin viewed by long-range side-scan sonar. 1: Gulf of Cadiz. *Oceanologica Acta*, **6**(3), 245–254.

GONTHIER, E., FAUGÈRES, J. C. & STOW, D. A. V. 1984. Contourite facies of the Faro Drift, Gulf of Cadiz. *In*: STOW, D. A. V. & PIPER, D. J. W. (eds), Fine-Grained Sediments: Deep-Water Processes and Facies. Geological Society, London, Special Publications, **15**, 775–797.

GRACIA, E., DAÑOOBEITIA, J., VERGÉS, J. & BARTOLOMÉ, R. 2003. Crustal architecture and tectonic evolution of the Gulf of Cadiz (SW Iberian margin) at the convergence of the Eurasian and African plates. *Tectonics*, **22**(4), 1033, doi:10.1029/2001TC901045.

GROUSSET, F., JORON, J. L., BISCAYE, P. E., *ET AL.* 1988. Mediterranean outflow through the Strait of Gibraltar since 18,000 years B.P. Mineralogical and geochemical agreement. *Geo-Marine Letters*, **8**(1), 25–34.

GUTSCHER, M.-A. 2004. What caused the Great Lisbon Earthquake? *Science*, **305**, 1247–1248.

GUTSCHER, M.-A., MALOD, J., REHAULT, J.-P., CONTRUCCI, I., KLINGELHOEFER, F., MENDES-VICTOR, L. & SPAKMAN, W. 2002. Evidence for active subduction beneath Gibraltar. *Geology*, **30**, 1071–1074.

HABGOOD, E. L., KENYON, N. H., MASSON, D. G., AKHMETZHANOV, A., WEAVER, P. P. E., GARDNER, J. & MULDER, T. 2003. Deep-water sediment wave fields, bottom current sand channels and gravity flow channel-lobe system: Gulf of Cadiz, NE Atlantic. *Sedimentology*, **50**, 483–510.

HEEZEN, B. C. & JOHNSON, G. L. 1969. Mediterranean undercurrent and microphysiography west of Gibraltar. *Bulletin de l'Institut Océanographique Monaco*, **67**(1382), 1–95.

HERNÁNDEZ-MOLINA, F. J., SOMOZA, L., REY, J. & POMAR, L. 1994. Late Pleistocene–Holocene sediments on the Spanish continental shelves: model for very high resolution sequence stratigraphy. *Marine Geology*, **120**, 129–174.

HERNÁNDEZ-MOLINA, F. J., SOMOZA, L. & LOBO, F. J. 2000. Seismic stratigraphy of the Gulf of Cadiz continental shelf: a model for Late Quaternary very high-resolution sequence stratigraphy and response to sea-level fall. *In*: HUNT, D. & GAWTHORPE, R. L. G. (eds) *Sedimentary Responses to Forced Regressions*. Geological Society, London, Special Publications, **172**, 329–361.

HERNÁNDEZ-MOLINA, F. J., SOMOZA, L., VÁZQUEZ, J. T., LOBO, F., FERNÁNDEZ-PUGA, M. C., LLAVE, E. & DÍAZ DEL RÍO, V. 2002. Quaternary stratigraphic stacking patterns on the continental shelves of the southern Iberian Peninsula: their relationship with global climate and palaeoceanographic changes. *Quaternary International*, **92**(1), 5–23.

HERNÁNDEZ-MOLINA, F. J., LLAVE, E., SOMOZA, L., *ET AL.* 2003. Looking for clues to paleoceanographic imprints: a diagnosis of the Gulf of Cadiz contourite depositional systems. *Geology*, **31**(1), 19–22.

HERNÁNDEZ-MOLINA, F. J., LLAVE, E., STOW, D. A. V., *ET AL.* 2006. The contourite depositional system of the Gulf of Cadiz: a sedimentary model related to the bottom current activity of the Mediterranean Outflow Water and the continental margin characteristics. *Deep-Sea Research*, **53**, 1420–1463.

KENYON, N. H. & BELDERSON, R. H. 1973. Bed forms of the Mediterranean undercurrent observed with side-scan sonar. *Sedimentary Geology*, **9**, 77–99.

KNAUSS, J. A. 1978. *Introduction to Physical Oceanography*. Prentice-Hall, Englewood Cliffs, NJ.

LEBREIRO, S. 1995. *Sedimentation history off Iberia: Tore Seamount, Tagus and Horseshoe Abyssal Plains*. PhD thesis, University of Cambridge.

LEBREIRO, S. M., MCCAVE, I. N. & WEAVER, P. P. E. 1997. Late Quaternary turbidite emplacement on the Horseshoe Abyssal Plain (Iberian margin). *Journal of Sedimentary Research*, **67**(5), 856–870.

LLAVE, E. 2003. *Análisis morfosedimentario y estratigráfico de los depósitos contorníticos del Golfo de Cádiz: implicaciones paleoceanográficas.* Tesis Doctoral, University of Cadiz.

LLAVE, E., HERNÁNDEZ-MOLINA, F. J., SOMOZA, L., DÍAZ-DEL-RÍO, V., STOW, D. A. V., MAESTRO, A. & ALVEIRINHO DIAS, J. M. 2001. Seismic stacking pattern of the Faro–Albufeira contourite system (Gulf of Cadiz): a Quaternary record of paleoceanographic and tectonic influences. *Marine Geophysical Researches*, **22**, 487–508.

LLAVE, E., HERNÁNDEZ-MOLINA, F. J., FERNÁNDEZ-PUGA, M. C., *ET AL.* 2003*a*. A fossil mounded elongate and separated drifts on the Middle Slope of the Gulf of Cadiz; paleoceanographic significance. *Deep Water Processes in Modern and Ancient Environments, Barcelona and Ainsa (Spain), 15–19. Abstracts Volume*, 25.

LLAVE, E., HERNÁNDEZ-MOLINA, F. J., FERNÁNDEZ-PUGA, M. C., ET AL. 2003*b*. Fossil mounded elongated and separated drift on the slope of the Gulf of Cadiz as the tectonics stress over the MOW paleoceanography. *Thalassas*, **19**(2a), 60–62.

LLAVE, E., SCHÖNFELD, J., HERNÁNDEZ-MOLINA, F. J., MULDER, T., SOMOZA, L. & DIAZ DEL RÍO, V. 2004*a*. Arquitectura estratigráfica de los depósitos contorníticos del Pleistoceno superior del Golfo de Cádiz: implicaciones paleoceanográficas de los eventos de Heinrich. *Geotemas*, **6**(5), 187–190.

LLAVE, E., FLORES, J. A., HERNÁNDEZ-MOLINA, F. J., SIERRO, F. J., SOMOZA, L., DÍAZ-DEL-RÍO, V. & MARTÍNEZ DEL OLMO, W. 2004*b*. Cronoestratigrafía de los depósitos contorníticos del talud continental del Golfo de Cádiz a partir del análisis de nanofósiles calcáreos. *Geotemas*, **6**(5), 183–186.

LLAVE, E., SCHÖNFELD, J., HERNÁNDEZ-MOLINA, F. J., MULDER, T., SOMOZA, L. & DIAZ DEL RÍO, V. 2006. High-resolution stratigraphy of the Mediterranean outflow contourite system in the Gulf of Cadiz during the Late Pleistocene: the impact of Heinrich events. *Marine Geology*, **227**, 241–262.

LOBO, F. J. 1995. *Estructuración y evolución morfosedimentaria de un sector del margen continental septentrional del Golfo de Cádiz durante el Cuaternario terminal.* Dissertation, University of Cadiz.

LOBO, F. J. 2000. *Estratigrafía de alta resolución y cambios del nivel del mar durante el Cuaternario del margen continental del Golfo de Cádiz (S de Iberia) y del Roussillon (S de Francia): estudio comparativo.* PhD thesis, University of Cadiz.

LOBO, F. J., HERNÁNDEZ-MOLINA, F. J., SOMOZA, L. & DÍAZ DEL RÍO, V. 2001. The sedimentary record of the post-glacial transgression on the Gulf of Cadiz Continental Shelf (Southwest Spain). *Marine Geology*, **178**, 171–195.

LOBO, F. J., HERNÁNDEZ-MOLINA, F. J., SOMOZA, L., DÍAZ DEL RÍO, V. & ALVEIRINHO DIAS, J. M. 2002. Seismic stratigraphic evidence of a upper Pleistocene TST to HST complex on the Gulf of Cádiz continental shelf (southwest Iberian peninsula). *Geo-Marine Letters*, **22**, 95–107.

LOBO, F. J., DIAS, J. M. A., GONZÁLEZ, R., HERNÁNDEZ-MOLINA, F. J., MORALES, J. A. & DÍAZ DEL RÍO, V. 2003. High-resolution seismic stratigraphy of a narrow, bedrock-controlled estuary: the Guadiana estuarine system, SW Iberia. *Journal of Sedimentary Research*, **73**(6), 973–986.

LOBO, F. J., DIAS, J. M. A., HERNÁNDEZ-MOLINA, F. J., GONZÁLEZ, R., FERNÁNDEZ-SALAS, L. M. & DÍAZ DEL RÍO, V. 2005*a*. Shelf-margin wedges and lateral accretion of upper slopes in the Gulf of Cadiz margin (southwest Iberian Peninsula). *In*: HODGSON, D. M. & FLINT, S. S. (eds) *Submarine Slope Systems: Processes and Products* Geologial Society, London, Special Publications. **244**, 7–25.

LOBO, F. J., FERNÁNDEZ-SALAS, L. M., HERNÁNDEZ-MOLINA, F. J., GONZÁLEZ, R., DIAS, J. M. A., DÍAZ DEL RÍO, V. & SOMOZA, L. 2005*b*. Holocene highstand deposits in the Gulf of Cadiz, SW Iberian Peninsula: a high-resolution record of hierarchical environmental changes. *Marine Geology*, **219**, 109–131.

MADELAIN, F. 1970. Influence de la topographie du fond sur l'écoulement méditerranéen entre le Detroit de Gibraltar et le Cap Saint-Vincent. *Cahiers Océanographiques*, **22**, 43–61.

MAESTRO, A., SOMOZA, L., DÍAZ DEL RÍO, V., ET AL. 1998. Neotectónica transpresiva en la plataforma continental Suribérica Atlántica. *Geogaceta*, **24**, 203–206.

MAESTRO, A., SOMOZA, L., MEDIALDEA, T., TALBOT, C. J., LOWRIE, A., VÁZQUEZ, J. T. & DÍAZ-DEL-RÍO, V. 2003. Large-scale slope failure involving Triassic and Middle Miocene salt and shale in the Gulf of Cadiz (Atlantic Iberian Margin). *Terra-nova*, **15**, 380–391.

MALDONADO, A., SOMOZA, L. & PALLARÉS, L. 1999. The Betic orogen and the Iberian–African boundary in the Gulf of Cadiz: geological evolution (central North Atlantic). *Marine Geology*, **155**, 9–43.

MALOD, J. 1982. *Comparaison de l'evolution des marges continentales au Nord et au Sud de la Peninsula Iberique.* Thèse Doctorat d'Etat, Université Paris VI.

MALOD, J. A. & MAUFFRET, A. 1990. Iberian plate motions during the Mesozoic. *Tectonophysics*, **184**, 261–278.

MALOD, J. A. & MOUGENOT, D. 1979. L'histoire géologique néogène du Golfe de Cadix. *Bulletin de la Société Geologique de France*, **7**(21), 603–611.

MATTHIESEN, S. & HAINES, K. 1998. Influence of the Strait of Gibraltar on past changes in Mediterranean thermohaline circulation. *Poster, VI International Conference on Palaeoceanography (ICP)*, Lisbon.

MEDIALDEA, T., VEGAS, R., SOMOZA, L., ET AL. 2004. Structure and evolution of the 'Olistostrome' complex of the Gibraltar Arc in the Gulf of Cadiz (eastern Central Atlantic): evidence from two long seismic cross-sections. *Marine Geology*, **209**, 173–198.

MÉLIÈRES, F. 1974. *Recherches sur la dynamique sédimentaire du Golfe de Cadix (Espagne).* Thèse Doctoral d'Etat, Université Paris A.

MOUGENOT, D. 1988. *Géologie de la Marge Portugaise.* These de Doctorat d'État, Université Paris VI.

MOUGENOT, D. & VANNEY, J. R. 1982. Les Rides de contourites Plio-Quaternaires de la pente continentale sud Portugaise. *Bulletin de la Société Géologique de Bassin Aquitaine*, **31**, 131–139.

MULDER, T., LECROART, P., VOISSET, M., ET AL. 2002. Past deep-ocean circulation and the paleoclimate record—Gulf of Cadiz. *EOS Transactions, American Geophysical Union*, **83**(43), 481–488.

Mulder, T., Voisset, M., Lecroart, P., *et al.* 2003. The Gulf of Cadiz: an unstable giant contouritic levee. *Geo-Marine Letters*, **23**(1), 7–18.

Murillas, J., Mougenot, D., Boillot, G., Comas, M. C., Banda, E. & Mauffret, A. 1990. Structure and evolution of the Galicia Interior Basin (Atlantic western Iberian continental margin). *Tectonophysics*, **184**, 297–319.

Nelson, C. H., Baraza, J. & Maldonado, A. 1993. Mediterranean undercurrent sandy contourites, Gulf of Cadiz, Spain. *Sedimentary Geology*, **82**, 103–131.

Nelson, C. H., Baraza, J., Maldonado, A., Rodero, J., Escutia, C. & Barber, J. H. 1999. Influence of the Atlantic inflow and Mediterranean outflow currents on Late Quaternary sedimentary facies of the Gulf of Cadiz continental margin. *Marine Geology*, **155**, 99–129.

Ochoa, J. & Bray, N. A. 1991. Water mass exchange in the Gulf of Cadiz. *Deep-Sea Research*, **38**(1), S465–S503.

Pailler, D. & Bard, E. 2002. High frequency paleoceanographic changes during the past 140,000 years recorded by the organic matter in sediments off the Iberian Margin. *Palaeogeography, Palaeoclimatology, Palaeoecology*, **181**, 431–452.

Paterne, M., Guichard, F., Labeyrie, J., Gillot, P. Y. & Duplessy, J. C. 1986. Tyrrhenian Sea tephrochronology of the oxygen isotope record for the past 60,000 years. *Marine Geology*, **72**, 259–285.

Perconig, E. 1960–1962. Sur la constitution geólogique de l'Andalousie Occidentale, en particulier du bassin du Guadalquivir (Espagne meridionale). *In*: *Livre Mémoire du Professeur Paul Fallot. Mémories Hors-Série, Société Géologique de France*, **1**, 229–256.

Pérez-Fernández, L. M. 1997. *Evolución del diapirismo en el margen continental del Golfo de Cádiz y su relación con la sedimentación Cuaternaria.* Dissertation, University of Granada.

Pinheiro, L. M., Wilson, R. C. L., Reis, R. P., Whitmarsh, R. B. & Ribeiro, A. 1996. The western Iberian margin: a geophysical and geological overview. *In*: Sawyer, D. S., Whitmarsh, R. B., Klaus, A. *et al.* (eds) *Proceedings of the Ocean Drilling Program, Scientific Results*, **149**. Ocean Drilling Program, College Station, TX, 3–23.

Reid, J. L. 1994. On the total contribution of the Mediterranean Sea outflow to the Norwegian–Greenland Sea. *Deep-Sea Research*, **26A**, 1199–1223.

Riaza, C. & Martínez del Olmo, W. 1996. Depositional model of the Guadalquivir–Gulf of Cadiz Tertiary basin. *In*: Friend, P. F. & Dabrio, C. J. (eds) *Tertiary Basins of Spain. The Stratigraphic Record of Crustal Kinematics*. Cambridge University Press, 330–338.

Ribeiro, A., Kullberg, M. C., Kullberg, J. C., Manuppella, G. & Phipps, S. 1990. A review of Alpine tectonics in Portugal: foreland detachment in basement and cover rocks. *Tectonophysics*, **184**, 357–366.

Rodero, J. 1999. *Dinámica sedimentaria y modelo evolutivo del margen continental suroriental del Golfo de Cádiz durante el Cuaternario Superior (Pleistoceno Medio-Holoceno).* Doctoral thesis, University of Granada.

Rodero, J., Pallarés, L. & Maldonado, A. 1999. Late Quaternary seismic facies of the Gulf of Cadiz Spanish margin: depositional processes influenced by sea-level change and tectonic controls. *Marine Geology*, **155**, 131–156.

Rogerson, M. 2002. Climatic influence on sediment transport in the Mediterranean outflow current (Gulf of Cadiz, Spain). *Newsletter of Micropalaeontology*, **66**, 16–17.

Rohling, E. J., Hayes, A., De Rijk, S., Kroon, D., Zachariasse, W. J. & Eisma, D. 1998. Abrupt cold spells in the northwest Mediterranean. *Paleoceanography*, **13**, 316–322.

Schönfeld, J. 1997. The impact of the Mediterranean Outflow Water (MOW) on benthic foraminiferal assemblages and surface sediments at the southern Portuguese continental margin. *Marine Micropaleontology*, **29**, 211–236.

Schönfeld, J. & Zahn, R. 2000. Late Glacial to Holocene history of the Mediterranean Outflow. Evidence from benthic foraminiferal assemblages and stable isotopes at the Portuguese margin. *Palaeogeography, Palaeoclimatology, Palaeoecology*, **159**, 85–111.

Shanmugan, G. 2000. 50 years of the turbidite paradigm (1950s–1990s): deep-water processes and facies models—a critical perspective. *Marine and Petroleum Geology*, **17**, 285–342.

Sierro, F. J., Flores, J. A. & Baraza, J. 1999. Late glacial to recent paleoenvironmental changes in the Gulf of Cadiz and formation of sandy contourite layers. *Marine Geology*, **155**, 157–172.

Smith, W. H. F. & Sandwell, D. T. 1997. Global sea floor topography from satellite altimetry and ship depth soundings. *Science*, **277**, 1956–1962.

Somoza, L., Hernández-Molina, F. J., De Andrés, J. R. & Rey, J. 1997. Continental shelf architecture and sea-level cycles: Late Quaternary high-resolution stratigraphy of the Gulf of Cádiz, Spain. *Geo-Marine Letters*, **17**, 133–139.

Somoza, L., Lowrie, A., Maestro, A., *et al.* 1999. Gravity spreading and sliding of mobile shales and salt sheets along the slope of the Gulf of Cadiz (Atlantic Iberian Margin). *Meeting of Association of European Geological Societies, Abstracts Volume*.

Somoza, L., Gardner, J. M., Díaz-del-Río, V., Vázquez, T., Pinheiro, L., Hernández-Molina, F. J. & TASYO/ANASTASYA Shipboard Scientific Parties. 2002. Numerous methane gas related seafloor structures identified in the Gulf of Cádiz. *EOS Transactions, American Geophysical Union*, **83**(47), 541–547.

Somoza, L., Díaz-del-Rio, V., León, R., *et al.* 2003. Seabed morphology and hydrocarbon seepage in the Gulf of Cadiz mud volcano area: acoustic imagery, multibeam and ultra-high resolution data. *Marine Geology*, **195**, 157–176.

SRIVASTAVA, S. P., SCHOUTEN, H., ROEST, W. R., KLITGORD, K. D., KOVACS, L. C., VERHOEF, J. & MACNAB, R. 1990. Iberian plate kinematics: a jumping plate boundary between Eurasia and Africa. *Nature*, **344**, 756–759.

STOW, D. A. V. 1979. Distinguishing between fine-grained turbidites and contourites on the Nova Scotian deep water margin. *Sedimentology*, **26**, 371–87.

STOW, D. A. V. 1982. Bottom currents and contourites in the North Atlantic. *Bulletin de l'Institut Géologique de Bassin Aquitaine*, **31**, 151–166.

STOW, D. A. V. & FAUGÈRES, J. C. (eds) 1993. Contourites and Bottom Currents. *Sedimentary Geology, Special Issue*, **82**.

STOW, D. A. V. & FAUGÈRES, J. C. (eds) 1998. Contourites, Turbidites and Processes Interaction. *Sedimentary Geology, Special Issue*, **82**.

STOW, D. A. V. & MAYALL, M. 2000. Deep-water sedimentary systems: New models for the 21st century. *Marine and Petroleum Geology*, **17**, 125–135.

STOW, D. A. V., FAUGÈRES, J. C. & GONTHIER, E. 1986. Facies distribution and textural variations in Faro drifts contourites: velocity fluctuation and drift growth. *Marine Geology*, **72**, 71–100.

STOW, D. A. V., READING, H. G. & COLLINSON, J. 1996. Deep seas. *In*: READING, H. G. (ed.) *Sedimentary Environments: Processes, Facies and Stratigraphy*. Blackwell, Oxford, 395–453.

STOW, D. A. V., FAUGÈRES, J. C., GONTHIER, E., ET AL. 2002. Faro–Albufeira drift complex, Northern Gulf of Cadiz. *In*: STOW, D. A. V., PUDSEY, C. J., HOWE, J. & FAUGERES, J. C. (eds) *Deep-Water Contourite Systems: Modern Drifts and Ancient Series, Seismic and Sedimentary Characteristics*. Geological Society, London, Special Publications, **22**, 137–154.

TERRINHA, P. A., RIBEIRO, C., KULLBERG, J. C., LOPES, C., ROCHA, R. & RIBEIRO, A. 2002. Compressive episodes and faunal isolation during rifting, Southwest Iberia. *Journal of Geology*, **110**, 101–113.

THOMSON, J., NIXON, S., SUMMERHAYES, C. P., SCHÖNFELD, J., ZAHN, R. & GROOTES, P. 1999. Implications for sedimentation changes on the Iberian margin over the last two glacial/interglacial transitions from $(^{230}Th_{excess})_0$ systematics. *Earth and Planetary Science Letters*, **165**, 255–270.

THORPE, S. A. 1975. Variability of the Mediterranean undercurrent in the Gulf of Cadiz. *Deep-Sea Research*, **23**, 711–727.

THUNELL, R. C. & WILLIAMS, D. F. 1989. Glacial–Holocene salinity changes in the Mediterranean Sea: hydrographic and depositional effects. *Nature*, **338**, 493–496.

THUNELL, R., RIO, D., SPROVIERI, R. & VERGNAUD-GRAZZINI, C. 1991. An overview of the post-Messinian paleoenvironmental history of the Mediterranean. *Palaeoceanography*, **6**(1), 143–164.

TORELLI, L., SARTORI, R. & ZITELLINI, N. 1997. The giant chaotic body in the Atlantic Ocean off Gibraltar: new results from a deep seismic reflection survey. *Marine and Petroleum Geology*, **14**, 125–138.

TORTELLA, D., TORNÉ, M. & PÉREZ-ESTAÚN, A. 1996. Evolución geodinámica del limite de placas entre Eurasia y África en la zona del Banco de Gorringe y Golfo de Cadiz. *Geogaceta*, **20**(4), 958–961.

VERNAUD-GRAZZINI, C., CARALP, M.-H., FAUGÈRES, J. C., GONTHIER, E., GROUSSET, F., PUJOL, C. & SALIÈGE, J. F. 1989. Mediterranean outflow throught the Strait of Gibraltar since 18,000 years B.P. *Oceanologica Acta*, **12**, 305–324.

WILSON, R. C. L., SAWYER, D. S., WHITMARSH, R. B., ZERONG, J. & CARBONELL, J. 1996. Seismic stratigraphy and tectonic history of the Iberian Abyssal Plain. *In*: SAWYER, D. S., WHITMARSH, R. B., KLAUS, A., ET AL. (eds) *Proceedings of the Ocean Drilling Program, Scientific Results*, **149**. Ocean Drilling Program, College Station, TX, 617–630.

ZAHN, R. 1997. *North Atlantic thermohaline circulation during the last glacial period: evidence for coupling between meltwater events and convective instability, GEOMAR Report*, **63**.

ZAHN, R., SARNTHEIN, M. & ERLENKEUSER, H. 1987. Benthic isotope evidence for changes of the Mediterranean outflow during the late Quaternary. *Paleoceanography*, **2**, 543–559.

ZAZO, C., SILVA, P. G., GOY, J. L., ET AL. 1999. Coastal uplift in continental collision plate boundaries: data from the Last Interglacial marine terraces of the Gibraltar Strait area (south Spain). *Tectonophysics*, **301**, 95–109.

ZENK, W. 1970. On the temperature and salinity structure of the Mediterranean water in the Northeast Atlantic. *Deep-Sea Research*, **17**, 627–631.

ZENK, W. 1975. On the Mediterranean outflow west of Gibraltar. *'Meteor' Forschungs ergebunisse, Reihe, A*, **16**, 23–34.

ZENK, W. & ARMI, L. 1990. The complex spreading patterns of Mediterranean water off the Portuguese continental slope. *Deep-Sea Research*, **37**, 1805–1823.

Bottom-current reworked Palaeocene–Eocene deep-water reservoirs of the Campos Basin, Brazil

MARCO A. S. MORAES[1], WALTER B. MACIEL[2], MARIO SÉRGIO S. BRAGA[2] & ADRIANO R. VIANA[2]

[1]*Petrobras Research & Development Centre, Rio de Janeiro, RJ 21941-598, Brazil (e-mail: masmoraes@petrobras.com.br)*

[2]*Petrobras Exploration and Production, Rio de Janeiro, RJ 20031-912, Brazil*

Abstract: Deep-water reservoirs consisting of turbiditic sandstones moderately to heavily reworked by bottom currents are common in canyon- and trough-filling deep-water (bathyal) Palaeocene–Eocene sequences of the Campos Basin, offshore southeastern Brazil. A number of wells with conventional logs, together with cores, provided the database for the study. Seismic data provide additional support, but low resolution and noise hamper detailed analysis. The sandstones presenting better reservoir quality in these sequences are interpreted as being deposited by turbidity currents, as suggested by the dominance of unstratified normally graded sandstones, with grain sizes ranging from fine to coarse sand, and low clay-matrix content. Sandstones interpreted as bottom-current deposits (mid-water contourites) form poor-quality reservoirs, or baffles and barriers. These rocks are commonly moderately to heavily bioturbated, with variable, frequently high, clay-matrix content. Common trace fossils include *Planolites*, *Palaeophycus* and *Zoophycos*. Locally, these sandstones show faint horizontal stratification and planar cross-stratification. Contourites with thickness ranging from a few decimetres to several metres occur intercalated with turbiditic sandstones. Because they present distinct reservoir qualities, the mapping of the limits between turbidites and contourites is critical for adequate reservoir characterization. Most of this mapping has been performed using well information, constrained by outcrop analogues. The currents responsible for reworking turbiditic sands are interpreted to be deviated geostrophic currents, with velocity enhanced in narrow canyons and troughs.

Bottom currents, generated by thermohaline circulation or pressure gradients, or deep tidal forces, are considered, in addition to gravitational flows (debris flows and turbiditic currents), to be among the most important processes of sediment transfer and distribution to deep ocean realms (Stow & Faugères 1998; Stow *et al.* 2002). Although there is growing evidence that bottom currents affect the distribution of sandy (and hence reservoir prone) material in deep marine settings (Viana *et al.* 1998), there is a lack of convincing descriptions of hydrocarbon reservoirs significantly composed of, or associated with, deep-water bottom-current deposits. The few published examples (see Shanmugan *et al.* 1993, and references therein) have been considered at least dubious and reinterpreted as turbidites or shallower deposits (Stow *et al.* 1998). The present paper describes a Palaeocene–Eocene sequence in the Campos Basin, offshore SE Brazil, comprising billion-barrel hydrocarbon reservoirs, which are mostly composed of massive to normally graded unstratified sandstones, interpreted as turbiditic sandstones, interbedded with moderately to heavily bioturbated sandstones, interpreted as bottom-current deposits. Turbiditic sandstones form high-quality reservoirs, whereas bottom-current deposits form low-quality reservoirs, or permeability baffles and barriers. For simplicity, these two types of sandstones will be named herein turbidites and contourites (mid-water contourites of Viana *et al.* 1998; see discussion below).

The most significant evidence for the bottom-current origin of the Campos Basin deep-water Palaeocene–Eocene bioturbated sandstones includes: (1) the intense and homogeneous nature of bioturbation, indicating slow aggradation rates; (2) the absence of well-defined grain-size trends, suggesting more sustained currents; (3) the abrupt contact between turbiditic deposits and inferred bottom-current deposits, indicating removal of turbidite tops by bottom currents; (4) the thickness of homogeneously bioturbated sandstones, indicating long-term processes. Recognition of bottom-current deposits and prediction of their distribution are essential for the adequate characterization of these important reservoirs. The criteria developed herein may be applicable to other sequences of similar origin, and perhaps contribute to the better understanding of deep-water processes in the Campos Basin and elsewhere.

From: VIANA, A. R. & REBESCO, M. (eds) *Economic and Palaeoceanographic Significance of Contourite Deposits*. Geological Society, London, Special Publications, **276**, 81–94.
0305-8719/07/$15.00

Geological setting

The Campos Basin (Fig. 1) is the most prolific of several eastern Brazilian continental-margin basins. These basins developed since the early Cretaceous break-up of Gondwana (Guardado *et al.* 1990; Bruhn 1998). The sedimentary prism that forms the Campos Basin is divided into five megasequences (Fig. 2), named, from older to younger: (1) continental rift (Neocomian); (2) transitional (evaporitic) (Aptian); (3) shallow marine (Albian); (4) marine transgressive (Cenomanian to Upper Eocene); (5) marine regressive (Upper Eocene to Present) megasequences. The Palaeocene–Eocene reservoirs occur mostly within the transgressive megasequence.

The deep-water Palaeocene–Eocene sandstones occur mostly confined within elongated troughs generated by salt tectonics (Fig. 3). In some troughs, only Eocene sandstones occur, which reflects a more restricted Palaeocene sandstone distribution. The troughs are typically several kilometres wide, and filled with up to a few hundred metres of sand-rich deposits. Commonly, within a given third-order stratigraphic sequence, the deposits are strongly confined at the base and less confined toward the top, suggesting that trough geometry was established by relatively short tectonic phases, followed by passive filling phases. Internally, trough-filling sequences consist mostly of sandstone with variable mud content (mostly in the form of detrital matrix within the bioturbated sandstones). Mudstone intercalations are rare, although they become more common at the margins and upper parts of the troughs. Most mud-rich intervals consist of heterolithic beds composed of centimetre-scale intercalations of bioturbated mudstones, bioturbated marls and

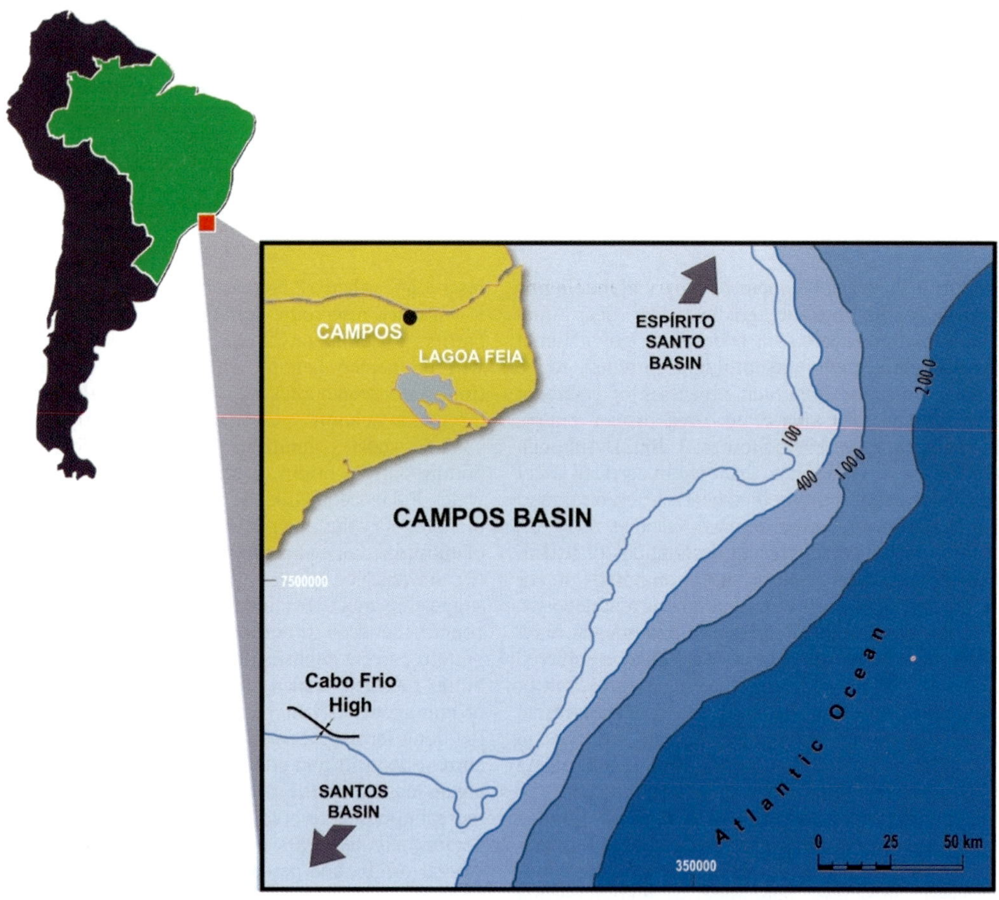

Fig. 1. Location map of the Campos Basin, off SE Brazil. Isobaths are in metres. The study area is located beneath the present middle and lower slope area in the central part of the figure.

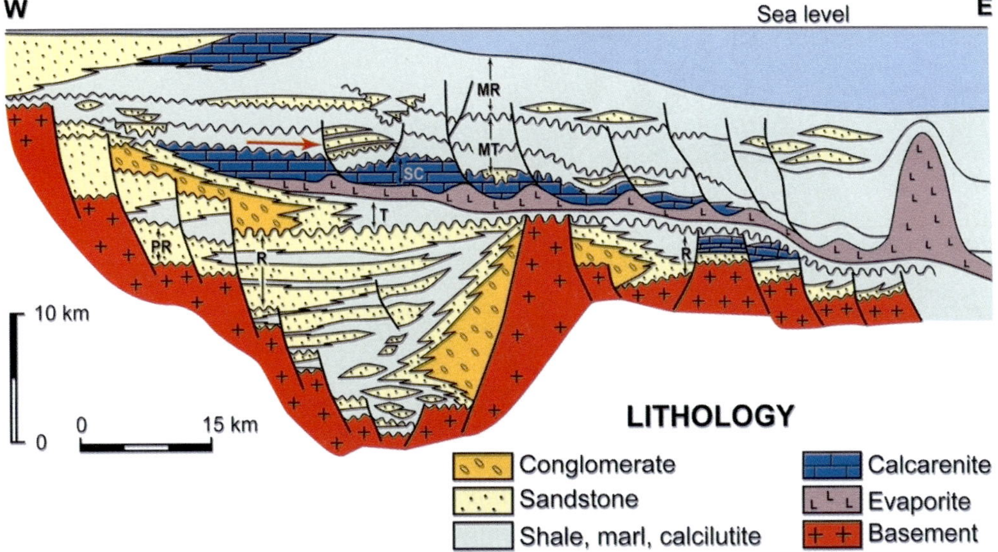

Fig. 2. Generalized geological section for the Eastern Brazilian continental margin basins, showing the main megasequences: PR, pre-rift (which does not occur in the Campos Basin); R, rift; T, transitional (which includes the evaporate section); SC, shallow carbonate; MT, marine transgressive; MR, marine regressive (modified from Bruhn 1998). The red arrow indicates the Palaeocene–Eocene reservoirs.

heavily bioturbated sandstones. Some of these intervals appear to be laterally continuous, and are used to separate and correlate internal stratigraphic units.

The Palaeocene–Eocene boundary of the Campos Basin is marked by a significant unconformity. In most of the basin the Lower Eocene sequence is entirely absent. Eocene reservoir-prone sequences occur in both the Middle and Upper Eocene deposits. In the Upper Eocene sequence, a regional mass-wasting complex (informally named 'Pebbly'), consisting of several widespread lenses of carbonate-rich debris flows, was formed, apparently in response to an intensification of tectonic activity that affected all the South Atlantic basins. The 'Pebbly' episode generated new relief on the ocean bottom. The reservoirs found intercalated with the debrites are again confined. The transition from Eocene to Oligocene is also marked by an unconformity, albeit not as significant as that of the Palaeocene–Eocene boundary.

The deposits discussed herein are located seaward of the Eocene reservoirs described by Mutti *et al.* (1980), which were interpreted as deep-water gravitational flows winnowed by bottom-current deposits. Although it is still accepted that gravitational flows and bottom currents were involved in their formation, these deposits are now considered to be mostly outer-shelf or upper-slope deposits. The same holds for Eocene reservoirs interpreted as bottom-current deposits by D'Avila *et al.* (1994).

Facies associations

The reservoir sequences, both in Palaeocene and Eocene units, are characterized by (1) relatively monotonous centimetre- to decimetre-scale intercalations of massive to normally graded, only locally stratified (horizontal laminations to ripples) clean sandstones, which are interpreted as turbidites, and (2) moderately to heavily bioturbated clay-rich sandstones, interpreted as bottom-current deposits (contourites; Fig. 4). The contourites described herein were formed in mid- to lower bathyal environments (M. R. Becker *et al.* unpubl. data), and, if their origin as bottom-current deposits is accepted, they could be included in the mid-water (300–2000 m) contourites of the classification of Viana *et al.* (1998).

Grain size in both cases varies from localized granules to dominantly fine to medium sand, with coarse or very fine sand as secondary components. The top of the turbidite beds is commonly bioturbated, although the contact between turbidites and bioturbated sandstones is in most cases abrupt (Fig. 5). This seems to indicate that the upper part of the turbidites was often removed by bottom currents before deposition started again. Bioturbation

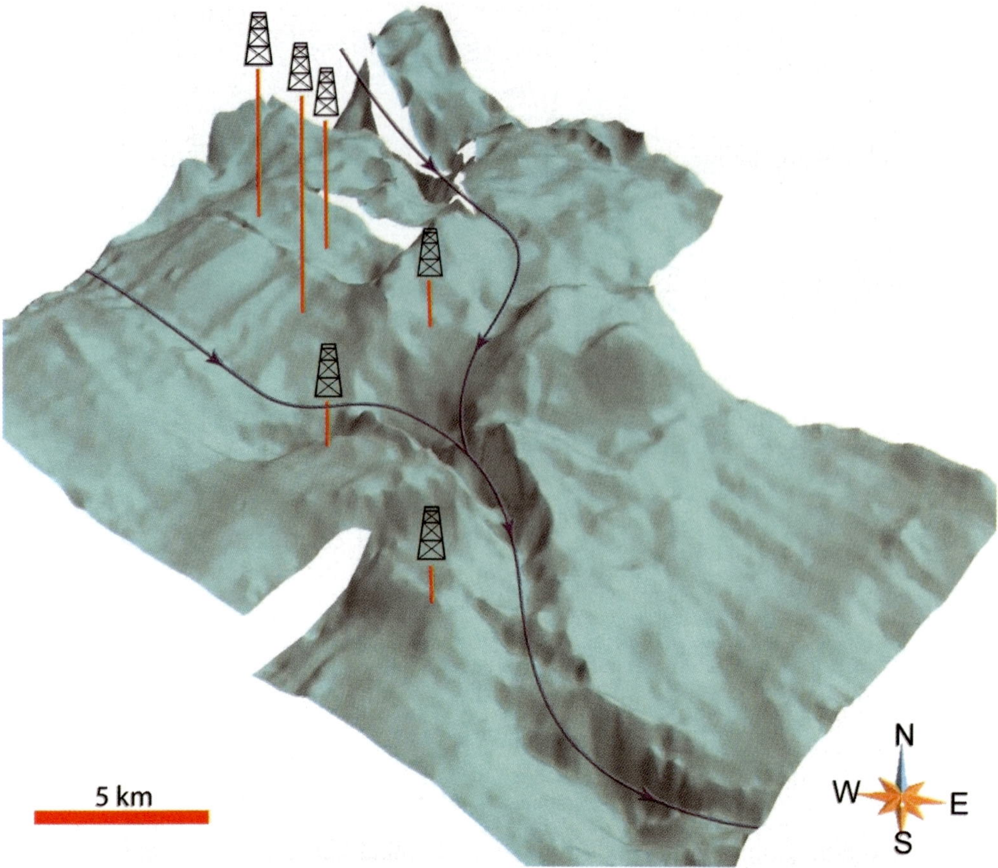

Fig. 3. A 3D view of the present top of the Cretaceous sequence (amplitude reflector) in the study area. The present morphology maintains the basic topographic elements (highs and troughs, generated by salt tectonics) that controlled the distribution of Palaeocene and Eocene sandstones. Positions of control wells are represented by red lines. Flow lines indicate main sand pathways. It should be noted that some wells were drilled at the margins of the troughs (structural highs).

is pervasive and homogeneously distributed within the bioturbated sandstones (Fig. 6). The most common trace fossils include *Planolites*, *Palaeophycus* and *Zoophycos*. This trace fossil assemblage is typical of deep-ocean realms (Pemberton *et al.* 1992). Commonly, there are no clear vertical trends in the degree of bioturbation, although the intensity can vary in different parts of the beds. No vertical trace fossils, which could suggest the presence of suspension feeders, and hence shallower waters, are observed. This is in agreement with palaeoecological estimates of mid- to lower bathyal depths. Locally, where bioturbation is less intense, sandstones may show faint horizontal stratification and planar cross-stratification.

In terms of texture, turbidites are commonly graded (coarse-tail grading) and are moderately to well sorted. In some cases, intraclast shale breccias occur at the base of the turbidites. Contourites commonly show poor sorting. Coarse sand and even granules are commonly found in different positions within the bed, which also show a variable mud content. The presence of dispersed coarse grains seems to indicate that the beds aggraded slowly. Bed aggradation was probably dominated by traction, and probably led to development of stratification, which was then obliterated by burrowing. The detrital matrix distribution is also controlled by bioturbation, occurring either as lined burrows or as former mud-rich intervals that are strongly obliterated by bioturbation.

Contourites are volumetrically more important in the Palaeocene units, where they can represent up to 50% of the sandstone sequences. In fact, in the Upper Palaeocene sequences, in the interval

Fig. 4. Core photograph showing the intercalation of centimetre-scale to decimetre-scale beds of graded, unstratified and horizontally stratified sandstones, interpreted as turbiditic deposits, with bioturbated sandstone beds, interpreted as bottom-current deposits (contourites). Darker-coloured zones correspond to an increase in oil saturation. The abrupt contacts at the base and the internal homogeneity of the contourites should be noted. Core top is to the left (depth is in metres).

Fig. 5. Detail of core sample showing the contact between a turbiditic sandstone showing Ta and Tb divisions, at the base, and a bioturbated sandstone (contourite) at the top. It should be noted that although the uppermost part of the turbidite shows a number of burrows (*Zoophycos*; Z), the contact (C) between the two beds is represented by an abrupt and irregular surface, indicating that a time gap, with some erosion, occurred before the deposition of the contourite bed.

Fig. 6. Detail of core sample showing the fabric of bioturbated sandstone. It should be noted that bioturbation (mostly *Planolites*, *Palaeophycus* and *Zoophycos*) is homogeneously distributed throughout the bed. Dispersed coarse sand grains and granules are also visible.

known as the 'sterile zone', contourites form as much as 70% of the sequence. In the Eocene units, turbidites are thicker, with better preserved fining-upward sequences. Contourites are less common, forming around 20% of the sequences. This change is interpreted as resulting from higher sedimentation rates during the Eocene. Debrites, as discussed above, are more common in the Upper Eocene sequences. In fact, the so-called 'Pebbly' unit is represented by carbonate-rich intervals (mostly red-algae rodoliths) consisting of debrites and turbidites (Fig. 7). The debrites are composed of shallow-water carbonate clasts (rodoliths) dispersed in a siliciclastic mud matrix. The carbonate turbidites consist of graded beds composed of rodoliths and bivalve shells, with low matrix content. In some areas, siliciclastic turbidites appear associated with those beds.

As mentioned above, shale beds are extremely rare within the sandstone sequences that fill the troughs. Most of the clay-rich intervals are heavily bioturbated sandstones. Commonly only a

Fig. 7. Core photograph showing the characteristics of the carbonate-rich debris-flow and turbidite deposits (the 'Pebbly' interval) of the Campos Basin Upper Eocene sequences. Carbonate fragments are mostly rodoliths, bivalves and equinoids. Bioclastic turbidites (whitish) are cemented by calcite. Core top is to the left (depth is in metres).

few shale beds, 10–20 cm thick, are found when examining a well. Despite their thin nature, most shales seem to be laterally continuous, representing third- or fourth-order condensed zones, as will be discussed in the next section.

Palaeocene–Eocene turbidites form high-quality reservoirs in numerous fields in the deep-water areas of the Campos Basin. Typical porosity values range from 25 to 30%, with permeabilities ranging from hundreds of millidarcies to a few darcies (Fig. 8). Contourites form poor-quality reservoirs, with porosities ranging from 20 to 30% and permeability of tens of millidarcies, or baffles and barriers with permeabilities of a few millidarcies or less. Reservoir quality in the turbidites depends upon the amount of calcite cement, which is more abundant in the Palaeocene units, whereas in the contourites it depends upon the amount of clay matrix.

Stratigraphic framework

As shales are relatively rare, and reworking and bioturbation common, the reservoir sequences are characterized by relatively low biostratigraphic-resolution. Essentially, biostratigraphic zonation permits the recognition of three units (Fig. 9), namely Lower–Middle Palaeocene, Middle Eocene, and Middle–Upper Eocene. However, vertical trends and/or variations in facies proportions, along with a few condensed layers, permit the recognition of a series of third- and fourth-order stratigraphic sequences (Fig. 9). Figure 10 shows that the stratigraphic units can be correlated using criteria such as: (1) facies proportions (identified in cores but also reflected by log patterns); (2) grain-size average and trends (also defined in cores and possibly recognizable in density logs); (3) shaliness (gamma-ray background); (4) in some cases, presence of condensed layers (high gamma-ray peaks). The correlation of such stratigraphic units revealed that, despite the trough-filling and relatively high-energy (inferred based on the grain size) nature of the sequences, there is good lateral continuity within the troughs. As a result, a high-resolution stratigraphic framework could be established (Figs 9 and 10), including three third-order and six fourth-order sequences in the Palaeocene, and two third-order and five fourth-order sequences in the Eocene. Such stratigraphic units have been used to define production zones and subzones, which are used to guide the reservoir management.

The stratigraphic units seem to present a subparallel geometry, and to onlap onto the border of the troughs. Such features suggest that there was little deformation coeval with the filling of the troughs. The occurrence of significant internal erosion

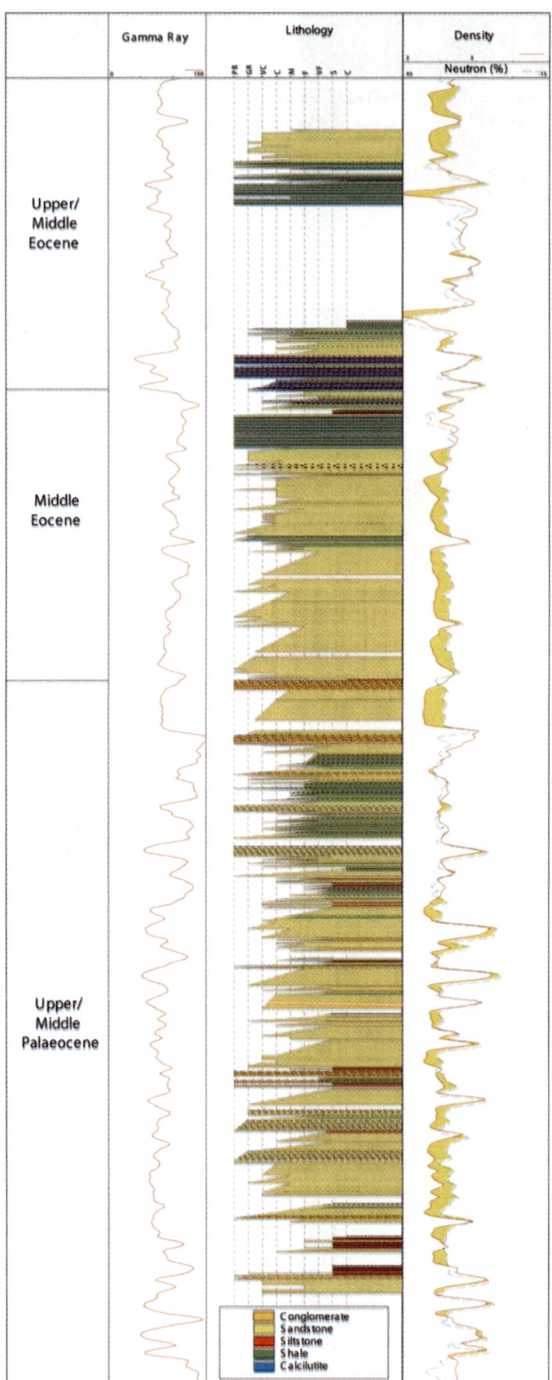

Fig. 8. Vertical section within the Campos Basin Palaeocene–Eocene section showing the typical well response and characteristic petrophysical properties of the reservoirs. Low gamma-ray, high porosity zones correspond to turbiditic sandstones, whereas higher mud content (higher gamma-ray), lower porosity zones commonly correspond to bioturbated sandstones (contourites).

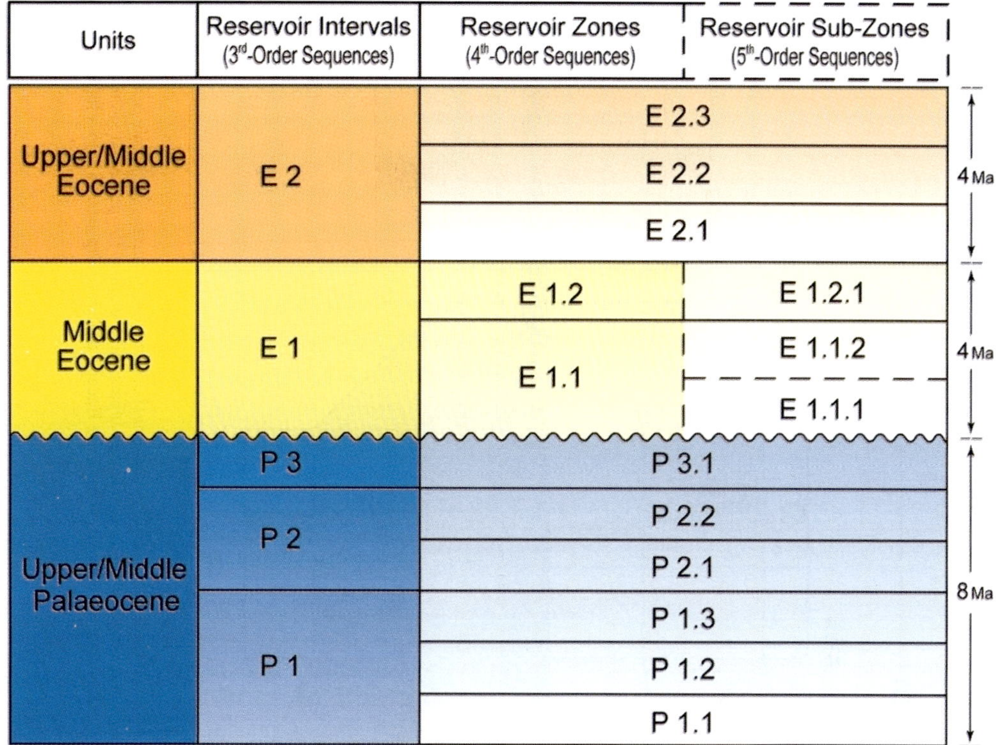

Fig. 9. Schematic stratigraphic chart of the Palaeocene–Eocene section of the deep-water Campos Basin showing low- and high-resolution sequences. The relationship between low- and high-resolution sequences and reservoir zones should be noted. Unit names are codes used for this study and do not imply any specific field or area.

surfaces within the reservoir sequences was considered as a possibility but, with the exception of the Lower Eocene unconformity, was ruled out based on the correlatable pattern found between wells. Subsurface correlation, especially when it is based only on lithostratigraphy, as shown in Figure 10, involves a high degree of interpretation and, of course, uncertainty. However, the features listed above, especially the striking similarities of log pattern (hence facies proportions) shown by the same unit at different wells, strongly suggest good lateral continuity. This seems to be a feature most commonly found in the Palaeocene–Eocene sequences. In most Campos Basin trough-filling sequences of other ages (Bruhn 1998; Moraes *et al.* 2000, 2004), such a repetitive pattern is not found, especially in the lower parts of the trough fills, suggesting the occurrence of internal erosive surfaces.

Depositional model

The characteristics of the lithofacies described above, along with observations of well-log patterns, correlation between wells, and regional seismic data, complemented by outcrop and recent analogue examples, led to the formulation of the depositional model presented in Figure 11. In this model, the turbiditic sandstones are interpreted as deposited by a series of shallow, probably braided, channels, with subsequent reworking, to variable degrees, by bottom currents. The channelized nature of the system is inferred based on: (1) grain size, which indicates high-energy currents, capable of suspending particles up to granule size, and hence promoting erosion of finer particles; (2) a certain degree of turbidite bed amalgamation (erosive contacts, some with intraclast shale breccia), which suggests that erosional surfaces were occasionally formed; (3) the apparent interconnection of the troughs, which would avoid complete ponding and the development of lobes. On the other hand, there is no evidence of deep erosional cuts. The system aggraded, preserving the continuity of the stratigraphic units. Such a pattern of channel complex aggradation was probably controlled by terminal deposition (ponding) in mini-basins located basinward. Local sand deposition was controlled by

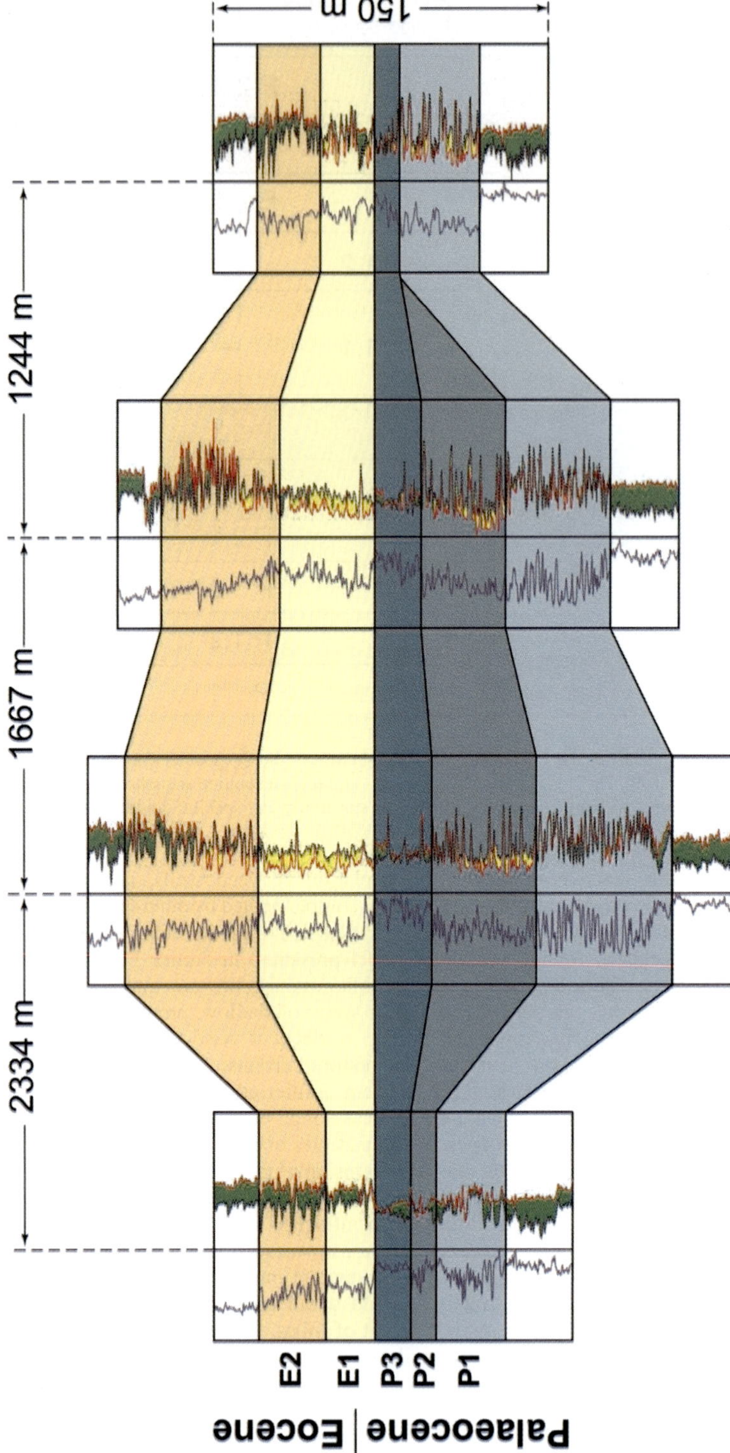

Fig. 10. Well-log correlation showing the main criteria used to define the stratigraphic units, which included (1) facies proportions (i.e. distinct log patterns), (2) grain-size average and trends (recognized in density logs), (3) shaliness (gamma-ray background), and, in some cases, (4) presence of condensed layers (high gamma-ray peaks). It should be noted that the datum in the Palaeocene–Eocene boundary preserves the trough shape of the Palaeocene section but distorts the Eocene trough-filling section into a high.

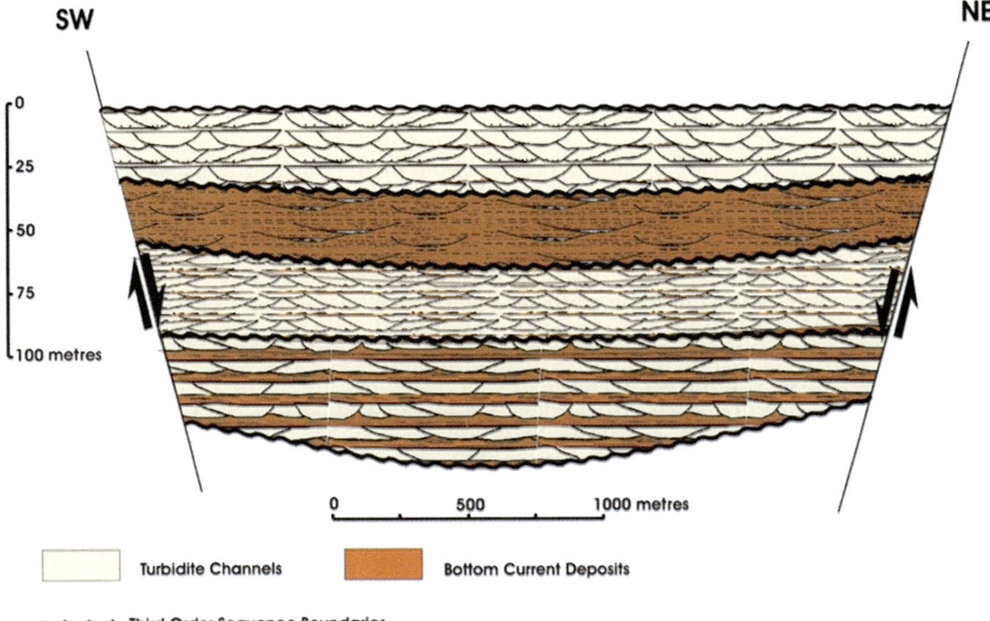

Fig. 11. Depositional model for Palaeocene–Eocene deep-water, trough-filling sandstones of the Campos Basin, showing typical facies proportions and dimensions inferred for the sandstone bodies. Dimensions of turbiditic channels and contourites are based on well observations and analogue data.

changes in the gradient of the trough, as sand filling is localized, and do not occur throughout the trough.

Several observations led to the interpretation that the bioturbated sandstones were produced by bottom-current reworking and redeposition of turbiditic sand. These include the following.

(1) *Abrupt contacts*: although the tops of many turbidite beds are marked by progressive bioturbation, even in these beds there is an abrupt passage from a progressively burrowed sandstone, commonly with primary structures still preserved, to a intensely and homogeneously bioturbated bed (Fig. 5). Such an abrupt passage indicates that some turbiditic (and possibly some hemipelagic) material was removed from the top of the previously deposited bed before the onset of deposition by bottom currents.

(2) *Homogeneous fabric*: whereas turbidite beds are commonly graded, bioturbated sandstones are very homogeneous (Figs 5 and 6). The beds show poor sorting, and coarse sand grains and even granules can be found at any position within the bed, mixed with a variety of grain size, including clay. Bioturbation is also homogeneously distributed. Although the intensity of bioturbation can vary throughout the bed, such variations are not abrupt, and do not show any regular vertical trend. Such homogeneity indicates very slow rates of deposition. Probably different grain sizes were emplaced as distinct laminae, but at such a slow rate that bioturbation was capable of mixing all the grain sizes. In this context, clay was probably also originally deposited in discrete layers, which was also mixed with sand by the action of organisms.

(3) *Lack of turbidite or hemipelagic fines*: as shale, marls or other fine-grained deposits, related to low-energy environments, are extremely rare, this seems to indicate a constant action of currents sweeping the bottom, and preventing the accumulation and/or preservation of the condensed layers that are very common in other turbidite sequences. On the other hand, part of the clay material preserved within the bioturbated sandstone is interpreted as emplaced by low-energy processes and may have originally formed thin layers or drapes. However, none of the original structure was preserved.

(4) *Thickness of bioturbated sandstones*: although most bioturbated sandstones have a centimetre- or decimetre-scale thickness, a number of them are on a metre, or even decametre scale. This, together with the homogeneous nature of the sandstones and the abrupt contacts, suggests that bioturbated sandstones were not formed by the simple reworking of the top of turbidite beds by

organisms, but rather by a more complex process of reworking and redeposition by bottom currents.

These observations are consistent with facies models for contourites proposed in the literature (Stow *et al.* 1998; Viana *et al.* 1998), although the sandstones described here are coarser grained than those commonly presented in the classical models.

In terms of reservoir architecture, it is clear from the model in Figure 11 that the delineation of the geometry of channels and associated contourite zones is critical for predicting reservoir performance. As there is no seismic resolution to determine directly the geometry of turbidite channels and contourite beds, and as well spacing (typically on the order of a few kilometres) is too large to allow a detailed assessment using well correlation, the dimensions of sandstone bodies have to be inferred based on dimensions observed in analogue systems (see data on channel dimensions, of Moraes *et al.* 2000, 2004). As such, the Palaeocene–Eocene trough-filling system of the Campos Basin was inferred, as shown in Figure 11, to consist of individual channels a few hundred metres wide and a few metres thick, which would stack laterally and vertically to form composite bodies several hundred metres wide and several metres thick. The stacking of composite channels would then produce channel complexes a few kilometres wide and several tens of metres thick.

Distribution and preservation of contourite beds is inferred to be dependent on the frequency and intensity of turbidity current activity. As bottom-current activity is more or less continuous and widespread, and turbidity current activity much more episodic and localized, the deposition and preservation of contourites would be favoured in areas with less frequent turbidite deposition. For example, because the volume of contourites is greater in the Palaeocene sequences, it is inferred that turbidity currents were less frequent in that period than in the Eocene, where contourite volume is smaller and turbidite amalgamation more common. In addition to temporal controls, there is also concentration of turbidites in the trough axis, at a large scale, and in channel (composite and individual) axis, at a smaller scale. The trough axis can be identified by determining facies proportions in well logs (see Fig. 8), and even using seismic data, where, despite low resolution and noise, areas with a higher proportion of clean sandstones can be related to zones with lower amplitude or impedance. Finally, at the regional scale, there should be expected, nevertheless, some relationship between the occurrence of contourites and turbidites, as the latter are the sources of the material reworked by bottom currents. The regional distribution of the contourites has not been determined yet, because of the lack of well data and the low seismic resolution.

The architectural pattern defined by turbiditic channels and contourite zones develops within the limits of the high-resolution stratigraphic units. As mentioned above, stratigraphic unit limits are considered to be continuous surfaces, with the units being distinguished from one another by differences in facies proportions, bed thickness trends, and other factors. As such, it is inferred that the third- and fourth-order units may each probably show a preferential axis location, which can be determined by careful well-log correlations or improved seismic-attribute mapping.

Palaeohydrological considerations

The origin of the bottom currents that formed the contourite sequences is not clear. As such currents developed within canyons and troughs, deep tidal currents or deviated geostrophic currents seem to be the most probable. Isotopic data suggest that, in the western South Atlantic, the modern pattern of regional thermohaline circulation could have been established during the Oligocene, when the Drake Strait was open (Viana & Faugères 1998). Thus, it is not probable that such ocean-wide currents would be active during the deposition of the studied deposits, and hence available to be deviated into the canyons and troughs. On the other hand, Viana & Faugères (1998) showed that surface slope boundary geostrophic currents, formed basically in response to hemispheric wind pressure, are very active today on the Campos Basin shelf, and have an effect down to the present upper slope. It seems that such an effect is favoured by the convex form of the Campos Basin shelf, which funnelled the northern-derived currents on the outer shelf, producing currents with measured speeds up to 50 cm s^{-1} (possibly with higher peaks in a longer period). Although this system affects today basically the shelf and the upper slope, a similar system, operating on the much smaller shelf of the Palaeocene–Eocene, might have been able to affect greater depths, especially during times of lowered sea level. The occurrence of Eocene outer shelf and upper slope reservoirs affected by bottom currents (Mutti *et al.* 1980; C. E. Souza Cruz *et al.* unpubl. data), is further evidence of the existence of such a system.

In the Campos Basin, contourites, on the other hand, are not common in the Cretaceous or the Oligocene–Miocene sequences. Their common occurrence in the Palaeocene–Eocene sequences may be due to two basic factors. First, the Cretaceous, despite the narrowness of the proto-South Atlantic, and of its shelf, was a greenhouse period, when a high sea level prevailed, which might have prevented or diminished the impact or duration of bottom currents in the deep realms. Second, in the Oligocene–Miocene, the Atlantic had already developed into a large ocean, with a marked shelf expansion. In such a large system, the action of bottom currents capable of reworking sand, even during times of lowered sea level, probably was restricted to the shelf and upper slope, never reaching the base-of-slope or basinal sectors where the turbidites are found.

Deep tides are also a possibility. Although there is no direct evidence of strong deep tidal currents in the Campos Basin, it is well known (Shepard *et al.* 1979, pp. 3–13; Zhenzhong *et al.* 1998) that deep tidal currents, funnelled in straits or canyons, can reach velocities high enough to move sand (see the simulation performed by Lima *et al.* 2007, for a modern submarine canyon in the Campos Basin). In the case presented here, the lack of preservation of primary structures makes it difficult to recognize features that would be diagnostic of tidal action, such as thick mud drapes, or bi-directional features. Moreover, there are no observations of significant occurrences of tidal deposits in the ancient or recent record of the Campos Basin. In addition, Eocene deposits located landward (beneath the present continental shelf; described by Mutti *et al.* 1980) seem to be deposits of shelf to upper slope geostrophic currents, without a major influence of tidal effects. On the other hand, tidal forces could serve to enhance geostrophic currents, allowing them to reach peak velocities high enough to cause sand reworking (Viana *et al.* 1998).

Economic relevance

Concerning the economic impact of the contourites, it is intuitive that the conceptual understanding of the bottom-current processes and their interaction with previously deposited gravitational-flow sediments, along with the resultant reservoir architecture, is crucial for hydrocarbon production optimization. Regarding the example presented herein, there are two different roles that contourites can assume: (1) the lower quality, more heavily biotubated or cemented sandstones can act as barrier and baffles to hydrocarbon flow through much higher permeability turbiditic sandstones; (2) the higher quality, less bioturbated and less cemented sandstones can host significant amounts of hydrocarbons, although they have lower permeability values than the turbidites. It is well known that, mainly where hydrocarbon recovery depends on water flooding, the sweep efficiency will be very much dependent on the reservoir architecture and the distribution of petrophysical properties distribution within the architectural elements. Thus, very important issues related to the drainage strategy of hydrocarbon reservoirs (e.g. well type definition, well positioning and well completion strategy) can be optimized with the correct understanding of reservoir architecture. This is relevant at any phase of a given field development. It is critical in the early phases, when the data are sparse and decisions have huge impacts on the project results, but it can also be important in more mature phases. In mature phases, time lapse seismic data (4D seismic), for instance, can show bypassed oil areas. Application of 4D seismic results, nevertheless, requires sound geological understanding to properly carry out the work-overs, infill drillings and flow rate tunnings.

In any case, using a consistent geological model to support the engineering project can lead to significant increases in the recovery and in the net revenues. The correlation technique shown in Figure 10 and the architectural model presented in Figure 11 represent the first steps toward such a goal. Given the limitations of the seismic method in providing a 3D image of the studied reservoirs, often with low impedance contrast, it can be very useful to perform geological investigations using modern surface and ancient outcrop analogues and physical modelling. Such data can fill the gaps in the sparse subsurface well data, improving the understanding of the spatial distribution patterns shown by the facies associations and related architectural elements.

Conclusions

Whatever the origin of the bottom currents that affected the Palaeocene–Eocene trough-filling sediments of the Campos Basin, they strongly affected the nature and the architecture of the reservoirs. The proposed depositional model predicts that the turbidites occur as a series of channel-fill sandstone bodies, and that contourites, represented by the bioturbated sandstone facies, occur as sheets and lenses that become more important volumetrically in areas away from the channel and channel complex axis, and in sequences with a lower turbidite sedimentation rate. The thickness of the deposits can be determined from the wells. The geometry and dimensions of the sand bodies, however, because of the lack of seismic resolution, are inferred from outcrop analogues, in the case of turbidite channels, and based on well correlation, for contourites. This last procedure is much more prone to uncertainty. However, the recognition and mapping of contourite zones, even with some degree of uncertainty, is essential for adequate reservoir management. The models presented herein represent important guides for building reservoir models, especially for predicting interwell heterogeneity, as they present not only the geometry, but also the dimensions of the various reservoir elements. At a regional scale, sandy contourites should be expected to occur associated with areas where turbidites occur, as sand-rich turbidites are necessary to provide the sandy material to be reworked.

We thank J. J. Marques, M. M. S. d'Oliveira, R. M. T. Silva, M. R. Becker and C. E. Souza Cruz for discussions and technical collaboration. Thanks also go to M. M. Alves, S. M. C. Anjos and João C. J. Conceição for management support. H. S. Barbosa provided drafting support. We also thank A. H. Bouma and D. Stow for their constructive reviews.

References

BRUHN, C. H. L. 1998. *Deep-water reservoirs from the eastern Brazilian rift and passive-margin basins.* Course No. 6, Part 2, AAPG International Conference and Exhibition, Rio de Janeiro.

D'AVILA, R. S. F., CADDAH, L. F. G. & GRASSI, A. A. 1994. Reconstrução paleoambiental de uma seção do paleógeno da Bacia de Campos, uma contribuição com base no estudo de traços fósseis. *São Leopoldo, XIII Congresso Brasileiro de Paleontologia, 1993, Anais. Acta Geologica Leopoldensia*, **XVII**, 15–28.

GUARDADO, L. R., GAMBOA, L. A. P. & LUCCHESI, C. F. 1990. Petroleum geology of the Campos Basin, Brazil: a model for a producing Atlantic-type basin. *In*: EDWARDS, J. D. & SANTOGROSSI, P. A. (eds) *Divergent/Passive Margin Basins.* AAPG Memoirs, **48**, 3–79.

LIMA, J. A. M., MOLLER, O. O. JR, VIANA, A. R. & PIOVESAN, R. 2007. Hydrodynamic modelling of bottom currents and sediment transport in the Canyon São Tomé (Brazil). *In*: VIANA, A. R. & REBESCO, M. (eds) *Economic and Palaeoceanographic Significance of Contourite Deposits.* Geological Society, London, Special Publications, **276**, 329–342.

MORAES, M. A. S., BECKER, M. R., MONTEIRO, M. C. & NETTO, S. L. A. 2000. Using outcrop analogs to improve 3D heterogeneity modeling of Brazilian sand-rich turbidite reservoirs. *In*: WEIMER, P., SLATT, R. M. & COLEMAN, J. ET AL. (eds) *Deep-Water Reservoirs of the World.* GCSSEPM Foundation, Houston, 587–605.

MORAES, M. A. S., BECKER, M. R., BLASKOVSKI, P. R. & JOSEPH, P. 2004. The Grès d'Annot as an analogue for Brazilian Cretaceous sandstone reservoirs: comparing convergent to passive-margin confined turbidites. *In*: JOSEPH, P. & LOMAS, S. (eds) *Deep-Water Sedimentation in the Alpine Basin of SE France: New Perspectives on the Grès d'Annot and Related Systems.* Geological Society, London, Special Publications, **221**, 419–436.

MUTTI, E., BARROS, M. C., POSSATO, S. & GUARDADO, L. R. 1980. *Deep-sea fan turbidite systems winnowed by bottom currents in the Eocene of the Campos Basin, Brazilian offshore.* 1st European Meeting of the International Association of Sedimentologists (Abstracts), 114.

PEMBERTON, S. G., MACEACHERN, J. A. & FREY, R. W. 1992. Trace fossil facies models: environmental and allostratigraphic significance. *In*: WALKER, R. G. & JAMES, N. E. (eds) *Facies Models: Response to Sea-level Change.* Geological Association of Canada, St. Johns, 47–72.

SHANMUGAM, G., SPALDING, T. D. & ROFHEART, D. H. 1993. Process sedimentology and reservoir quality of deep-marine bottom-current reworked sands (sandy contourites): an example from the Gulf of Mexico. *AAPG Bulletin*, **78**, 910–937.

SHEPARD, F. P., MARSHALL, N. F., MCLOUGHLIN, P. A. & SULLIVAN, G. G. 1979. *Currents in submarine canyons and other seavalleys.* AAPG Studies in Geology, **8**.

STOW, D. A. V. & FAUGÈRES, J.-C. (eds) 1998. Contourites, turbidites and process interaction. *Sedimentary Geology, Special Issue*, **115**.

STOW, D. A. V., FAUGÈRES, J. C., VIANA, A. R. & GONTHIER, E. 1998. Fossil contourites: a critical review. *Sedimentary Geology*, **115**, 3–31.

STOW, D. A. V., FAUGÈRES, J.-C., HOWE, J. A., PUDSEY, C. J. & VIANA, A. R. 2002. Bottom currents, contourites and deep-sea sediment drifts: current state-of-the-art. *In*: STOW, D. A. V., PUDSEY, C. J., HOWE, J. A., FAUGÈRES, J.-C. & VIANA, A. R. (eds) *Deep-Water Contourites: Modern Drifts and Ancient Series, Seismic and Sedimentary Characteristics.* Geological Society, London, Memoirs, **22**, 7–20.

VIANA, A. R. & FAUGÈRES, J. C. 1998. Upper slope sand deposits: the example of Campos Basin, a latest Pleistocene–Holocene record of the interaction between alongslope and downslope currents. *In*: STOKER, M. S., EVANS, D. & CRAMP, A. (eds) *Geological Processes on Continental Margins: Sedimentation, Mass Wasting and Stability.* Geological Society, London, Special Publications, **129**, 287–316.

VIANA, A. R., FAUGÈRES, J. C. & STOW, D. A. V. 1998. Bottom-current-controlled sand deposits—a review of modern shallow- to deep-water environments. *Sedimentary Geology*, **115**, 53–80.

ZHENZHONG, G., ERIKSSON, K. A., YOUBIN, H., SHUNSHE, L. & JIANHUA, G. 1998. *Deep-Water Traction Current Deposits: a Study of Internal Tides, Internal Waves, Contour Currents and Their Deposits.* Science Press, Beijing; VSP, Utrecht.

Interaction of processes and importance of contourites: insights from the detailed morphology of sediment Drift 7, Antarctica

M. REBESCO[1], A. CAMERLENGHI[2], V. VOLPI[1], C. NEAGU[1], D. ACCETTELLA[1], B. LINDBERG[3], A. COVA[1], F. ZGUR[1] & the MAGICO PARTY

[1]*Istituto Nazionale di Oceanografia e di Geofisica Sperimentale (OGS), Borgo Grotta Gigante 42/c, 34010 Sgonico (Trieste), Italy (e-mail: mrebesco@ogs.trieste.it)*

[2]*ICREA, Institució Catalana de Recerca i Estudis Avançats, c/o GRC Geociències Marines, Departament d'Estratigrafia, P. i Geociències Marines, Universitat de Barcelona, Facultat de Geologia, C/ Martí i Franquès, s/n, E-08028 Barcelona, Spain*

[3]*Department of Geology, University of Tromsø, Dramsveien 201, NO-9037 Tromsø, Norway*

Abstract: As the definition of contourites has widened to embrace a large spectrum of sediments in so-called mixed systems, the distinction between contourites and turbidites has become at times vague. The case history of sediment Drift 7 off the Antarctic Peninsula is analysed in this paper in the light of newly acquired swath bathymetry data. The co-existence of various sedimentary processes is reflected in a complex morphology: erosional gullies produced by debris flows on the upper part of the continental slope; deeply incised channels at the slope base; main trunk-type inter-drift turbidity channels separating the drifts; slide scars; undulating depositional bedforms interpreted as bottom-current sediment waves; fluid escape structures perhaps associated with deep-water coral bioherms. The data suggest that Drift 7 is a genuine sediment drift in which bottom currents pirate the sediment of the turbidity currents. Finally, we propose that the control on location and elongation of the drift is inherited from an older margin structure. The relationships between bottom current and deposition are investigated through a comparison with the SE Greenland continental margin, an analogous counterpart in the northern hemisphere.

Contourites are sediments deposited or substantially reworked by the action of bottom currents (see recent summaries by Stow & Faugères 1993, 1998; Faugères *et al.* 1999; Rebesco & Stow 2001; Stow *et al.* 2002*a*; Rebesco 2005). The term was originally introduced by Heezen *et al.* (1966) to define sediments that were deposited in the deep sea by contour currents (i.e. alongslope thermohaline bottom currents). The term was subsequently widened to embrace a larger spectrum of sediments in a wide range of water depths such as on continental shelves and slopes (e.g. Camerlenghi *et al.* 2001; Harris *et al.* 2001; Laberg *et al.* 2001), in depositional environments such as lakes (e.g. Ceramicola *et al.* 2001), formed by different type of currents (e.g. tides along canyons, Shanmugam 2003), and including newly defined sedimentary processes (e.g. sea-floor polishing, Viana *et al.* 1998). The definition of contourites applies to very large parts of the modern ocean floor and to a variety of sedimentary deposits. The present-day discussion on contourites also embraces some terminological issues. Although possibly too ample, the long-established term 'contourites' should be maintained (at least for non-specific use). Strictly speaking, the term 'contour currents' should be applied only to contour-parallel thermohaline currents (and bottom currents in other cases). The term 'contourite drifts' should be restricted to sediment accumulations formed principally by contour currents, and 'sediment drifts' for those formed principally by other types of bottom current. Depth definition has to be considered where available, but it is often of limited value for fossil contourites. Mixed systems are the norm rather than the exception for contourites.

If applied to contourite systems, the terms closely associated with downslope processes (such as fan, channel, etc.) have to be preceded by the prefix 'contourite'. The employment of modifying terms to qualify the deposit in terms of current action, depth, current type and interacting process would hence be recommended (see Rebesco 2005).

From: Viana, A. R. & Rebesco, M. (eds) *Economic and Palaeoceanographic Significance of Contourite Deposits.* Geological Society, London, Special Publications, **276**, 95–110.
0305-8719/07/$15.00

Importance of contourites

The study of contourites is important in at least three respects: palaeoceanography, hydrocarbon exploration, and slope stability.

The depositional environment of the sediment drifts generally allows a continuous record with relatively high resolution providing priceless information for palaeoceanography and palaeoclimatology, especially regarding variability in circulation pattern, current velocity, oceanographic history and basin interconnectivity (e.g. Grützner *et al.* 2003; Hernández-Molina *et al.* 2003).

Oil and gas reservoir development in the deep sea is generally considered a prerogative of the turbidite systems. Nevertheless, the action of bottom currents (or the interaction between bottom and turbidity currents) may be crucial where weak flows favour the accumulation of source rocks and robust flows form 'clean' sands (e.g. Shanmugam 2000; Viana 2001, 2007).

Nowadays, the critical role played by fine-grained contourites in slope stability is being perceived. Potential overpressured glide planes may be provided when the high water content of these sediments is loaded, and escape is prevented by low permeability. The most representative examples lie in areas of episodically high sediment input of glacigenic sediments, such as the Norwegian margin (Laberg & Vorren 2000; Laberg *et al.* 2001, 2002), although the Antarctic Peninsula sediment drifts also appear to be prone to gravitational instability (Volpi *et al.* 2003).

Despite their significance, contourites are still relatively poorly known, especially because the nature of these complex deposits is not clear cut, and because of the dominance of the turbidite paradigm. The distinction between contourites and turbidites (Fig. 1) appears reasonably easy when the general, large-scale aspects of the end-members are considered (e.g. Faugères *et al.* 1999; Rebesco & Stow 2001; Stow *et al.* 2002*b*; Rebesco 2005). However, end-members are seldom found in the deep sea, where interaction of processes is the norm rather than the exception. For this reason, interesting insights may arise from the case history of the sediment Drift 7 off the Antarctic Peninsula (Fig. 2), where hybrid contourites, with unusual well-preserved lamination, are recognized (Lucchi & Rebesco 2006). The Antarctic Peninsula drifts are well-studied examples (McGinnis & Hayes 1995; McGinnis *et al.* 1997; Rebesco *et al.* 1997, 2002; Pudsey 2000; Lucchi *et al.* 2002; Dowdeswell *et al.* 2004; Amblas *et al.* 2006) that are relevant to palaeoceanography (Grützner *et al.* 2004; Hernández-Molina *et al.* 2004) and slope stability (Volpi *et al.* 2003). Moreover, they constitute a complement for hydrocarbon exploration by providing instructive comparison for exploration–production cases.

In this paper we analyse the detailed morphology of Drift 7 and integrate the pre-existing lithological, seismic and oceanographic data with the aim of contributing to the understanding of contourite system behaviour, with a special focus on the interaction of down- and along-slope processes.

Dataset and knowledge development

The sediment Drift 7 (Fig. 2), also named the Alexander sedimentary mound (McGinnis & Hayes 1995), was initially identified on the basis of multichannel seismic reflection profiles (Rebesco *et al.* 1994). Notwithstanding the evidence for both turbidite and contourite processes in the development of this and similar adjacent sedimentary bodies (McGinnis & Hayes 1995; McGinnis *et al.* 1997), Drift 7 was interpreted as a sediment drift according to the large-scale asymmetric geometry of the sediment wave type, in contrast to the symmetrical gull-wing geometry typical of channel–levee systems (Rebesco *et al.* 1996). The analysis of six seismostratigraphic sequences allowed the identification of three main evolutionary stages of the sedimentary history of the continental rise from the Mid-Miocene to the present (Rebesco *et al.* 1997). Successively, oceanographic data confirmed the previously inferred present-day occurrence in this area of a weak, SW-flowing, contour-following bottom current (Camerlenghi *et al.* 1997*a*; Giorgetti *et al.* 2003). The compilation of existing soundings in the area (Rebesco *et al.* 1998) allowed a better, but still rough, definition of the bathymetry and morphology of the drift and of the relationships with the adjacent morphological elements (deep-sea channels, prograding slope wedges and glacial shelf troughs). Core data allowed the detection in the Pleistocene of two alternating lithological facies. That deposited during interglacials consists of brown, bioturbated, diatom-bearing mud with Foraminifera. Conversely, that deposited during glacial stages consists of grey mud, essentially barren and non-bioturbated, with very fine laminated intervals without the appreciable normal grading and/or the systematic succession of sedimentary facies typical of turbidites (Pudsey & Camerlenghi 1998; Pudsey 2000; Lucchi *et al.* 2002; Villa *et al.* 2003). Scientific drilling (Ocean Drilling Program (ODP) Leg 178) showed that the alternation of lithofacies is less regular and evident in the Pliocene and Miocene than in the Pleistocene, and that silty and muddy turbidites are present within the glacial facies in the distal part of sediment Drift 7 (Barker & Camerlenghi 2002). The low energy of this depositional environment is confirmed by the fine grain size throughout.

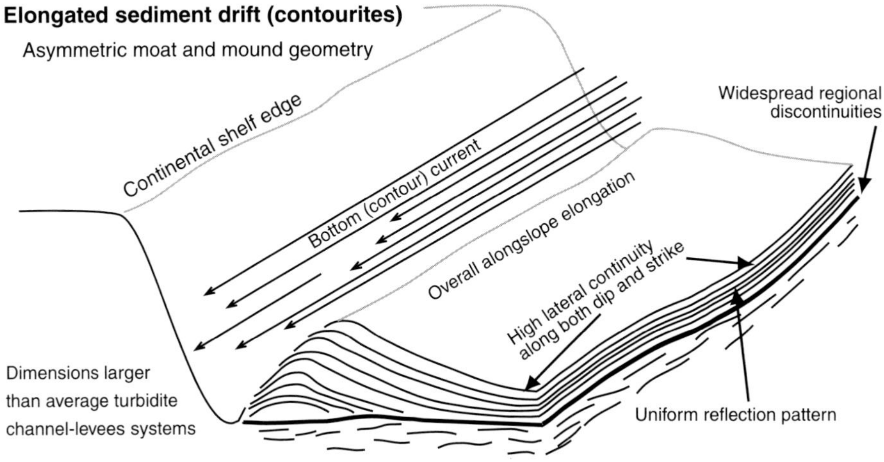

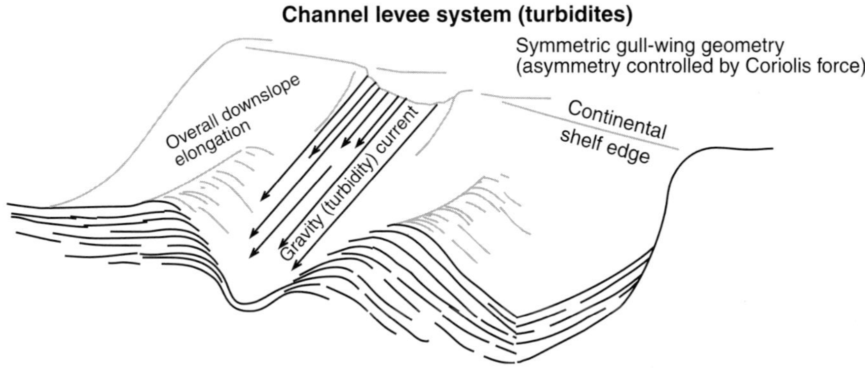

Fig. 1. Schematic model showing ideal, large-scale differences between sediment drifts and channel–levee systems (modified after Rebesco 2005; with permission from Elsevier).

The general depositional model was proposed by Rebesco *et al.* (1997) and it is still widely accepted (Amblas *et al.* 2006). It implies an interpretation of sediment Drift 7 as the product of weak SW-flowing bottom currents pirating (and depositing downstream) the suspended fine material from turbidity currents flowing during glacial stages in the adjacent channel to the NE. Enhanced deposition of the entrained sediment on the NE, gentler, upstream side facing the feeder channel (and slope instability on the SW, steeper, downstream side) generates a moderate upstream migration of the asymmetric drift (similarly to a huge sediment wave).

Methods

New swath bathymetric and sub-bottom acoustic data were acquired during the IX Antarctic Geophysical Cruise, which was part of the XIX expedition of PNRA (Programma Nazionale di Ricerche in Antartide). The MAGICO (Multibeam Antarctic Glacial system Integral COverage) cruise took place on the R.V. *OGS-Explora* from Ushuaia (Argentina), from 17 January, to the same port on 17 February 2004. The swath bathymetry of the whole sediment Drift 7 (nearly 37 000 km^2) and over 4000 km of sub-bottom profiles (Fig. 2) were collected using a Reson SeaBat 8150 multibeam echosounder and a Benthos CAP-6600 Chirp sonar, respectively.

The ship speed during acquisition was generally about 8 knots (*c.* 15 km/h), but it varied depending on weather and sea conditions. The 12 kHz multibeam system employed 234 beams with a total swath of 150°. The resulting horizontal resolution varied with the water depth and the ship speed.

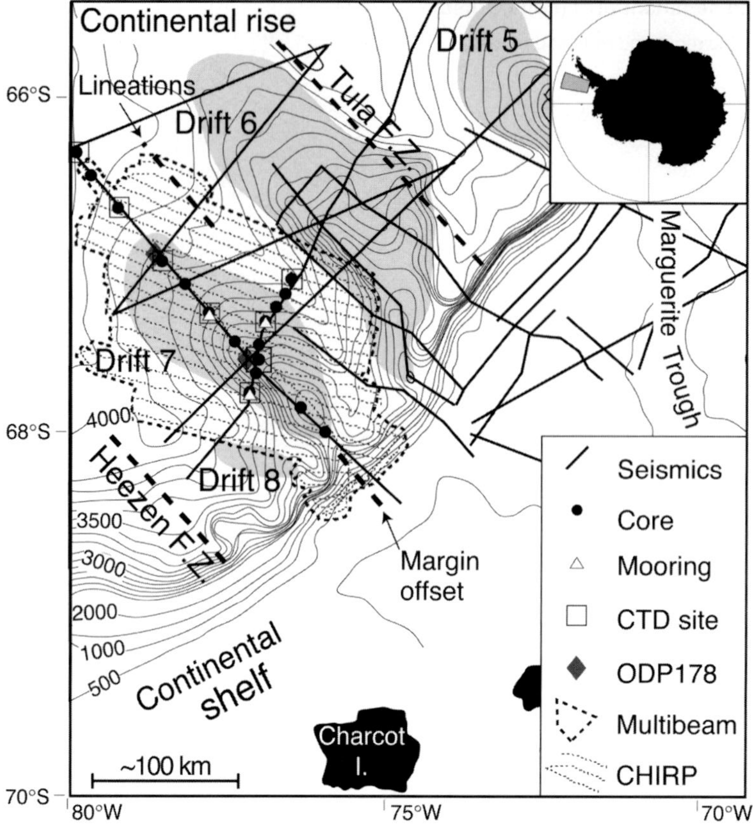

Fig. 2. Position map of the case study of sediment Drift 7 with location of PNRA data.

The maximum resolution attained during the survey in the deepest water depth (over 4000 m depth) was about 100 m on average (depending on sea and weather conditions and the ship's speed). The sub-bottom profiler employed 16 transducers with a sweep of 2–7 kHz. The trigger rate was 8 s and the recording rate 8192 records per ping (1024 record s^{-1}). The swath bathymetry data were first edited onboard to produce preliminary maps and then reprocessed at the Istituto Nazionale di Oceanografia e di Geofisica Sperimentale (OGS). Only display-during-acquisition (both on paper and on screen) was performed onboard for sub-bottom data, which were then replotted at OGS using the SEISPRO software.

Morphology

A shaded relief display of the multibeam dataset collected on sediment Drift 7 (Fig. 3) reveals a complex morphology, which provides new insights into the recent sedimentary processes that have shaped this part of the Antarctic Peninsula Pacific margin.

Continental slope

As in many other parts of the Antarctic Peninsula Pacific margin (e.g. Ambals *et al.* 2006), the continental slope is very steep (up to >13°), and it is not incised by any canyons or major channels or by recent large-scale slide scars. Conversely, similarly to the Marguerite Bay margin (Dowdeswell *et al.* 2004), a series of large straight subparallel gullies cut the upper slope (Fig. 3) just beyond the shelf edge. Gullies are V-shaped, about 100 m deep on average and up to about 500 m wide. At their down-slope mouth, at about mid-slope they fade into a smooth sea floor where there is no evidence, at the scale of this survey, of sediment accumulation (such as debris lobes or mid-slope fans) resulting from the deposition of sediments transported in the gullies. The base of the continental slope has a rough topography, being incised by a large number of channels within a dendritic network.

The channels converge in the upper rise to form single, trunk-type, deep-sea channels such as the Alexander and Charcot channels (Fig. 3). An analogous pattern was described in the northern part of the Pacific margin of the Antarctic Peninsula on the basis of swath bathymetry (Rebesco *et al.* 2002) and long-range sidescan sonar data (Tomlinson *et al.* 1992).

The MAGICO survey outlines the presence of a sharp offset of the continental slope facing Drift 7 (Fig. 3). The straight slope, trending NE–SW for the length imaged in the survey, is offset by about 3 km for its entire extent from its base to the shelf edge.

The proximal part of the drift

The SE, proximal part of the drift approximately occupies the upper continental rise and includes the transition between the sediment drift and the continental slope (Fig. 3). Closer to the slope, the drift undergoes erosion on both sides through a dendritc pattern of small channels converging in two major the channels, the Charcot and Alexander channels, bounding the drift to the SW and NE, respectively. Further oceanwards, the drift enlarges and develops the typical asymmetric cross-section, with a SW rough, eroded side and a NE smooth, depositional side. Two arcuate slide scars occur on the NE gentle side, close to the top of the drift. One of the scars erodes the sediment surface up to the drift crest. The slide deposit is not obvious in the morphology or in the seismic record, although sediment blocks are present downslope from the scars. It is clear that any sedimentary product of these slides has fed the Alexander Channel. At the moment we have no elements to determine age and volume of these slides, but the freshness of the morphology suggests a relatively recent age.

Alexander Channel

The depressed area between drifts 6 and 7 is incised by the Alexander Channel (Fig. 3). The channel originates at the base of the continental slope from the convergence of a dendritic network of tributary channels conveying sediments from the base of the continental slope as well as from the proximal parts of Drift 7 and, probably, Drift 6. In the upper rise the main channel is flat floored and about 10 km wide. About 150 km seaward from the base of the continental slope, in about 3800 m water depth, there is a slight change in the sea-floor gradient that delineates the transition between the upper and lower rise. Here the depression between the drifts widens and the channel loses its morphological expression. This is where the transition between confined and unconfined gravity flows is suggested to have occurred in the past (Lucchi *et al.* 2002; Diviacco *et al.* 2006). Chirp profiles and coring data suggest that this channel has been inactive recently, at least since the last interglacial period (Marine Isotope Stage 5, Lucchi *et al.* 2002). The Charcot Channel, bounding Drift 7 on the SW side, originates at the base of the continental slope similarly to the Alexander Channel and has a comparable width, but is shorter and more sinuous.

Mass wasting on the steep side of the drift

Widespread sediment mass wasting affects the steep side of Drift 7 in the form of several subparallel erosional channels and gullies cutting into the sediment drift sequence up to the sharp crest separating the SW steep flank from the NE gentler one (Figs 3 and 4). This crest is formed by coalescent slide scars that give it an overall regularly undulated shape with concavity consistently oriented towards the SW. Sediment drainage is towards the Charcot Channel.

Undulating depositional bedforms on the gentler side

The gentle NE side of the drift is characterized by the presence of sediment waves and previously unreported sedimentary ridges. The sediment wave field to the north of ODP Site 1095, already observed in seismic profiles (Camerlenghi *et al.* 1997*b*), is located in the distal part of the drift, separated from the Alexander Channel by an area about 30 km wide in which waves are not recognized. The geometry of the sediment waves is variable, with average height less than 20 m and length less than 2 km. The orientation, barely visible in the map of Figure 3, is SW–NE, subparallel to the sea-floor slope. Most (although not all) of them show an apparent upslope migration (Fig. 5). Depositional ridges (Figs 3, 6 and 7), which are 5–10 km wide, a few tens of kilometres long and up to 100 m high, occur with orientation oblique (SSW–NNE) to the drift main axis of elongation. The flanks are mostly even, or concave upwards, and no appreciable migration is visible in the sub-bottom profiles. They are subparallel, at times bifurcated, and they lie in the most elevated part of the drift. Their dimensions are peculiar, being smaller than those typical of sediment drifts and channel–levees, but definitely larger than those of turbidity current sediment waves, and just at the upper limit of those of bottom-current sediment waves. Bottom-current related bedforms of comparable dimensions have been described to occur as large sediment waves (Flood *et al.* 1993), as series of elongated mounds (e.g. Bulat & Long 2001), or within a multicrested drift (Lu *et al.* 2003).

Lineations beside the Alexander Channel

A series of subparallel NW–SE-trending lineations occurs beyond the distal part of the Alexander Channel (Fig. 3). Their origin is unclear. Some Chirp data across these lineations show the presence of sharp steps where the sea floor and the underlying reflectors are offset by a few tens of metres (see also the 18 kHz data of Fig. 8). Conversely, the seismic data crossing the southernmost part of these lineations show the presence of undulations of the sea floor about 1 km in wavelength and a few tens of metres high. Only the shallowest reflectors appear to be affected by these undulations (in the form of terminations and/or changes in amplitude of the reflectors), whereas the reflectors beneath are not affected.

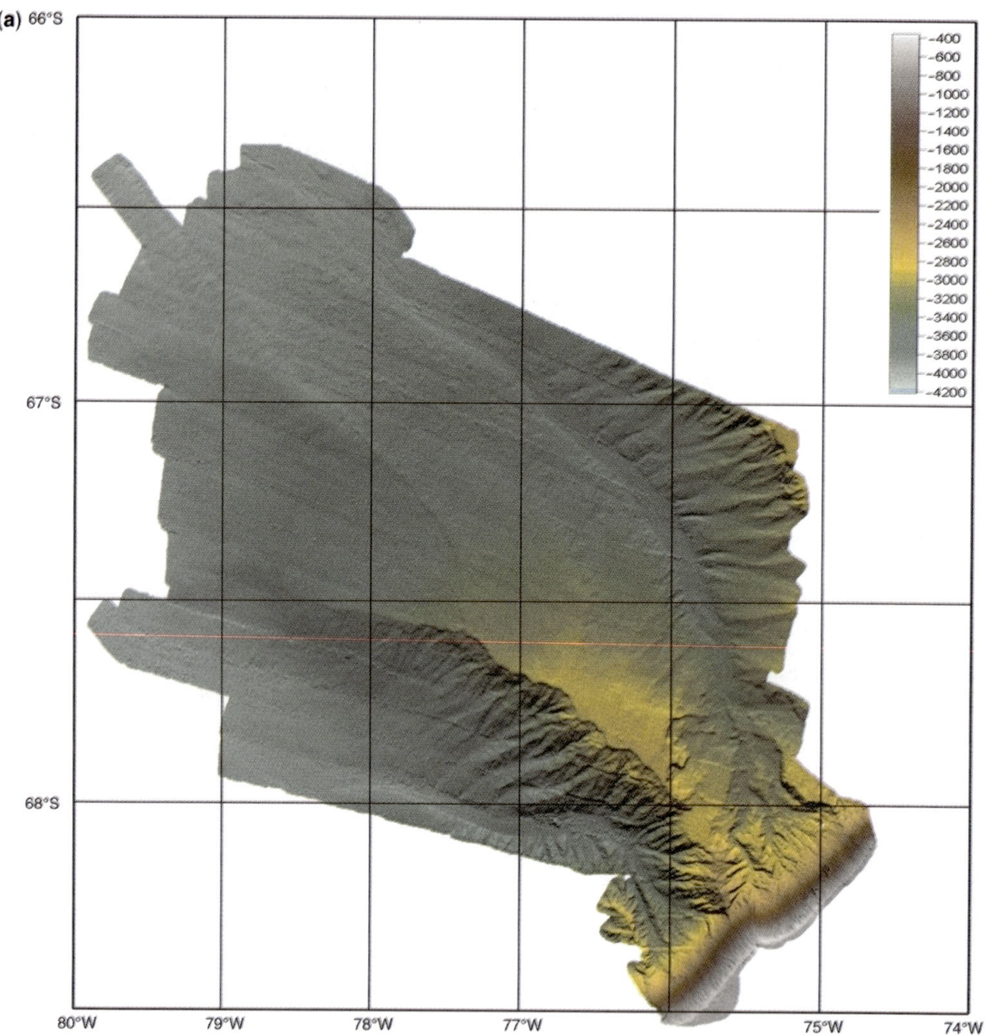

Fig. 3. Colour shaded relief bathymetric map produced from a regularly spaced 200 m grid on the basis of the multibeam data collected within the MAGICO project. Mercator projection, WGS 1984 spheroid, true-scale latitude 65°S, vertical exaggeration ×7.5, illumination from north. (**a**) Uninterpreted; (**b**) with superimposed interpretation and data location.

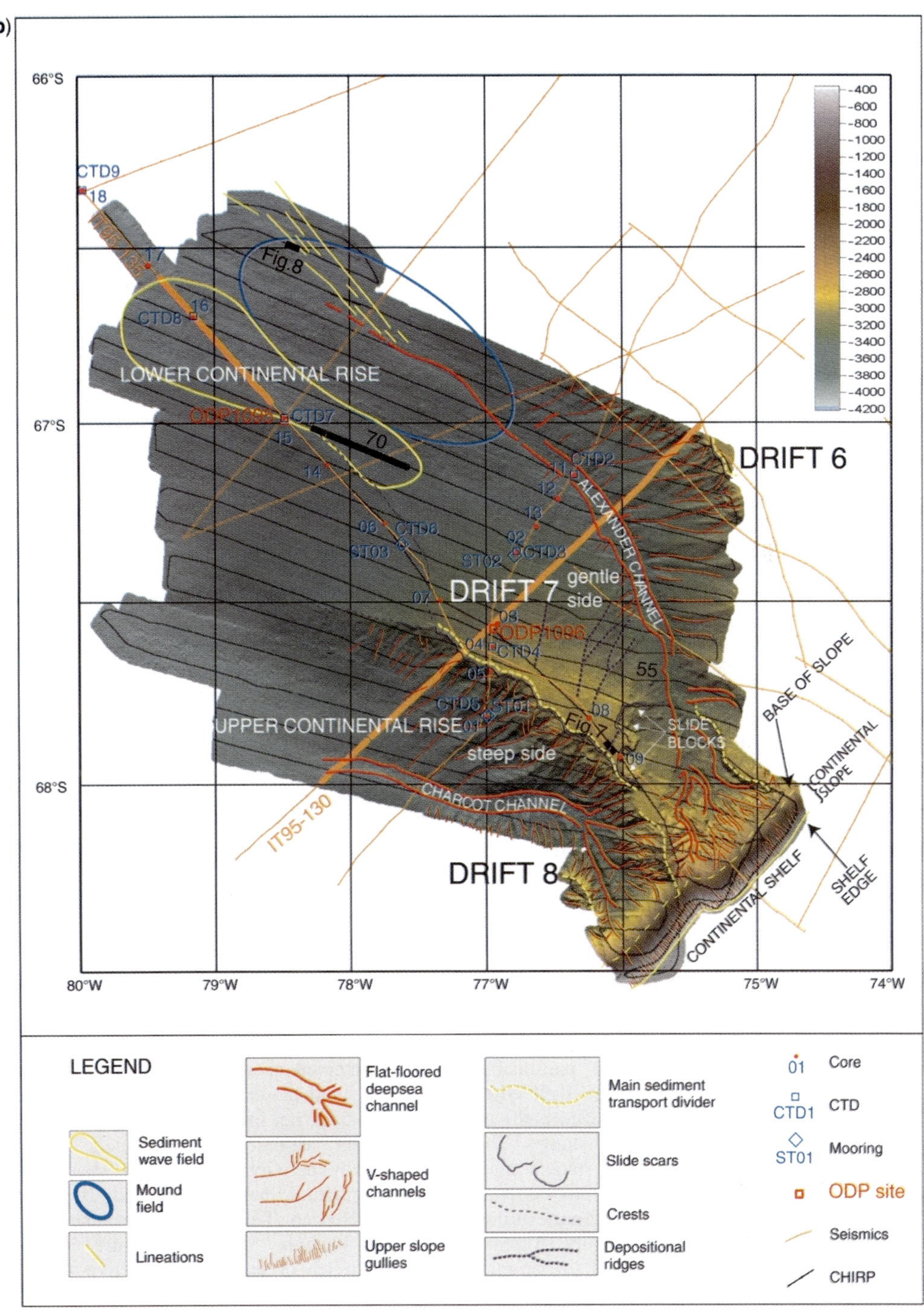

Fig. 3. *Continued*

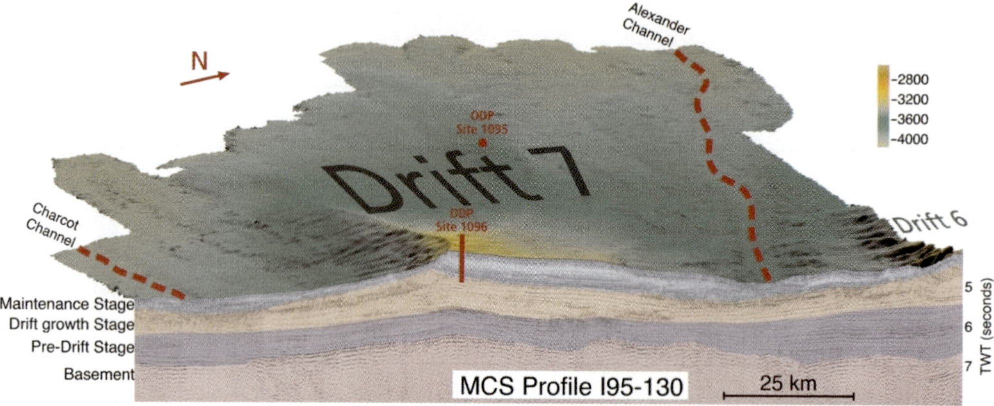

Fig. 4. Perspective view of multibeam data from the distal part of Drift 7, cut by seismic profile I95-130A (location shown in Fig. 3b). TWT, two-way travel time.

Acoustically transparent, cone-shaped mounds

In the deepest part of the MAGICO survey, a number of acoustically transparent, cone-shaped mounds were imaged on both sub-bottom profiler and 18 kHz echosounder data (Fig. 8). The mounds are variable in dimensions and geometry, but are generally subcircular, several hundred metres wide and a few tens to over 70 m high. They were observed to occur in two different settings: aligned along the faults beside the Alexander Channel and scattered in a vast field in the most distal part of Drift 7 (Fig. 3).

Interpretation and discussion

Downslope sediment transfer on the Antarctic Peninsula Pacific margin is inferred to occur by numerous small-scale slumps on the upper slope, and debris flows and turbidity currents on the lower slope and continental rise (Larter & Cunningham 1993; Rebesco *et al.* 1997). The gullies on the upper slope (Fig. 3) are therefore the expression of sediment transport from the shelf, occurring presumably during stages of ice grounding at the shelf edge. They may be interpreted as channelized paths of debris flows or as a series of small slumps. The disappearance of the gullies in the smooth mid-slope may suggest that the erosive process ends there, whereas sediment transport continues further down to the base of the continental slope. Alternatively, channelized erosion from shelf to rise may have been continuous from gullies to deep-water channels earlier in the history of this margin, and the gullies on the mid-slope may have been infilled in later stages (during ice sheet retreat). None the less, the roughness at the base of slope suggests that the debris flows evolve downslope into density flows (turbidity currents) that erode the substrate to form deep-sea channels in a dendritic centripetal pattern. Moreover, it has been recently observed that the stiff glacial diamictons of the steep prograding wedge produced during the last 3 Ma of polar history of the margin show a significant downlap over a regional surface, eroding the underlying sediments deposited during the earlier polythermal history of the glacial margin (Rebesco *et al.* 2006). Channel incision at the base of the slope may hence be inferred to develop at the transition from the stiff glacial diamictons to the softer, more erodible polythermal sediment, which can be exposed, or subcropping only at the base of the slope.

The origin of the offset in the straight margin (Fig. 3) is puzzling and probably is not identifiable in the erosional processes affecting the base of the slope. In fact, either retrogressive erosion at the head of a deep-sea channel or a slide scar would have resulted in a semicircular scar with relatively sharp flanks. An interpretation of the displacement of the slope as one flank of a mega slide scar would require the existence of an opposite flank and an intervening semicircular shape. Furthermore, the available deep penetration multichannel seismic data do not favour this hypothesis.

A recent tectonic origin appears unlikely, because any activity in this area should have ceased sometime between 30 and 20 Ma ago when the Heezen–Tula ridge-crest segment of the Antarctic–Phoenix spreading centre migrated into the trench of the formerly active Antarctic Peninsula margin (Larter *et al.* 1997).

More reasonably, this offset could be derived by horizontal displacement between plates produced between 30 and 20 Ma ago. The dextral

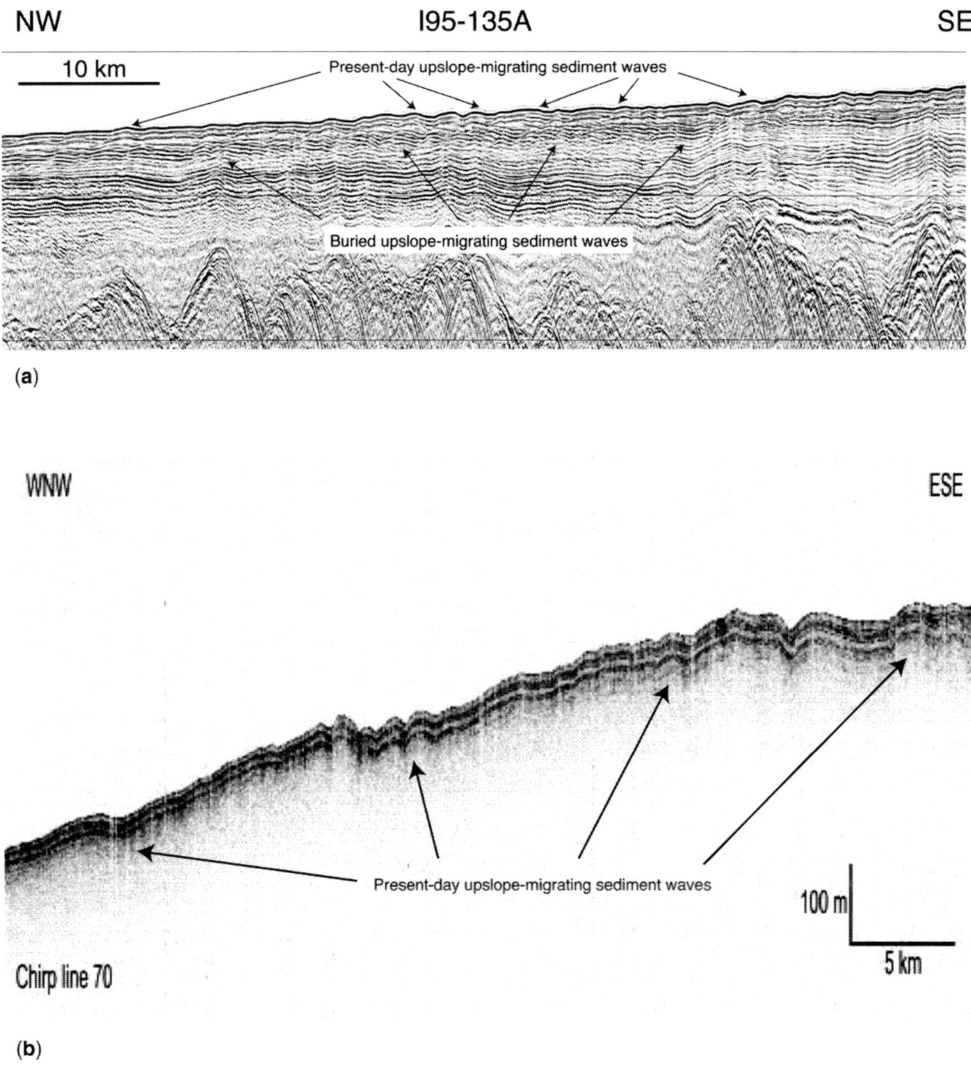

Fig. 5. Examples of buried and outcropping upslope-migrating sediment waves. (**a**) Seismic profile I95-135A; (**b**) sub-bottom profile 70. (Locations shown in Fig. 3b.)

displacement and the orientation of the fault plane (roughly perpendicular to the margin) are consistent with the orientation and horizontal displacement of the Antarctic Plate fracture zones. Moreover, the presence of a fracture zone in the oceanic basement below Drift 7 was suggested by deep penetration multichannel seismic data showing moderate oceanic basement offsets (McGinnis & Hayes 1995). The lineations beyond the distal part of the Alexander Channel correspond to this fracture zone and are oriented subparallel to it and to the Tula and Heezen fracture zones (Fig. 2). These lineations, if interpreted as faults, may have been produced by remobilization, subsequent to debris-flow loading, of the sedimentary succession overlying this fracture zone (Diviacco *et al.* 2006). Alternatively, they may be interpreted as gravity faults at the base of Drift 6 induced by underconsolidation of biogenic silica-containing sediments (Volpi *et al.* 2003), or as a result of unknown neo-tectonic deformation. Nevertheless, the seismic data available in the southern part of these lineations do not show any displacement of the underlying sedimentary succession, hence suggesting a sedimentary origin as flow structures related to the Alexander Channel. It is also possible

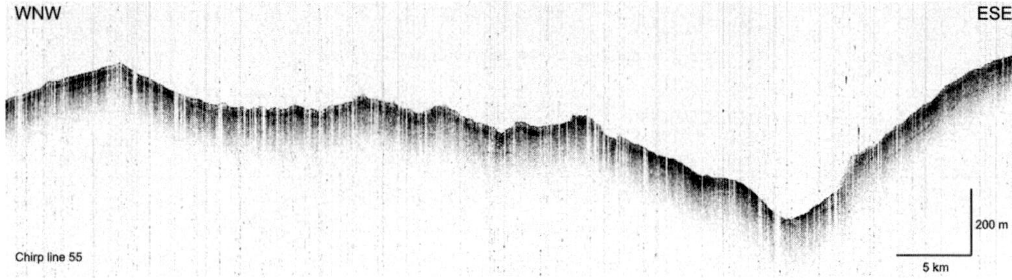

Fig. 6. Sub-bottom Profile 55 (location shown in Fig. 3b) showing the depositional ridges on the gentler side of Drift 7, elevated above the adjacent thalweg of the Alexander Channel.

that some of these lineations (the southern ones) are of sedimentary origin and others (the northern ones) are of structural origin.

The present margin offset, which lies approximately on the landward projection of the fracture zone identified by McGinnis & Hayes (1995) and of the lineations beyond the Alexander Channel (Fig. 3), may hence be reasonably considered as the remnant of a larger ancestral offset at the ocean–continental transition, now buried by the Plio-Pleistocene glacial wedge development.

The presence of such an offset and of the fracture zone may be relevant for the understanding of the processes that led to the drift growth in the Miocene. A comparison may be attempted with other changes in the margin's trend that have been considered relevant in controlling the direction of elongation of sediment drifts. This direction is generally subparallel to the margin, but the development of drifts with a trend perpendicular to the margin may result in response to a change in the margin's trend (Faugères *et al.* 1999), as in the case of the Eirik drift (Arthur *et al.* 1989) or the Greater Antilles Outer Ridge (Tucholke & Ewing 1974). Moreover, the orientation of Drift 7 may have been controlled by the underlying fracture zone, similarly to the drift that developed corresponding to the subducted oceanic slab off Peter I Island in the Bellingshausen Sea (Nitsche *et al.* 2000). The answer to the hitherto unsolved questions regarding location and orientation of Drift 7 (whose contouritic origin was challenged by some because of its elongation perpendicular to the margin) may lie in the presence of the offset of the margin and/or of the fracture zone.

The lesson that may come from the more than 10 years of persistent research surrounding Drift 7 is that a reliable understanding of deep-sea processes may come only from complementary datasets and multidisciplinary approaches. Similarly, the answer to the question of the direction of elongation and location of the other drifts may come after the acquisition of additional, more detailed datasets.

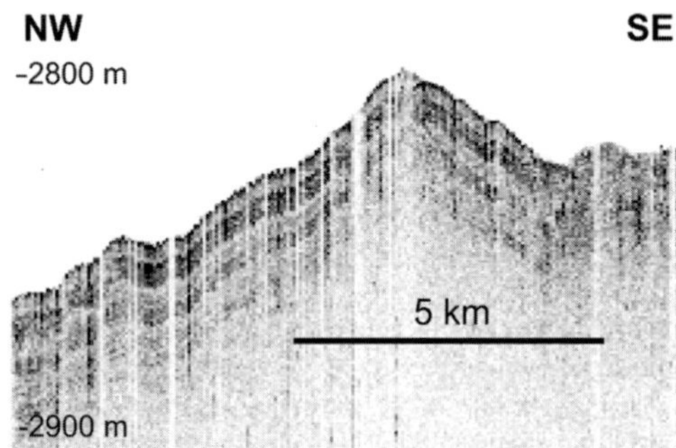

Fig. 7. Sub-bottom Profile 04 (location shown in Fig. 3b) showing how sub-bottom reflections are continuous and parallel to the sea floor on both sides of the ridges. The ridges are therefore interpreted as depositional features rather than being related to fault activity or to erosional processes.

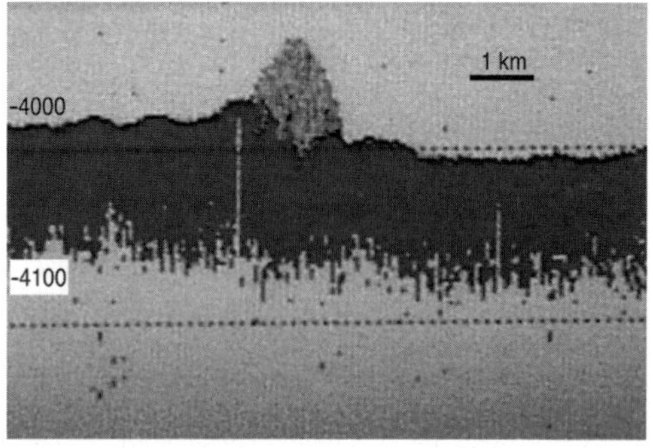

Fig. 8. An 18 kHz echosounder profile showing a conical diffraction about 75 m high and several hundreds of meters wide (location shown in Fig. 3b).

The very rough morphology of the proximal part of the drift (Fig. 3) clearly suggests that it is affected by sediment mass wasting that originated on the continental slope. However, the smooth, depressed, depositional nature of the gentle side in this most proximal part of the drift strongly suggests reduced deposition with respect to the top of the drift in consequence of a flow of bottom current confined along slope by the Coriolis force (i.e. SW of the depositional ridge of Fig. 7). This part of the drift hence appears to result from the combined action of reduced deposition by a bottom current, and sediment mass wasting, with the former being more evident south of the top of the drift and the latter prevailing closer to the slope. Therefore the action of bottom currents is one of the most effective processes on the continental rise, masked by mass wasting only in the most proximal part of the drift.

The erosion of the steep, SW side of the drift (Fig. 3) was attributed before the MAGICO survey to the gullies and small tributary channels feeding the Charcot Channel (Rebesco *et al.* 2002). In other words, the process controlling the morphology of this side was inferred to be that operated by turbidity flows undercutting and deepening the base of the SW slope of the drift. However, this process is justified by swath bathymetry only in the proximal part of the SW side of the drift, where the gullies coalesce and reach the thalweg of the Charcot Channel. Conversely, the incisions affecting the distal part are too far and disconnected from the Charcot Channel for us to infer a genetic connection (Fig. 3). The reasons for the instability have hence to be sought in other processes.

By analogy with smaller-scale, fine-grained sediment waves that originated from bottom currents, non-deposition or erosion on the steep side of the drift may occur as a consequence of the greater flow speed (supposed to be about >16 cm s^{-1}) on the downstream side (Wynn & Stow 2002). The contrast between relief building by sediment deposition of the gentle side and non-deposition or erosion on the steep side favours the progressive steepening of the steep side until gravitational instability is generated. Drift 7 sediments, which contain large amounts of biogenic silica, have anomalously high porosities at depth (Hillenbrand & Fütterer 2001) because of the low compressibility of the skeletons of the siliceous microfossils (Volpi *et al.* 2003). The fine grain size and high porosity at depth favour a reduced shear resistance to applied stresses and enhanced gravitational instability. The average gradient of the steep slope of the drift is 2°, close to the 3–4° average slope gradient of low-latitude continental slopes at which sediment mass wasting occurs normally throughout the evolution of a continental margin. The occasional character of the erosion on the SW side is indicated by the non-steady migration of the crest of the initially symmetrical drift visible on seismic profiles (Fig. 4).

The causes of the instability are therefore inherent in the bottom-current processes that control the growth of the drift: (1) persisting enhanced accumulation of opal-rich fine-grained sediment on the upstream (NE) side; (2) reduced accumulation and occasional erosion on the downstream (SW) side. Both processes have produced the northeastward migration of the crest and the present asymmetric morphology.

The origin of the sediment failure on the proximal gentle side of the drift (Fig. 3), where the sea-floor gradient is in the range of 0–1°, appears to be, as in the case of the steep side, the high porosity at depth produced by siliceous microfossils (Volpi *et al.* 2003) combined with the relatively high accumulation rate, which has been shown to decrease from proximal to distal parts of the drift (Lucchi *et al.* 2002). The downslope dip of the depositional surfaces, which are parallel to the sea floor throughout the sedimentary succession, may also contribute to sliding along weak surfaces. The blocks visible within the slide scar close to the slide headwall suggest that the lower sea-floor gradient of these slides has resulted in a shorter run-out compared with the slides of the steep side of the drift. Probably, the rest of the slide sediments were transferred to the deepest part of the rise along with the density flows moving within the Alexander Channel.

According to their dimensions and characteristics, the sediment waves in the distal part of the drift (Figs 3 and 5) may be produced by both turbidity current and bottom-current processes. However, turbidity current-generated waves usually show greater regularity, a downslope trend of decreasing dimensions, and a relationship with the parent turbidity channel (Wynn & Stow 2002). As none of these criteria are observed here, a contour-current origin is preferentially envisaged. Moreover, because of their dimension, orientation and setting, we interpret the depositional ridges in the most elevated part of the drift (Figs 3, 6 and 7) as a novel type of sharp-crested large fine-grained bottom-current sediment wave.

It must be noted, however, that Volpi *et al.* (2003) described long-wavelength folds induced by creep of the sedimentary sequence of Drift 7 above the biogenic opal-A to opal-CT diagenetic boundary, outlined in seismic profile by a bottom-simulating reflector. We think that these folds, undoubtedly affecting a few hundred metres of sedimentary sections, may overlap and imprint the depositional ridges only SE of the location of ODP Site 1095.

Apart from the locally overlapping creep effect, the most likely process controlling the production of sediment waves on the distal NW side and depositional ridges on the gentle NE side is hence that of bottom currents.

The acoustically transparent, cone-shaped mounds in the deepest part of the MAGICO survey (Figs 3 and 8) are interpreted as fluid escape structures indicating the presence of overpressured fluids within the sedimentary column. In this preliminary phase of the study, where we await additional information from future investigation, the origin of the overpressured fluids can be generally identified in two types of mechanisms: (1) sediment compaction at the Opal A–CT transition at depth; (2) shallow depth sediment overloading by large debris-flow accumulations between drifts in the distal part of the continental rise. The former hypothesis was introduced by Volpi *et al.* (2003). The latter originated in a recent study through reprocessing of multichannel seismic reflection profiles between the distal part of Drifts 6 and 7, which identified large debris-flow deposits that originated during the Late Pliocene associated with seismic evidence of fluid mobilization and venting (Diviacco *et al.* 2006).

Some of these mounds may be alternatively interpreted as deep, cold-water coral bioherms, as their acoustic signature is similar to that of the Norwegian margin (e.g. Lindberg *et al.* 2006). However, framework or reef-building corals have not previously been described in similar water depths. The possibility that local favourable conditions are provided by the observed fluid escapes will form the focus of further research.

Relationship between bottom current and deposition

Considering the numerous features that suggest the predominant action of bottom-current processes, Drift 7 is interpreted as a genuine sediment drift in which bottom currents pirate the sediment of turbidity currents.

The newly collected swath and sub-bottom data confirm the interpretation of Drift 7 as having been produced by the interactions of dip- and along-slope processes. The model proposing that turbidity currents provide the sedimentary input and bottom currents control the geometry and morphology of the deposit is consistent with the new information. Turbidity currents alone cannot explain the undulating depositional bedforms on the gentler side (especially those towards its top), whereas bottom-current deflection produced by the margin offset and/or by the fracture zone may explain the dipslope elongation of the sedimentary body. Moreover, the mass wasting on the steep side is certainly not related to turbidity-current undercutting, but is probably inherent in the bottom-current controlled depositional process.

Given the essence of sediment Drift 7, it is interesting to analyse the relationship between bottom current and deposition and to attempt a comparison with an analogous high-latitude counterpart. Within the northern hemisphere many analogies are found on the SE Greenland continental margin, where a series of sedimentary mounds have been interpreted to result from the interplay between turbidity and bottom currents (Rasmussen *et al.* 2003). In both cases very large asymmetric sedimentary mounds separated by deep-sea channels lie within an overall alongslope-flowing contour current (Fig. 9). However, whereas the setting of the SE Greenland mounds is consistent with a channel–levee model in which the asymmetry may be explained by the Coriolis force alone (towards the right in the northern hemisphere), that of the Antarctic Peninsula drifts requires a bottom-current influence (as discussed by Rebesco *et al.* 1997). The stronger influence of turbidity currents in the case of the SE Greenland margin was acknowledged by Rasmussen *et al.* (2003). Moreover, in both cases the mounds are dipslope elongated, slightly oblique with respect to the trend of the margin (Fig. 9), but whereas the Antarctic Peninsula examples lean and prograde upstream (like huge sediment waves), the SE Greenland mounds lean and prograde downstream (like current-deflected levees). Similarly, the orientation of the steep side of the mound is opposite (downstream in the Antarctic Peninsula and upstream in SE Greenland).

This comparison hence suggests that the sediment drifts of the Antarctic Peninsula, although similar in shape to asymmetric channel–levees, behave differently and require the action of bottom currents for their growth. Moreover, a comparative examination of the geometry and morphology of deep-sea deposits may allow the assessment of the relative importance of bottom and turbidity currents.

Conclusions

(1) Studies in the Antarctic Peninsula Pacific margin confirm once again that a reliable understanding of deep-sea processes may come only from complementary datasets and multidisciplinary approaches, and that the assessment of the relative importance of bottom and turbidity currents may be effectively assisted by an accurate examination of the geometry and morphology of deep-sea deposits.

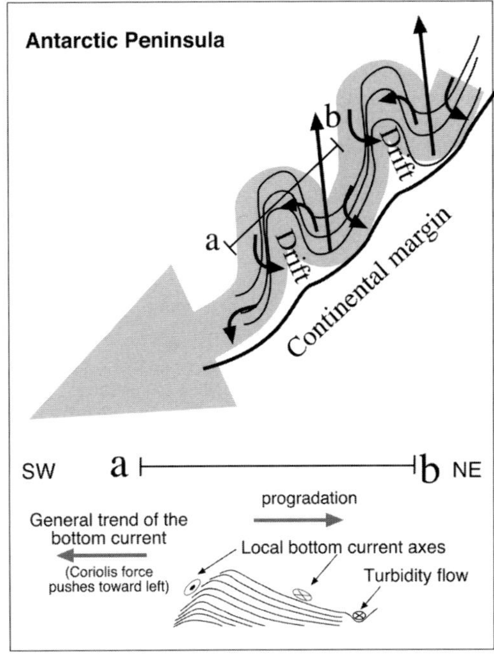

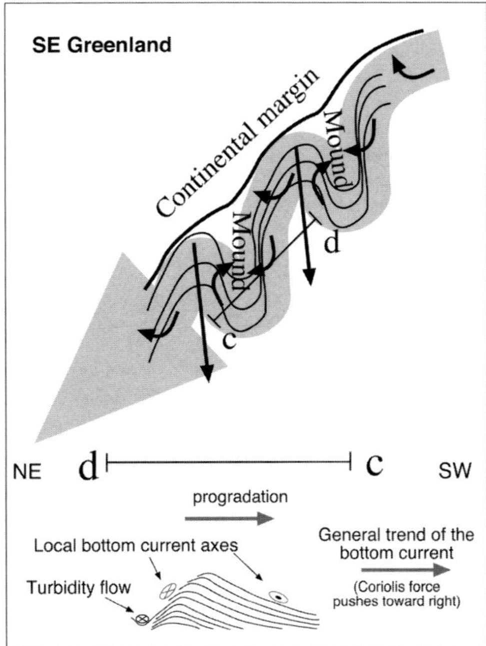

Fig. 9. Schematic diagram showing the comparison between the sedimentary mounds of the Pacific margin of the Antarctic Peninsula and those of the SE Greenland margin. The wide grey arrow shows the general path of the bottom current and the small black arrows show the deviating trend of the Coriolis force.

(2) In particular, the genuine sediment drift nature of Drift 7, in which bottom currents pirate and control the deposition of the sediment of turbidity currents, is suggested by a number of features: (a) the location and elongation perpendicular to the margin of Drift 7 can be explained by the presence of a fracture zone beneath it and of an associated offset in the margin in front of it; (b) the geometry of the proximal, depressed part of the drift suggests the combined action of reduced deposition by bottom currents confined along slope by the Coriolis force, and sediment mass wasting; (c) the persisting accumulation on the upstream side and reduced accumulation or occasional erosion on the downstream side (inherent in the bottom-current process) have produced the asymmetric morphology and caused of the instability of the steep downstream side; (d) sediment waves on the distal NW side and depositional ridges on the gentle NE side are probably produced by bottom currents.

(3) Additional features imaged by the multibeam survey are: the whole extent of the Alexander and Charcot channels; previously unreported lineations of enigmatic origin in the distal part of the drift; acoustically transparent, cone-shaped mounds in the distal part of the drift interpreted as fluid escape structures.

(4) Comparison with an analogous high-latitude counterpart suggests that the bottom currents appear to have a crucial influence on the geometry and progradation direction of Drift 7.

Further developments in the understanding of contourite system behaviour and of the interaction of down- and along-slope processes, relevant for palaeoclimatic studies and for oil exploration, will come from comprehensive datasets from well-studied examples.

We thank the captain and crew of the R.V. *OGS-Explora* for their help during the field data acquisition. Funding for this study was provided by PNRA (Programma Nazionale di Ricerche in Antartide) through the MAGICO (Multibeam Antarctic Glacial system Integral COverage) project. This work is also a contribution to the EC Research and Training Network EURODOM, which funded the activity of B.L. and C.N. We thank B. Kuvaas and C. Escutia for their critical suggestions, which helped improve the initial manuscript. We are also grateful to L. Gasperini for providing a preliminary version of the SEISPRO software used to replot the sub-bottom record.

References

AMBLAS, D., URGELES, R., CANALS, M., CALAFAT, A. M., REBESCO, M., CAMERLENGHI, A., ESTRADA, F., DE-BATIST, M. & HUGHES-CLARKE, J. E. 2006. Relationship between continental rise development and palaeo-ice sheet dynamics, northern Antarctic Peninsula Pacific margin. *Quaternary Science Reviews*, **25**, 933–944.

ARTHUR, M. A., SRIVASTAVA, S. P., KAMINSKI, M., JARRARD, R. & OSLER, J. 1989. Seismic stratigraphy and history of deep circulation and sediment drift development in Baffin Bay and the Labrador Sea. *In*: SRIVASTAVA, S. P., ARTHUR, M. A., CLEMENT, B. M., *ET AL.* (eds) *Proceedings of the Ocean Drilling Program, Scientific Results, 105*. Ocean Drilling Program, College Station, TX, 957–988.

BARKER, P. F. & CAMERLENGHI, A. 2002. Synthesis of Leg 178 results: glacial history of the Antarctic Peninsula from Pacific margin sediments. *In*: BARKER, P. F., CAMERLENGHI, A., ACTON, G. D. & RAMSAY, A. T. S. (eds) *Proceedings of the Ocean Drilling Program, Scientific Results, 178*. [Online.] World Wide Web address: http://www-odp.tamu.edu/publications/178_sr/volume/synth/synth.pdf.

BULAT, J. & LONG, D. 2001. Images of the seabed in the Faroe–Shetland Channel from commercial 3D seismic data. *Marine Geophysical Researches*, **22**, 345–367.

CAMERLENGHI, A., CRISE, A., PUDSEY, C. J., ACCERBONI, E., LATERZA, R., & REBESCO, M. 1997*a*. Ten-month observation of the bottom current regime across a sediment drift of the Pacific margin of the Antarctic Peninsula. *Antarctic Sciences*, **9**, 424–431.

CAMERLENGHI, A., REBESCO, M. & PUDSEY, C. J. 1997*b*. High resolution terrigenous sedimentary record of a sediment drift on the Antarctic Peninsula Pacific margin (initial results of the 'SEDANO' Program). *In*: RICCI, C. A. (ed.) *The Antarctic Region: Geological Evolution and Processes, Proceedings of the VII ISAES (International Symposium of Antarctic Earth Sciences), Siena (Italy), August 1995*. Terra Antartica, Siena, 705–710.

CAMERLENGHI, A., DOMACK, E., REBESCO, M., *ET AL.* 2001. Glacial morphology and post-glacial contourites in northern Prince Gustav Channel (NW Weddell Sea, Antarctica). *Marine Geophysical Researches*, **22**, 417–443.

CERAMICOLA, S., REBESCO, M., DE BATIST, M. & KHLYSTOV, O. 2001 Seismic evidence of small-scale lacustrine drifts in Lake Baikal (Russia). *Marine Geophysical Researches*, **22**, 445–464.

DIVIACCO, P., REBESCO, M. & CAMERLENGHI, A. 2006. Late Pliocene mega debris flow deposit and related fluid escapes identified on the Antarctic continental margin by seismic reflection data analysis. *Marine Geophysical Researches*, **27**, 109–128.

DOWDESWELL, J. A., O'COFAIGH, C. & PUDSEY, C. J. 2004. Continental slope morphology and sedimentary processes at the mouth of an Antarctic palaeo-ice stream. *Marine Geology*, **204**, 203–214.

FAUGÈRES, J. C., STOW, D. A. V., IMBERT, P. & VIANA, A. 1999. Seismic features diagnostic of contourite drifts. *Marine Geology*, **162**, 1–38.

FLOOD, R. D., SHOR, A. N. & MANLEY, P. L. 1993. Morphology of abyssal mudwaves at Project MUDWAVES sites in the Argentine Basin. *Deep-Sea Research II*, **40**, 859–888.

GIORGETTI, A., CRISE, A., LATERZA, R., PERINI, L., REBESCO, M. & CAMERLENGHI, A. 2003. Water masses and bottom boundary layer dynamics above a sediment drift of the Antarctic Peninsula Pacific margin. *Antarctic Science*, **15**, 537–546.

GRÜTZNER, J., REBESCO, M., COOPER, A. K., FORSBERG, C. F., KRYC, K. A. & WEFER, G. 2003. Evidence for orbitally controlled size variations of the East Antarctic Ice Sheet during the late Miocene. *Geology*, **31**, 777–780.

GRÜTZNER, J., HILLENBRAND, C.-D. & REBESCO, M. 2004. Terrigenous flux and biogenic silica deposition at the Antarctic continental rise during the late Miocene to early Pliocene: implications for ice sheet stability and sea ice coverage. *Global and Planetary Change*, **45**(1–3), 131–149.

HARRIS, P. T., BRANCOLINI, G., ARMAND, L., *ET AL.* 2001. Continental shelf drift deposit indicates non-steady state Antarctic bottom water production in the Holocene. *Marine Geology*, **179**, 1–8.

HEEZEN, B. C., HOLLISTER, C. D. & RUDDIMAN, W. F. 1966. Shaping of the continental rise by deep geostrophic contour currents. *Science*, **152**, 502–508.

HERNANDEZ-MOLINA, J., LLAVE, E., SOMOZA, L. *ET AL.* 2003. Looking for clues to paleoceanographic imprints; a diagnosis of the Gulf of Cadiz contourite depositional systems. *Geology*, **31**, 19–22.

HERNANDEZ-MOLINA, J., LARTER, R., REBESCO, M. & MALDONADO, A. 2004. Miocene changes in bottom current regime recorded in continental rise sediments on the Pacific margin of the Antarctic Peninsula. *Geophysical Research Letters*, **31**, L22606.

HILLENBRAND, C.-D. & FÜTTERER, D. K. 2001. Neogene to Quaternary deposition of opal on the continental rise west of the Antarctic Peninsula, ODP Leg 178, Sites 1095, 1096, and 1101. *In*: BARKER, P. F., CAMERLENGHI, A., ACTON, G. D. & RAMSAY, A. T. S. (eds) *Proceedings of the Ocean Drilling Program, Scientific Results, 178*. [Online.] World Wide Web address: http://www-odp.tamu.edu/publications/178_SR/chap_23/chap_23.htm.

LABERG, J. S. & VORREN, T. O. 2000. The Traenadjupet Slide, offshore Norway; morphology, evacuation and triggering mechanisms. *Marine Geology*, **171**, 95–114.

LABERG, J. S., DAHLGREN, T., VORREN, T. O., HAFLIDASON, H. & BRYN, P. 2001. Seismic analyses of Cenozoic contourite drift development in the northern Norwegian Sea. *Marine Geophysical Researches*, **22**, 401–416.

LABERG, J. S., VORREN, T. O., MIENERT, J., ET AL. 2002. Late Quaternary palaeoenvironment and chronology in the Traenadjupet Slide area offshore Norway. *Marine Geology*, **188**, 35–60.

LARTER, R. D. & CUNNINGHAM, A. P. 1993. The depositional pattern and distribution of glacial–interglacial sequences on the Antarctic Peninsula Pacific Margin. *Marine Geology*, **109**, 203–219.

LARTER, R. D., REBESCO, M., VANNESTE, L. E., GAMBOA, L. A. P. & BARKER, P. F. 1997. Cenozoic tectonic, sedimentary and glacial history of the continental shelf west of Graham Land, Antarctic Peninsula. *In*: COOPER, A. K. & BARKER, P. F. (eds) *Geology and Seismic Stratigraphy of the Antarctic Margin, Part 2*. American Geophysical Union, Antarctic Research Series, **71**, 1–27.

LINDBERG, B., BERNDT, C. & MIENERT, J. 2006. The Fugløy Reef at 70°N; acoustic signature, geologic, geomorphologic and oceanographic setting. *International Journal of Earth Sciences*. doi:10.1007/s00531-005-0495-y.

LU, H., FULTHORPE, C. & MANN, P. 2003. Three-dimensional architecture of shelf-building sediment drifts in the offshore Canterbury Basin, New Zealand. *Marine Geology*, **193**, 19–47.

LUCCHI, R. G. & REBESCO, M. 2007. Glacial contourites on the Antarctic Peninsula margin: insight for palaeoenvironmental and palaeoclimatic conditions. *In*: VIANA, A. R. & REBESCO, M. (eds) *Economic and Palaeoceanographic Signifiance of Contourite Deposits*. Geological Society, London, Special Publications, **276**, 111–127.

LUCCHI, R. G., REBESCO, M., CAMERLENGHI, A., ET AL. 2002. Mid–Late Pleistocene glacimarine sedimentary processes of a high-latitude, deep-sea sediment drift (Antarctic Peninsula Pacific Margin). *Marine Geology*, **189**, 343–370.

MCGINNIS, J. P. & HAYES, D. E. 1995. The roles of downslope and along-slope depositional processes: Southern Antarctic Peninsula continental rise. *In*: COOPER, A. K., BARKER, P. F. & BRANCOLINI, G. (eds) *Geology and Seismic Stratigraphy of the Antarctic Margin*. American Geophysical Union, Antarctic Research Series, **68**, 141–156.

MCGINNIS, J. P., HAYES, D. E. & DRISCOLL, N. W. 1997. Sedimentary processes across the continental rise of the southern Antarctic Peninsula. *Marine Geology*, **141**, 91–109.

NITSCHE, F. O., CUNNINGHAM, A. P., LARTER, R. D. & GOHL, K. 2000. Geometry and development of glacial continental margin depositional systems in the Bellingshausen Sea. *Marine Geology*, **162**, 277–302.

PUDSEY, C. J. 2000. Sedimentation on the continental rise west of the Antarctic Peninsula over the last three glacial cycles. *Marine Geology*, **167**, 313–338.

PUDSEY, C. J. & CAMERLENGHI, A. 1998. Glacial–interglacial deposition on a sediment drift on the Pacific margin of the Antarctic Peninsula. *Antarctic Science*, **10**(3), 286–308.

RASMUSSEN, S., LYKKE-ANDERSEN, H., KUIJPERS, A. & TROELSTRA, S. R. 2003. Post-Miocene sedimentation at the continental rise of Southeast Greenland: the interplay between turbidity and contour currents. *Marine Geology*, **196**, 37–52.

REBESCO, M. 2005. Contourites. *In*: RICHARD, C. SELLEY, R. C., COCKS, L. R. M. & PLIMER, I. R. (eds) *Encyclopedia of Geology, Vol. 4*. Elsevier, Amsterdam, 513–527.

REBESCO, M. & STOW, D. A. V. 2001. Seismic expression of contourites and related deposits: a preface. *Marine Geophysical Researches*, **22**, 303–308.

REBESCO, M., LARTER, R. D., BARKER, P. F., CAMERLENGHI, A. & VANNESTE, L. E. 1994. The history of sedimentation on the continental rise west of the Antarctic Peninsula. *Terra Antarctica*, **1**, 277–279.

REBESCO, M., LARTER, R. D., CAMERLENGHI, A. & BARKER, P. F. 1996. Giant sediment drifts on the continental rise west of the Antarctic Peninsula. *Geo-Marine Letters*, **16**, 65–75.

REBESCO, M., LARTER, R. D., BARKER, P. F., CAMERLENGHI, A. & VANNESTE, L. E. 1997. The history of sedimentation on the continental rise west of the Antarctic Peninsula. *In*: COOPER, A. K. & BARKER, P. F. (eds) *Geology and Seismic Stratigraphy of the Antarctic Margin 2*. American Geophysical Union, Antarctic Research Series, **71**, 29–50.

REBESCO, M., CAMERLENGHI, A. & ZANOLLA, C. 1998. Bathymetry and morphogenesis of the continental margin west of the Antarctic Peninsula. *Terra Antarctica*, **5**, 715–725.

REBESCO, M., PUDSEY, C., CANALS, M., CAMERLENGHI, A., BARKER, P., ESTRADA, F. & GIORGETTI, A. 2002. Sediment drift and deep-sea channel systems, Antarctic Peninsula Pacific Margin. *In*: STOW, D. A. V., PUDSEY, C. J., HOWE, J. A., FAUGÉRES, J. C. & VIANA, A. R. (eds) *Deep-Water Contourite Systems: Modern Drifts and Ancient Series, Seismic and Sedimentary Characteristics*. Geological Society, London, Memoirs, **22**, 353–371.

REBESCO, M., CAMERLENGHI, A., GELETTI, R. & CANALS, M. 2006. Margin architecture reveals the transition to the modern Antarctic ice sheet ca. 3 Ma. *Geology*, **34**, 301–304.

SHANMUGAM, G. 2000. 50 years of the turbidite paradigm (1950s–1990s): deep-water processes and facies models—a critical perspective. *Marine and Petroleum Geology*, **17**, 285–342.

SHANMUGAM, G. 2003. Deep-marine tidal bottom currents and their reworked sands in modern and ancient submarine canyons. *Marine and Petroleum Geology*, **20**, 471–491.

STOW, D. A. V. & FAUGÈRES, J. C. (eds) 1993. Contourites and bottom currents. *Sedimentary Geology*, **82**.

STOW, D. A. V. & FAUGÈRES, J. C. (eds) 1998. Contourites, turbidites and process interaction. *Sedimentary Geology*, **115**.

STOW, D. A. V., FAUGÈRES, J. C., HOWE, J., PUDSEY, C. J. & VIANA. A. (eds) 2002*a*. *Deep-Water Contourite Systems: Modern Drifts and Ancient Series, Seismic and Sedimentary Characteristics*. Geological Society, London, Memoirs, **22**.

STOW, D. A. V., FAUGÈRES, J. C., HOWE, J., PUDSEY, C. J. & VIANA, A. 2002*b*. Bottom currents, contourites and deep-sea sediment drifts: current state of the art. *In*: STOW, D. A. V. & FAUGÈRES, J. C., HOWE, J., PUDSEY, C. J. & VIANA. A. (eds), *Deep-Water Contourite Systems: Modern Drifts and Ancient Series, Seismic and Sedimentary Characteristics*. Geological Society, London, Memoirs, **22**, 7–20.

TOMLINSON, J. S., PUDSEY, C. J., LIVERMORE, R. A., LARTER, R. D. & BARKER, P. F. 1992. Long-Range Sidescan Sonar (GLORIA) Survey of the Antarctic Peninsula Pacific Margin. *In*: YOSHIDA, Y., KAMINUMA, K. & SHIRAISHI, K. (eds) *Recent Progress in Antarctic Earth Science: 6th International Symposium on Antarctic Earth Science, Tokyo, Proceedings*, 423–429.

TUCHOLKE, B. E. & EWING, J. I. 1974. Bathymetry and sediment geometry of the Greater Antilles Outer Ridge and vicinity. *Geological Society of America Bulletin*, **85**, 1789–1802.

VIANA, A. 2001. Seismic expression of shallow- to deep-water contourites along the south-eastern Brazilian margin. *Marine Geophysical Researches*, **22**, 509–521.

VIANA, A. R., ALMEIDA, W. JR, NUNES, M. C. V. & BULHÕES, E. M. 2007. The economic importance of contourites. *In*: VIANA, A. R. & REBESCO, M. (eds) *Economic and Palaeoceanographic Signifiance of Contourite Deposits*. Geological Society, London, Special Publications, **276**, 1–24.

VIANA, A., FAUGÈRES, J. C., KOWSMANN, R. O., LIMA, J. A. M., CADDAH, L. F. G. & RIZZO, J. G. 1998. Hydrology, morphology and sedimentology of the Campos continental margin, offshore Brazil. *Sedimentary Geology*, **115**, 133–157.

VILLA, G., PERSICO, D., BONCI, M. C., LUCCHI, R. G., MORIGI, C. & REBESCO, M. 2003. Biostratigraphic characterization and Quaternary microfossil palaeoecology in sediment drifts west of the Antarctic Peninsula—implications for cyclic glacial–interglacial deposition. *Palaeogeography, Palaeoclimatology, Palaeoecology*, **198**, 237–263.

VOLPI, V., CAMERLENGHI, A., HILLENBRAND, C.-D., REBESCO, M. & IVALDI, R. 2003. The effect of biogenic silica on sediment consolidation and slope instability, Pacific margin of the Antarctic Peninsula. *Basin Research*, **15**, 339–354.

WYNN, R. B. & STOW, D. A. V. 2002. Classification and characterisation of deep-water sediment waves. *Marine Geology*, **192**, 7–22.

Glacial contourites on the Antarctic Peninsula margin: insight for palaeoenvironmental and palaeoclimatic conditions

RENATA G. LUCCHI[1,2] & MICHELE REBESCO[1]

[1]*Istituto Nazionale di Oceanografia e di Geofisica Sperimentale (OGS), Borgo Grotta Gigante 42/c, 34010 Sgonico (Trieste), Italy*

[2]*Present address: GRC Geociències Marines, Departament d'Estratigrafia, P. i Geociències Marines, Universitat de Barcelona, Facultat de Geologia, C/Martí i Franquès, s/n, E-08028 Barcelona, Spain (e-mail: rglucchi@ub.edu)*

Abstract: Deep-sea finely laminated and barren glacial sediments occur in the sediment drift field offshore the Pacific margin of the Antarctic Peninsula where a weak contour current flows at present to the SW. Atypical sedimentary facies were related to the coexistence and interaction of different sedimentary processes. Three 'end-members' of radiograph facies were defined to represent the sedimentary sequences controlled by a dominant process, as follows. (1) Direct influence of turbidity currents on sedimentation is observed in the area surrounding the Alexander Channel system with silty layers interbedded with laminated mud free of ice-rafted debris (IRD). (2) Distal meltwater turbid flows dominate the more proximal area of the top plateau with structureless and coarser-grained sediments containing IRD. (3) Along the crest of the drift, persistent weak bottom currents control the deposition of fine-grained sediments conveyed into the system through other processes. These laminated sediments contain IRD and are, atypically, not bioturbated, because of unusual, climatically related, environmental conditions of suppressed primary productivity and oxygen-reduced deep waters. These glacial contourites were observed on most of the Antarctic margin with the exception of the areas in which polynyas were maintained during the glacial stages. Glacial contourites can be used as a proxy to define temporal and spatial extension of the Antarctic sea-ice.

Deep-sea sediments of the sediment drifts offshore the Pacific margin of the Antarctic Peninsula were investigated within the SEDANO (SEdiment Drifts of the ANtarctic Offshore) and MOGAM (Morphology and Geology of Antarctic Margins (Antarctic Peninsula and Wilkes Land)) projects of the Italian Programma Nazionale di Ricerche in Antartide (Fig. 1). The continental shelf of the Pacific margin of the Antarctic Peninsula is dissected by a number of modern, U-shaped troughs that reflect the Neogene path of large ice-streams (Rebesco *et al.* 1998, 2002). Sediments across the continental shelf were transported mainly by ice-streams and delivered to the continental slope and rise during the periods of maximum glacial expansion. The continental rise hosts a number of large sediment drifts separated by dendritic deep-sea channel systems that originated on the lower continental slope in alignment with the reaches of each glacial trough (Amblas *et al.* 2006; Rebesco *et al.* 2007).

Two main alternating lithologies related to climatic changes were initially distinguished (Pudsey & Camerlenghi 1998; Pudsey 2000): interglacial brown bioturbated mud containing sparse ice-rafted debris (IRD), and glacial grey laminated mud. Further detailed studies consisting of over 1000 sedimentological, compositional and biostratigraphic analyses (diatoms, foraminifers, and nannofossils) allowed Lucchi *et al.* (2002*a*) to distinguish four distinctive sets of sedimentary facies connected to a sequence of four climatic stages, each characterized by the onset of typical sedimentary processes (Fig. 2), as follows. (1) Strongly bioturbated mud with abundant IRD and aeolian sediment input was produced during the interglacial phase (Fig. 2a). The increased bioactivity is matched by insignificant down- and along-slope processes. Negligible suspended load and sediment transport by the southwesterly flowing bottom currents is occurring at present on the continental rise (Giorgetti *et al.* 2003). (2) Terrigenous structureless mud including a series of close-spaced IRD layers was produced during the deglaciation (transition from glacial to interglacial conditions) by the disintegration of the retreating grounded ice sheet (Fig. 2b). The absence of calcareous microfossils is related to strong carbonate dissolution and rise of the carbonate compensation depth (CCD) as a result of major meltwater discharge, which in turn caused water stratification (with less oxygenated and nutrient-enriched conditions at the sea bottom). (3) A large variety of

From: VIANA, A. R. & REBESCO, M. (eds) *Economic and Palaeoceanographic Significance of Contourite Deposits*. Geological Society, London, Special Publications, **276**, 111–127.
0305-8719/07/$15.00

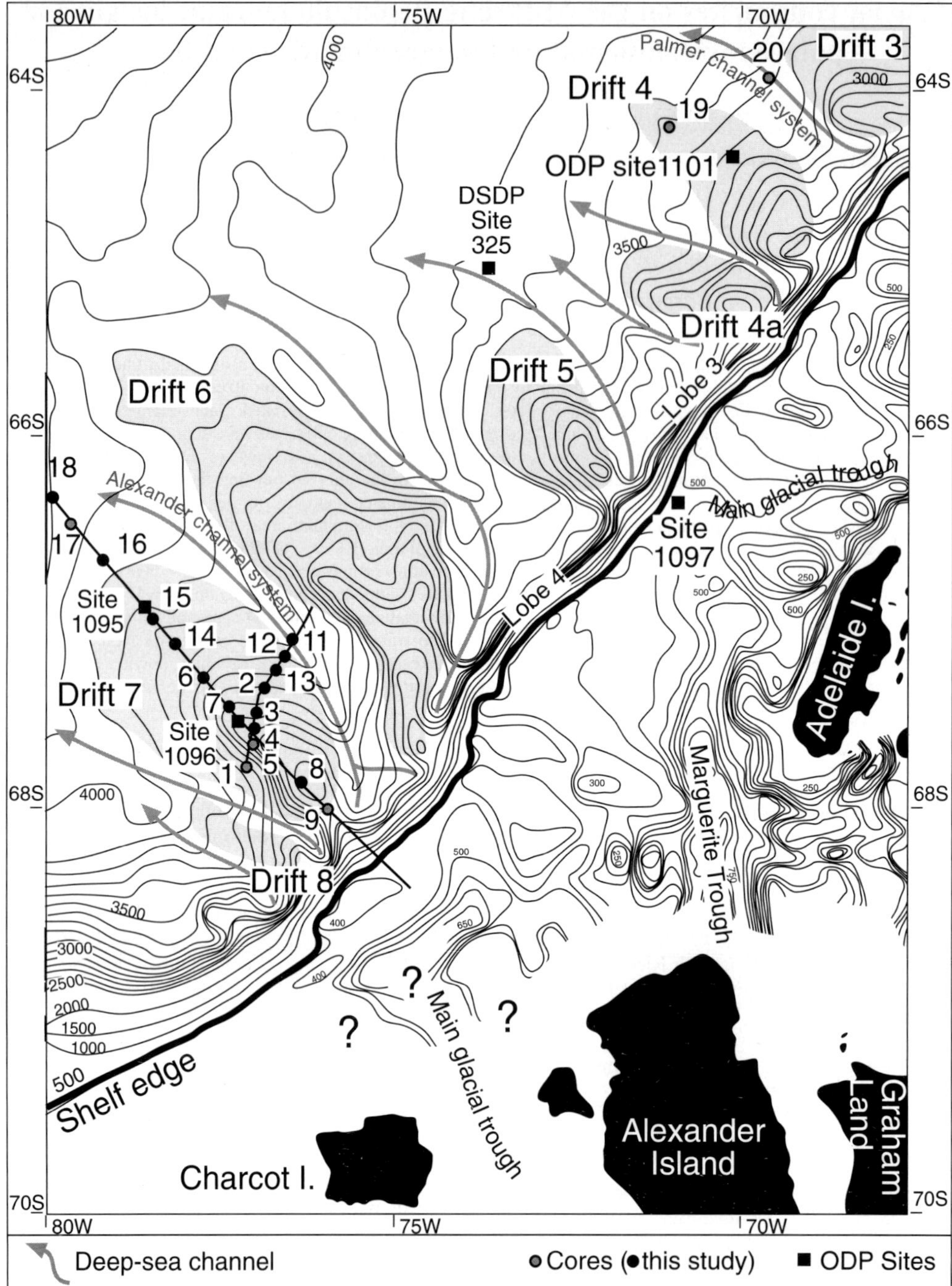

Fig. 1. Location map of the study area (bathymetry after Rebesco *et al.* 1998). The whole set of SEDANO cores is shown along the strike- and dip-oriented seismic lines. ●, Cores used for this study; ■, ODP Leg 178 drilling sites.

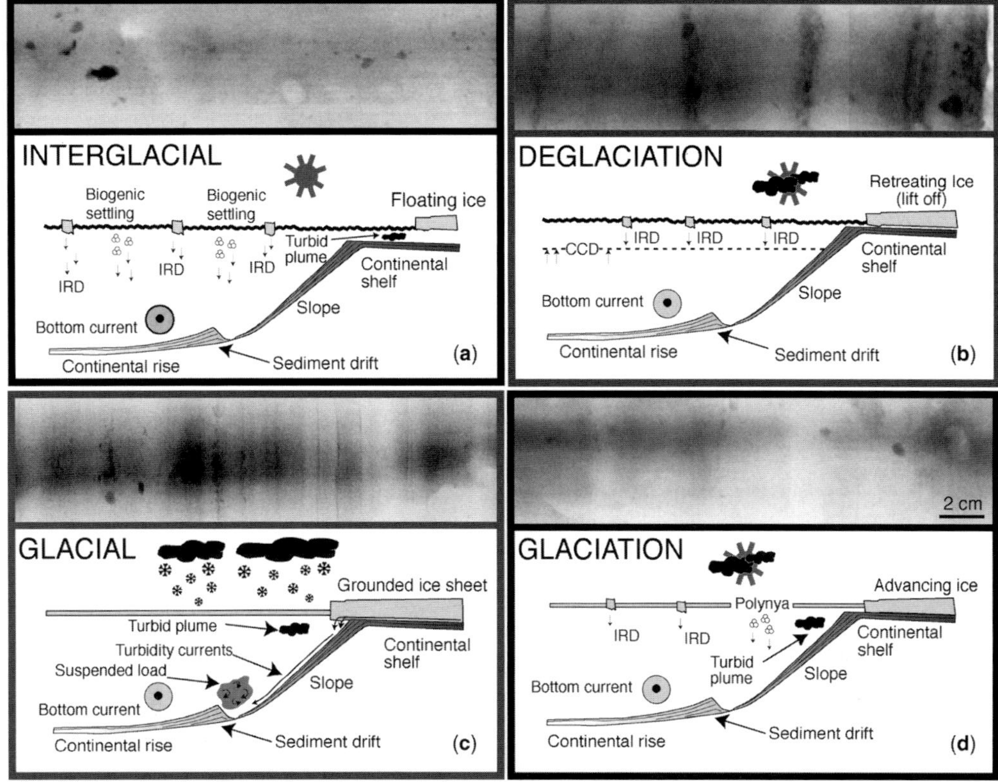

Fig. 2. Schematic model of the four climatic stages proposed to explain the repeated succession of four distinctive sedimentary facies related to the onset of different sedimentary processes (after Lucchi *et al.* 2002*a*).

almost barren facies was produced during the glacial conditions, suggesting the presence of different interacting sedimentary processes. Sediments appear generally laminated (Fig. 2c) with planar and/or ripple-like lamination, and can contain silty layers, laminae and lenses, and IRD layers or sparse pebbles. Typical mass gravity flow facies are also present, including slumps and turbidites. (4) Terrigenous structureless mud with rare and sparse IRD and bioclasts was produced during the glaciation phase (transition from interglacial to glacial conditions; Fig. 2d). Sediment input probably occurred through meltwater turbid plumes, wind and ice-rafted detritus, with biological activity reduced to polynyas.

Several researchers have advocated the possible concurrence of turbiditic and contouritic processes on the Antarctic Peninsula continental slope and rise as well as meltwater turbid plumes and the existence of benthic nepheloid layers at the sea bottom (Pudsey & Camerlenghi 1998; Pudsey 2000; Lucchi *et al.* 2002*a*). Nevertheless, sedimentary processes have not been associated with the corresponding facies, which in this area appear significantly different from those defined by the 'classic' tests.

In this study we investigate in detail the last glacial interval corresponding to isotopic stages 2–4 (the above set of facies (3)) in order to: (1) specify the type of sedimentary process, or coupled sedimentary processes, that were responsible for the large variety of sub-facies observed in our sediment cores; (2) define the environmental and palaeoclimatic conditions during last glacial maximum with a possible extrapolation to the older glacial intervals; (3) compare the glacial sedimentation on the Pacific margin of the Antarctic Peninsula with other Antarctic margins.

Materials and methods

This study has been carried out on the last glacial interval (IS 2–4, spanning 74–13 ka) identified by Lucchi *et al.* (2002*a*) on the basis of biostratigraphy, magnetic susceptibility logs, clay mineral assemblage, sediment facies and textural characteristics (Fig. 3). The stratigraphic scheme employed has been developed without the aid of isotope

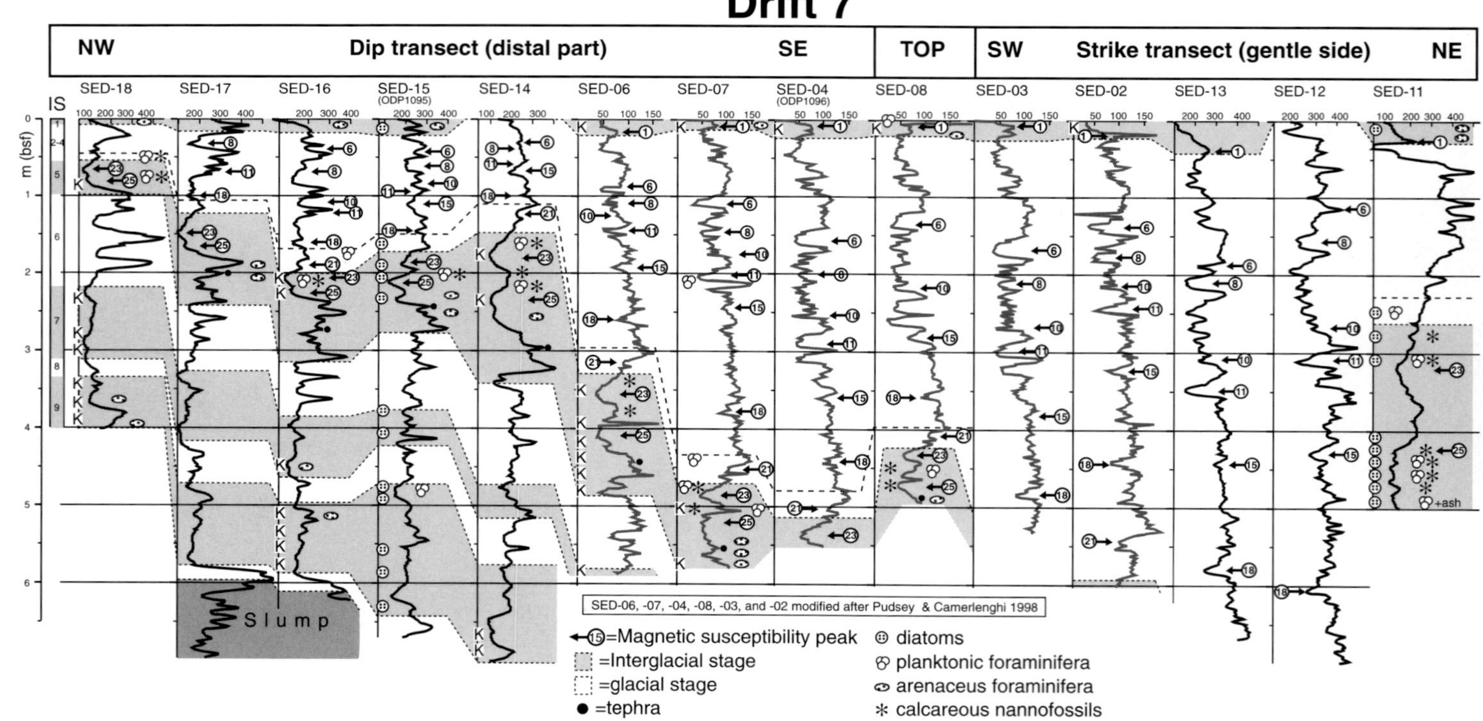

Fig. 3. Magnetic susceptibility logs with correlation of the SEDANO cores based on magnetic susceptibility peaks, biostratigraphy, visual core description, X-radiograph facies, and clay mineral analyses (core location is shown in Fig. 1). In this study we investigated last glacial sediments corresponding to Isotopic Stages 2–4 spanning 74–13 ka. Stratigraphic framework and core logs have been given by Lucchi *et al.* (2002*a*).

Table 1. *Core location and sedimentological investigation*

Core	Length (m)	Water depth (m)	Location within or around Drift 7	Sedimentology	
				Grain size	X-ray
SED-02	6.11	3313	Mid-part of the NE gentle side		•
SED-03	5.32	3176	Upper part of the NE gentle side		•
SED-04	5.52	3066	Proximal to the crest of Drift 7		•
SED-06	5.89	3606	Distal part of Drift 7		•
SED-07	5.79	3266	Distal part of Drift 7	•	•
SED-08	4.95	2750	Crest of drift 7	•	•
SED-11	5.00	3725	Upper reaches of Alexander Channel	•	•*
SED-12	6.97	3620	Base of the NE gentle side		•
SED-13	6.67	3559	Lower part of the NE gentle side	•	
SED-14	6.98	3768	Distal part of Drift 7		•
SED-15	6.68	3880	Distal part of Drift 7	•	
SED-16	6.31	4055	Distal part of Drift 7	•	
SED-17	6.60	4130	Distal part of Drift 7		
SED-18	4.00	4215	Lower reaches of Alexander Channel	•	•*

*On significant sections only.

stratigraphy because the foraminiferal record is discontinuous in the sedimentary record (Barker *et al.* 2001). However, it has been confirmed by the chronology derived from the palaeointensity of the magnetic record (Sagnotti *et al.* 2001).

From the original set of 17 SEDANO cores collected in and around sediment Drift 7, we selected 14 cores located along two orthogonal transects: a strike line through the NE gentle side of Drift 7, facing the Alexander Channel system, and a dip line along the NW-oriented crest of Drift 7 (Fig. 1 and Table 1). None of the cores collected on the steep sides of Drift 7 have been considered for this study as their sedimentary sequences are possibly affected by local sediment instability and hence are not unquestionably continuous.

Glacial sediments have been studied for lithological properties and sedimentary structures by visual description of the fresh core surface and by use of X-radiographs.

The textural characteristics of sediments have been studied by grain-size analysis of 168 samples at 10 cm spacing. The sediments were initially dried at 30 °C to determine the total weight, and then were left to disaggregate for 24 h in a 0.05% Calgon solution. Sand and silt–clay (mud) fractions were separated by wet sieving at 63 μm, and the fine fraction was analysed using a Sedigraph™ 5100. The analytical results were compared using cluster analysis with *k*-means.

Results

Glacial sediments are grey in colour (typically 5Y 5/1 on the Munsell Soil Colour Chart), very fine grained (dominated by mud), and they can occasionally contain both silty layers or laminae and IRD layers (unsorted sand or gravelly mud layers with clasts usually >63 μm and <1 cm).

A statistical approach to grain-size analysis allowed a initial distinction of three sediment types forming glacial sediments (Fig. 4a): (1) *terrigenous mud*, which forms the bulk of glacial sediments, is characterized by the highest clay content (over 65%) and the lowest sand content (<0.5%); (2) *silty layers*, which have a mode within the coarse silt range (5–6ϕ); (3) *gravelly mud layers* (IRD layers), which contain a significant amount (up to 40%) of sand, gravel and rare pebbles, with a minimum of very coarse silt (4–5ϕ) and a large tail of fine-grained sediments.

The *terrigenous mud* is generally poorly sorted and negatively skewed. Figure 5a and b and Table 2 indicate the textural variability of the *terrigenous mud* along the dip transect. Core SED-08, located at the top of Drift 7, contains the 'coarsest' sediments, with a relatively higher percentage of non-cohesive fraction (from very fine sand to coarse silt). These values exceed those observed in core SED-13 located on the strike transect in the area close to the Alexander Channel system (Table 2). In the distal area (SED-15 and -16), some intervals have a relatively high fine sand content compared with the proximal cores. Nevertheless, the overall grain-size distribution of the *terrigenous mud* fines oceanward, where sediments have the mean size of clay (>9ϕ) and appear better sorted (Fig. 5b).

A statistical investigation conducted on the silt fraction of the *terrigenous mud* indicated the presence of four sediment groups (clusters), named

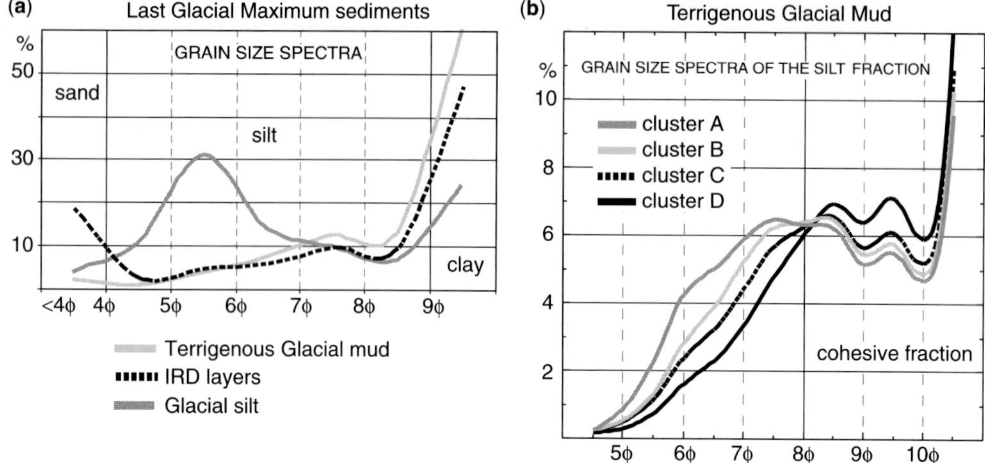

Fig. 4. Textural characterization of last glacial sediments. (**a**) Distinctive grain-size spectra of the three sediment types forming the glacial sequences; (**b**) grain-size spectra of the silty fraction of the terrigenous mud forming the bulk of glacial sediments.

A–D, that reflect a progressive decrease of environmental energy and/or dynamics of sedimentary processes (Fig. 4b). According to Table 3, which shows the distribution of the four classes of cluster, cores SED-13 (close to the channel system) and SED-08 (top of the drift) are very similar. These cores contain a few samples belonging to cluster A (medium–high energy) with a predominance of medium–low (cluster B) and low-energy (cluster C) related deposits. Core SED-07 (crest of Drift 7) contains samples belonging to only clusters B and C, whereas in the distal area core SED-15 contains all classes of clusters with predominance of cluster D, which forms the entire glacial interval of core SED-16.

Visual core description and X-radiograph analysis indicated that on the NE gentle side of Drift 7 towards the Alexander Channel system, the last glacial interval is over 7 m thick close to the Alexander Channel (core SED-12) and thins

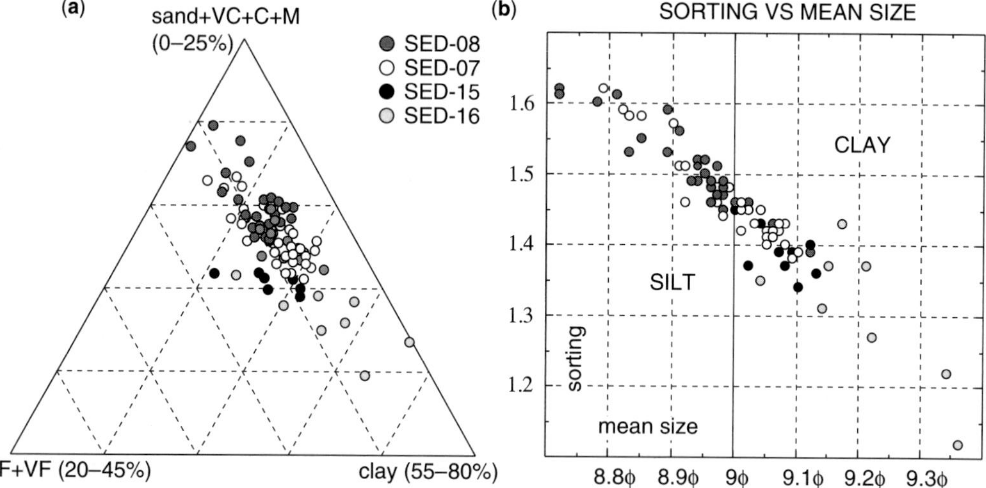

Fig. 5. Grain-size characteristics of the terrigenous mud using (**a**) the triangle plot for fine-grained sediments suggested by Lucchi *et al.* (2002*a*) (values in wt%), and (**b**) diagram of sorting v. mean grain size. VC, very coarse silt; C, coarse silt; M, medium silt; F, fine silt; VF, very fine silt.

Table 2. *Grain-size averages of last glacial sediments (wt%)*

Cores	Sand	VC silt	C silt	M silt	F silt	VF silt	Clay
SED-13	0.11	0.75	3.84	8.69	12.72	11.74	62.17
SED-08	0.22	0.83	4.70	8.68	12.28	11.87	61.42
SED-07	0.15	0.69	4.01	8.30	12.19	11.93	62.74
SED-15	0.45	0.36	3.03	7.40	12.46	13.71	62.59
SED-16	0.43	0.50	2.27	6.20	11.35	12.75	66.50

VC, very coarse; C, coarse; M, medium; F, fine; VF, very fine.

toward the crest of the drift. Pervasive lamination is produced by faintly laminating mud alternating with closely spaced silty laminae, lenses (millimetre-thick) and layers (up to 4 cm thick; Fig. 6a). These coarser-grained deposits drastically decrease in thickness and frequency towards the crest of the drift (from SED-12 to -02, Table 4) where silty laminae are detectable only in radiographs (SED-03, Fig. 6a). On the lower part of the NE gentle slope, centimetre-thick silty layers appear massive, without lamination or bioturbation, often with irregular or erosive bases (SED-12). Moving upslope toward the crest of the drift, the silty layers are finely laminated with sharp not erosive bases (core SED-13, not illustrated).

At the plateau on top of Drift 7 (SED-08) the last glacial sediments consist of homogeneous layers (2–10 cm thick) and laminae of alternating dark and light grey mud (5Y 5/1 grey and 5Y 4/1 dark grey). This sediment arrangement can locally appear on the fresh surface as a faint lamination, whereas in X-radiographs it is structureless with alternating dark and light grey intervals separated by sharp and irregular boundaries (Fig. 6b). The X-radiographs also indicate the presence of 1–2 cm thick intervals of scattered fine-grained sand and silt.

Along the crest of Drift 7, the thickness of the last glacial interval increases from the centre of the plateau topping of Drift 7 (SED-08, 4 m thick) to its distal part (SED-04, 5 m thick), and then decreases oceanward (Fig. 3). Also along the dip-transect, millimetre-thick silty laminae are common. Their occurrence increases from SED-06 oceanward to SED-15. Beyond this point the silty laminae decrease in frequency but increase in thickness (centimetre-thick silty layers were found in the most distal cores, see Table 4).

Table 3. *Occurrence of the silty fraction clusters (percentages) in the terrigenous mud*

Cores	Cluster A	Cluster B	Cluster C	Cluster D
SED-13	11	59	29	2
SED-08	14	50	36	–
SED-07	–	49	51	–
SED-15	10	10	30	60
SED-16	–	10	10	80

The last glacial sediments of cores SED-11 and SED-18, located within the Alexander Channel system, were a series of coarse-grained deposits, often graded and/or laminated, of unequivocal turbiditic origin. Core SED-11 (upper reaches of the channel) contains silty layers 4 cm thick and a 20 cm thick turbidite with an erosive base overlain by 4 cm of coarse and massive sand. The upper part of this turbidite is faintly laminated and contains a sequence of four medium-grained sandy intervals topped by very fine sand, suggesting scattering of the main turbidity flow (the overall grain size fines upward). Core SED-18 (lower reaches of the channel) contains a 19 cm thick turbidite and at least three amalgamated turbidity intervals, each formed of graded and laminated fine sand with a mud cap. The 19 cm thick turbidite has a sharp non-erosive base and inverse–normal grading within medium-grained sands.

Discussion

Last glacial sedimentary facies and depositional processes

The last glacial sediments of Drift 7 indicate a mixed system responsible for atypical sedimentary facies related to the coexistence of different depositional processes. Three of the facies we identified were called 'end-members' (turbidites, contourites, plumites), as we relate them to the predominance of a single process (Fig. 7). The degree of interplay between these processes is then responsible for the large variety of sub-facies observed within Drift 7.

The Alexander Channel system and the NE gentle side of Drift 7: turbidites v. contourites. The Alexander Channel system is the area directly influenced by turbidity currents. Decimetre-thick turbidites have been recovered from both the upper (SED-11) and lower (SED-18) reaches. Sedimentary structures and textural characteristics of these

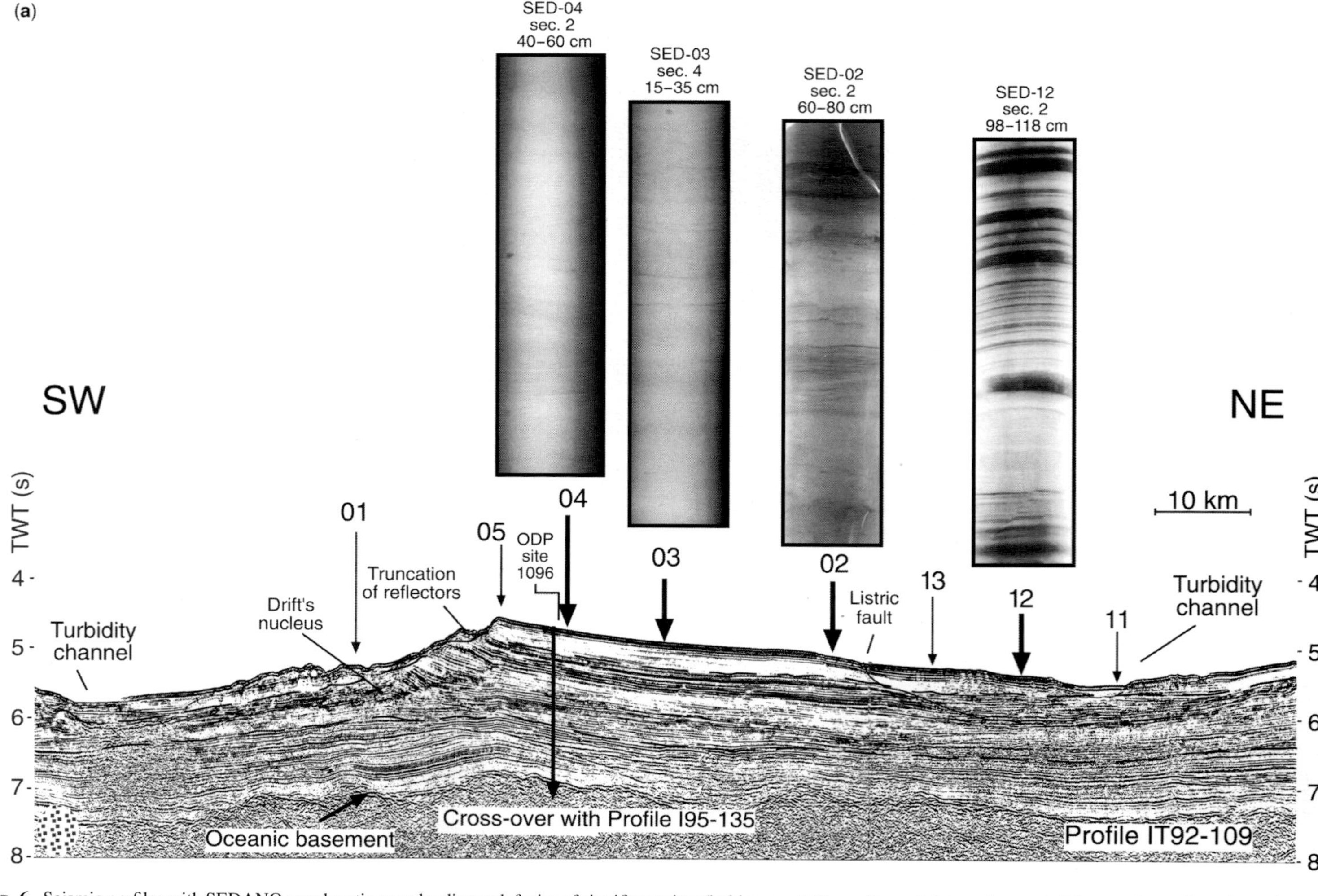

Fig. 6. Seismic profiles with SEDANO core locations and radiograph facies of significant sites (bold arrows). The radiograph intervals reported from each core have been chosen to represent the same chrono-stratigraphic interval. However, for each core the radiograph pattern is consistent within last glacial sediments. (**a**) Strike transect (OGS seismic line IT92-109); (**b**) dip transect (OGS seismic line I95-135).

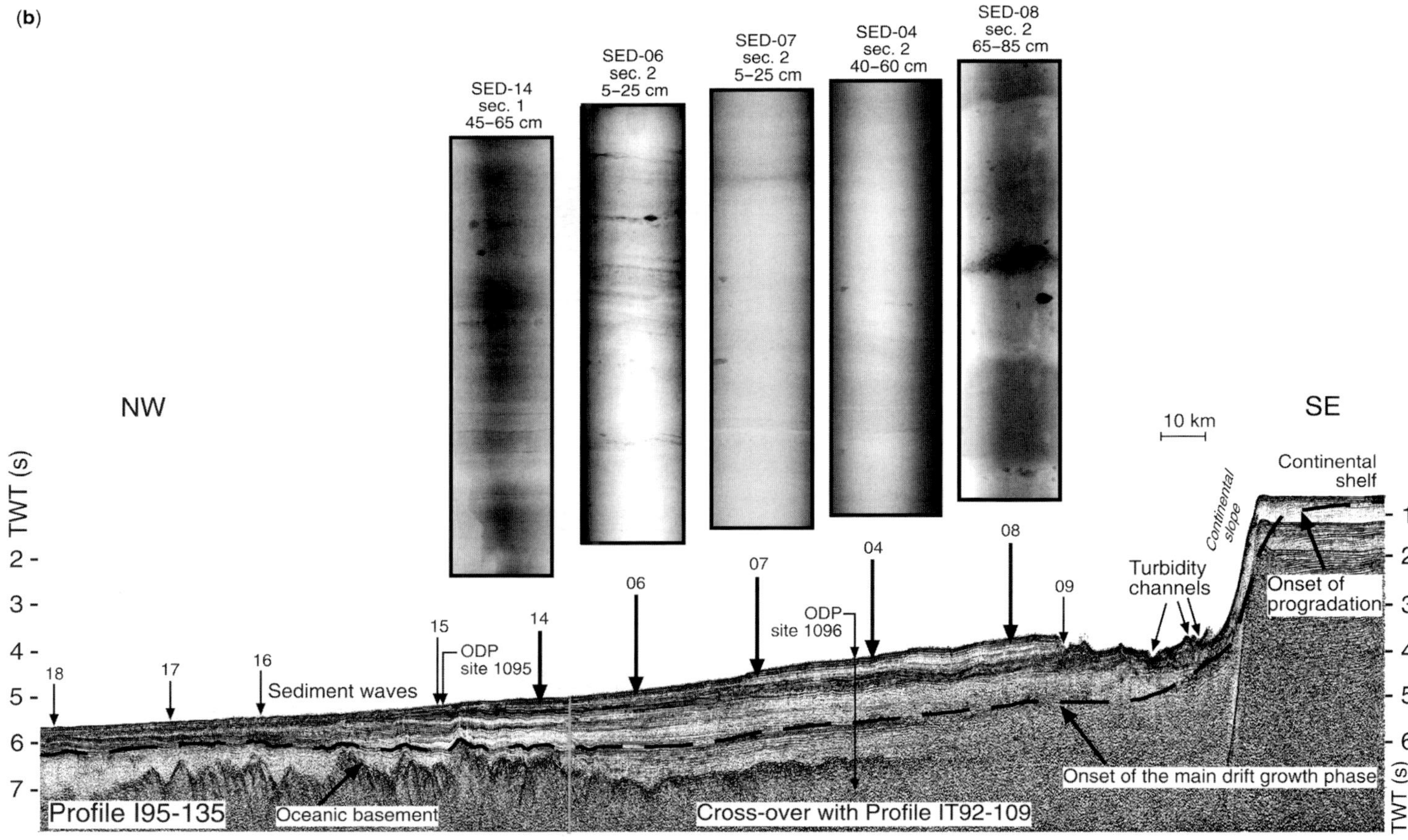

Fig. 6. *Continued.*

Table 4. *Frequency and thickness of silty deposits within last glacial sediments*

Core	Water depth (m)	Number of laminae (mm thick)	Number of layers and lenses (cm thick)
SED-11	3725	Amalgamated turbidites decimetre thick	
SED-12	3620	164	36 (up to 4 cm thick)
SED-13	3559	70	2 (up to 1 cm thick)
SED-02	3313	8	–
SED-03	3176	A few laminae visible only on X-radiographs	
SED-08	2750	–	–
SED-04	3066	–	–
SED-07	3266	–	–
SED-06	3606	6	–
SED-14	3768	10	–
SED-15	3880	20	5 (1 cm thick)
SED-16	4055	5	3 (up to 3 cm thick)
SED-17	4130	–	>10 (up to 3 cm thick)
SED-18	4215	Amalgamated turbidites decimetre thick	

deposits suggest scattering of the original flow, and high efficiency of the system.

In core SED-11 the presence of a 1 m thick sequence of structureless or thinly laminated mud containing sparse IRD pebbles can be related to sediment settling through nepheloid layers with suspended sediment reworking by weak local bottom currents. Analogous facies have been described by Gilbert *et al.* (1998) for continental rise deposits in the NW Weddell Sea. We believe that during

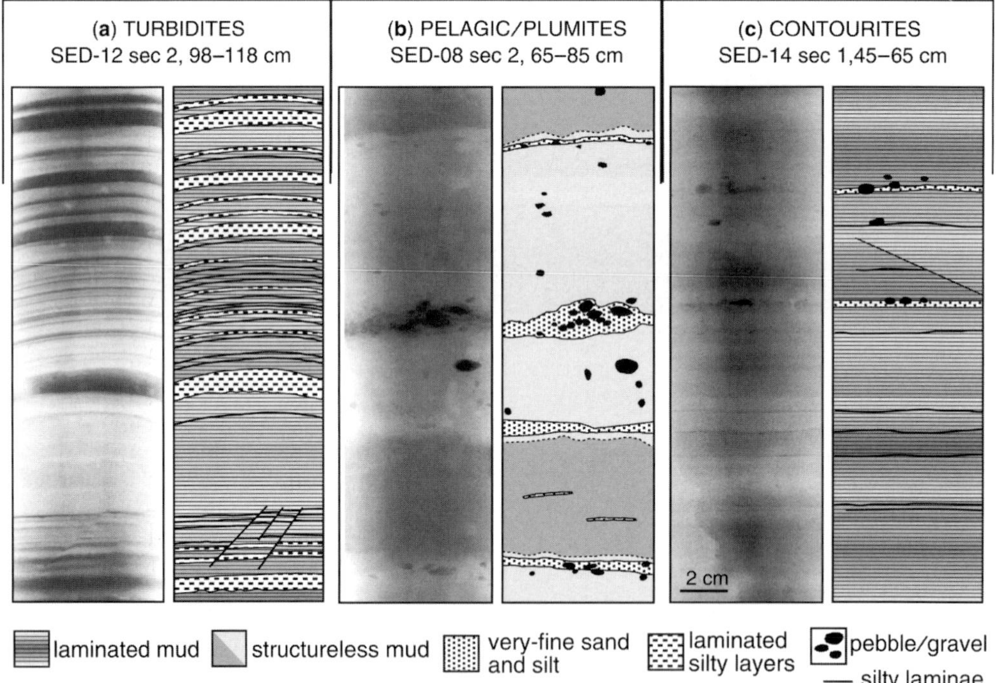

Fig. 7. 'End-members' of radiograph patterns, and interpreted illustrations, related to the predominance of one sedimentary process: (**a**) turbidity flows producing laminated mud with IRD-free silty layers; (**b**) meltwater plumes generating structureless mud with IRD layers, sparse pebbles, and coarse-silt-rich layers; (**c**) contour currents producing finely laminated mud with IRD layers and sparse pebbles (see text for discussion).

glacial periods the area of the Alexander Channel system as well as the surrounding areas were overrun by a thick persistent nepheloid layer in which turbidity currents, ice rafting and meltwater plumes periodically contributed sediment. Persistent weak bottom currents controlled the deposition of this episodically nourished nepheloid layer, resulting in faintly laminated sediments.

On the strike transect along the NE gentle side of Drift 7 the sedimentary facies in the area close to the Alexander Channel system consist of laminated mud interbedded with very fine-grained sand and silty layers (Fig. 7a). We infer that the silty layers recovered in cores SED-12 and -13 originated from spillout of the turbidity currents that moved along the Alexander Channel system enhanced by both the Coriolis force and the NW-flowing contour current. This hypothesis is consistent with the sedimentological characteristics of the silty layers, including textural variation along the NE gentle slope (fining upslope) and the frequency and thickness of the layers.

On the lower slope (core SED-12, *c*. 100 m above the channel's thalweg), silty laminae are common and centimetre-thick layers appear massive, often with erosive bases (proximal area of spillout of turbid flows). Here, the presence of cross-stratified mud (ripple-like lamination) suggests a direct influence of turbidity currents on sedimentation. In core SED-13 (only 60 m more elevated than core SED-12) the occurrence of silty layers and laminae is drastically reduced (Table 4), and they have sharp but not erosive bases (reduced energy). The frequency and thickness of silty layers, hence, rapidly decreases towards the crest of the drift, where silty laminae are detectable only on radiographs (core SED-03, Fig. 6a).

Silty layers appear on X-radiographs to contain no ice-rafted detritus, suggesting that the silt settled under a fast depositional process such as a turbidity current. Moreover, the grain-size spectra of these deposits are comparable with those of the turbiditic sand units described from the Weddell Sea slope and rise by Gilbert *et al.* (1998). However, the sediments described in Drift 7 contain a lower percentage of sand with a consistent fine-grained tail suggesting low-density turbidity currents.

The predominance of turbidity flow processes in the sedimentation in the area surrounding the channel system is shown also by the variation in thickness of the last glacial sediment across the continental rise, which exceeds 7 m near the Alexander Channel system and thins up the drift slope to 5.50 m (Fig. 3). In core SED-12 (base of NE slope) the presence of finely laminated, IRD-free mud can be related to turbiditic events, whereas in core SED-03 (summit of NE slope) laminated fine-grained sediments containing IRD layers and sparse pebbles are interpreted as contourites, as deposition of an IRD layer is not compatible with the relatively fast deposition from a turbidity flow.

Crest of Drift 7: plumites v. contourites. The structureless, colour banded, coarser-grained sediments described at the topmost, proximal part of Drift 7 (core SED-08) are interpreted to originate from settling of suspended particles of very distal overflowing plumes undisturbed by bottom currents (Fig. 7b).

This interpretation is consistent with the textural and compositional characteristics of these deposits. The most elevated and proximal areas of Drift 7 contain the highest percentages of very-coarse and coarse silt (5–6ϕ) and the clay mineral assemblage is dominated by illite, whose crystallinity index indicates a generally low alteration consistent with a continental provenance of detritus trough meltwater interflows and/or overflowing turbid plumes (Lucchi *et al.* 2002*a*). Nevertheless, the sedimentary facies of core SED-08 sediments largely differ from the plumite facies described by several researchers in Quaternary and modern sediments from polar and sub-polar margins (e.g. Mackiewicz *et al.* 1984; Cowan & Powell 1990; Cowan *et al.* 1997; Hesse *et al.* 1997, 1999; Yoon *et al.* 1997, 1998; Ó-Cofaigh *et al.* 2001; Evans *et al.* 2002; Curran *et al.* 2004). These deposits appear as cyclic couplets of a thin fine-grained sand or silt lamina that grades normally into a thicker poorly to very poorly sorted mud lamina. The laminations are defined by the alternation of coarse- and fine-grained sediments forming interlaminated deposits (textural laminations). This facies has been described for sediments recovered at up to 20 km distance from glacial outlet sources (Hesse *et al.* 1997). Mackiewicz *et al.* (1984), however, indicated that fine-grained muddy sediments dominate the distal deposition in fjords at 50 km from the efflux point. These sediments derived from overflows have a mean grain size of fine to very fine silt and are structureless or thinly interlaminated.

We argue that the absence of the 'typical' plumite facies in our deposits can be attributed to their distal characteristics (mean grain size of 8.7–9.4ϕ corresponding to very fine silt and clay). Very fine sand and coarse silt particles in overflowing plumes settle as individual grains whereas fine silt and clay particles, in a marine environment, aggregate to form floccules that settle as discrete turbid layers (Syvitski *et al.* 1985; Cowan & Powell 1990; Curran *et al.* 2004). Each turbid layer represents a pulsing event of a sediment-laden meltwater plume. In the sediments studied here, the faint colour contrast on the fresh sediment surface defines horizontal laminations for which we cannot find a textural correspondence on the fresh surface or a structural correspondence on X-radiographs. These types of lamination may represent separated sediment incursions by very distal turbid overflows depositing very fine-grained sediments, whereas textural interlaminations cannot be observed because of the homogeneity of the distal very fine-grained sediments involved. Further in support of this hypothesis, we observed that the grain-size spectra of our sediments present some similarities to the upper finer-grained interval of the individual plumites studied in the Labrador Sea (Hesse *et al.* 1997). In that area, deposits with the seismic characteristics of plumes have been recognized up to 130 km seawards from the Hudson Strait as a result of the composition of the suspended sediments, which suppressed flocculation.

It is proposed that around the Drift 7 area overflowing plumes, during glacial periods, were able to run for long distances, transporting large quantities of fine-grained particulate matter (glacial flour) before settling, as a result of their physical and compositional properties and uncommon environmental conditions. In the marine environment, fine-grained sediment flocculation and settling within an overflowing plume is enhanced by the presence of organic matter and/or salt (Syvitski *et al.* 1985; Curran *et al.* 2004). Overflows in Antarctica consist of fresh waters with almost no biogenic detritus suspended with the particulate terrigenous matter (Yoon *et al.* 1998). Moreover, the presence of a thick oceanwards extended sea-ice coverage during glacial periods inhibited the surface water mixing normally caused by katabatic winds and/or tidal action.

In core SED-08 the lack of laminations suggests insignificant reworking by bottom currents. This idea is also supported by the clay mineral assemblage, which contains only traces of smectite supposed to be transported by the SW-flowing bottom currents (Hillenbrand & Ehrmann 2001; Lucchi *et al.* 2002*a*; Hillenbrand *et al.* 2003). Conversely, laminations become progressively more evident oceanwards along the dip transect of Drift 7 (cores SED-04, -07, -06, -14; see Fig. 6b) where persistent bottom currents are supposed to have been present (Fig. 7c). Here very low-energy (cluster D), poorly sorted and negatively skewed laminated mud containing sparse IRD has been related to contour-current deposition, whereas the presence of millimetre-thick, faintly laminated silty layers with sharp non-erosive bases is attributed to pulsing events of weak very distal turbid flows (oceanward decrease in silty layers frequency and mean grain size).

'Glacial' contourites: palaeoenvironmental and palaeoclimatic significance

In the discussion above, we argue for a contouritic origin of most of the sediments recovered along the crest of Drift 7 and in its distal area. This type of contourite has been defined as 'hybrid' by Stow *et al.* (2002) because it has characteristics attributable to both contour and turbidity currents. In contrast to the 'classic' contouritic facies described by several workers (e.g. Stow & Lovell 1979; McCave *et al.* 1980; Stow 1982; Stow & Holbrook 1984; Wang & Hesse 1996; Armishaw *et al.* 2000), our sediments do not show bioturbation, and the lamination is generally more pronounced and laterally continuous. These characters are generally used to indicate turbidity current related deposits (Stow 1979). Piper & Brisco (1975) described the facies of contour-current deposits in Antarctica as indistinctly layered bioturbated mud. We relate the preservation of lamination in our study case to the absence of bioturbation. Reduced bioturbation in contourites has been attributed to both high-energy conditions at the sea bottom (Yoon & Chough 1993) and a high sedimentation rate (e.g. 190 cm ka^{-1} of the Lofoten Contourite Drift, Laberg & Vorren 2004). We believe that neither of these causes can be applied to our study case. Low-velocity bottom current are currently measured in the area of Drift 7 (Camerlenghi *et al.* 1997; Giorgetti *et al.* 2003), and a low velocity as been inferred for glacial stages on the basis of the grain size of the sediment forming Drift 7 (Pudsey & Camerlenghi 1998; Pudsey 2000, 2001; Lucchi *et al.* 2002*a*). Moreover, the higher sedimentation rate recorded in the Drift 7 glacial sediments (*c.* 12 cm ka^{-1} in core SED-12) with respect to the interglacial stages (2–3.5 cm ka^{-1}) is still a much lower value compared with those of the Weichselian contourites described from the Lofoten Contourite Drift (Norway continental slope) by Laberg & Vorren (2004).

Laminated, not bioturbated contourites have been described for shallow-water organic-rich muddy contourites from the Baltic Sea, where high biological productivity associated with periodic stagnation of near-bottom waters generated an anoxic condition at the sea bottom (Sivkov *et al.* 2002). In Drift 7, sediments contain a low percentage of organic carbon (usually $<0.2\%$, Lucchi *et al.* 2002*b*) compatible with other oceanic contourites, and are almost exclusively barren (both benthic and planktonic fauna are absent). We attribute the lack of bioturbation in Drift 7 glacial sediments to the absence of biological activity at the sea bottom.

We argue that during the last glacial stage the presence of a long-lasting sea-ice cover suppressed planktonic biological activity by preventing sunlight from reaching the upper part of the water column and gas exchanges (O_2 and CO_2) with the atmosphere (Fig. 8). This would have caused a drastic reduction of nutrient supply through the water column, which is necessary to sustain the benthic fauna.

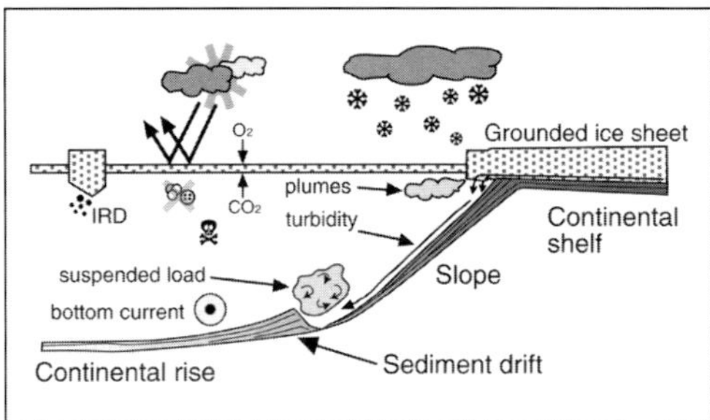

Fig. 8. Schematic representation of the environmental conditions during glacial periods. The absence of bioturbation in barren sediments has been related to the presence of long-lasting sea-ice that prevents the sunlight from reaching the upper part of the water column and gas exchange with the atmosphere. In these conditions the phytoplankton (i.e. diatoms), and consequently the zooplankton, is unable to survive. This would have caused a drastic reduction of nutrient supply, necessary to sustain the benthic fauna, throughout the water column. Life at the sea bottom is also limited by oxygen-reducing conditions (presence of pyrrhotite in the sediments) and possibly nutrient-depleted bottom currents (low organic carbon in the sediments).

The distribution of diatoms, which form the siliceous Antarctic phytoplankton (primary productivity), is sensitive to small changes in environmental parameters including light, temperature, water circulation and sea-ice cover (Dunbar *et al.* 1985; Leventer & Harwood 1993; Cunningham & Leventer 1998). Abelmann & Gersonde (1991), using time-series sediment traps, observed that at present the surface planktonic productivity is restricted to ice-free periods (peaks of vertical particle flux), whereas under sea-ice cover the vertical flux of siliceous organisms is extremely low.

Evidence of reduced or low-diversity and/or suppressed planktonic biological activity related to sea-ice extension has been reported from modern and old sedimentary records of northern and southern high-latitude margins (Barcena *et al.* 1998; Armand 2000; Pudsey 2001; Barcena *et al.* 2002; Wlodarska-Kowalczuk *et al.* 2004; Gersonde *et al.* 2005). Dowdeswell *et al.* (2000) reported barren neoglacial sediments in East Greenland as the result of multi-year shorefast sea-ice. In particular, the sediments that they attributed to the 'Little Ice Age' (Lamb 1965) were deposited during a period of several hundred years when temperatures were often a few degrees lower than today. Evidence of this period in Antarctica has been found in marine sediments by Leventer & Dunbar (1988) and Leventer *et al.* (1996), and in ice cores by Barnola *et al.* (1995). This means that severe low temperatures are not needed to maintain the sea-ice cover.

The presence of pervasive bioturbation is considered as one of the diagnostic characteristics for the identification of muddy contourites, as bottom currents are supposed to provide a continuous supply of oxygen and nutrients favourable to benthic proliferation (Stow & Holbrook 1984; Chough & Hesse 1985). The absence of benthos in the Antarctic Peninsula Pacific margin sediments is related to decreased nutrient supply through the water column but also possibly to oxygen-reduced conditions and nutrient-depleted bottom currents.

The lack of sapropelic sediments excludes anoxic conditions at the sea bottom during glacial periods (low organic carbon in the sediments). Nevertheless, the presence of pyrrhotite in the sediments suggests oxygen-reduced environmental conditions, as it represents one of the intermediate products along the pathway of pyrite formation (Sagnotti *et al.* 2001; Macrì 2004). This hypothesis is consistent with the calculation model for Antarctic glacial–interglacial atmospheric CO_2 variations of Stephens & Keeling (2000). They stated that because dissolved O_2 equilibrates with the atmosphere more rapidly than CO_2, the gas-exchange limitation imposed by the glacial sea-ice would have a lesser effect on O_2 entering the deep ocean than on CO_2 leaving, and thus would not produce deep-water anoxia. Low glacial atmospheric CO_2 concentrations might result from reduced deep-water ventilation associated with environmental and/or physical conditions that prevented air–sea gas exchanges.

The atypical facies of the muddy contourites described within the last glacial interval of Drift 7 is present in the older glacial stages (down to *c.* 340 ka of IS 10) of the sediment cores recovered from this area of the Pacific margin of the Antarctica Peninsula (Lucchi *et al.* 2002*a*), and it has also been described from other areas of deep-sea Antarctic margins (e.g. Anderson *et al.* 1979; Pudsey *et al.* 1988; Mackensen *et al.*, 1989; Grobe & Mackensen 1992; Pudsey 1992; Gilbert *et al.* 1998; Anderson 1999; Yoon *et al.* 2000). In contrast, in Wilkes Land (East Antarctica), the last glacial sediments recovered from the Wega Channel and surrounding sediment mounds are faintly laminated, bioturbated (Busetti *et al.* 2003; Caburlotto 2003), and contain maghemite (an oxidized form of magnetite), suggesting the presence of oxygen during deposition (Macrì, 2004; Macrì *et al.* 2005). At present, this area is characterized by the presence of a long-lasting polynya, west of the Mertz glacier, that is responsible for the production of High Salinity Shelf Waters (HSSW, Bindoff *et al.* 2000). The HSSW mix with the Modified Circumpolar Deep Waters in the proximity of the shelf break, producing the cold and saline Antarctic Bottom Water (AABW), which transports oxygen and nutrients downslope towards deeper environments. We believe that during glacial periods, oxygen- and nutrient-rich deep waters had only a local impact on sedimentation, and their biological content was rapidly depleted during the motion around the continent driven by bottom-current flow. As the Antarctic margins are overrun by circumpolar currents, we believe that reduced oxygen and nutrient conditions in deep-sea environments had a circum-Antarctic relevance during glacial periods, except in the areas where a polynya was maintained during sea-ice conditions.

Conclusion

Detailed study of sediments from the last glacial interval has shown a large variety of coexisting sedimentary processes in sediment Drift 7 on the Pacific margin of the Antarctic Peninsula. Despite the variety of sedimentary facies described, three end-members were identified, representing turbidity flows, meltwater plumes and contour-current deposits.

Low-density turbidity flows moved along the Alexander Channel system and contributed fine-grained sediments to the nepheloid layer. These gravity-driven currents were concurrently affected by both the Coriolis force and NW-flowing contour currents, and their direct influence on sedimentation rapidly decreased with distance and elevation from the thalweg of the Alexander Channel system. These deposits appear laminated and non-bioturbated, and do not contain ice-rafted debris, as this is not compatible with their fast deposition.

Meltwater sediment-laden interflows and/or overflowing turbid plumes represented in West Antarctica an important sedimentary process conveying large amounts of fine-grained particulate matter into the depositional system as far as very distal areas. These deposits appear structureless in X-radiographs but finely laminated on a fresh sediment surface, and can contain sparse IRD.

The presence of IRD layers and sparse pebbles within finely laminated, non-bioturbated glacial sediments of Drift 7 allowed the identification of the 'glacial contourites'. This atypical conturitic facies originated from the piracy of the sediments delivered into the system by other sedimentary processes and indicates unusual, climatically related, environmental conditions of suppressed primary productivity and oxygen-reduced deep waters. The presence of this facies in most of the Antarctic margins indicates that this unusual condition had a circumpolar importance, except in those areas where polynyas were maintained during the glacial stages.

Barren, non-bioturbated Antarctic contourites (*glacial contourites*) can be used as a proxy to define the temporal and spatial extension of the Antarctic sea-ice.

The authors would like to thank Salvi, G., Salvi, C. and Bandelli, N. of the Antarctic Museum in Trieste for prompt support with X-radiograph analyses on two key cores. G. Fontolan assisted with statistical analyses. M. A. Barcena and R. Dunbar helped with discussion and suggestions on diatom ecology. Camerlenghi, A. contributed with suggestions and informal revision of the manuscript. Cowan, E. A. and Laberg, J. S. are acknowledged for helpful advice and revision of the manuscript. The present research has been funded by the PNRA-MOGAM (MOrphology and Geology of Antarctic Margins) project.

References

ABELMANN, A. & GERSONDE, R. 1991. Biosiliceous particle flux in the southern ocean. *Marine Chemistry*, **35**, 503–536.

AMBLAS, D., URGELES, R., CANALS, M., *ET AL.* 2006. Relationship between continental rise development and palaeo-ice sheet dynamics, Northern Antarctic Peninsula Pacific margin. *Quaternary Science Reviews*, **25**, 933–944.

ANDERSON, J. B. 1999. *Antarctic Marine Geology*. Cambridge University Press, Cambridge.

ANDERSON, J. B., KURTZ, D. D. & WEAVER, F. M. 1979. Sedimentation on the Antarctic continental slope. *In*: DOYLE, L. J. & PILKEY, O. (eds) *Geology of Continental Slopes*. Society of Economic Paleontologists and Mineralogists, Special Publications, **27**, 265–283.

ARMAND, L. K. 2000. An ocean of ice—advances in the estimation of past sea ice in the southern ocean. *GSA Today*, **10**, 1–2.

ARMISHAW, J. E., HOLMES, R. W. & STOW, D. A. V. 2000. The Barra Fan: a bottom-current reworked, glacially-fed submarine fan system. *Marine and Petroleum Geology*, **17**, 219–238.

BARCENA, M. A., GERSONDE, R., LEDESMA, S. *ET AL.* 1998. Record of Holocene glacial oscillations in Bransfield Basin as revealed by siliceous microfossil assemblages. *Antarctic Sciences*, **10**, 269–285.

BARCENA, M. A., ISLA, E., PLAZA, A., *ET AL.* 2002. Bioaccumulation record and palaeoclimatic significance in the Western Bransfield Strait. The last 2000 years. *Deep-Sea Research II*, **49**, 935–950.

BARKER, P. F., OSTERMAN, L. E. & HALL, M. A. 2001. Data report: Oxygen and carbon isotope measurements on *Neogloboquadrina pachyderma*(*s*) from Holes 1096B and 1101A, Antarctic Peninsula margin, Leg 178. *In*: BARKER, P. F., CAMERLENGHI, A., ACTON, G. D. & RAMSAY, A. T. S. (eds) *Proceedings of the Ocean Drilling Program, Scientific Results*, *178*. [Online.] World Wide Web address: http://www-odp.tamu.edu/publications/178_SR/chap_20/chap_20.htm.

BARNOLA, J. M., ANKLIN, M., PORCHERON, J., RAYNAUD, D., SCHWANDER, J. & STAUFFER, B. 1995. CO_2 evolution during the last millennium as recorded by Antarctic and Greenland ice. *Tellus*, **47B**, 264–272.

BINDOFF, N. L., ROSEMBERG, M. A. & WARNER, M. J. 2000. On the circulation and water masses over the Antarctic continental slope and rise between 80 and 150E. *Deep-Sea Research II*, **47**, 2299–2326.

BUSETTI, M., CABURLOTTO, A., ARMAND, L., *ET AL.* 2003. Plio-Quaternary sedimentation on the Wilkes Land continental rise: preliminary results. *Deep-Sea Research II*, **50**, 1529–1562.

CABURLOTTO A. 2003. *Processi sedimentari di mare profondo sul Margine Continentale Glaciale Antartico*. PhD thesis, Università degli Studi di Siena.

CAMERLENGHI, A., CRISE, A., PUDSEY, C. J., ACCERBONI, E., LATERZA, R. & REBESCO, M. 1997. Ten-month observation on the bottom current regime across a sediment drift of the Pacific margin of the Antarctic Peninsula. *Antarctic Science*, **9**, 426–433.

CHOUGH, S. K. & HESSE, R. 1985. Contourites from Eirik Ridge, South of Greenland. *Sedimentary Geology*, **41**, 185–199.

COWAN, E. A. & POWELL, R. D. 1990. Suspended sediment transport and deposition of cyclically interlaminated sediment in a temperate glacial fjord, Alaska, U.S.A. *In*: DOWDESWELL, J. A. & SCOURSE, J. D. (eds) *Glacimarine Environments: Processes and Sediments*. Geological Society, London, Special Publications, **53**, 75–89.

COWAN, E. A., CAI, J., POWELL, R. D., CLARK, J. D. & PITCHER, J. N. 1997. Temperate glacimarine varves: an example from Disenchantment Bay, Southern Alaska. *Journal of Sedimentary Research*, **67**(3), 536–549.

CUNNINGHAM, W. L. & LEVENTER, A. 1998. Diatom assemblages in surface sediments of the Ross Sea: relationship to present oceanographic conditions. *Antarctic Science*, **10**(2), 134–146.

CURRAN, K. J., HILL, P. S., MILLIGAN, T. G., COWAN, E. A., SYVITSKI, J. P. M. & KONINGS, S. M. 2004. Fine-grained sediment flocculation below the Hubbard Glacier meltwater plume, Disenchantment Bay, Alaska. *Marine Geology*, **203**, 83–94.

DOWDESWELL, J. A., WHITTINGTON, R. J., JENNINGS, A. E., ANDREWS, J. T., MACKENSEN, A. & MARIENFELD, P. 2000. An origin for laminated glacimarine sediments through sea-ice build-up and suppressed iceberg rafting. *Sedimentology*, **47**, 557–576.

DUNBAR, R. B., ANDERSON, J. B., DOMACK, E. W. & JACOBS, S. S. 1985. Oceanographic influences on sedimentation along the Antarctic continental shelf. *Antarctic Research Series*, **43**, 291–312.

EVANS, J., DOWDESWELL, J. A., GROBE, H., NIESSEN, F., STEIN, R., HUBBERTEN, H. W. & WHITTINGTON, R. J. 2002. Late Quaternary sedimentation in Kejser Franz Josephs Fjord and the continental margin of East Greenland. *In*: DOWDESWELL, J. A. & Ó-COFAIGH, C. (eds) *Glacier-Influenced Sedimentation on High-Latitude Continental Margins*. Geological Society, London, Special Publications, **203**, 149–179.

GERSONDE, R., CROSTA, X., ABELMANN, A. & ARMAND, L. 2005. Sea-surface temperature and sea ice distribution of the Southern Ocean at the EPILOG Last Glacial Maximum: a circum-Antarctic view based on siliceous microfossil records. *Quaternary Science Reviews*, **24**, 869–896.

GILBERT, I. M., PUDSEY, C. J. & MURRAY, J. W. 1998. A sediment record of cyclic bottom-current variability from the northwest Weddell Sea. *Sedimentary Geology*, **115**, 185–214.

GIORGETTI, A., CRISE, A., LATERZA, R., PERINI, L., REBESCO, M. & CAMERLENGHI, A. 2003. Water masses and bottom boundary layer dynamics above a sediment drift of the Antarctic Peninsula Pacific margin. *Antarctic Science*, **15**(4), 537–546.

GROBE, H. & MACKENSEN, A. 1992. Late Quaternary climate cycles as recorded in sediments from the Antarctic continental margins. *In*: KENNATT, J. P. & WATKINS, N. B. (eds) *The Antarctic Paleoenvironment*. American Geophysical Union, Antarctic Research Series, **56**, 349–376.

HESSE, R., KHODABAKHSH, S., KLAUCK, I. & RYAN, W. B. F. 1997. Asymmetrical turbid surface-plume deposition near ice-outlets of the Pleistocene Laurentide ice sheet in the Labrador Sea. *Geo-Marine Letters*, **17**, 179–187.

HESSE, R., KLAUCK, I., KHODABAKHSH, S. & PIPER, D. 1999. Continental slope sedimentation adjacent to an ice margin. III. The upper Labrador Slope. *Marine Geology*, **155**, 249–276.

HILLENBRAND, C.-D. & EHRMANN, W. 2001. Distribution of clay minerals in drift sediments on the continental rise west of the Antarctic Peninsula, ODP Leg 178, Sites 1095 and 1096. *In*: BARKER, P. F., CAMERLENGHI, A., ACTON, G. D. & RAMSAY, A. T. S. (eds) *Proceedings of the Ocean Drilling Program Scientific Results, 178*. [Online.] World Wide Web address: http://www-odp.tamu.edu/publications/178_SR/chap_08/chap_08.htm.

HILLENBRAND, C.-D., GROBE, H., DIEKMANN, G. K. & FÜTTERER, D. K. 2003. Distribution of clay minerals and proxies for productivity in surface sediments of the Bellingshausen and Amundsen seas (West Antarctica)—relation to moden environmental conditions. *Marine Geology*, **193**, 253–271.

LABERG, J. S. & VORREN, T. O. 2004. Weichselian and Holocene growth of the northern high-latitude Lofoten Contourite Drift on the continental slope of Norway. *Sedimentary Geology*, **164**, 1–17.

LAMB, H. H. 1965. The early medieval warm epoch and its sequel. *Palaeogeography, Palaeoclimatology, Palaeoecology*, **1**, 13–37.

LEVENTER, A. & DUNBAR, R. B. 1988. Recent diatom record of McMurdo Sound, South Antarctica: implications for history of sea-ice extent. *Palaeoceanography*, **3**, 259–274.

LEVENTER, A. & HARWOOD, D. M. 1993. The geologic use of polar marine diatoms. *In*: BRYAN, J. R. (ed.) *Workshop on Antarctic Glacial Marine and*

Biogenic Sedimentation: Notes for a Short Course. Antarctic Marine Geology Research Facility, Florida State University, Sedimentology Research Laboratory Contribution no. 57, 134–238.

LEVENTER, A., DOMACK, E., ISHMAN, S. E., BRACHFELD, S., MCCLENNEN, C. E. & MANLEY, P. 1996. Productivity cycles of 200–300 years in the Antarctic Peninsula region: understanding linkages among the sun, atmosphere, oceans, sea-ice, and biota. *Geological Society of America Bulletin*, **108**, 1626–1644.

LUCCHI, R. G., REBESCO, M., CAMERLENGHI, A., ET AL. 2002*a*. Mid–late Pleistocene glacimarine sedimentary processes of a high-latitude, deep-sea sediment drift (Antarctic Peninsula Pacific Margin). *Marine Geology*, **189**, 343–370.

LUCCHI, R. G., REBESCO, M., BUSETTI, M., CABURLOTTO, A., COLIZZA, E. & FONTOLAN, G. 2002*b*. Sedimentary processes and glacial cycles on the sediment drifts of the Antarctic Peninsula Pacific Margin: preliminary results of SEDANO-II Project. *Royal Society of New Zealand Bulletin*, **35**, 275–280.

MACKENSEN, A., GROBE, H., HUBBERTEN, H.-W., SPIESS, V. & FÜTTERER, D. 1989. Stable isotope stratigraphy from the Antarctic continental margin during the last one million years. *Marine Geology*, **87**, 315–321.

MACKIEWICZ, N. E., POWELL, R. D., CARLSON, P. R. & MOLNIA, B. F. 1984. Interlaminated ice-proximal glacimarine sediments in Muir inlet, Alaska. *Marine Geology*, **57**, 113–147.

MACRÌ, P. 2004. *Magnetostratigrafia e magnetismo ambientale di sedimenti pleistocenici dai margini continentali antartici.* PhD thesis, Università di Bologna.

MACRÌ, P., SAGNOTTI, L., DINARÈS-TURELL, J. & CABURLOTTO, A. 2005. A composite record of Late Pleistocene relative geomagnetic paleointensity from the Wilkes Land Basin (Antarctica). *Physics of the Earth and Planetary Interiors*, **151**, 223–242.

MCCAVE, I. N., LONSDALE, P. F., HOLLISTER, C. D. & GARDNER, W. D. 1980. Sediment transport over the Hatton and Gardar contourite drifts. *Journal of Sedimentary Petrology*, **50**, 1049–1062.

Ó-COFAIGH, C., DOWDESWELL, J. A. & GROBE, H. 2001. Holocene glacimarine sedimentation, inner Scoresby Sund, East Greenland; the influence of fast-flowing ice-sheet outlet glaciers. *Marine Geology*, **175**, 103–129.

PIPER, D. J. W. & BRISCO, C. D. 1975. Deep water continental margin sedimentation. *In*: HAYES, D. E., FRAKES, L. A., ET AL. (eds) Initial Reports of the Deep-Sea Drilling Project, *28*. US Government Printing Office, Washington, DC, 727–755.

PUDSEY, C. J. 1992. Late Quaternary changes in Antarctic Bottom Water velocity inferred from sediment grain size in the northern Weddell Sea. *Marine Geology*, **107**, 9–33.

PUDSEY, C. J. 2000. Sedimentation on the continental rise west of the Antarctic Peninsula over the last three glacial cycles. *Marine Geology*, **167**, 313–338.

PUDSEY, C. J. 2001. Neogene record of Antarctic Peninsula glaciation in continental rise sediments: ODP Leg 178, Site 1095. *In*: BARKER, P. F., CAMERLENGHI, A., ACTON, G. D. & RAMSAY, A. T. S. (eds) *Proceedings of the Ocean Drilling Programs, Scientific Results, 178*. [Online.] World Wide Web address: http://www-odp.tamu.edu/publications/178_SR/chap_25/chap_25.htm [cited 2005-07-15].

PUDSEY, C. J. & CAMERLENGHI, A. 1998. Glacial–interglacial deposition on a sediment drift on the Pacific margin of the Antarctic Peninsula. *Antarctic Science*, **10**(3), 286–308.

PUDSEY, C. J., BARKER, P. F. & HAMILTON, N. 1988. Weddell Sea abyssal sediments: a record of Antarctic Bottom Water flow. *Marine Geology*, **81**, 289–314.

REBESCO, M., CAMERLENGHI, A. & ZANOLLA, C. 1998. Bathymetry and morphogenesis of the continental margin west of Antarctic Peninsula. *Terra Antartica*, **5**, 715–725.

REBESCO, M., PUDSEY, C. J., CANALS, M., CAMERLENGHI, A., BARKER, P. F., ESTRADA, F. & GIORGETTI, A. 2002. Sediment drift and deep-sea channel systems, Antarctic Peninsula Pacific margin. *In*: STOW, D. A. V., PUDSEY, C. J., HOWE, J., FAUGÈRES, J. C. & VIANA, A. (eds) *Deep-water contourite systems: Modern Drifts and Ancient Series, Seismic and Sedimentary Characteristics*. Geological Society, London, Memoirs, **22**, 353–371.

REBESCO, M., CAMERLENGHI, A., VOLPI, V., ET AL. 2007. Interaction of processes and importance of contourites: insights from the detailed morphology of sediment Drift 7, Antarctica. *In*: VIANA, A. R. & REBESCO, M. (eds) *Economic and Palaeoceanographic Significance of Contourite Deposits*. Geological Society, London, Special Publications, **276**, 95–110.

SAGNOTTI, L., MACRÌ, P., CAMERLENGHI, A. & REBESCO, M. 2001. Environmental magnetism of Antarctic Pleistocene sediments and interhemispheric correlation of climatic events. *Earth and Planetary Science Letters*, **192**, 65–80.

SIVKOV, V., GORBATSKIY, V., KULESHOV, A. & ZHUROV, Y. 2002. Muddy contourites in the Baltic Sea: an example of a shallow-water contourite system. *In*: STOW, D. A. V., PUDSEY, C. J., HOWE, J. A., FAUGÈRES, J.-C. & VIANA, A. R. (eds) *Deep-Water Contourite Systems: Modern Drifts and Ancient Series, Seismic and Sedimentary Characteristics*. Geological Society, London, Memoirs, **22**, 121–136.

STEPHENS, B. B. & KEELING, R. F. 2000. The influence of Antarctic sea ice on glacial–interglacial CO_2 variations. *Nature*, **404**, 171–174.

STOW, D. A. V. 1979. Distinguishing between fine-grained turbidites and contourites on the Nova Scotia outer continental margin. *Sedimentology*, **26**, 371–385.

STOW, D. A. V. 1982. Bottom currents and contourites in the North Atlantic. *Bulletin de l'Institut Géologique du Bassin d'Aquitaine*, **31**, 151–166.

STOW, D. A. V. & LOVELL, J. P. B. 1979. Contourites: their recognition in modern and ancient sediments. *Earth-Science Reviews*, **14**, 251–291.

STOW, D. A. V. & HOLBROOK, J. A. 1984. North Atlantic contourites: an overview. *In*: STOW, D. A. V. & PIPER, D. (eds) *Fine-grained Sediments: Deep-Water Processes and Facies*. Geological Society, London, Special Publications, **15**, 121–136.

STOW, D. A. V., FAUGÈRES, J.-C., HOWE, J. A., PUDSEY, C. J. & VIANA, A. R. 2002. Bottom currents, contourites and deep-sea sediment drifts: current state-of-the-art. *In*: STOW, D. A. V., PUDSEY, C. J., HOWE, J. A., FAUGÈRES, J.-C. & VIANA, A. R. (eds) *Deep-Water Contourite Systems: Modern Drifts and Ancient Series, Seismic and Sedimentary Characteristics*. Geological Society, London, Memoirs, **22**, 7–20.

SYVITSKI, J. P. M., ASPREY, K. W., CLATTENBURG, D. A. & HODGE, G. D. 1985. The prodelta environment of a fjord: suspended particle dynamics. *Sedimentology*, **32**, 83–107.

WANG, D. & HESSE, R. 1996. Continental slope sedimentation adjacent to ice-margin. II. Glaciomarine depositional facies on Labrador Slope and glacial cycles. *Marine Geology*, **135**, 65–96.

WLODARSKA-KOWALCZUK, M., KENDALL, M. A., WESLAWSKI, J. M., KLAGES, M. & SOLTWEDEL, T. 2004. Depth gradients of benthic standing stock and diversity on the continental margin at a high-latitude ice-free site (off Spitsbergen, 79°N). *Deep-Sea Research I*, **51**, 1903–1914.

YOON, H. I., HAN, M. W., PARK, B.-K., OH, J.-K. & CHANG, S.-K. 1997. Glaciomarine sedimentation and palaeo-glacial setting of Maxwell Bay and its tributary embayment, Marian Cove, South Shetland Islands, West Antarctica. *Marine Geology*, **140**, 265–282.

YOON, H. I., PARK, B. K., DOMACK, E. W. & KIM, Y. 1998. Distribution and dispersal pattern of suspended particulate matter in Maxwell Bay and its tributary, Marian Cove, in the South Shetland Islands, West Antarctica. *Marine Geology*, **152**, 261–275.

YOON, H. I., PARK, B. K., KIM, Y. & KIM, D. 2000. Glaciomarine sedimentation and its paleoceanographic implications along the fjord margins in the South Shetland Islands, Antarctica during the last 6000 years. *Palaeogeography, Palaeoclimatology, Palaeoecology*, **157**, 189–211.

YOON, S. H. & CHOUGH, S. K. 1993. Sedimentary characteristics of Late Pleistocene bottom-current deposits, Barents Sea slope off northern Norway. *Sedimentary Geology*, **82**, 33–45.

The role of intermediate-depth currents in continental shelf–slope accretion: Canterbury Drifts, SW Pacific Ocean

R. M. CARTER

Marine Geophysical Laboratory, James Cook University, Townsville, Qld. 4811, Australia (e-mail: bob.carter@jcu.edu.au)

Abstract: The Late Oligocene to Recent Canterbury Drifts were deposited in water depths between *c.* 400 and *c.* 1500 m by northward-flowing, cold, intermediate-depth water masses: Subantarctic Mode Water (SAMW), Antarctic Intermediate Water (AAIW) and their predecessor current flows. Drift accumulation started at *c.* 24 Ma, fed by terrigenous sediment derived from the newly rising Alpine Fault plate boundary in the west, which has built a progradational shelf–slope sediment prism up to 130 km wide at rates of eastward advance of up to 5.4 km Ma^{-1}. Gentle uplift associated with the nearby plate boundary has exposed older Late Oligocene and Miocene drifts onland (Bluecliffs Formation). Ocean Drilling Program Site 1119, located 100 km offshore at a water depth of 394 m, penetrated a 428 m thickness of mid-Pliocene to Pleistocene (0–3.9 Ma) drift located just seaward of the eastern South Island shelf edge. Uniquely, these large (>60 000 km^3), regionally extensive, intermediate-depth sediment drifts can be examined in outcrop, in marine drill-core and at the modern sea bed. The drifts comprise planar-bedded units up to several metres thick. Some sand intervals have sharp, erosive bases and normally graded tops into overlying siltstone; others are symmetrically graded with reverse-graded bases and normally graded tops. Bioturbation is moderate and rarely destroys the pervasive background, centimetre-scale, planar or wispy alternation of muddy and sandy silts displayed by Formation Micro-Scanner imagery. These features are consistent with deposition from rhythmically fluctuating bottom currents. Texturally, the drifts are polymodal quartzofeldspathic silty sands, sandy silts, silts and silty clays, with varying admixtures of benthic and biopelagic carbonate and silica. Miocene samples are mostly dominated by coarse silt (45–60 μm) and very fine sand (70–105 μm) grain-size modes, whereas strong fine silt (11–13 μm) and very fine silt–clay (<5 μm) modes become dominant after *c.* 3.1 Ma in the Late Pliocene, consistent with an increasing input of glacially ground material. Over the Plio-Pleistocene part of the succession, the sand–silt lithological rhythmicity occurs in synchroneity with Milankovitch-scale climate cycling, with periods of inferred faster current flow (sand intervals) mostly corresponding to warm, interglacial periods. Northward drift dispersal has helped cause the seaward growth of the eastern South Island shelf–slope system since the Late Oligocene probably by clinoform progradation and by episodic welding of mounded slope drifts onto the pre-existing sediment prism. Such along-slope, contourite drift accumulation occurs even in the absence of mounded drifts on seismic profiles, and represents a previously underemphasized mechanism for the progradation of shelf–slope clinoforms, worldwide. The Canterbury Drifts vary in thickness from *c.* 300 m near the early Miocene shoreline, where they were accumulating in limited shallow-water accommodation, to *c.* 2000 m under the modern shelf edge. Mounded drifts first occur in the Middle Miocene, at *c.* 15 Ma, their appearance perhaps reflecting more vigorous intermediate water flow consequent upon the worldwide climatic deterioration between 15 and 13 Ma. At Site 1119, a further change from large (>10 km wide) to smaller (1–3 km wide) mounded slope drifts occurs at *c.* 3.1 Ma, marking further cooling and perhaps the inception of discrete SAMW flows and initiation of the Subantarctic Front.

The concept that deep ocean currents play a major role in shaping the continental slope originated from seismic observations of migratory abyssal sediment waves (Ballard 1966; Lonsdale & Hollister 1979), and sea-bed photographs of nearby current-influenced bedforms. Similar features were later shown to occur worldwide beneath the path of contour-hugging thermohaline currents (e.g. Ewing *et al.* 1971; Hollister *et al.* 1974; Gardner & Kidd 1987; Howe *et al.* 1997), and also laterally to turbidity-current pathways (e.g. Damuth 1979; Normark *et al.* 1980; Carter *et al.* 1990). The accompanying 'contourite' sediments, with characteristic sand–mud sedimentary structures and textures (Stow & Lovell 1979), have been described both from sea-bed cores and from ancient sedimentary basins (summarized by Stow *et al.* 1998). Long Deep Sea Drilling Project (DSDP) core samples through well-known sediment drifts, the Hatton, Gardar and Feni sediment drifts in the North Atlantic, were described by Laughton *et al.* (1972), Montadert *et al.* (1979), McCave *et al.* (1980) and Kidd & Hill (1987).

From: VIANA, A. R. & REBESCO, M. (eds) *Economic and Palaeoceanographic Significance of Contourite Deposits*. Geological Society, London, Special Publications, **276**, 129–154.
0305-8719/07/$15.00

The 1980s saw a widening interest in marine sediment drifts, extending to include those developed in shallower intermediate water depths. The Faro Drift, a 300 m thick, 50 km long body located at depths of 500–700 m along the path of the deep Mediterranean outflow in the eastern Atlantic ocean (Faugères *et al.* 1984; Stow *et al.* 1986; Nelson *et al.* 1993), was a subject of particular study about the time that the first detailed facies models for drift sediments were emerging (Stow *et al.* 1998; Viana *et al.* 1998). Although it was implicit in many papers that the deposition of sediment drifts helped build up the continental margin (for instance, McCave & Tucholke (1986) referred to the 'plastering and decorating' of the sides of the North Atlantic Ocean basin), most previous research has been focused on the description of drift geometry and sedimentary facies, and the inference of current pathways.

Latterly, it has become apparent that some drift deposits play a determining role in the progradational building of the continental shelf and slope (Fulthorpe & Carter 1991; Seranne & Abeigne 1999). This paper discusses the sedimentary texture, composition, structure and origin of one such field of drifts, cored to a depth of 513 m (*c.* 3.9 Ma) offshore at Ocean Drilling Program (ODP) Site 1119 (Carter *et al.* 1999). Site 1119 is located at 394 m water depth on the upper continental slope, about 100 km east of Timaru, New Zealand (Fig. 1a and b). The sediments there represent the most recent part of a long-lived, *c.* 24–0 Ma, succession of terrigenous drifts that underlie the eastern South Island coastal plain–shelf–slope sediment prism (Fig. 2) (Carter, R. M., *et al.* 1996) and form an important part of the Eastern New Zealand Oceanic Sedimentary System (ENZOSS; Carter, L., *et al.* 1996).

This paper summarizes the available published information on the Canterbury Drifts, both offshore (ODP Site 1119) and onland (Bluecliffs Formation). New sediment textural analyses provide insights into the evolution of the drift succession since the Early Miocene, and comparison between onshore and offshore sites contributes to our understanding of the climatic and oceanographic history of the region.

Sediment analyses were conducted according to the laboratory protocols described in the Appendix. For discussion, textural data have been aggregated into three grain-size classes, cohesive mud (cM; <8.70 or <9.48 μm), sortable silt (sZ; <60.65 or <56.09 μm) and sand (sS; >60.65 or >56.09 μm). These classes represent the bin-boundary grain diameters on the laser particle sizer that, for the 2000 or 600 mm lenses, respectively, approximate to the conventional cohesive–noncohesive and silt–sand boundaries of *c.* 10 and 62.5 μm. It should be noted also that the non-specific allocation of clay–very fine silt modes to <5 or <10 μm is because these grain sizes fall at the lower end of the Mastersizer grain-size spectrum (using, respectively, the 600 or 2000 mm lens), the accuracy of mode identification in these ranges being degraded by instrumental edge effects.

Previous research on the Canterbury Drifts

Seismic delineation

The Pliocene–Recent Canterbury Drifts (Fulthorpe & Carter 1991) were deposited on the mid–upper slope east of South Island New Zealand, in depths of *c.* 400–1500 m from north-travelling Subantarctic Mode Water (SAMW) and Antarctic Intermediate Water (AAIW), and their predecessor water masses (Fig. 2). The drifts were first imaged on reconnaissance petroleum exploration seismic lines shot by Gulf (1973), and their widespread regional distribution was revealed by a detailed seismic survey of the offshore Canterbury Basin (Western Geophysical 1982). Small, upslope-migrating sediment drifts are also known to occur on the modern sea bed in the head of the Bounty Trough (Fig. 1c), but remain undescribed in detail.

Using the Western Geophysical seismic database, Fulthorpe & Carter (1991) showed that deposition of the offshore Canterbury Drifts has played a pivotal role in the seaward progradation of the continental shelf sediment prism since at least the Late Miocene. Drifts initially form on the slope at depths several hundred metres below the shelf edge, from which they are separated by a channel-like, longitudinal gutter, and build up alongshope and shorewards (west) under the influence of the Coriolis effect. Once the crest of a drift approaches the same depth as the continental shelf edge, sediment aggradation fills the intervening gutter. Thereby, the seaward part of the crest of the drift comes to mark a new shelf edge. By this process, the shelf edge intermittently vaults seawards across the gutter and half the width of the newly emplaced drift (5–15 km) in a single step (see Fig. 3a).

An extensive grid of new high-resolution multichannel seismic profiles through the Canterbury Drifts was collected during R.V. *Maurice Ewing* cruise EW00-01 in 2000. These data, the upper parts of which have a resolution of *c.* 5 m, have been analysed by Lu *et al.* (2003) and Lu & Fulthorpe (2004), who described the 3D shape of what they termed elongate, complex and subsidiary drifts. Those workers recognized 11 major drifts within the younger, Late Miocene–Pleistocene part of the Otakou Group, and established a detailed model for their deposition.

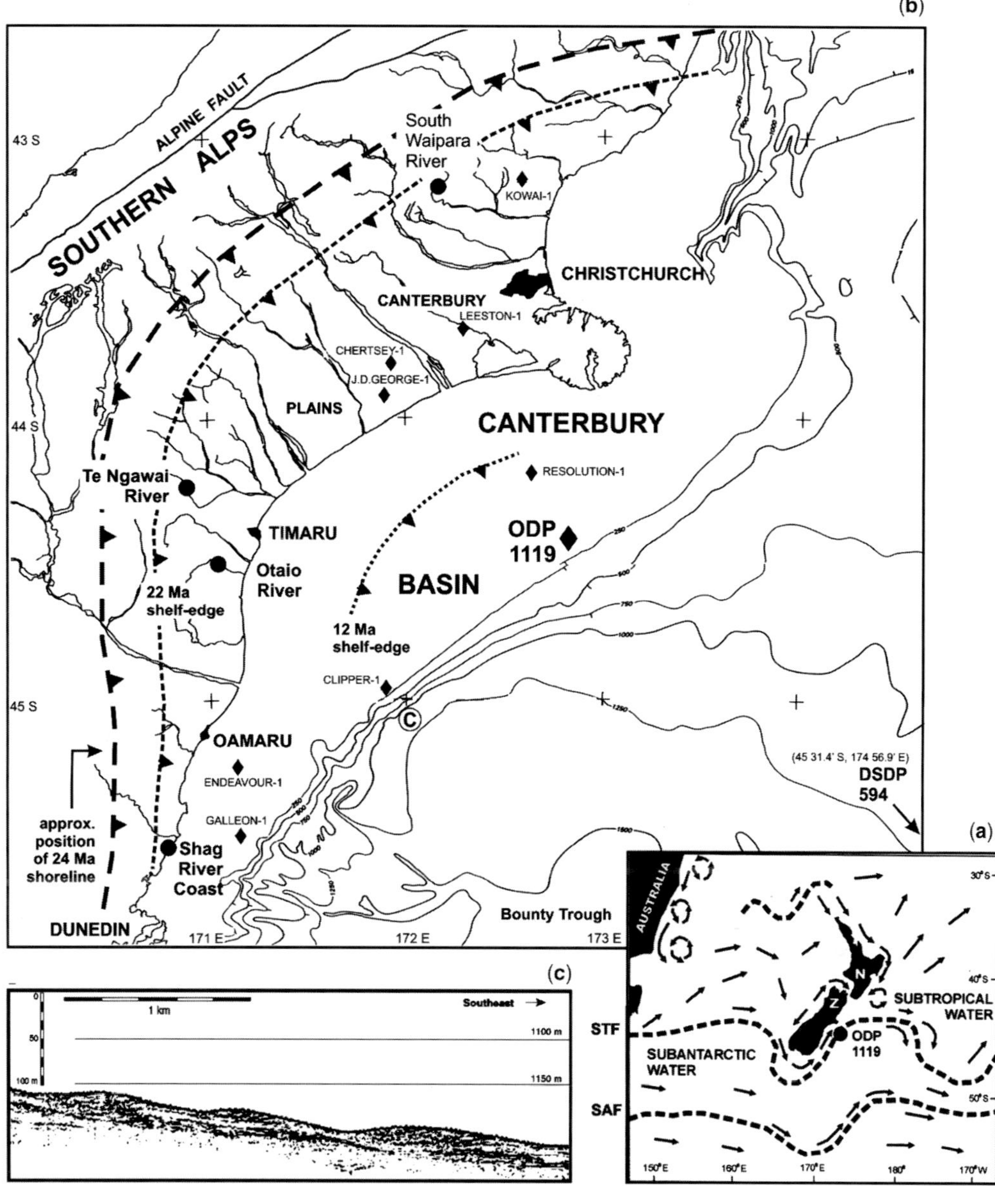

Fig. 1. (**a**) Regional oceanographic setting for the New Zealand region, showing the disposition of modern surface current flows and fronts. STF, Subtropical Front; SAF, Subantarctic Front. (**b**) Central and eastern South Island, showing location of the Canterbury Basin, ODP Site 1119, petroleum drill-holes and important localities for onland drifts of the Bluecliffs Formation. Bold dashed line indicates the Late Oligocene (24 Ma) position of the western edge (shoreline) of the Canterbury Basin at the commencement of regression and eastward shelf–slope progradation; the approximate position of the shelf-edge at *c*. 22 and *c*. 12 Ma are also indicated. (**c**) A 3.5 kHz profile downslope and SW of ODP Site 1119, showing small (1 km spaced) modern sediment drifts at a water depth of 1100 m (location marked, as **c**, in (**b**)).

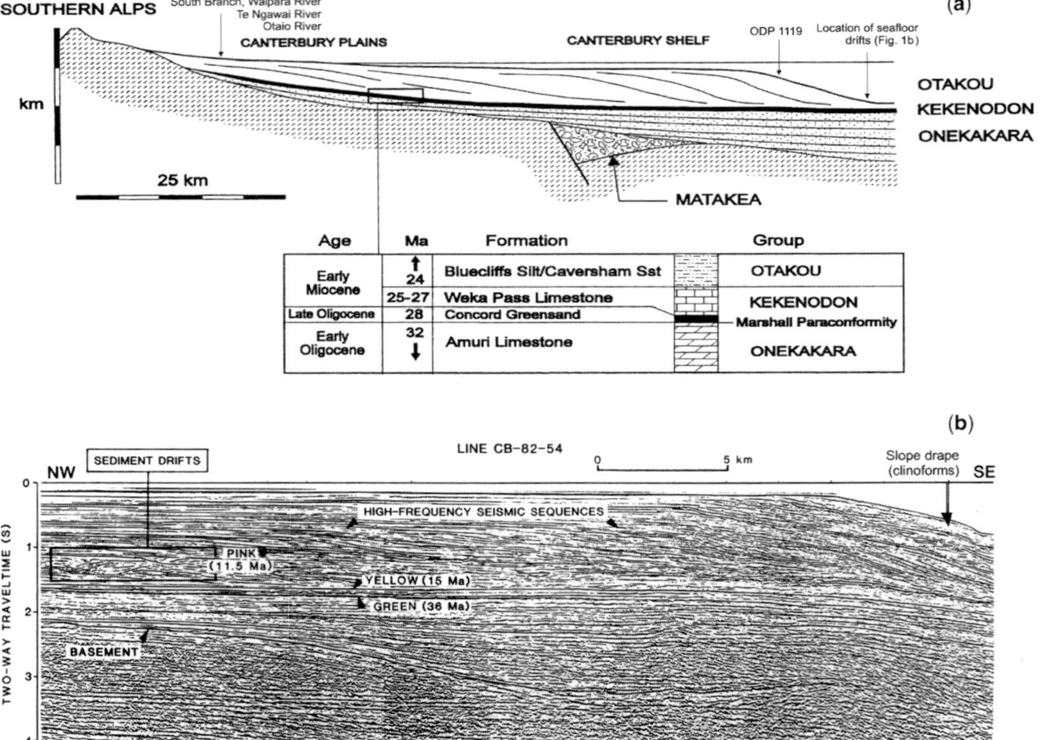

Fig. 2. (**a**) Schematic west–east profile across the Canterbury Basin. Late Cretaceous, post-Gondwanan rifting (rift-fill Matakea Group) was followed by thermal subsidence and marine transgression until the Early Oligocene (onlapping Onekakara Group). At *c*. 33.5 Ma, opening of the Tasman Rise gateway far to the SW caused the establishment of strong proto-Circum Antarctic Current flows across the largely (Fleming 1962) or completely (LeMasurier & Landis 1996) submergent New Zealand Plateau, causing the formation of a regional paraconformity (Marshall Paraconformity) followed by current-emplaced deposition of greensands and bioclastic carbonates (oceanic plateau, Kekenodon Group). Uplift on the Alpine Fault plate boundary commenced in the Late Oligocene (*c*. 25 Ma), the yield of new terrigenous sediment causing the eastward progradation of the shelf–slope prism (Otakou Group). Clinoform foresets of the Otakou Group correspond onland to the Bluecliffs Formation. The Pliocene to Recent Canterbury Drifts are centred under the outer shelf and slope in the vicinity of ODP Site 1119, but mounded drifts as old as Late Miocene occur within the Otakou Group, and sediments with drift characteristics occur down to the base of the group onland. (**b**) Commercial (low-resolution) seismic line through the outer shelf and upper slope, central Canterbury Basin, showing geometry of the major stratigraphic units. Kekenodon Group is delimited by the green (below) and yellow (above) seismic reflectors (after Fulthorpe & Carter 1991).

ODP Site 1119

ODP Site 1119 provided the first drill-core through the younger part of the Canterbury Drifts, penetrating down as far as the top of Drift 10 of Lu & Fulthorpe (2004) with an age of *c.* 3.9 Ma. Carter *et al.* (1999) briefly characterized the drift succession as alternating thicker (intervals up to 10 m thick) muds and intervening thinner (intervals up to 3 m thick) sand beds. Most of the sand intervals, some of which are carbonate rich, correspond to warmer, interglacial periods when the water at the site was *c.* 100 m deeper than at glacial lowstand and the shoreline was located many tens of kilometres inboard (west) of the shelf edge. Carter *et al.* (2004*a*) showed that a number of smaller late Pliocene to Pleistocene drifts overlie the large, early Pliocene Drift D10. They interpreted the change in drift size and seismic geometry that occurred at *c.* 3.1 Ma as reflecting intensified intermediate water flow, consequent upon climatic deterioration and the development of Northern Hemisphere glaciation.

Regional distribution of drifts

Plate boundary uplift during the later Cenozoic has caused the gentle eversion of the western (landward) parts of the Canterbury Basin succession, such that Early to Middle Miocene equivalents of the Site 1119 drifts are today exposed onland as the regressive Bluecliffs Silt Formation (Gair 1959; Carter 1988). These older drifts are widely present beneath the continental shelf and coastal plain of eastern South Island, which they have materially built (Fulthorpe & Carter 1991; Lu *et al.* 2003) (see Fig. 2a). Outcrops show that the lowest part of the terrigenous drift succession is marly, with an enhanced content of biopelagic tests, consistent with its deposition immediately above, and transition from, a mid–late Oligocene greensand–carbonate facies that was deposited widely on the New Zealand mid-Cenozoic oceanic plateau (Kekenodon Group; see Ward & Lewis 1975). The marls represent the distal (seaward) toes or bottomsets of the eastward-advancing prism of terrigenous sediment.

The term Canterbury Drifts is here extended to encompass all the terrigenous sediment drifts that underlie the eastern South Island coastal plain, shelf and slope, and that range from latest Oligocene to Recent in age. These sediments display a mineralogy that reflects their origin from South Island Rangitata Synthem rocks, especially the Haast Schist (chlorite–biotite zone), and its north-flanking Torlesse metagreywacke (prehnite–pumpellyite zone) belt (e.g. Reed 1957; Craw 1984). Terrigenous sand and silt petrography is dominated by quartz, albite, epidote–clinozoisite, sericite and chlorite.

Terrigenous-silt-dominated units of Early Miocene to Pliocene age have been mapped widely across the Canterbury Basin under local names that include the Grey Marl (Hector 1877), Mt. Harris Beds (Hector 1884), Goodwood Limestone (Service 1934), Tokama Siltstone (Mason 1941), Awamoa Beds (Gage 1957), Waikari Formation (Andrews 1963), Conway Formation (Warren & Speden 1978), Waima Siltstone (Lewis *et al.* 1980) and Caversham Sandstone (Bishop & Turnbull 1996). These units, and probably also some of the younger Pliocene Greta (Hutton 1888) and Motunau (Buchanan 1870) siltstones and their equivalents, are part of the regional Bluecliffs Formation of Carter (1988) and the Canterbury Drifts as interpreted here.

Canterbury Drifts offshore (ODP Site 1119)

Stratigraphy and lithology

The topmost interval at Site 1119, unit A, 0–86.19 m composite depth (mcd), comprises sediment that has been bypassed seawards across the shelf and shelf edge to accumulate as a clinoform drape on the upper slope (Carter *et al.* 2004*b*) (Fig. 3a and b). This surficial 86 m thick interval of terrigenous muddy siltstone and muddy very fine sandstone has a sand:mud bed ratio of 1:23 and an average sedimentation rate of 36 cm ka^{-1} (Unit A; 0–0.252 Ma; MIS 1–7). Below a small 25 ka long unconformity in MIS 8, Site 1119 penetrated 428 m of similar sediments, which represent the Canterbury Drift succession *sensu stricto* (Units B–C; 0.28–3.92 Ma; MIS 8–Gi 11) (Carter & Gammon 2004). Units B2–3 exhibit lesser sedimentation rates (20 and 21 cm ka^{-1}, respectively) and higher sand:mud bed ratios (1:8 and 1:7, respectively) than the clinoform strata above. Lower in the section still, Units B4–C2 are characterized by remarkably constant sedimentation rates of *c.* 10 cm ka^{-1}, and a sand:mud bed ratio either between 1:6 and 1:8 (Units B4–B5c) or an absence of sand beds altogether (Units C1–2). Overall, Site 1119 records a total sand bed thickness of 53.92 m out of a total thickness of 513.5 m of sediment, for an average sand:mud bed ratio of 1:9.5, and the succession was deposited at water depths that shoaled from *c.* 900 to 400 m over the last 3.9 Ma.

Site 1119 cores and Formation Micro-Scanner (FMS) images (Figs 3c and 4) show that Units B and C comprise centimetre-scale-bedded, silty mudstone and muddy and sandy siltstone, variously

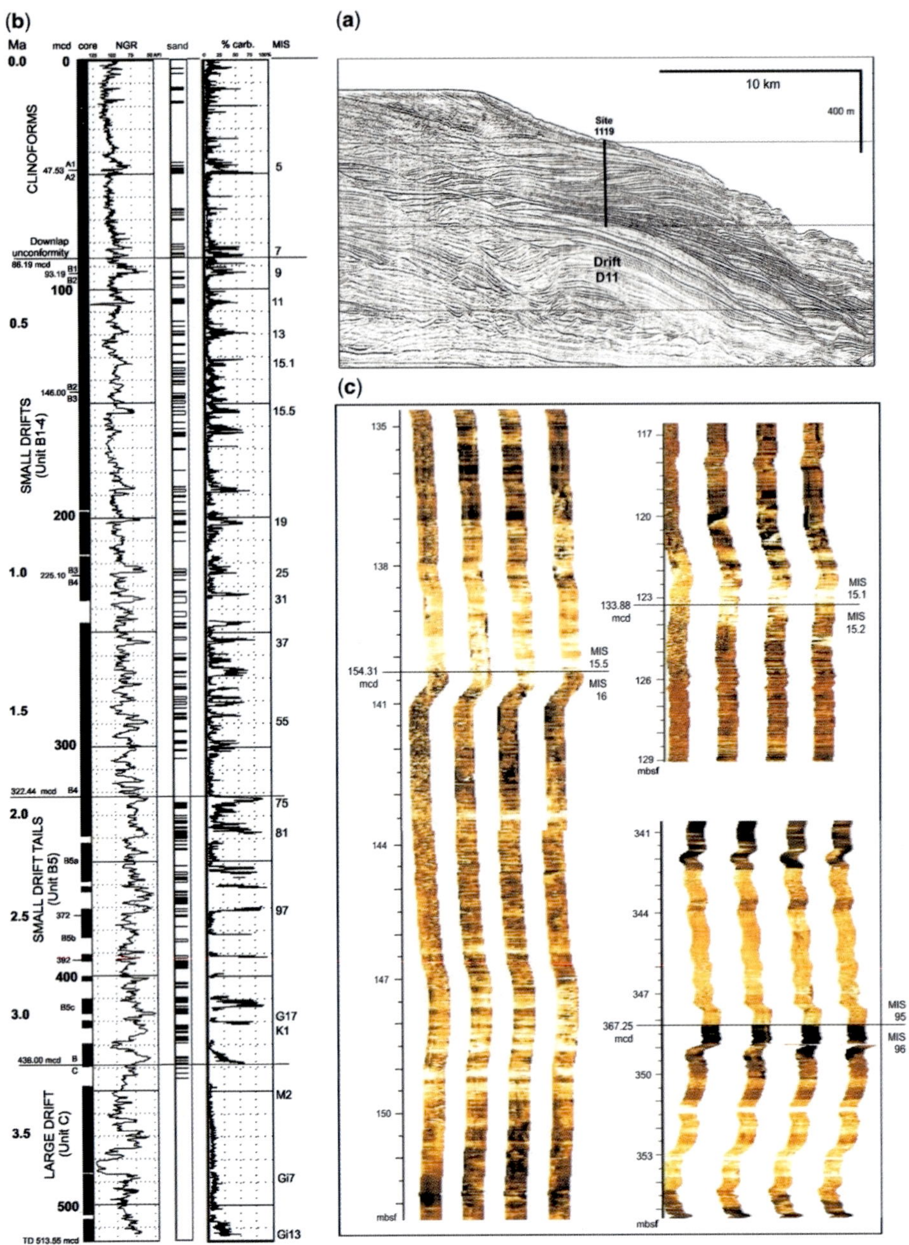

Fig. 3. (**a**) High-resolution seismic section through the Canterbury Drifts at ODP Site 1119 (courtesy C. Fulthorpe). (**b**) Stratigraphy of Site 1119. Columns, from the left, are: age (Ma), core depth in metres composite depth (mcd), core recovery (core), natural gamma-ray intensity (NGR), sand log (sand), grey-scale reflectance (% carb.) and selected Marine Isotope Stages (MIS; identified from the NGR record, after Carter & Gammon 2004). (**c**) Formation Microscanner (FMS) resistivity images of drift intervals 117–129 (unit B2), 135–152 (units B1–B2) and 341–355 m below sea floor (mbsf) (unit B5a); white and yellow indicate high resistivity (concretionary layers, sands); orange to brown, lower resistivity (sandy silts, silts); black, borehole washout. Noteworthy features area: (1) the presence of a sharp-based, normally graded sand near 140 mbsf; a sharp-based and sharp-topped sand between 343 and 348 mbsf; rhythmic double-graded sand intervals centred near 123, 143 and 148 mbsf; (2) the pervasive background presence of undulose, centimetre-scale, interbedded coarser and finer silts.

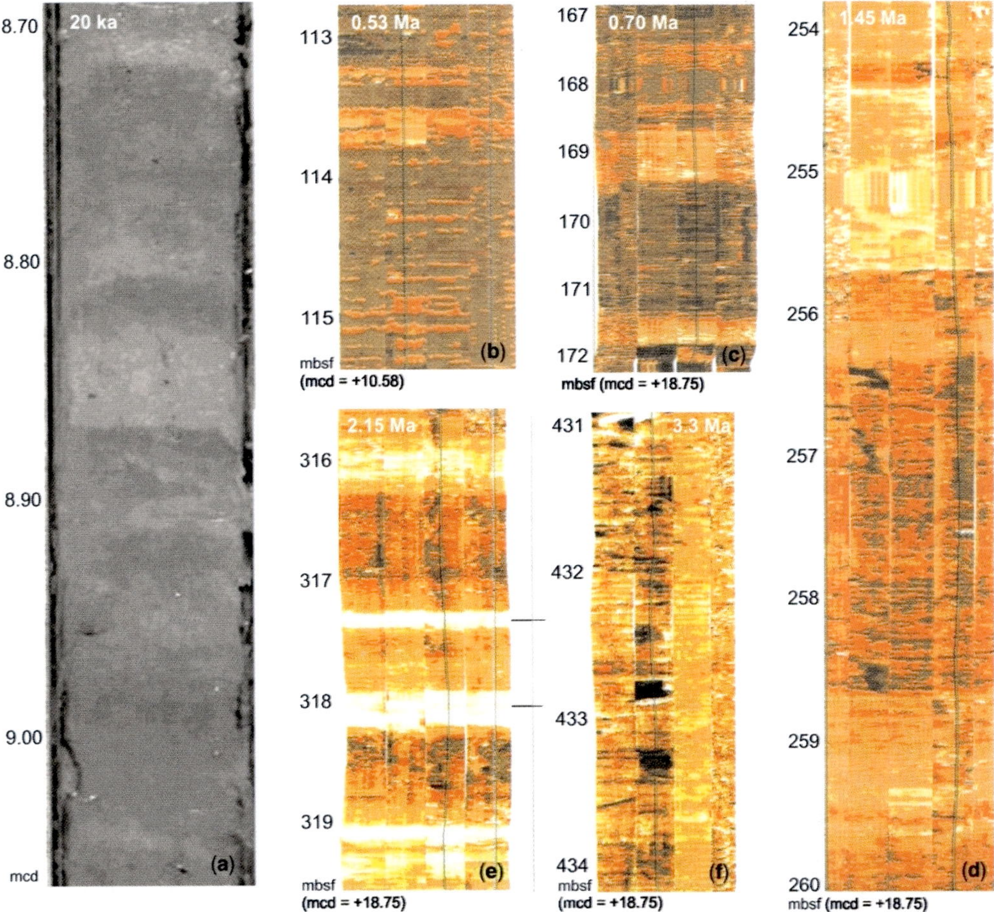

Fig. 4. Images from core 1119C. (**a**) Core photograph from the interval 8.70–8.10 mcd, showing interbedded, slightly burrowed, light and dark muds deposited during the last glacial maximum. The lighter muds have a higher carbonate and sand content, the darker muds are clay-rich, and their interbedding is inferred to represent the settlement of seasonal flood plumes (dark grey) against a background of more general hemipelagic mud (light grey). This bedded-mud facies is clay-rich (see Fig. 6a–c), mainly restricted to cold climate parts of Unit A (when the shoreline was nearby), and does not occur within the underlying Canterbury Drifts. (**b**–**f**) composite FMS resistivity images of selected drift intervals; white and yellow–gold indicate high resistivity (concretionary layers, sands); orange to brown, lower resistivity (sandy silts, silts); black, borehole washout. The variety of style in the *c*. 30–110 cm thick sand beds should be noted; some are both sharp-based (**c**, 171.9 mbsf; **d**, 255.8 mbsf) and sharp-topped (**b**, 113.5 mbsf), others are gradationally based (**b**, 113.8 mbsf; **d**, 259.9 mbsf) and gradationally topped (**c**, 171.4 mbsf; **e**, 317 mbsf). In the more deeply buried drifts, incipient cementation may give rise to high resistivity in sand-bed cores (arrowed; **e**, 317.3, 318.1 mbsf). The sands are set in a background of pervasively interlaminated, laterally pinching and swelling, silty sand (lighter) and sandy silt (darker) on scales of *c*. 1–10 cm. Some intervals of background mud are sand-rich (e.g. **b**, 113–115 mbsf), others mud-rich (e.g. **c**, 167–172 mbsf). The absence of discrete, thicker sand beds from drift Unit C (**f**, 431–434 mbsf), but retention of the thin-bedded, current-drift fabric in the silty sands and sandy silts, should also be noted.

burrowed or bioturbated, and punctuated by 2–300 cm thick beds of muddy, very fine-grained quartzofeldspathic sand. Internally, some of the thicker sand units display pulsatory changes in sand or mud content; organized Bouma turbidite structures (see Stow *et al.* 1998) are not present. Some sands have both gradational bottoms and tops, and comprise a reverse- to normal-graded mud–sand–mud triplet similar to descriptions of other sediment drift deposits (Stoker *et al.* 1998) and to the idealized drift facies model of Faugères *et al.* (1984) and Stow *et al.* (1998). The overall characteristics of the Site 1119 muds, silts and sands below 86 mcd coincide with those first described for sediment drifts from the Feni Drift by Faugères *et al.* (1984), namely, 'contacts between the different facies may be more or less planar and sharp, distinctly erosional, or completely gradational, and all three types are equally common at the base or at the top of separate beds ... Mottling occurs throughout ... together with other more or less distinct and isolated pockets and streaks.' Virtually all of the cored sediments of the Canterbury Drifts, between 86 and 514 mcd (TD at Site 1119), therefore, belong to the 'silt and sand' and 'mottled silt and mud' sediment drift facies of Faugères *et al.* (1984) and Stow *et al.* (1998).

Finally, the 328 m drilled thickness of Units B and C at Site 1119 appears to contain only two thin (<1 cm thick) beds of graded sand that might be interpreted as slope turbidites (Carter *et al.* 1999). Overall, therefore, the lithology of the dominantly siltstone and mudstone sediment pile (which is best revealed by FMS images; Fig. 4) is consistent with the seismic geometry of the deposits (see Fig. 3a) and with contourite deposition from systematically fluctuating bottom currents.

Grain-size textural distributions

Grain-size distributions have been grouped according to stratigraphic unit, in order downwards from the sea bed (Table 1). Samples from units A and B are similar in their lithological range, comprising a background of silty mudstone–muddy or sandy siltstone with distinct modes of very fine (<5 μm), fine (8–16 μm) and coarse (35–60 μm) silt, and interspersed beds of muddy, very fine- or medium-grained sandstone (modes 100–130 and 235–440 μm) (Figs 5a–f and 6b,c). Unit C (large drift D11) differs in its lack of sand interbeds and in having a dominant coarse, rather than very fine or fine, silt mode (Figs 5g and 6a). For graphical presentation, samples have been divided into two main facies groups, silty mudstone + muddy siltstone and silty sandstone + sandy siltstone, respectively. Generally, these groups share common modes though with different proportions of each, as represented by average cM:sZ:sS ratios of 45:47:9 and 14:24:63 for the mudstone and sandstone facies, respectively (Table 1).

Sediments of grain sizes 12, 35 and 110 μm diameter become entrained at current speeds of *c.* 11, *c.* 15 and *c.* 25 cm s^{-1}, respectively (Miller *et al.* 1977; McCave *et al.* 1996; McCave 2005; Table 1). Against such a background, the sands that occur interspersed throughout the Unit B drifts commonly display medium to coarse sand modes (235–440 μm), indicative of enhanced current strengths up to 50 cm s^{-1}. That some sand beds have sharp bases indicates that in such cases even stronger currents eroded the underlying muddy sea bed prior to waning to allow sand deposition.

Depth plots of sediment texture

Despite a variable sample spacing (Fig. 7a) and core gaps (Fig. 3), depth-ordered, contiguous-sample plots of the cM:sZ:sS ratio and mean and modal grain sizes reveal useful information about grain-size trends against stratigraphic height (Fig. 7b and c).

The succession is punctuated by episodic sand beds, which occur within a silt- and cohesive mud-dominated sediment background. A long-term trend of increasing cohesive mud and decreasing sand and silt starts at *c.* 390 mcd (*c.* 2.75 Ma). This trend passes up across the downlap unconformity and into the clinoform downlapping sediments of unit A, where cM:sZ:sS ratios as mud-rich as 54:40:7 occur. Although a single trend line can be fitted to this long-term mud increase, the data could also be interpreted as showing stepped increases in mud content at *c.* 350 mcd (2.3 Ma) and *c.* 150 mcd (0.7 Ma).

Above 320 mcd (unit B4 and above; small drifts), sand is mostly represented by terrigenous grains, dominantly quartz; between 320 and 440 m (unit B5; drift tails), sands also contain large numbers of calcareous microfossils and shell fragments (Carter *et al.* 1999). The background silts and muds of drift units B1–B5 have average cM:sZ:sS ratios of 41:50:9, in comparison with 21:52:28 for unit C (large drift D11). In parallel with the trend of increasing mud content upwards, the dominant grain-size mode decreases from coarse silt (40–55 μm) in unit C, to fine silt (9–16 μm) in units B4–A2, to very fine silt (<5 μm) in unit A1 (Table 1). The two dominant mean sand grain sizes, similar across the succession, are very fine (100–130 μm) and medium (250–440 μm) sand, with a concentration of the latter, coarser mode in small drifts B3 and B4 (Fig. 7c, Table 1).

Canterbury Drifts onland (Bluecliffs Silt Formation)

A nearly complete sequence through the Bluecliffs Silt Formation occurs at its type locality in the Otaio River, a little inland from Timaru, central South Island (Figs 1b and 8). Samples were collected at Otaio River through the middle part of the formation (*c.* 50–320 m above the base), above its poorly outcropping basal part. Samples were also collected from two other well-exposed sections through Bluecliffs Formation (Te Ngawai River; central basin, inland of Otaio River, Fig. 8b; south branch of Waipara River; northern

Table 1. *Summary table of grain-size modes for the Canterbury Drifts at Site 1119 and onland exposures of the Bluecliffs Formation*

		ODP Site 1119 (all samples)													
		Cohesive	Sortable Silt	Sand	*n*	%'mud'	%'sand'	S:M							
< 20% sand	M	44.74	46.73	8.54	115	91.46	8.54	0.09							
20–50% sand	mS	24.71	40.97	34.32	50	65.68	34.32	0.52							
> 50% sand	S	13.75	23.53	62.72	38	37.28	62.72	1.68							
Drift 11	zC	20.61	51.68	27.72	17	72.28	27.72	0.38							
		ODP Site 1119 (mudstone facies only)							Modal Grain Diameters						
		Cohesive	Sortable Silt	Sand	*n*	%'mud'	%'sand'	S:M	vfz	1st fz	2nd mz	3rd cz	4th vfs	fs	ms–cs
	Unit A1	46.39	45.99	7.62	29	92.38	7.62	0.08	<5	8–16		40–50	100–117		or 350
	Unit A2	42.83	49.64	7.53	21	92.47	7.53	0.08	<5	9–13		37–42	100–130		or 300–355
	Unit B1	31.44	55.58	13	5	87.02	13	0.15	<5	11–13		35–50	100–110		or 350
	Unit B2	33.48	51.09	15.42	29	84.57	15.42	0.18	<5	10–13		35–50	100–130		or 285–440
	Unit B3	27.8	37.99	34.2	11	65.8	34.2	0.52	<5	9–16		35–65	110–125	or 180–275	or 265–330
	Unit B4	25.21	40.45	34.33	13	65.66	34.33	0.52	<5	12–16		35–60			235–405
	Unit B5	22.34	43.22	34.43	8	65.56	34.43	0.53	<5		16–28	or 35–55		125–270	
	Unit C	55.07	8.91	0.12	17	63.97	0.12	0	<5			40–55		170–260	
	All (except C)	32.79	46.28	20.93	116	69.18	18.32	0.26							
	All (including C)	35.57	41.61	18.33	133	77.18	18.33	0.24							
				Total samples	133										
Bluecliffs Silt onland															
Walpara River	All samples	12.66	46.4	40.94	20	59.06	40.94	0.69	<10		27	or 55–65	or 80–140		
Te Ngawai River	All samples	7.96	41.12	50.92	36	49.08	50.92	1.04	<10				80–105		700–1200
Shag River mouth	All samples	7.45	45.96	46.59	16	53.4	46.59	0.87	<10			45–50	or 80–105		300–800
Otaio River	All samples	5.67	46.94	47.39	36	52.61	47.39	0.9							
	Silt facies	6.97	51.62	41.42	17	58.58	41.42	0.71	<10				70–100		300–800
	Sand facies	4.51	42.75	52.74	19	47.26	52.74	1.12	<10				70–100		720–960
	All onland	8.43	45.1	46.46	108	53.54	46.46	0.87							
Grain diameter (µm)									<10	12	30	60	110	250	450
Entrainment velocities (cm s^{-1})									cohesive	11	13	17	25	38	50

Inferred current speed figures are estimated for 1 m above the bed, based upon Miller et al. (1977) and McCave et al. (1995).

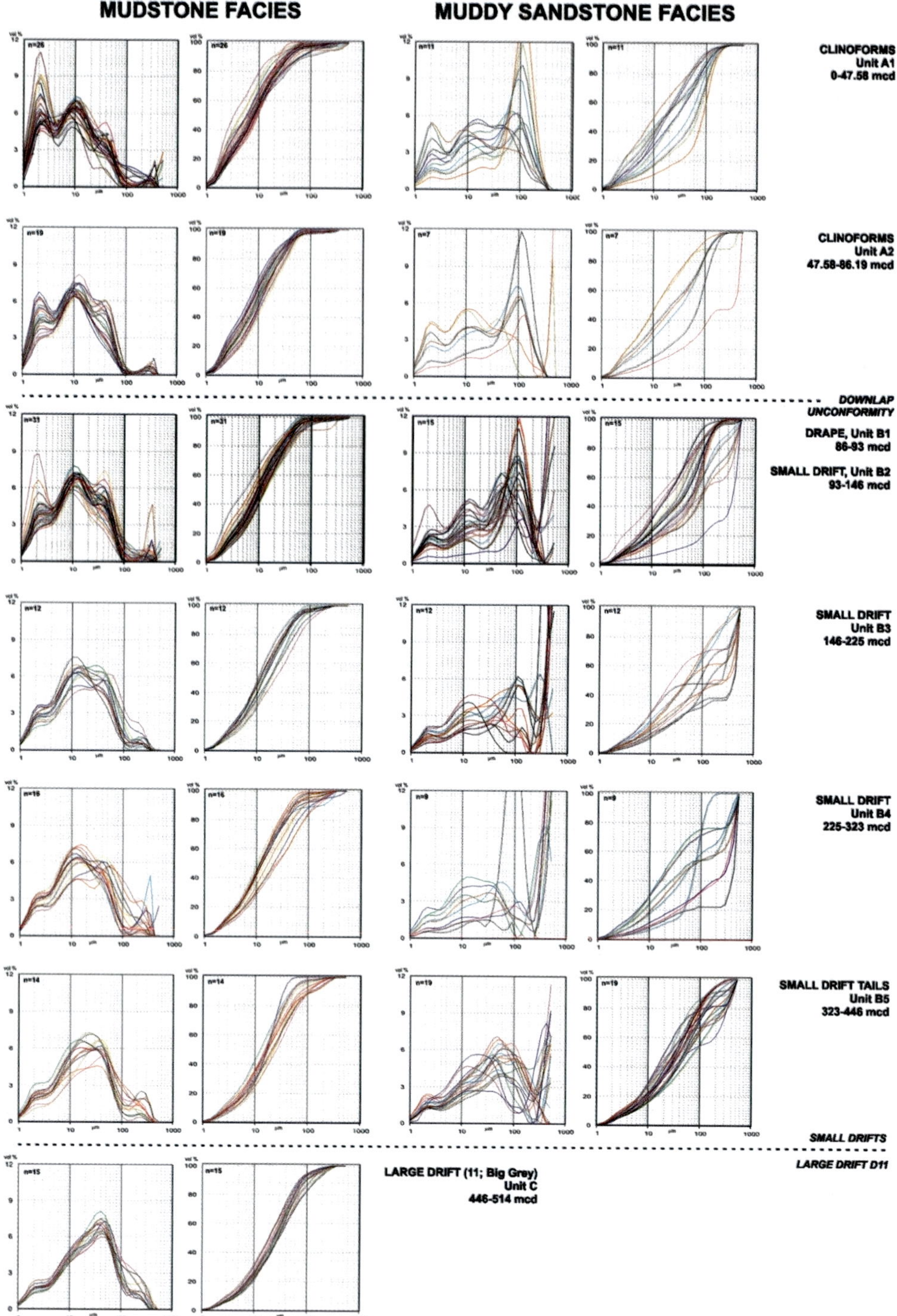

Fig. 5. Grain-size distributions for ODP Site 1119 sediments, determined using a Malvern laser particle sizer fitted with a 1.2–600 μm lens. Left column graph-pair and right column graph-pair comprise histogram and cumulative frequency plots for the mudstone and muddy sandstone facies from each specified stratigraphic interval, respectively. Data plotted on a logarithmic x-axis for all graphs; y-axes indicate volume percentage and cumulative volume percentage, respectively.

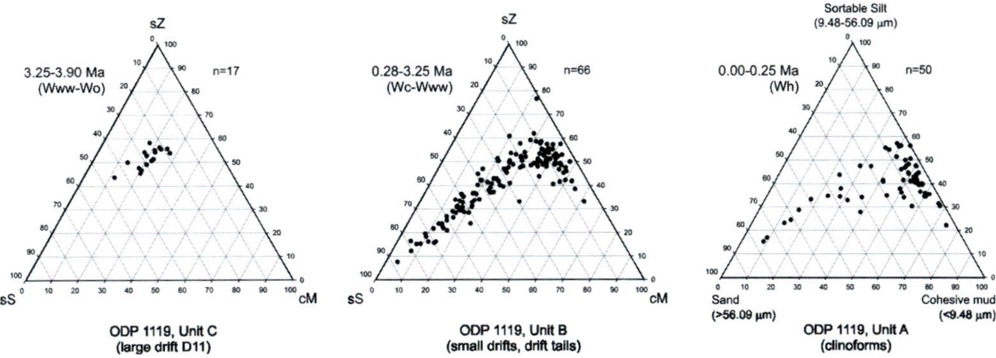

Fig. 6. Tri-plots for cohesive mud:sortable silt:sortable sand (sS) (cM:sZ:sS) for sediments from units A, B and C at Site 1119.

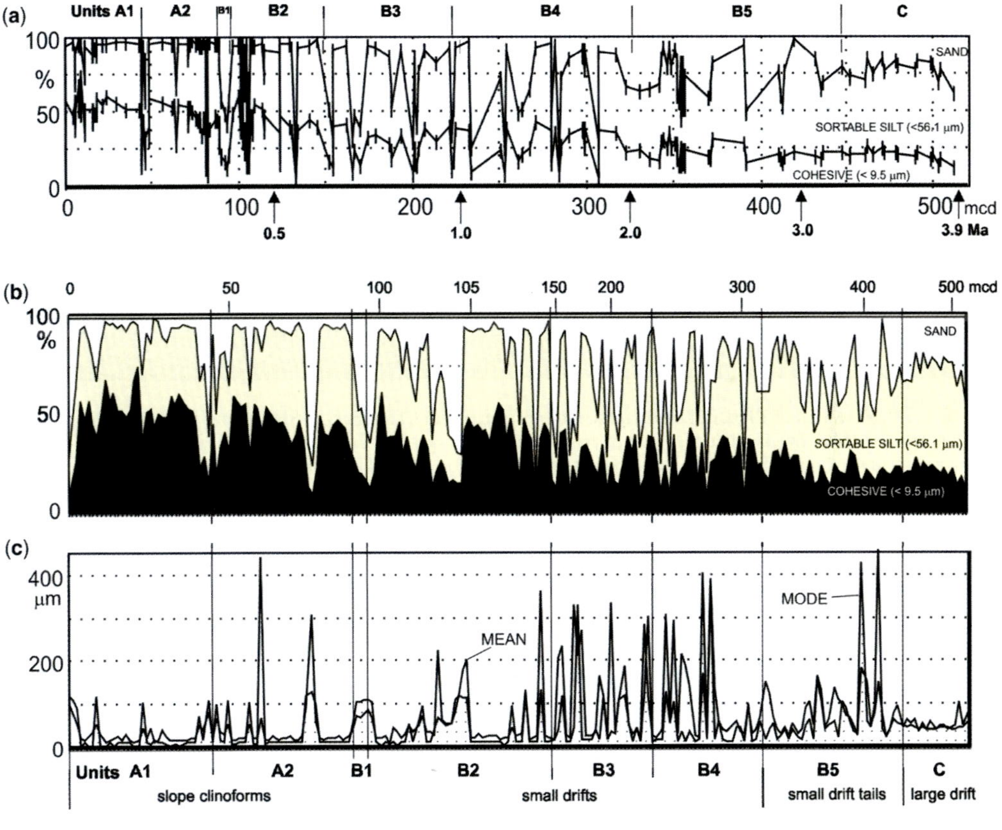

Fig. 7. (**a**) Sediment textural data for ODP Site 1119 plotted against metres composite depth, and (**b**) collapsed depth (i.e. samples plotted left to right in depth order and contiguous to each adjacent sample, regardless of the true interval that separates them). Percentage amounts of cohesive mud (black in **b**), sortable silt and sortable sand are plotted throughout Site 1119. (**c**) Grain-size mean and primary mode against collapsed depth (note the correspondence in pattern between these two measures, and the correlation of coarser peaks with the sand-rich intervals delineated in (**a**) and (**b**)).

Fig. 8. (**a**) Type section of Bluecliffs Formation, Otaio River, South Canterbury, beds dipping 20° ESE. (Note the *c.* 20 cm thick cemented layers located about 3 m apart, which lie at and just above intervals of sand, some sharp-based and normally graded, others gradationally based with reverse-to-normal grading.) (**b**) Outcrop transition between Weka Pass Limestone and Bluecliffs Formation Siltstone, Te Ngawai River, South Canterbury. The gradational change between the two units, indicated by the argillaceous upper limestone and marly lower siltstone intervals, should be noted. (**c**) Photomicrographs (×80, plane-polarized light) of (above) Goodwood Limestone, Shag River Coast and (below) Bluecliffs Formation, Te Ngawai River, basal marly siltstone. The poorly sorted nature should be noted; clayey matrix contains silt-sized grains of clear, angular quartz (q), sericite (s), pelleted glauconite (g), sponge spicules (arrowed, upper photograph) and planktic foraminifers (arrowed, lower photograph). Scale bars represent 100 μm.

basin), and through a coastal outcrop of the upper part of the formation on the coast south of Shag River ('Goodwood limestone'; southern basin).

Stratigraphy and lithology

The *Otaio River section* is the type for both the Bluecliffs Formation and the local New Zealand Early Miocene Otaian biostratigraphic stage (23.2–21.7 Ma), and because of this has recently been subjected to detailed study (Morgans *et al.* 1999). The section comprises a *c.* 330 m thick succession of regressive, terrigenous siltstone and sandstone, marly in its lower part, which succeeds an offshore carbonate platform deposit (Weka Pass Limestone; *c.* 30 m thick) and is followed by a shoaling shelf to shoreface, sand-dominated unit (Southburn Sand; 150+m thick). The section represents the eastward passage of the slope–shelf succession, as it built eastwards and downlapped onto the pre-existing, quasi-horizontal, regional Oligocene carbonate platform (Fig. 2a). That the succession in Otaio River is only *c.* 330 m thick, compared with *c.* 2000 m thick for its equivalent beneath the modern outer shelf and slope, is consistent with its accumulation in a relatively nearshore geographical position in initial platform water depths of perhaps 450–750 m.

Bluecliffs Silt comprises a background continuum of sandy siltstone–silty very fine sandstone, punctuated by 1–5 m spaced beds or intervals of muddy, medium- to coarse-grained sandstone up to 30 cm thick (Figs 8a and 9b). The sand beds are often concretionary, and sometimes glauconitic or contain *in situ* bivalved venerids. Some sandstone beds are sharp-based and grade up into overlying siltstone, others are double-graded against both underlying and overlying siltstone. Small-scale bioturbation, often marked by pyritized burrows, is pervasive, but not to a degree that homogenizes the thicker sand–silt rhythmicity. Rare hydroplastic deformation in some beds attests to a sea-floor slope. Fossils include scattered molluscs (*Limopsis, Zeacolpus, Stiracolpus, Palomelon, Dentalium*), shell fragments, annelids and solitary corals (*Flabellum, Notocyathus*), all of bathyal aspect, and abundant microfossils and nannofossils. In their detailed study of the Otaio River section, Morgans *et al.* (1999) showed that sediment cyclicity occurs as *c.* 5 m spaced peaks in foraminiferal abundance which co-vary with decreases in coarse sediment (>74 μm) content. Planktic taxa make up 45–85% of total Foraminifera, and nannofossils up to *c.* 50% of the sediment matrix. Foraminiferal census counts suggest deposition of the lower part of the section at *c.* 600 m water depth, shoaling to depths of *c.* 100 m in the upper part, where the molluscan fauna becomes more diverse and contains taxa of uppermost bathyal to outer shelf aspect. The composition of nannofossil and foraminiferal assemblages indicates deposition from a cool-temperate water mass that was eutrophic, vertically well mixed and located in the transition zone between coastal and fully oceanic water.

The above description of Bluecliffs Silt at its type locality, and inferences therefrom, applies with only minor local modification to other outcrops of Bluecliffs Formation throughout the basin.

In the south branch of the Waipara River (Fig. 1b), the formation is 300 m thick and again early Miocene (Otaian) in age. The basal few metres (Pahau Siltstone of Andrews 1963) are calcareous and mildly glauconitic, marking the transition from the underlying carbonate platform deposit (Weka Pass Limestone). A conspicuous interval of well-sorted sands occurs between *c.* 50 and 100 m above the base (the Crantara Sandstone of Andrews 1963), above which again the formation becomes sandier upwards towards the shoreface sands of the overlying Southburn Sand (=Mt. Brown Beds). The Crantara Sandstone contains about six sharp-based, 1–3 m thick units of massive sand, which grade up to siltier intervals of either bioturbated sandy silt or centimetre-scale sand–silt flaser and lenticular bedding. These facies are clearly both rhythmic and strongly bottom-current influenced.

At *Te Ngawai River* (Fig. 1b), the Bluecliffs Silt is 324 m thick (Douglas 1975), with a slightly calcareous base that is transitional from the underlying limestone. Foraminifera indicate an age of either latest Oligocene (Waitakian Stage, *c.* 22.7–25.2 Ma; Finlay 1953) or earliest Miocene (Otaian Stage, *c.* 19.0–22.7 Ma; Vella, *in* Eade & Kennett 1962). Well-washed outcrops display an indistinct, 1.5–5 m spaced, rhythmicity of 20–40 cm thick slightly sandier layers separated by less sandy siltstone. The formation is bioturbated and slightly glauconitic throughout (Fig. 8c). The upper 15 m of the formation (separated by Douglas (1975) as the Upper Member) is sandier, more glauconitic, more fossiliferous and more bioturbated as it approaches the contact with the overlying shallow shelf–shoreface Southburn Sand. In his detailed study, Douglas (1975) showed that (1) the ratio of planktic:benthic Foraminifera is highest (between 60:40 and 75:25) in the lowest 30 m of the formation, averages 55:45 between 30 and 290 m, and falls to 20:80 over the top 35 m, and (2) between 5 and 30% of the sediment is a non-terrigenous coarse fraction, comprising foraminifers, molluscan shell fragments and glauconite. These characteristics are closely similar to those described by Morgans *et al.* (1999) for the Bluecliffs Silt at Otaio River, and indicate again that the formation represents shoaling (*c.* 600–100 m palaeodepth) oceanic slope environments adjacent to a narrow, sandy shelf–shoreface system to the west.

At the southern end of the Canterbury Basin, the early Neogene terrigenous succession is unusually sandy, cropping out around Dunedin as the muddy Caversham Sandstone (Bishop & Turnbull 1996). A similar facies occurs also offshore in exploration well Endeavour-1, off Oamaru. North of Dunedin, on the coast south of the Shag River mouth (Fig. 1b), the upper part of the Caversham Sandstone interval is represented by rhythmically bedded, Bluecliffs-facies bioturbated silts and muddy sands, 10–20 m thick packets of which show lensing, lateral onlap and other evidence of current-influenced deposition (Carter, R., *et al.* 1996, fig. 7b). The 20–40 cm thick muddy sandstone beds at the base of individual sedimentation units, which sometimes contain abundant siliceous sponge spicules (Fig. 8c) and diatom frustrules, often carry a concretionary calcite cement, causing the rather misleading name Goodwood

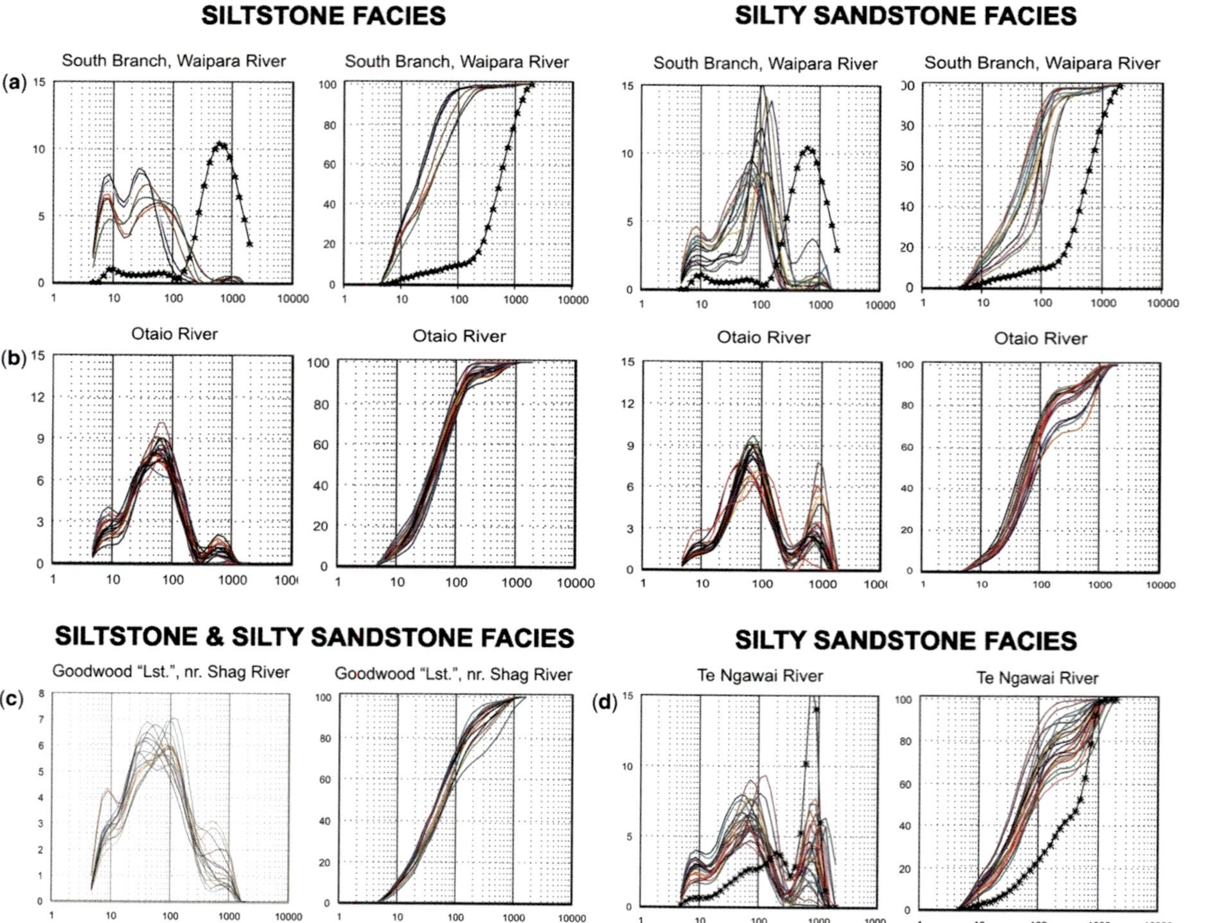

Fig. 9. Grain-size distributions for the onland Bluecliffs Formation from: (**a**) South Branch, Waipara River; (**b**) Otaio River; (**c**) coast near Shag River mouth; (**d**) Te Ngawai River. Determined using a Malvern laser particle sizer fitted with a 4–2000 μm lens. Left column graph-pair and right column graph-pair for (**a**) and (**b**) comprise histogram and cumulative frequency plots for the siltstone and sandstone facies from each locality, respectively. (**c**) and (**d**) are similar, but samples are not subdivided into silt and sand facies at each locality. Data plotted on a logarithmic x-axis for all graphs; y-axes indicate volume percentage and cumulative volume percentage, respectively.

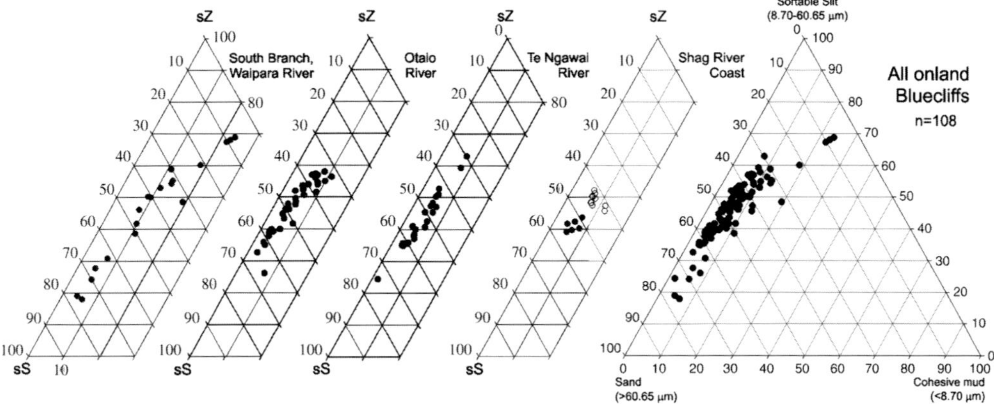

Fig. 10. Tri-plots for cohesive mud:sortable silt:sortable sand (sS) (cM:sZ:sS) for the onland Canterbury Drifts at (left to right): South Branch, Waipara River, Otaio River, Te Ngawai River, and coast south of Shag River mouth. Samples are replotted as a single, composite set on the full triangle (right).

Limestone (Service 1934) to be assigned to them. The 'limestone' succession is of early Miocene age (Otaian to Altonian Stages; *c.* 16–23 Ma; Benson, *in* Fleming 1955).

Grain-size textural distributions

Bluecliffs Formation sediments from all the onland Early Miocene localities are closely similar in their sedimentary characteristics and grain-size texture. They comprise a continuum of moderately sorted, silty sands and sandy silts with three dominant modes of very fine silt (<10 μm), very fine sand (70–105 μm) and medium–coarse sand (300–800 μm) (Figs 9 and 10; Table 1). As at Site 1119, the grain-size modes are common to both the sandy and silty facies, but differ almost symmetrically between facies in their relative proportions. For instance, at Otaio River, sands and silts have cM:sZ:sS ratios of 5:43:53 and 7:52:42, respectively, compared with a ratio of 8:45:46 for the average of all 108 onland samples (Table 1). The great majority of samples contain <12% mud, apart from four samples from the Pahau member at the base of the Bluecliffs Formation at Waipara River, which have mud contents of 18–23%.

Depth plots of sediment texture

Depth-ordered, contiguous-sample plots of the cM:sZ:sS ratio and mean and modal grain sizes demonstrate the stratigraphic and regional homogeneity of the Bluecliffs Formation, and reveal useful information about grain-size trends (Fig. 11).

The lower *c.* 50 m of the formation at Waipara and Te Ngawai is noticeably richer in mud (Fig. 11a and b), and all sections are characterized by decametre-wavelength cycles of fluctuating sZ:sS ratio, which are often paralleled by changes in the mean and modal grain size. The main sediment mode increases in size upwards through the basal parts of the Waipara River and Te Ngawai sections and thereafter fluctuates slightly in sympathy with the changing mean grain size. At Otaio River (section between 50 and 220 m) the very fine sand mode (70–100 μm) remains remarkably constant throughout fluctuations in mean grain size, which represent the intermittent presence of the medium sand mode.

Comparison between offshore and onshore drifts

The offshore Plio-Pleistocene strata from Site 1119 and the onshore, Late Oligocene and younger Bluecliffs Formation belong to the same stratigraphic body, which is the prograding prism of upper slope sediment that I have termed the Canterbury Drifts (Fig. 2).

The average compositions of sediment samples from different stratigraphic levels are plotted on a cM:sZ:sS textural diagram in Figure 12, which summarizes the same dataset that is plotted as individual points in Figures 6 and 10. A striking pattern emerges. Early Miocene sediments are low in cohesive mud, and form a tight-knit cluster of sandy silts and silty sands. In contrast, samples from mid-Pliocene drift, Unit C, contain on average *c.* 15% less sand and *c.* 15% more cohesive mud than the onland Bluecliffs Formation, and are intermediate between the Early Miocene samples and Late Pliocene–Pleistocene samples from Unit B5 at Site 1119. Successively younger Site 1119 intervals,

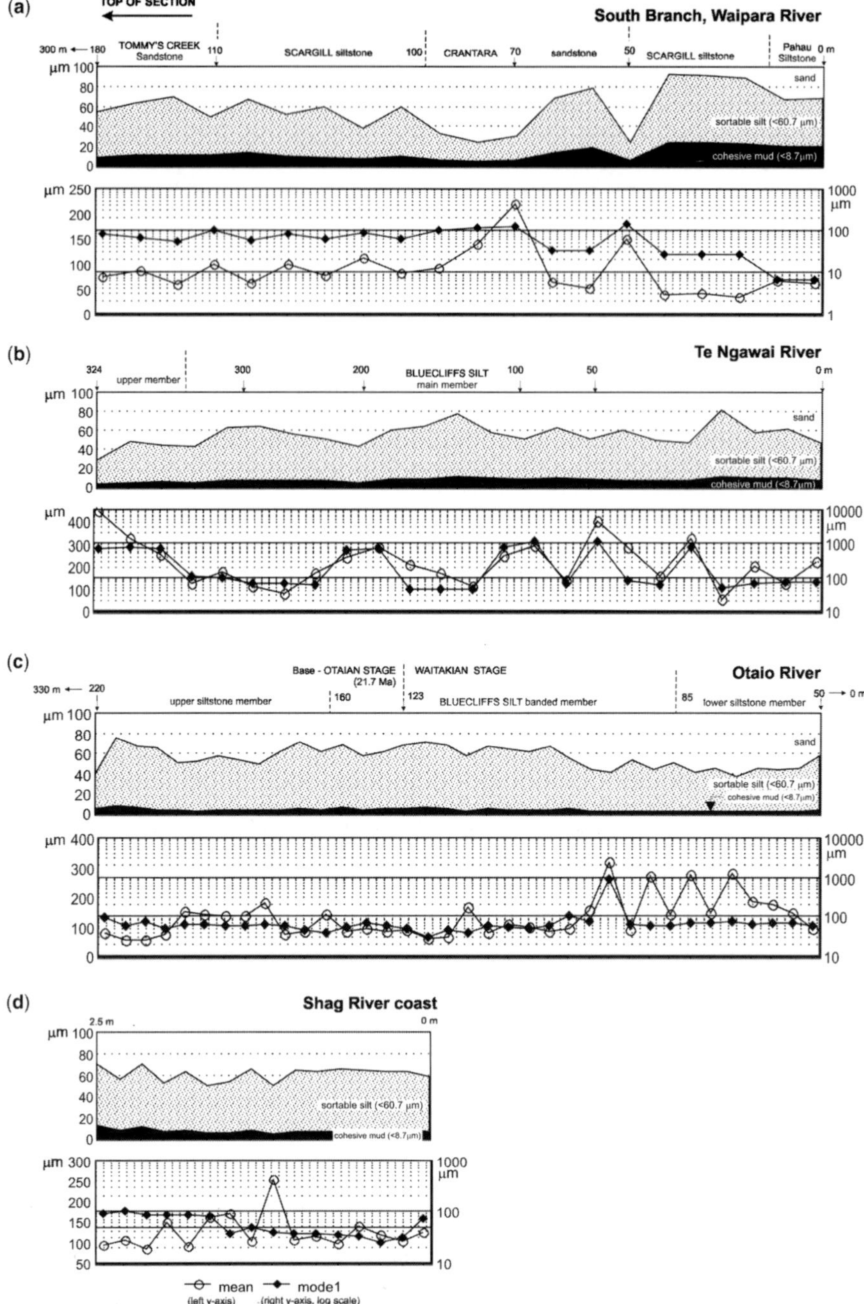

Fig. 11. Sediment textural data for onland Bluecliffs Formation localities plotted against collapsed depth (i.e. samples plotted left to right in depth order and contiguous to each adjacent sample, regardless of the true interval that separates them). Upper graph of each pair plots percentages of cohesive mud (cM; black), sortable silt (sZ; shaded) and sortable sand (sS; white); lower graphs plot the mean (left-axis arithmetic scale) and modal (right-axis logarithmic scale) grain size. (**a**) South Branch, Waipara River; lower and middle parts of Bluecliffs Formation (0–180 m of the total 300 m); (**b**) Te Ngawai River; complete section through Bluecliffs Formation (0–324 m); (**c**) Otaio River; middle parts of Bluecliffs Formation (50–220 m of the total 330 m); (**d**) coast south of Shag River mouth; small, characteristic interval of Bluecliffs Formation (3.5 m thickness).

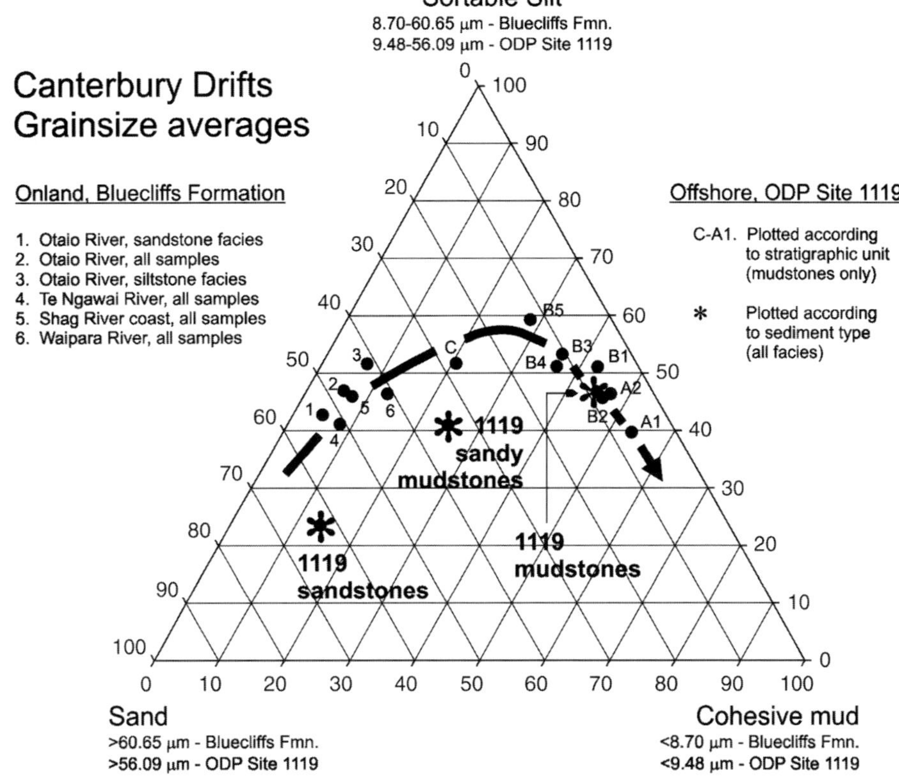

Fig. 12. Tri-plot for values of cohesive mud:sortable silt:sortable sand (sS) (cM:sZ:sS) for Canterbury Drift sediments from both offshore (ODP Site 1119) and onshore (Bluecliffs Formation). The average value is plotted for sediments from each locality (see Fig. 10) or indicated stratigraphic interval (see Fig. 6), and also for the three commonest facies variants at Site 1119, namely sandstone, sandy mudstones and mudstone.

B4 to B1, lie along a path of increasing cohesive mud, which is 5% higher than Unit C in Unit B5 and increases to 25% higher in the topmost drift, Unit B2. The trend to higher mud continues through into the clinoform Unit A, with Unit A1 samples containing 20% less sand and 33% more cohesive mud than Unit C samples.

Synthesis

The previous part of this paper contains a synopsis of the regional nature of the Late Oligocene to Recent Canterbury Drifts, and presents new data to help delineate their character. This information, in turn, points to new oceanographic, climate-history and stratigraphic perspectives, as discussed below.

Time period of drift deposition

The accumulation of substantial sediment drifts requires an extended period of stability, both geological (controlling, via tectonics, the source of sediment, the transport pathway and the site of deposition) and oceanographic (controlling, via current flows, the mechanism of deposition).

Morgans *et al.* (1999) showed that the base of the Early Miocene Otaian stage lies at 73 m above the base of the Bluecliffs Formation as exposed at Otaio River, and estimated its age as 21.7 Ma. A further *c.* 50 m of poorly outcropping calcareous siltstone (Waitoura Marl, here treated as a basal member of Bluecliffs Formation) lies beneath this section and above the top of the Weka Pass Limestone, indicating that the incoming of post-limestone terrigenous sediment, and the true base of the Bluecliffs Formation, lies *c.* 123 m below the base-Otaian boundary. Assuming a sedimentation rate of 10 cm ka^{-1}, this thickness of sediment corresponds to 1.23 Ma, which yields an age estimate of 22.93 Ma for the local base of the Bluecliffs Formation. Thus at Otaio River terrigenous sediment first appeared in the Early Miocene, about 0.87 Ma later than the 23.80 Ma Oligocene–Miocene boundary.

The eastward progradational nature of the Bluecliffs Formation implies that successions along the western edge of the Canterbury Basin will be both thinner and older at their base than is the section at Otaio River. Consistent with this inference, some earlier workers inferred an age of Waitakian (25.2–22.7 Ma) for outcrops of Bluecliffs Formation at Te Ngawai River (see Fig. 1). The earliest Bluecliffs Formation onland, and the start of regional terrigenous drift sedimentation in eastern New Zealand, therefore, dates from *c.* 24 Ma in the Late Oligocene. Because the Canterbury Drifts everywhere overlie older, *c.* 28–24 Ma, greensand–carbonate platform drifts of the Kekenodon Group, it is clear that the *c.* 24 Ma date corresponds not to the start of drift sedimentation *per se*, but rather to the start of supply of terrigenous sediment from the western plate boundary. The almost simultaneous appearance of terrigenous sediment at *c.* 24.0 Ma in offshore abyssal drifts of ENZOSS (ODP Site 1124; Joseph *et al.* 2004) and terrigenous turbidites of similar age in the Waiau Basin (Carter *et al.* 2005) confirms the Late Oligocene–Early Miocene appearance of a new, voluminous and widespread source.

In summary, the terrigenous Canterbury Drifts have a longevity of *c.* 24 Ma. This results from their deposition along a passive continental margin, which, over that period of time was (1) tectonically stable, and (2) consistently fed with sediment from a nearby, uplifting plate boundary.

Transport pathways for drift-feeding sediment

Since the Late Oligocene, terrigenous sediment has been eroded from mountains aligned along the western South Island plate boundary and delivered by east-flowing rivers to the coast. The mechanism whereby this terrigenous sediment was then moved across the shelf and onto the upper slope is less clear.

Along eastern South Island today, modern terrigenous sand is restricted to the shoreface and inner shelf, and the middle and outer shelf are either mud-covered (Herzer 1981) or, under the influence of the Southland Current, comprise shelly and bryozoan-rich facies (Carter *et al.* 1985). Although storm-engendered, cross-shelf mud-hopping is a feasible mechanism for the delivery of clay and very fine silt to the slope, it is much less likely for medium- to coarse-grained sand grain sizes. How, then, has such sand been delivered to the Canterbury Drifts in the past, and how does it cross the shelf today? There are at least three possible answers to these questions.

First, the width of the early Neogene shelf was much less than it is today (e.g. Fulthorpe & Carter 1991, fig. 4b), making direct cross-shelf sand transport more feasible. *Second*, since about 3.1 Ma, and to a lesser extent before, eustatic sea-level falls of up to 130 m during cold or glacial periods have intermittently narrowed the shelf by exposing its inner portion and moving the shoreline seaward up to many tens of kilometres, which again would have enhanced the chance of cross-shelf sand transport and deposition around the shelf-edge (Carter *et al.* 2004*b*). *Third*, in the southern part of the basin the slope and shelf-edge are dissected by numerous submarine channels and canyons (Krause 1966; Herzer 1979) around the headward part of the Bounty Trough. Turbidity currents travelling down these canyons and into the Bounty Channel system are deflected left by the Coriolis effect, resulting in preferential overspilling on the left (north) bank (Carter & Carter 1988) and the delivery of abundant terrigenous sediment along the north-flowing paths of both intermediate AAIW and abyssal CDW. Although similar ancient (buried) canyon systems are rarely imaged on seismic profiles through the offshore basin, they do occur (e.g. Herzer & Lewis 1979; Lu & Fulthorpe 2004).

In summary, the bulk of the sediment that now comprises the Canterbury Drifts was probably transported to the continental slope through shelf-edge indenting canyons, aided by the direct dumping of shoreface sediments around the shelf-edge and canyon heads during sea-level lowstands.

Sedimentary structures accompanying drift deposition

In outcrop, the Bluecliffs Silt comprises mildly bioturbated, poorly bedded terrigenous siltstone and very fine sandstone, punctuated by beds of very fine or medium- to coarse-grained muddy sand, sometimes concretionary. Sedimentary structures are usually not conspicuous. However, data from ODP Site 1119 reveal the presence of abundant sedimentary structures in the Canterbury Drifts, most of which are also present in outcrop.

The offshore Canterbury Drifts are dominated by alternating plane-beds of sandy and muddy silt on a centimetre scale, with interspersed muddy sands up to several metres thick; millimetre-scale planar and undulose lamination occurs within some thicker sand beds. Cut-core and outcrop surfaces generally present as unstructured, massive, bioturbated muddy silt, sandy silt and silty sand, occasionally with subtle centimetre-scale layering of slightly different colours or grain sizes. Given the visual rarity of strong macroscopic layering in core and outcrop, it is surprising that FMS images of the borehole wall show the pervasive presence throughout the Site 1119 drillhole of resistivity layering at the centimetre-scale (Fig. 4). This layering reflects the presence of interbedded muddy and sandy silts, is often slightly undulose, consistent with the presence of ripple-like features, and is only mildly disturbed by bioturbation.

Thicker sand beds, up to 3 m, may have either sharp or gradational bases and tops. Most commonly, these beds are either normally graded with sharp, erosive bases against underlying siltstone, or are doubly graded, with a reverse-graded base and a normally graded top. Given sand grain sizes between 80 and *c*. 1000 μm, this implies (1) currents that are strong and then gradually wane, or that gradually increase and then gradually wane, respectively, and (2) current velocities in the range of 20–70 cm s^{-1} (Table 1).

The Canterbury Drifts were deposited from strong bottom currents that waxed and waned episodically at Milankovitch periodicity to deposit the thicker sand units (Carter *et al.* 2004*a*), and gentler currents (5–15 cm s^{-1}) that operated regularly at short periodicity to produce the pervasive, centimetre-scale, mud–silt, background layering.

In summary, both offshore and onland, the Canterbury Drifts contain a range of sedimentary structures indicative of contourite deposition. Where not homogenized by bioturbation, these structures include pervasive centimetre-scale interbeds of background muddy and sandy silt, in which occur interspersed thicker, sharp-based normally graded, and gradationally based double-graded sandstones.

Oceanographic and climatic controls on drift deposition

Deposition of a large sediment drift requires the stable operation of a significant geostrophic current over an extended period of time. Such major ocean currents represent flow responses to the particular configuration of an ocean basin. Changes to basin configuration of sufficient magnitude to reset geostrophic flows generally take place over periods of tens of millions of years or longer, unless rapid gateway opening or shutting occurs.

In the modern ocean, deep, cold water flows generated at high southern latitudes pass east of New Zealand to flow northward into the Pacific basin. The three main flows are: circumpolar deep water (2000–4500 m; generated by sinking waters around the periphery of Antarctica; Carter & McCave 1997); AAIW (800–1100 m; generated by subduction at the Antarctic Polar Front (APF); Lynch–Stieglitz *et al.* 1994); and SAMW (300–800 m; generated by subduction at the Subantarctic Front; Morris *et al.* 2001). Predecessor flows to these date back to the inception of the 'modern' stratified ocean at *c*. 33.5 Ma, when the Southern Ocean was created by the deep-water separation of Tasmania and Antarctica (Exon *et al.* 2001). Cold water flows of Southern Ocean derivation have therefore affected the deposition of the Canterbury Drifts throughout their 24 Ma history. Near Site 1119 the shallowest of the modern flows, SAMW, is augmented by an outer shelf–upper slope current jet (Southland Current; Heath 1972; Chiswell 1996; Sutton 2003), which follows the NE-oriented, margin-parallel path of the Subtropical Front off eastern South Island (Sutton 2003).

At 394 m water depth, Site 1119 lies a little seawards of the Southland Current core and at the top of the zone of north-flowing SAMW (Morris *et al.* 2001). Deeper still, at around 800 m water depth, SAMW is underlain by similarly north-flowing AAIW. In consequence, the upper clinoform and small-drift parts of Site 1119 were deposited from SAMW, whereas the basal *c*. 100 m of the core (big drift 11) was probably deposited under AAIW (Carter *et al.* 2004*a*) (Fig. 3b). The change in drift architecture observed in seismic profiles at *c*. 3.1 Ma may have resulted from the initiation then of the SAF, or the movement of the STF into its present postion, or both, the latter accompanied by SAMW generation and Southland Current flow. Alternatively, should both these fronts have already existed in approximately their modern positions, the change from large AAIW drifts to small SAMW drifts may simply reflect shoaling and accretion of the sea bed from AAIW to SAMW depths.

Similarly, the change in the Canterbury Drifts from simple slope clinoforms to clinoforms with large drifts at *c*. 15 Ma (Lu *et al.* 2003) is also likely to mark a significant shift in the dynamics of the northward along-slope current regime. Corresponding as it does to the worldwide Late Miocene cooling caused by growth of the East Antarctic ice sheet, this change from clinoform to large drifts may mark the inception of the APF and the first generation of steady AAIW circulation.

In summary, the Canterbury Drifts were deposited from persistent, north-directed, cold-water, intermediate-depth, geostrophic currents. Changes in their internal geometry, from clinoform to large to small mounded drifts, reflect changing current-flow regimes associated with frontal and water mass evolution, driven by global cooling since the Late Miocene.

Changes in drift texture, mineralogy and provenance

Changes in sediment supply to drift successions include rate changes, compositional changes and textural changes. Steadily increasing the geographical distance of a site of deposition from a source region will tend to result in a decreasing grain size through time. Changing climate or changing oceanography can result in either grain-size increase or decrease, depending upon the circumstances of a particular case.

The sediment within the Canterbury Drifts changes texture through time (Fig. 12). Between *c.* 21 and 4 Ma, sand content decreased by *c.* 20% at the expense of sortable silt, and cohesive mud increased by *c.* 10%. A further 15% decrease in sand occurred between 4 and 3.1 Ma, after which cohesive mud content increased by up to 33%, again mostly at the expense of sortable silt. Over the same period that these changes occurred, the main sediment mode declined in size from very fine sand in the Early Miocene to coarse silt in the Late Pliocene, and then to abundant very fine silt for the remainder of the succession (Table 1).

The decreasing size of the main sediment mode and increase in cohesive mud that occur through the Canterbury Drifts is consistent with the following known wider changes. First, sediment fining is expected to result from the increasing width of the shelf (transport path) through the Neogene. Second, sediment fining, and enrichment in micas, is expected to have accompanied Neogene uplift along the Alpine Fault plate boundary, which successively exposed higher grade, phyllosilicate-rich metamorphic rocks in the terrestrial source area for the Canterbury Drifts (Carter *et al.* 2004*c*). Third, accelerated uplift during the Early Pliocene, and climatic deterioration from the Late Pliocene on, acted together to encourage the development of montane glaciers in South Island, which greatly increased yields of very fine silt and clay after *c.* 3.1 Ma. This last change is marked also by a rapid rise in chlorite + illite and decrease in smectite within the clay assemblages of nearby DSDP Site 594 (Dersch & Stein 1991) and ODP Site 1123 (Winkler & Dullo 2004).

In summary, the long-term mineralogical and textural trends seen through the Canterbury Drifts were controlled by regional tectonic and climatic events. The trends are consistent with unroofing of basement Rangitata rocks along the South Island mountain spine since the Late Oligocene, which supplied sediment to a continually widening shelf, and with rapid climatic deterioration from the Late Pliocene onwards.

Stratigraphic architecture of the drift body: tectonic implications

The Canterbury Drifts comprise an eastward-thickening, shelf–slope sediment prism *c.* 130 km wide at their widest point in the central basin, *c.* 300 m or less thick at their inner western edge, 2000 m or more thick under the present-day shelf edge in the east and *c.* 400 km long at their seaward edge (Fig. 2a). The volume of this prism is *c.* 60 000 km^3, and since originating in the Late Oligocene (*c.* 24 Ma) it has built seawards at rates between 1 and 5 km Ma^{-1}. Eastward progradation took place into seaward-increasing accommodation (created by the gentle eastward dip of the subjacent Oligocene carbonate platform), and an approximately equal volume of sediment bypassed the inboard prism to be deposited within the Bounty Trough (Carter *et al.* 1994; see Adams 1980, fig. 2).

The Canterbury Drifts are built of sediment fed from mountains along the Alpine Fault plate boundary in western South Island, commencing at *c.* 25 Ma (Carter & Norris 1976). Prior to the Late Miocene, relative motion between the Pacific and Australian plates was dominantly strike-slip, with strong compression and continental collision starting at *c.* 6.4 Ma, since when about 90 km of shortening has occurred in the central Southern Alps (Walcott 1998). None the less, abundant terrigenous sediment was fed eastward from *c.* 24 Ma onwards, indicating that significant transpressive mountains were in existence along the plate boundary throughout the Neogene, despite assertions that development of the Southern Alps as a topographic feature did not occur until the Late Miocene or even Pliocene (Adams & Gabites 1985; Kamp *et al.* 1989; Chamberlain *et al.* 1999). Phases of Southern Alps mountain building can be inferred from peaks in clay mineral trends at 18, 8.5, 6.4, 2.1 and 1.2 Ma (Carter *et al.* 2004*c*, fig. F18).

Adams (1980) has estimated that today about 13×10^9 kg of modern sediment is deposited on the Canterbury shelf and slope each year; that is, *c.* 7.2×10^6 m^3 year^{-1}, assuming an average bulk sediment density of 1.8. Applying such a rate since the onset of plate collision at 6.4 Ma yields a sediment volume of 46 000 km^3, leaving *c.* 12 000 km^3 of the Canterbury Drifts prism to be accounted for by earlier deposition at a rate *c.* 0.68×10^6 m^3 year^{-1}. Outcrop and petroleum well-site information suggest that long-term rates of sediment deposition of 10–15 cm ka^{-1} are characteristic for the drifts that lie beneath the coastal plain and shelf (Carter 1988).

In summary, the Canterbury Drifts are a progradational shelf–slope sediment prism *c.* 400 km long, *c.* 130 km wide and up to 2000 m thick, with an overall volume of *c.* 60 000 km^3. The drifts represent the interaction of an abundant, plate boundary sediment source and major intermediate-depth current flows, and yield a high-resolution record of climatic, oceanographic and tectonic events.

Conclusions

The Canterbury Drifts are of particular interest because of their *c.* 24 Ma longevity, size, deposition at intermediate water depths, and southern mid-latitude location. Regional warping associated with plate boundary tectonics has caused gentle uplift of Late Oligocene–Miocene drifts onland along the western basin margin. Uniquely, therefore, the Canterbury Drifts can be studied in outcrop, in offshore petroleum wells on the shelf, at ODP Site 1119 and DSDP Site 594 on the upper and middle slope, respectively, and in the sea-floor sediments that lie beneath the path of the modern, north-travelling SAMW and AAIW.

Study of the Canterbury Drifts provides insight into the sedimentary characteristics of intermediate-depth drift systems. The drift sediments are dominated by structureless and plane-bedded, polymodal, quartzofeldspathic clayey silts, sandy silts and silty sands, with varying admixtures of benthic and biopelagic carbonate and silica, within which occur intermittent, sharp-based normal-graded and gradational-based double-graded muddy sands, which form up to 10% of the overall drift thickness. The appearance between *c.* 3.2 and 2.0 Ma of a distinctive 9–15 μm grain-size mode is consistent with its origin from glacial grinding processes ('rock-flour'), and the development at that time of uplift and significant montane glaciation on South Island.

The Canterbury Drifts exhibit three seismic signatures. Two display the characteristic mounded shapes and internal sigmoidal bedding that are known from other intermediate- and abyssal-depth sediment drifts (Fig. 3). The two types are differentiated mainly according to their size. A change from larger to smaller drifts at ODP Site 1119 occurred at *c.* 3.1 Ma; that is, at the time that global climate started to deteriorate towards its Pleistocene nadir. Such changes in drift architecture are likely to mark significant changes in oceanographic conditions, and in this case perhaps signal the inception of discrete SAMW flows generated from a newly formed Subantarctic Front located to the south of Site 1119.

Mounded drifts as old as Middle Miocene (*c.* 15 Ma) have been imaged on seismic profiles, and represent climatic deterioration and probably enhanced current flows after that time. The mounded drifts occur enclosed within apparently planar clinoforms of the prograding slope (see Fig. 2). At least some (those along depositional strike from a drift mound) and perhaps all of these clinoforms were deposited under the same northward-moving water masses that built the mounded drifts. Therefore, I infer that a third type of drift signature can be represented by simple clinoform bedding alone. Indeed, Lu *et al.* (2004, fig. 5a) have recently shown that in circumstances where industry-standard seismic profiles resolve only plain clinoform reflectors, higher resolution data yield abundant evidence for a lack of reflector continuity, with minor discordances and lensing consistent with current-influenced deposition. Lu *et al.* concluded (p. 1363) that 'along-strike transport could play a role in forming clinoform morphologies worldwide, even where (seismic) geometries diagnostic of current reworking are lacking'.

Finally, within the Canterbury Drifts the accretion of an individual mounded drift to the upper part of the slope, which occurs by the infilling of its delimiting landward gutter, causes the shelf-edge to 'vault' seawards by several to many kilometres at a time. This episodic type of shelf-edge progradation, together with the possibility that many slope clinoforms may represent contourite drift deposition rather than simple shelf mud-spillover progradation, suggests that a critical appraisal of mechanisms of shelf–slope progradation is overdue.

I thank A. Viana for the invitation to write this paper, and the members of the scientific and technical parties and the crew of R.V. *JOIDES Resolution* for the contributions they made towards the collection, logging and shipboard interpretation of Site 1119 cores on Leg 181. I also thank A. Orpin and C. McKeagney for assistance with grain-size analysis, C. Fulthorpe for providing the seismic lines reproduced in Figures 1 and 3, and J. Howe and J.-C. Faugères for their constructive criticisms of the draft manuscript. This research used samples and data provided by the Ocean Drilling Program (ODP). The ODP is sponsored by the US National Science Foundation (NSF) and participating countries under management of Joint Oceanographic Institutions (JOI), Inc. Financial support was provided by the Australian Research Council (ARC), Grant A-39805139.

Appendix

Grain-size analysis

A total of 220 samples were selected from throughout the Site 1119 succession to characterize the main lithologies and sediment facies present. As a result, sample spacing is at non-systematic intervals. Because of the presence also of core gaps, sample spacing therefore ranges from a few centimetres to 18 m, with an average of 2.33 m.

A total of 108 samples were collected also from the Early Miocene part of the Bluecliffs Formation, the onland representative of the Canterbury Drifts (R. Carter *et al.* 1996). Localities were selected in the south (coast south of Shag River mouth),

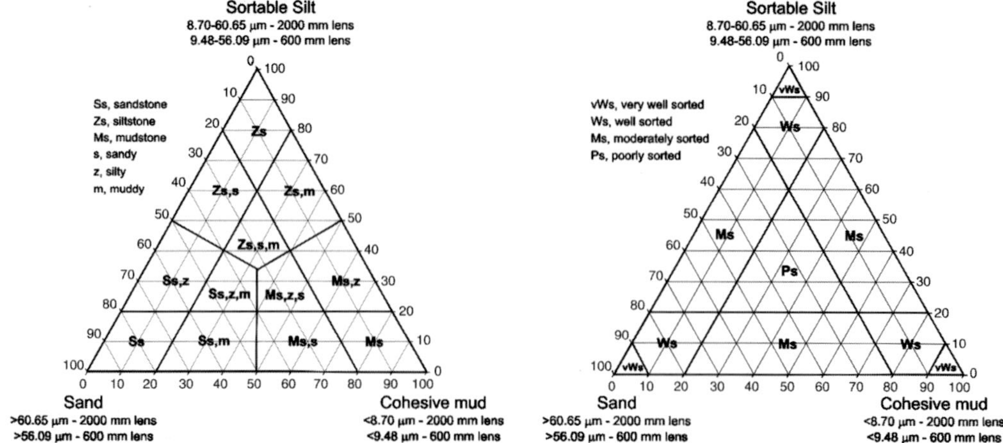

Fig. A1. Sediment type (left) and sorting terminology (right) used in this paper. The cM:sZ:sS compositional space for sediment type is divided into three equal mudstone, siltstone and sandstone fields, with qualifying adjectives delimited by the relevant 20th percentile dividers. Sorting terminology follows similar boundaries.

central (Otaio River; Te Ngawai River) and northern (South Waipara River) parts of the onland Canterbury Basin.

Procedure

Samples were analysed by laser diffraction using a Malvern MasterSizer-X long bed laser particle size analyser. The measurement routine analyses the sample 10 000 times and produces an averaged result processed using a polydisperse algorithm. Depending upon the lens fitted, the particle size distribution is constructed from 32 separate size classes within the range of 1.2–600 µm, or 28 separate size classes within the range of 5–2000 µm. Samples from Site 1119 were analysed using the 600 µm lens; samples from the onshore Bluecliffs Formation, using the 2000 µm lens.

Sediment from each 5 or 10 cm^3 push-core at Site 1119 was gently crumbled into a beaker of tap water, stirred and left to soak for 48 h to ensure complete disaggregation of all sediment particles. To test for complete dispersion of the finer sediment fractions, 15 of the samples between 86 and 115 metres composite depth (mcd), and most of the samples above 86 mcd, were pre-treated with *c.* 10 ml of 30% hydrogen peroxide to remove organic material and allowed to react for at least 36 h. Where appropriate, samples were warmed in a 60 °C water bath to confirm that the reaction had stopped. Samples were then treated again with *c.* 10 ml of 10% glacial acetic acid to remove carbonate material. These samples were left for at least 36 h to react, and, where appropriate, warmed to confirm that all the carbonate had been dissolved.

Samples were dispersed in tap water within a sample bath of *c.* 1 l in volume. When placing the samples into the Malvern sample bath, the entire sample was carefully washed from the beaker. Ultrasonic dispersion was not used. In most cases samples were subjected to vigorous stirring for around 5–10 min, or until the laser obscuration became stable. Laser obscuration was generally maintained in the range 20–50%. Occasional higher obscurations ranged to *c.* 70%, but the results indicate that, provided the samples were well dispersed, accurate and repeatable results are achieved.

A known defect of laser diffraction particle measurement is that erroneous measurement effects occur near the limits of the range of the chosen lens, particularly when a significant amount of the sample is finer-grained than the nominated lens lower limit. In such circumstances, the Malvern controlling software overcompensates by ‘squeezing’ diffraction data into the finest-grained channels for the lens concerned. This effect produces a spuriously strong sediment mode at around 7 µm for the 2000 mm lens and 2.5 µm for the 600 mm lens. When such spurious peaks are compared with the size distribution of the same sample run using a shorter lens (i.e. more accurate for determining the distribution of very fine silt and clay), the volume per cent of the sample less than 10 µm in diameter (the area beneath the size–frequency plot) is closely similar. Therefore, despite this type of edge effect the Malvern system can be used to make accurate measurements of the volume per cent of sediment <10 μm, which is why a cutoff as close as possible to that grain diameter was chosen to separate the cohesive mud (cM) and sortable silt (sZ) categories in this study, namely <8.70 or <9.48 μm for the 2000 and 600 mm lens, respectively.

The Bluecliffs Formation samples from onland South Island were subjected to the same laboratory procedure, but analysed using the 2000 μm lens. To allow comparison of grain-size distributions between onland and offshore samples, and because the 2000 μm lens does not give an accurate measurement of clay content, summary statistics for all samples have been calculated using the three categories cohesive mud:'sortable silt':sand (cM:sZ:sS) rather than the conventional clay:silt:sand grain-size classes. The boundary between the cohesive mud (which comprises clay plus very fine silt) and 'sortable silt' categories in nature lies at about 10 μm, corresponding to the grain-size diameter below which fine-grained sediment becomes cohesive and resistant to traction entrainment (McCave 2005). Because the size output classes from the Malvern system do not correspond exactly to traditional geometric grain-size boundaries, and in keeping with the nature of this study, the cohesive mud–sortable silt and sortable silt–sand boundaries were therefore approximated at 8.70 and 60.65 μm (2000 mm lens) and 9.48 and 56.09 μm (600 mm lens), respectively, rather than at the usual clay–silt and silt–sand boundaries of 3.9 and 62.5 μm. Based on these boundary conditions, a pragmatic mud:silt:sand terminology has been used to describe Canterbury Drift sediments (Fig. 12).

A silica 122 μm standard was run for each separate analysis session of around 20 samples to ensure consistent machine performance, and selected samples were reanalysed to ensure that acceptable levels of precision (i.e. closely matching size envelopes) were maintained throughout the periods of analysis.

References

Adams, C. J. & Gabites, J. E. 1985. Age of metamorphism and uplift in the Haast Schist Group at Haast Pass, Lake Wanaka and Lake Hawea, South Island, New Zealand. *New Zealand Journal of Geology and Geophysics*, **28**, 85–96.

Adams, J. 1980. Contemporary uplift and erosion of the Southern Alps, New Zealand. *Geological Society of America Bulletin*, **91**(1), 1–114.

Andrews, P. B. 1963. Stratigraphic nomenclature of the Omihi and Waikari Formations, north Canterbury. *New Zealand Journal of Geology and Geophysics*, **6**, 228–256.

Ballard, J. A. 1966. Structure of the lower continental rise hills of the western North Atlantic. *Geophysics*, **31**, 506–523.

Bishop, D. G. & Turnbull, I. M. (compilers) 1996. Geology of the Dunedin area. *Institute of Geological & Nuclear Sciences (Lower Hutt, New Zealand) 1:250 000 geological map*, **21**.

Buchanan, J. 1870. On the Wanganui Beds (upper Tertiary). *Transactions of the New Zealand Institute*, **2**, 163–166.

Carter, L. & Carter, R. M. 1988. Late Quaternary development of left-bank-dominant levees in the Bounty Trough, New Zealand. *Marine Geology*, **78**, 185–197.

Carter, L. & McCave, I. N. 1997. The sedimentary regime beneath the deep western boundary current inflow to the southwest Pacific Ocean. *Journal of Sedimentary Research*, **67**, 1005–1017.

Carter, L., Carter, R. M., McCave, I. N. & Gamble, J. 1996. Regional sediment recycling in the abyssal Southwest Pacific Ocean. *Geology*, **24**, 735–738.

Carter, L., Carter, R. M., Nelson, C. S., Fulthorpe, C. S. & Neil, H. L. 1990. Evolution of Pliocene to recent abyssal sediment waves on Bounty Channel levees, New Zealand. *Marine Geology*, **95**, 97–109.

Carter, R. M. 1988. Post-breakup stratigraphy of the Kaikoura Synthem (Cretaceous–Cenozoic), continental margin, southeastern New Zealand. *New Zealand Journal of Geology and Geophysics*, **31**, 405–429.

Carter, R. M. & Gammon, P. 2004. New Zealand maritime glaciation: millennial-scale southern climate change since 3.9 Ma. *Science*, **304**, 1659–1662.

Carter, R. M. & Norris, R. J. 1976. Cainozoic history of southern New Zealand: an accord between geological observations and plate-tectonic predictions. *Earth and Planetary Science Letters*, **31**, 85–94.

Carter, R. M., Carter, L. & Davy, B. 1994. Geologic and stratigraphic history of the Bounty Trough, southwestern Pacific Ocean. *Marine and Petroleum Geology*, **11**, 79–93.

Carter, R. M., Carter, L. & McCave, I. N. 1996. Current controlled sediment deposition from the shelf to the deep ocean: the Cenozoic evolution of circulation through the SW Pacific gateway. *Geologische Rundschau*, **85**, 438–451.

Carter, R. M., Carter, L., Williams, J. & Landis, C. A. 1985. *Modern and relict sedimentation on the Otago continental shelf*. New Zealand Oceanographic Institute, Memoir, **93**.

Carter, R. M., McCave, I. N., Richter, C., et al. 1999. Southwest Pacific Gateways, Sites 1119–1125. *In*: Richter, C., McCave, I. N., Carter, R. M. & Carter, L. et al. (eds) *Proceedings of the Ocean Drilling Program, Initial Reports, 181*, Ocean Drilling Program, (CD-ROM), College Station, TX, 1–80.

Carter, R. M., Fulthorpe, C. S. & Lu, H. 2004*a*. Canterbury Drifts at Ocean Drilling Program Site 1119: climatic modulation of southwest Pacific intermediate water flows since 3.9 Ma. *Geology*, **32**, 1005–1008.

CARTER, R. M., GAMMON, P. R. & MILLWOOD, L. 2004*b*. Glacial–interglacial (MIS 1–10) migrations of the Subtropical Front (STF) across ODP Site 1119, Canterbury Bight, Southwest Pacific Ocean. *Marine Geology*, **205**, 29–58.

CARTER, R. M., MCCAVE, I. N. & CARTER, L. 2004*c*. Fronts, flows, drifts, volcanoes, and the evolution of the southwestern gateway to the Pacific Ocean. *In*: RICHTER, C., MCCAVE, I. N., CARTER, R. M. & CARTER, L. ET AL. (eds) *Proceedings of the Ocean Drilling Program, Scientific Results*. Ocean Drilling Program, College Station, TX, 1–110.

CARTER, R. M., NORRIS, R. J. & TURNBULL, I. M. 2005. *The Geology of the Blackmount district, Te Anau–Waiau Basin, western Southland*. Institute of Geological and Nuclear Sciences, Report 2004/123.

CHAMBERLAIN, C. P., POAGE, M. A., CRAW, D. & REYNOLDS, R. C. 1999. Topographic development of the Southern Alps recorded by the isotopic composition of authigenic clay minerals, South Island, New Zealand. *Chemical Geology*, **155**, 279–294.

CHISWELL, S. M. 1996. Variability in the Southland Current, New Zealand. *New Zealand Journal of Marine and Freshwater Research*, **30**, 1–17.

CRAW, D. 1984. Lithologic variations in Otago Schist, Mt Aspiring area, northwest Otago, New Zealand. *New Zealand Journal of Geology and Geophysics*, **27**, 151–166.

DAMUTH, J. E. 1979. Migrating sediment waves created by turbidity currents in the northern South China Basin. *Geology*, **7**, 520–523.

DERSCH, M. & STEIN, R. 1991. Palaoklima und paloozeanische Verhaltnisse im SW-Pazifik wahrend der letzten 6 Millionen Jahre (DSDP—Site 594, Chatham Rucken, ostlich Neuseeland). *Geologische Rundschau*, **80**, 535–556.

DOUGLAS, B. J. 1975. *A paleoenvironmental analysis of a Lower Miocene regressive sequence, upper Tengawai River, South Canterbury, New Zealand*. MSc thesis, University of Otago, Dunedin.

EADE, J. & KENNETT, J. P. 1962. Tertiary sequence at upper Tengawai River, South Canterbury. *New Zealand Journal of Geology and Geophysics*, **5**, 163–174.

EWING, M., EITTREIM, S. A., EWING, J. L. & LE PICHON, S. 1971. Sediment transport and distribution in the Argentine Basin. 3. Nepheloid layer and processes of sedimentation. *Physics and Chemistry of the Earth*, **8**, 49–78.

EXON, N. F., KENNETT, J. P., MALONE, M. J. & Shipboard Scientists 2001. The Tasmanian Gateway: Cenozoic climatic and oceanographic development. *In*: EXON, N. F., KENNETT, J. P. & MALONE, M. J. ET AL. (eds) *Proceedings of the Ocean Drilling Program, Initial Reports, 189*. Ocean Drilling Program, College Station, TX, 1–98.

FAUGÈRES, J.-C., GONTHIER, E. & STOW, D. A. V. 1984. Contourite drift molded by deep Mediterranean outflow. *Geology*, **12**, 296–300.

FINLAY, H. J. 1953. Appendix II—Foraminifera. *In*: WELLMAN, H. W. (ed.) *The Geology of the Geraldine Subdivision, S102 Sheet District*, New Zealand Geological Survey, Bulletin, **50**, 47–51.

FLEMING, C. A. 1955. Oceania, Fascicule 4, New Zealand. *Lexique Stratigraphique International*. CNRS, Paris.

FLEMING, C. A. 1962. New Zealand biogeography: a paleontologist's approach. *Tuatara*, **10**(2), 53–108.

FULTHORPE, C. S. & CARTER, R. M. 1991. Continental shelf progradation by sediment drift accretion. *Geological Society of America Bulletin*, **103**, 300–309.

GAGE, M. 1957. *The Geology of the Waitaki Subdivision*. New Zealand Geological Survey, Bulletin, New Series, **55**.

GAIR, H. S. 1959. The Tertiary geology of the Pareora district, south Canterbury. *New Zealand Journal of Geology and Geophysics*, **2**, 265–296.

GARDNER, J. V. & KIDD, R. B. 1987. Sedimentary processes on the northwestern Iberian continental margin viewed by long-range side-scan sonar and seismic data. *Journal of Sedimentary Petrology*, **57**, 397–407.

GULF, 1973. *Basic geophysical data from M.V.* Gulfrex *cruises 94–98, October 1972–January 1973, New Zealand*. New Zealand Geological Survey Open-file Petroleum Report, **614**.

HEATH, R. A. 1972. The Southland Current. *New Zealand Journal of Marine and Freshwater Research*, **6**, 497–533.

HECTOR, J. 1877. *Progress Report*. New Zealand Geological Survey, Reports of Geological Exploration, **1876–77**, iv.

HECTOR, J. 1884. *Progress Report*. New Zealand Geological Survey, Reports of the Geological Explorations **1883–4**, xiii.

HERZER, R. H. 1979. Submarine slides and submarine canyons on the continental slope off Canterbury, New Zealand. *New Zealand Journal of Geology and Geophysics*, **22**, 391–406.

HERZER, R. H. 1981. *Late Quaternary stratigraphy and sedimentation of the Canterbury continental shelf, New Zealand*. New Zealand Oceanographic Institute, Memoirs, **89**.

HERZER, R. H. & LEWIS, D. W. 1979. Growth and burial of a submarine canyon off Motunau, North Canterbury, New Zealand. *Sedimentary Geology*, **24**, 69–83.

HOLLISTER, C. D., FLOOD, R. D., JOHNSON, D. A., LONSDALE, P. & SOUTHARD, J. B. ET AL. 1974. Abyssal furrows and hyperbolic echo traces on the Bahama Outer Ridge. *Geology*, **2**, 395–400.

HOWE, J. A., PUDSEY, C. J. & CUNNINGHAM, A. P. 1997. Pliocene–Holocene contourite deposition under the Antarctic Circumpolar Current, Western Falkland Trough, South Atlantic Ocean. *Marine Geology*, **138**, 27–50.

HUTTON, F. W. 1888. On some railway cuttings in the Weka Pass. *Transactions of the New Zealand Institute*, **20**, 262.

JOSEPH, L. H., REA, D. K. & VAN DER PLUIJM, B. A. 2004. Neogene history of the Deep Western Boundary Undercurrent at Rekohu Sediment Drift, Southwest Pacific (ODP Site 1124). *Marine Geology*, **205**, 185–206.

KAMP, P. J. J., GREEN, P. F. & WHITE, S. H. 1989. Fission track analysis reveals character of collisional tectonics in New Zealand. *Tectonics*, **8**, 169–195.

KIDD, R. B. & HILL, P. R. 1987. Sedimentation on Feni and Gardar sediment drifts. *In*: RUDDIMAN, W. F., KIDD, R. B., THOMAS, E., ET AL. (eds) *Initial Reports of the Deep Sea Drilling Project, 94*, US Government Printing Office, Washington, DC, 1217–1244.

KRAUSE, D. C. 1966. *Geology and geomagnetism of the Bounty region, east of South Island, New Zealand*. New Zealand Oceanographic Institute, Memoirs, **30**.

LAUGHTON, A. S., BERGGREN, W. A., ET AL. (eds) 1972. *Initial Reports of the Deep Sea Drilling Project, 12*, US Government Printing Office, Washington, DC.

LEMASURIER, W. E. & LANDIS, C. A. 1996. Mantle-plume activity recorded by low-relief erosion surfaces in West Antarctica and New Zealand. *Geological Society of America Bulletin*, **108**, 1450–1466.

LEWIS, D. W., LAIRD, M. G. & POWELL, R. D. 1980. Debris flow deposits of early Miocene age, Deadman Stream, Marlborough, New Zealand. *Sedimentary Geology*, **27**, 83–118.

LONSDALE, P. & HOLLISTER, C. D. 1979. A near-bottom traverse of Rockall Trough: hydrographic and geologic inferences. *Oceanologica Acta*, **2**, 91–106.

LU, H. & FULTHORPE, C. S. 2004. Controls on sequence stratigraphy of a middle Miocene–Holocene, current-swept, passive margin: offshore Canterbury Basin, New Zealand. *Geological Society of America Bulletin*, **116**, 1345–1366.

LU, H., FULTHORPE, C. S. & MANN, P. 2003. Three-dimensional architecture of shelf-building sediment drifts in the offshore Canterbury Basin, New Zealand. *Marine Geology*, **193**, 19–47.

LYNCH-STIEGLITZ, J., FAIRBANKS, R. G. & CHARLES, C. D. 1994. Glacial–interglacial history of Antarctic Intermediate Water: relative strengths of Antarctic versus Indian Ocean sources. *Paleoceanography*, **9**, 7–29.

MASON, B. H. 1941. The geology of Mount Grey District, North Canterbury. *Transactions of the Royal Society of New Zealand*, **71**, 120–122.

MCCAVE, I. N. 2005. Deposition from suspension. *In*: SELLEY, R. C., COCKS, R. & PLIMER, R. (eds) *Encyclopedia of Geology*, **5**. Elsevier, 8–17.

MCCAVE, I. N. & TUCHOLKE, B. E. 1986. Deep current controlled sedimentation in the Western North Atlantic. *In*: VOGT, P. R. & TUCHKOLKE, B. E. (eds) *The Geology of Western North America, V.M., The Western North Atlantic Region*. Geological Society of America, Boulder, CO, 451–468.

MCCAVE, I. N., LONSDALE, P. F., HOLLISTER, C. D. & GARDNER, W. D. 1980. Sediment transport over the Hatton and Gardar contourite drifts. *Journal of Sedimentary Petrology*, **50**, 1049–1062.

MCCAVE, I. N., MANIGHETTI, B. & BEVERIDGE, N. A. S. 1995. Circulation in the glacial North Atlantic inferred from grain-size measurements. *Nature*, **374**, 149–152.

MCCAVE, I. N., MANIGHETTI, B. & ROBINSON, S. G. 1996. Sortable silt and fine sediment size/composition slicing: parameters for palaeocurrent speed and palaeoceanography. *Paleoceanography*, **10**, 593–610.

MILLER, M. C., MCCAVE, I. N. & KOMAR, P. D. 1977. Threshold of sediment motion under unidirectional currents. *Sedimentology*, **24**, 507–528.

MONTADERT, L., ROBERTS, D. G., ET AL. (eds) 1979. *Initial Reports of the Deep Sea Drilling Project, 48*. US Government Printing Office, Washington, DC.

MORGANS, H. E. G., EDWARDS, A. R., SCOTT, G. H., ET AL. 1999. Integrated biostratigraphy of the Waitakian–Otaian boundary stratotype, Early Miocene, New Zealand. *New Zealand Journal of Geology and Geophysics*, **42**, 581–614.

MORRIS, M., STANTON, B. & NEIL, H. 2001. Subantarctic oceanography around New Zealand: preliminary results from an ongoing survey. *New Zealand Journal of Marine and Freshwater Research*, **35**, 499–519.

NELSON, C. H., BARAZA, J. & MALDONADO, A. 1993. Mediterranean undercurrent sandy contourites, Gulf of Cadiz, Spain. *Sedimentary Geology*, **82**, 103–131.

NORMARK, W. R., HESS, G. R., STOW, D. A. V. & BOWEN, A. J. 1980. Sediment waves on the Monterey Fan levee: a preliminary physical interpretation. *Marine Geology*, **37**, 1–18.

REED, J. J. 1957. *Petrology of the Lower Mesozoic rocks of the Wellington district*. New Zealand Geological Survey Bulletin, New Series **57**.

SERANNE, M. & ABEIGNE, C. R. N. 1999. Oligocene to Holocene sediment drifts and bottom currents on the slope of Gabon continental margin (west Africa). Consequences for sedimentation and southeast Atlantic upwelling. *Sedimentary Geology*, **128**, 179–199.

SERVICE, H. 1934. The geology of the Goodwood District, North-east Otago, New Zealand. *New Zealand Journal of Science and Technology*, **15**, 266–274.

STOKER, M. S., AKHURST, M. C., HOWE, J. A. & STOW, D. A. V. 1998. Sediment drifts and contourites on the continental margin off northwest Britain. *Sedimentary Geology*, **115**, 33–51.

STOW, D. A. V. & LOVELL, J. P. B. 1979. Contourites: their recognition in modern and ancient sediments. *Earth-Science Reviews*, **14**, 251–291.

STOW, D. A. V., FAUGÈRES, J.-C. & GONTHIER, E. 1986. Facies distribution and textural variation in Faro Drift contourites: velocity fluctuation and drift growth. *Marine Geology*, **72**, 71–100.

STOW, D. A. V., FAUGÈRES, J.-C., VIANA, A. & GONTHIER, E. 1998. Fossil contourites: a critical review. *Sedimentary Geology*, **115**, 3–31.

SUTTON, P. J. H. 2003. The Southland Current: a subantarctic current. *New Zealand Journal of Marine and Freshwater Research*, **37**, 645–652.

VIANA, A. R., FAUGÈRES, J.-C. & STOW, D. A. V. 1998. Bottom-current-controlled sand deposits—a review of modern shallow- to deep-water environments. *Sedimentary Geology*, **115**, 53–80.

WALCOTT, R. I. 1998. Modes of oblique compression: late Cenozoic tectonics of the South Island of New Zealand. *Reviews of Geophysics*, **36**, 1–26.

WARD, D. M. & LEWIS, D. W. 1975. Paleoenvironmental implications of storm-scoured ichnofossiliferous mid-Tertiary limestones, Waihao district, South Canterbury, New Zealand. *New Zealand Journal of Geology and Geophysics*, **18**, 881–908.

WARREN, G. & SPEDEN, I. 1978. *The Piripauan and Haumurian stratotypes (Mata Series, Upper Cretaceous) and correlative sequences in the Haumuri Bluff district, South Marlborough (S56)*. New Zealand Geological Survey Bulletin, **92**.

WESTERN GEOPHYSICAL. 1982. *Final operation report, Canterbury Bight. PPLs 38202 and 38203, BP, Shell, Todd (Canterbury) Services Limited*. New Zealand Geological Survey Open-file Petroleum Report, **898**.

WINKLER, A. & DULLO, W.-C. 2004. Data Report: Miocene to Pleistocene sedimentation pattern on the Chatham Rise, New Zealand. *In*: RICHEER, C., MCCAVE, I. N., CARTER, R. H. & CARTER, L. ET AL. (eds) *Proceedings of the Ocean Drilling Program, Scientific Results, 181*. Ocean Drilling Program, College Station, TX (CD-ROM).

Slope currents and contourites in an eastern boundary current regime: California Continental Borderland

REBECCA S. ROBINSON[1], JANETTE M. MURILLO DE NAVA[2] & DONN S. GORSLINE[3]

[1]*Department of Geological and Geophysical Sciences, Princeton University, Princeton, NJ 08544, USA (e-mail: rebeccar@gso.uri.edu)*

[2]*CICIMAR-IPN, Av. S/N Colonia Playa Palo de Santa Rita, La Paz, Baja California Sur, Mexico Cp 23096*

[3]*Department of Earth Sciences, University of Southern California, Los Angeles, CA 90089-0740, USA*

Abstract: Analysis of piston cores from lower slopes of central and outer basins of the California Continental Borderland shows the presence of structures ranging from starved silt ripples and lenses to erosion surfaces and truncated burrows to cross-bedded units that testify to reworking of the sediments by bottom currents. Adjacent basin-floor piston cores do not reveal these structures, but exhibit the usual bioturbated hemipelagic mud and turbidites. Slope sediments generally contain more silt than the adjacent basin-floor clay silts. The slope grain-size distributions are multimodal as a result of mixing of hemipelagic mud and thin (millimetre scale) silt layers as a consequence of the sampling intervals at centimetre scale. Carbonate and organic carbon contents tend to be low during the glacial periods but variations from this pattern occur that are probably related to the shift in the upwelling associated with the California Counter Current between interglacial and glacial conditions. The evidence for reworking is most abundant from late in Marine Isotope Stage 5 (MIS 5) to early Stage 4 (MIS 4), and it decreases into the Holocene. No structures are observed in the Holocene sections of the slope cores. The temporal distribution of the reworking structures can be explained either by changes in the degree of bioturbation (bottom-water oxygenation), which would work to erase the structures, or variation in the intensity of lower slope circulation through time. The observed pattern is in rough agreement with documented changes in bottom-water oxygenation conditions for these basins. However, the occurrence of reworking is also dominantly within glacial intervals, where lower sea levels produce stronger circulation within the basins as a result of exposure of banks and resulting restriction of cross-sections of the deeper flow pathways. These observations add to the increasing evidence that sediment transport by bottom currents is not restricted to the intensively studied western boundary current drift deposits. The action of bottom currents on eastern boundary slopes can introduce subtle effects on what are commonly assumed to be continuous records.

The wind-driven surface circulation systems in the oceans include narrow, high-velocity and deep western boundary current systems, which move poleward, swing eastward at the sub-polar margins of the hemispheres, then flow as broad slower equatorward eastern boundary current systems associated with strong upwelling. Both boundary current systems include complex local circulations that interact with the continental slopes. The western systems attain higher velocities and as a result produce high-energy flow regime sedimentary structures in the slope sedimentary deposits. Examples are the thick accumulations of large ripple sets and dunes observed in the major drifts of the North Atlantic (e.g. Bouma & Hollister 1963; Heezen & Hollister 1964; Keigwin & Jones 1989; reviews by Stow *et al.* 1998; Viana *et al.* 1998*a*, *b*). Another large-scale upslope process off the West African margin has been reported by Seranne & Abeigne (1999), who discussed upslope-directed large furrows produced by strong bottom flows. An alternative explanation of these structures was offered by Rasmussen, who noted the contributions of erosion and slope failures (Rasmussen 1994, 1997), and further debated by Rasmussen (2000) and Seranne *et al.* (2000).

Higher flow regimes also operate when flow is restricted by bathymetry as in Drake's Passage in the southern Circumpolar Current, and over gaps in the mid-ocean ridges and passages (e.g. Hollister *et al.* 1974; Lonsdale & Malfait 1974; Lonsdale 1981). These produce coarse-grained ripples in the axis of the passage.

The mechanics of fluid gravity flows (Middleton & Southard 1977) can be categorized on the basis of such factors as velocity of flow, thickness of flow, roughness of the surface over which the flow occurs, and fluid descriptors such as density

From: VIANA, A. R. & REBESCO, M. (eds) *Economic and Palaeoceanographic Significance of Contourite Deposits.* Geological Society, London, Special Publications, **276**, 155–169.
0305-8719/07/$15.00

and viscosity. These parameters can be related to sedimentary structures in flow regime diagrams where punctuated changes in turbulence to maintain quasi-equilibrium conditions occur at critical levels of flow. Each flow regime interacts with unconsolidated sediments to produce sedimentary structures that can be identified in many stratigraphic settings. Low-energy flow regimes may produce lineations, small-scale ripples, or isolated 'starved' ripples where sediment supply is limited. Higher velocities may produce large-scale ripples, dunes and antidunes. Sediment supply is an important factor, as well as grain size, water content, cohesiveness and roughness of the substrate.

For a given grain size and hydrodynamic condition, the resulting flow regime will produce the same sedimentary structures in stream beds, the deep ocean floor, or in dune fields and volcaniclastic sediments (Middleton & Southard 1977). Identifying the depositional environment requires additional information such as biogenic components, stratigraphic associations, deposit shape and dimensions, and sediment composition (Sloss 1962). Structures associated with the spectrum of slope currents in the ocean are governed by these same factors.

Whereas recent work has emphasized the occurrence of lateral, down-slope transport of sedimentary materials on the margins underlying eastern boundary current upwelling systems, sedimentary structures attributed to slope currents are singularly unreported in the literature (e.g. Mollenhauer *et al.* 2002). Lucchi & Rebesco (2004) reported sedimentary structures in fine slope sediments on the Antarctic continental margin, which appear to have many similarities to those reported here. That margin has weak countercurrent systems that are probably similar in scope to the borderland examples. Hams (1987) has described possible winnowing structures in one basin-slope box core in the California Borderland. This lack of observations is probably an artefact of sampling and economic interest. Coring has been concentrated in margin basin floors or in areas of thick potential petroleum reservoirs.

In this paper we present X-radiograph images and other data to demonstrate the presence of slope current-generated small-scale structures in an eastern boundary current regime. Estimates of downcore changes in the abundance of reworking structures are used to discuss the possible origins of the structures in the context of temporal variation of bottom-water velocities during times of lowered sea levels and increasing restriction of flow pathways.

Oceanographic and geological setting

The California Continental Borderland (Shepard & Emery 1941; Gorsline & Teng 1989) is a transform margin (Atwater 1989) that is composed of *c.* 21 margin basins arranged in three rows oriented NW–SE for *c.* 1600 km between Viscaino Bay, Baja California and Point Conception, California (Fig. 1). In this paper, we discuss sites in the low accumulation rate basins of the central and outer Borderland.

The major feature of the California Borderland circulation is the California Current System, which is a broad slow eastern boundary current (Hickey 1979, 1992). The core of maximum velocity is generally located about 300 km offshore. As the main California Current passes south of Point Conception, its inner waters turn shoreward and form a major gyre over the northern Borderland called the California Counter Current (Fig. 2a). This current flows north and turns west and flows along the south slope of the northern Channel Islands. The California Current and Counter Current transport Pacific Subarctic water southward, which is cold, low-salinity water with a high oxygen content. In the subsurface, the inner margin of the California Current is marked by the position of the poleward-flowing California Undercurrent (Emery 1960; Hickey 1992, 1993). This current has its source to the south in the Eastern Tropical South Pacific. It transports warm, saline and oxygen-poor water northward along the margin. At lower sea levels this current must move to a more seaward position as exposed banks block its present northward path. Deep-water flows into the basins at sill depth are predominantly from the south (Emery 1960; Fig. 2b).

A major oceanographic component of the eastern boundary current systems in the Pacific is a strongly developed Oxygen Minimum Zone (OMZ, Reid 1965). This zone, at present between 400 and 1000 m, restricts macro-benthic mixing of the slope and margin basin sediments in that depth range, preserving primary sedimentary structures (e.g. DeDiego & Douglas 1999). In the borderland, the Santa Barbara, Santa Monica and San Pedro Basins all have sills in the OMZ and exhibit preserved primary laminations (Emery & Hulsemann 1962; Christensen *et al.* 1994; Gorsline *et al.* 1996).

Sill depths of the individual basins have changed as sea levels rise and fall during climatic shifts of the Pleistocene. These climatic–oceanographic changes are also exhibited in changing oxygenation conditions over this period, as discussed by Stott *et al.* (2000). These changes produce two stratigraphic end-members: (1) the preserved varved sediments of anoxic basins such as the Cariaco Basin off Venezuela and the Santa Barbara and Santa Monica Basins in the borderland; (2), the homogeneous bioturbated grey–green muds that typify deep-sea sediment cores from sites bathed in oxygenated bottom waters. Within the California

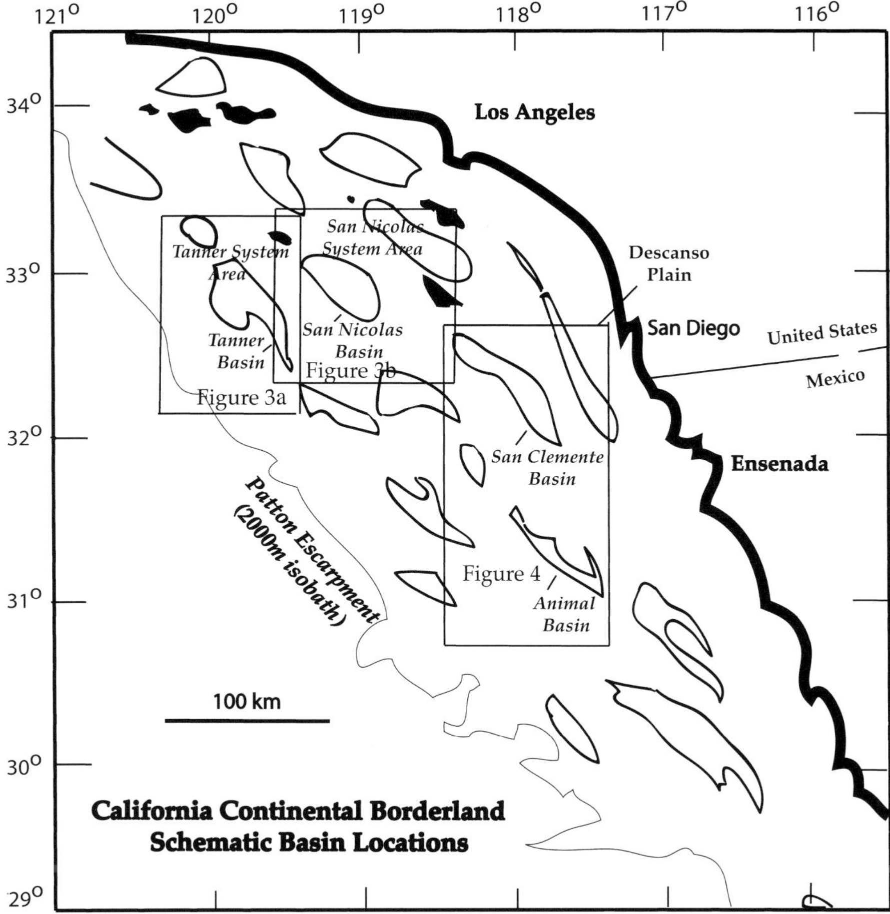

Fig. 1. A schematic view of the California Continental Borderland and the locations of more detailed maps (borderland basin identifications and locations from Emery 1960; Doyle 1973).

Continental Borderland, both the initial oxygen concentration and the export production and subsequent utilization of oxygen within the basins play a role in determining the local oxygenation conditions. There is evidence for a broad, regional shift in oxygenation conditions during the last glacial cycle (Behl 1995; Kennett & Ingram 1995; Behl & Kennett 1996; van Geen *et al.* 1996; Stott *et al.* 2000).

Berelson (1991) and Hickey (1992) have measured deep currents in inner basins of the Borderland. Hickey (1992), in an extensive multi-year study, observed prevailing current velocities over the lower slopes of the Santa Monica and San Pedro Basins of 1–3 cm s^{-1} with event velocities of more than 5 cm s^{-1}. Circulations in the deep basins driven by renewal of bottom water from annual to multi-year periods (Christensen *et al.* 1994) can generate brief episodes of stronger flows that can rework high water content surficial mud. These bottom circulation gyres, influenced by the Coriolis effect, produce highest velocities at the margins of the basin floors. Low-velocity cores of the gyres are centred in the basin floors (Hickey 1992).

During falling sea levels, basin sill cross-sectional areas decrease and flows will increase in strength as the cross-sections decrease. In the outer Borderland from the San Miguel Gap south through the Tanner Basin (Fig. 3a), a branch of the California Current is channelled through a smaller cross-section at lower sea levels when exposed banks inhibit the formation of the present major eddy over the central Borderland and increase bottom velocities in the Tanner Basin (Fig. 2b). It is also probable that the exposed

(a)

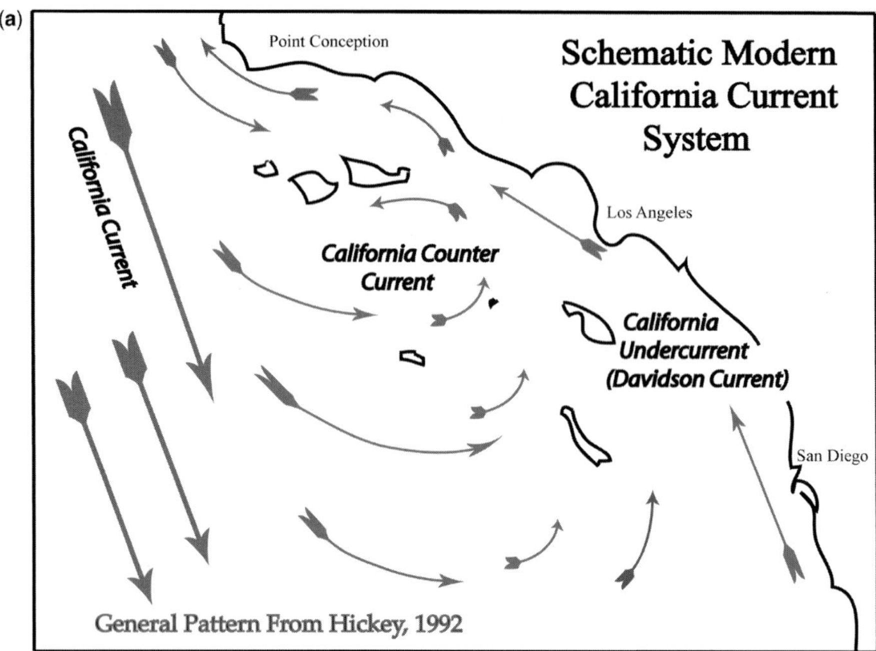

(b)

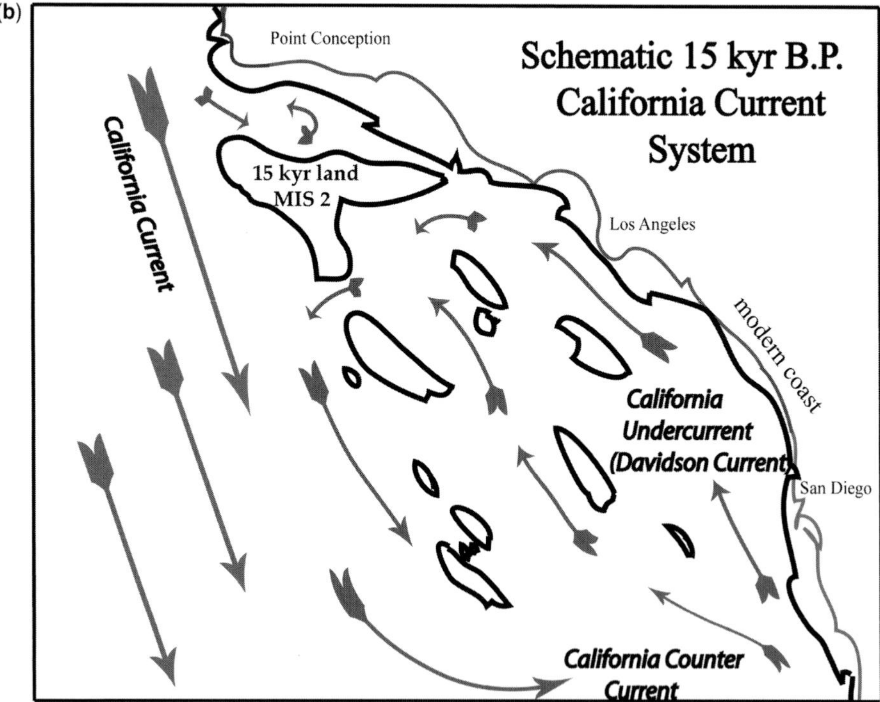

Fig. 2. (**a**) Schematic depiction of the main contemporary surface circulation elements in the California Borderland (Hickey 1992). The core of the Counter Current is at the label. (**b**) Schematic view of the probable general surface circulation at the time of the last glacial maximum. Shorelines based on the −150 m isobaths from the N Series bathymetric maps of the National Ocean Survey. The core of the Counter Current (the position of the label) may have moved south by as much as 200 km.

(a)

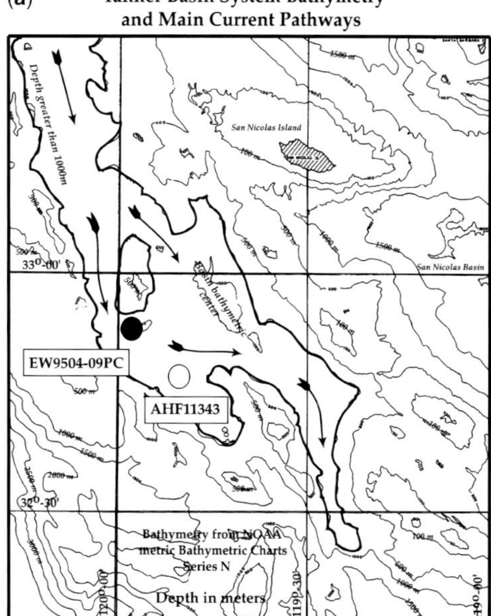

(b)

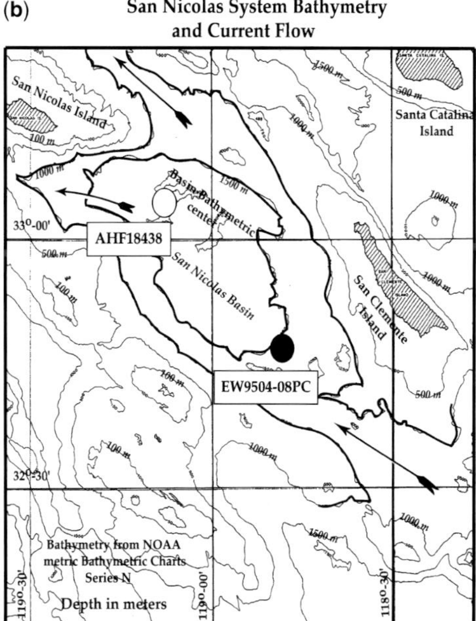

Fig. 3. (**a**) Detailed bathymetric map of the Tanner Basin showing location of cores discussed in the paper and flow lines of the inner branch of the California Current (see Fig. 1 for regional location). It should be noted that EW9504-09PC lies in the mouth of the constricted channel. (**b**) Detailed bathymetric map of the San Nicolas Basin showing locations of cores discussed in this paper. It should be noted that the California Undercurrent passes through a constriction and that EW9504-08PC is beyond the constriction and the most probable maximum velocity zone.

banks would deflect the California Undercurrent offshore away from the present mainland upper slope to a position farther offshore (Fig. 2b). This would increase the flow through the San Nicolas Basin (Fig. 3b), and the lowered sea levels would decrease the channel cross-section and increase flow velocities.

Methods and materials

Cores

Two sets of piston cores were used for this study. One set, ranging in length from 5 to 15 m, was collected during R.V. *Maurice Ewing* cruise EW9504 in April 1995 (Lyle *et al.* 1995). The cores were collected as part of a site survey for Ocean Drilling Program Leg 167 along the California margin. All cores were collected from lower basin slopes and open slope sites to minimize turbidite contribution and slide deposits, and to obtain longer stratigraphic records. Seven cores were collected from California Borderland basins; of these the Santa Monica core was excluded, as it sampled only about 6 ka of late Holocene time (Table 1). Five of the six cores are the basis for this study (Figs 3a, b and 4). Age dating was done using ^{210}Pb, accelerator mass spectrometry ^{14}C magnetostratigraphy and δ^{18}O stratigraphy by Stott *et al.* (2000). Their age models were used to calculate mass accumulation rates.

A second set of older piston cores was selected from the Allan Hancock Foundation archives at the University of Southern California (USC) (Table 1). These cores were collected aboard the R.V. *Velero IV* in the 1970s (Gorsline & Prensky 1975; Doyle & Gorsline 1977) from locations near *Ewing* slope sites (Figs 3b and 4; Table 1). We wished to determine if the structures seen in the slope cores were unique to the slope sediments in contrast to the continuously accumulating sediments of the central basin floors, which presumably were much less affected by bottom-water flows.

Velero cores are archived in dry storage at USC. *Ewing* cores and their subsamples were stored in a cold room at 13 °C to reduce drying rate during the period of analysis. *Ewing* archive core halves are stored in the Oregon State University Core Facility.

X-ray analysis

After cores were split in the laboratory, 2 cm thick longitudinal slabs were cut from the core halves and X-rayed using a Hewlett–Packard Faxitron industrial X-ray unit and a Penetrex industrial X-ray unit. X-ray negatives were digitized using a UMAX 2100 X-ray negative scanner. Images

Table 1. *List of piston core locations and core dimensions*

Core	Latitude (North)	Longitude (West)	Water depth (m)	Core length (cm)	Basin (see Fig. 1)	Age (ka),* base of core
Ewing cores						
EW9504-02PC	31°25.92′	117°35.09′	2040	720	Animal	190
EW9504-03PC	32°04.39′	117°21.85′	1300	400	Descanso	170
EW9504-04PC	32°17.01′	118°23.73′	1760	810	East Cortez	160
EW9504-05PC	32°28.55′	118°07.49′	1820	530	San Clemente	110
EW9504-08PC	32°48.05′	118°48.00′	1440	490	San Nicolas	125
EW9504-09PC	32°51.51′	119°57.48′	1195	650	Tanner	115
Hancock cores						
AHF11343	32°48.00′	119°51.22′	1170	790	Tanner	No data
AHF16833	32°00.00′	117°59.00′	1980	556	East Cortez	No data
AHF17803	32°30.00′	118°03.17′	1890	240	San Clemente	No data
AHF18017	32°17.00′	119°18.00′	1710	288	West Cortez	No data
AHF18438	33°05.33′	119°04.50′	1770	468	San Nicolas	No data

*Age dates from Stott *et al.* (2000).

were inverted to yield positive images, and contrast and brightness were balanced.

Study of the X-radiographs defined several types of reworking structures indicative of different flow regimes and bioturbation features. In order of increasing flow regime they are: type 1, bioturbated mud (also type 1 of Savrda *et al.* 1984; type 2, lenses and possible starved ripples; type 3, erosion surfaces with or without residual debris cover; type 4, cross-bedding; type 5, turbidites. Examples of these at different depths in the cores are illustrated in Figure 5a and b, and the relationship of the structure types to age is shown in Figure 6. As seen in Figure 6, the structures are concentrated in periods primarily of falling sea level (MIS 5–4, MIS 7–6).

Geochemical and isotopic studies

Particle size, carbonate, organic carbon, opal silica and clay mineralogy were studied by Robinson (1997). Carbonate content and total carbon were determined on powdered samples using a LECO gasometric carbon analyser modified following procedures described by Kolpack & Bell (1968). Organic matter was determined by the difference between carbonate carbon and total carbon. Opal silica was determined colorimetrically after sample digestion in D. E. Hammond's laboratory at USC. Clay mineralogy was determined by X-ray diffraction using methods of Fleischer (1970). Results of stable isotopic analyses have been given by Stott *et al.* (2000). Geochemical results are only briefly described in this paper to the degree that they explain changes in facies.

Textural analysis

Particle-size distributions were determined using a Beckman–Coulter Particle Size Analyzer Model LS 32 120 at Centro Interdisciplinario de Ciencias Marinas (CICMAR-IPN) in La Paz, Baja California Sur, Mexico. Dry sediment samples (0.25–0.3 g) were hydrated with 10 ml of distilled water, and rubbed lightly with the fingers to disaggregate large aggregates. Fifty ml of 5% hexametaphosphate solution, and 50 ml of 50% acetone solution were added to dissolve volatile organic matter and prevent flocculation, after which the suspensates were placed in an ultrasonic bath for 15 s for further disaggregation. After standing for 24 h, final disaggregation was completed by a 30 s treatment in a sonic bath. The suspensate samples were homogenized by using a magnetic agitator and an aliquot of *c.* 1 ml was pipetted into the sample receiver in the Beckman–Coulter Particle Size Analyser.

Statistical parameters in microns and in phi units were calculated from the measured volume per cent of particles in *c.* 120–150 windows from 0.04 μm to 2000 μm. Phi moment measures were calculated for use in textural diagrams. Size–frequency curves are plotted in terms of particle diameters in microns.

Sedimentology

The sediments in the cores are primarily hemipelagic clay silts and silts with mean diameters ranging from 4 to 10 μm (7–8ϕ). Winnowing of the fine sediments of the slope cores removes clays, and the lenses and lamina are silts with mean diameters of 15–60 μm (4–6ϕ). Figure 7 shows the sedimentary features of representative sampled intervals, and Figure 8a–d shows corresponding size–frequency curves for the hemipelagic mud, winnowed lamina, cross-beds and basin-floor distal turbidites. The plots are multimodal because of the difficulty of sampling specific thin (millimetre-scale) sediment lamina or lenses, and the specific

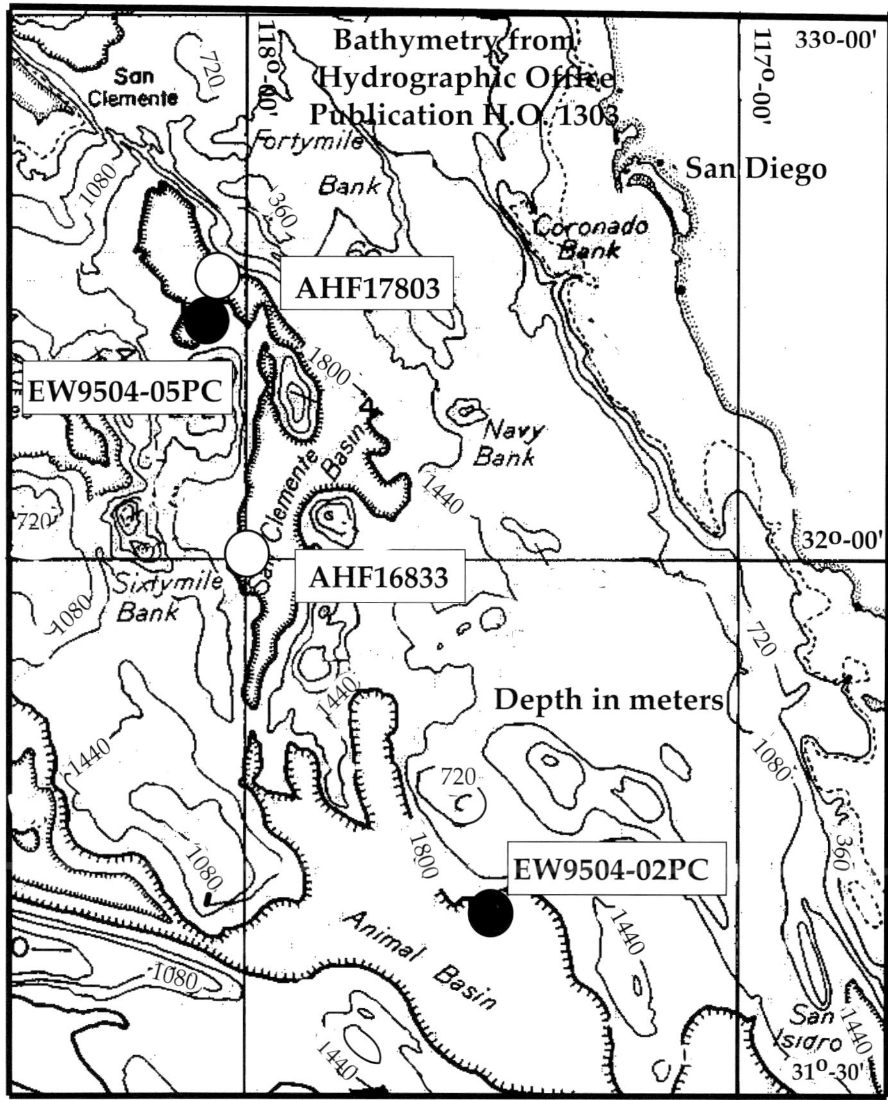

Fig. 4. Detailed bathymetric map of the San Clemente and Animal Basins showing positions of cores discussed in this paper (see Fig. 1 for location).

sediment types appear as modes in the composite samples.

The basin-slope and basin-floor sediments are from the distal slow accumulation rate basins of the Borderland and are mixtures of terrigenous particles with biogenous carbonate, silica and organic carbon. Characterization of the sediments is best shown in terms of mass accumulation rates (MAR). Total mass accumulation rates are low and range from about 5 to 8 mg cm^{-2} a^{-1} except where turbidite spikes appear in the records (Fig. 9). There is an increase in Holocene time except in Animal. San Nicolas exhibits values approaching 10 mg cm^{-2} a^{-1} in MIS 4 and 5 possibly related to the appearance of silt cross-bedding at that time. However, Tanner Basin slope shows no significant change in total MAR although that core has the best developed cross-bedded units at that time.

Figure 10a shows calcium carbonate MAR v. age. Most cores show lowest accumulation rates during glacial time. San Nicolas shows a significant

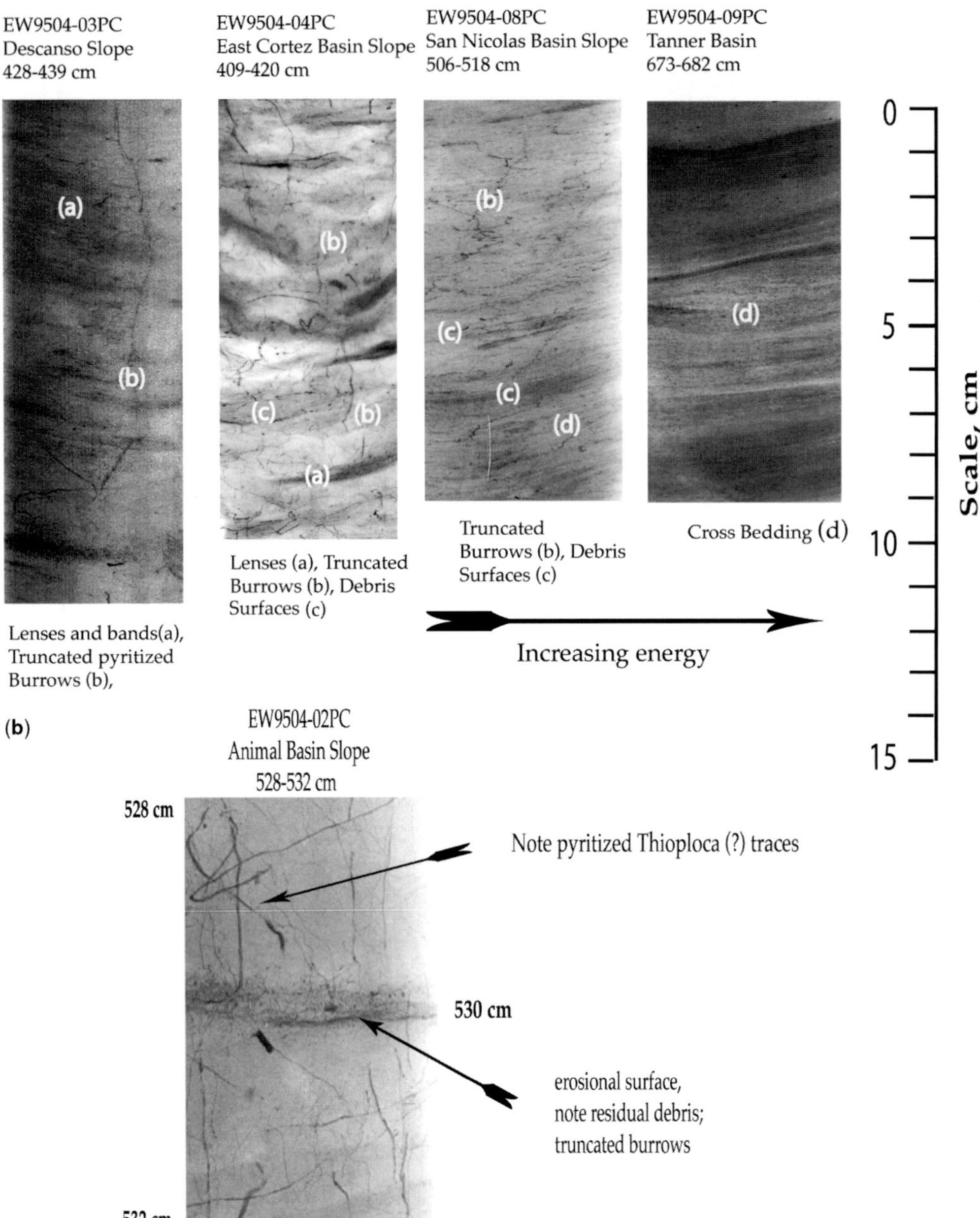

Fig. 5. (**a**) Examples of sedimentary structures showing the progressively higher flow regime structures. (See Figs 3–5 for core locations). (**b**) Enlarged image of an erosional surface with a thin layer of residual debris containing fragments of pyritized micro-burrows. (Note truncation of burrows.) One later burrow appears to have been stopped by the residual layer at the left side of the upper figure.

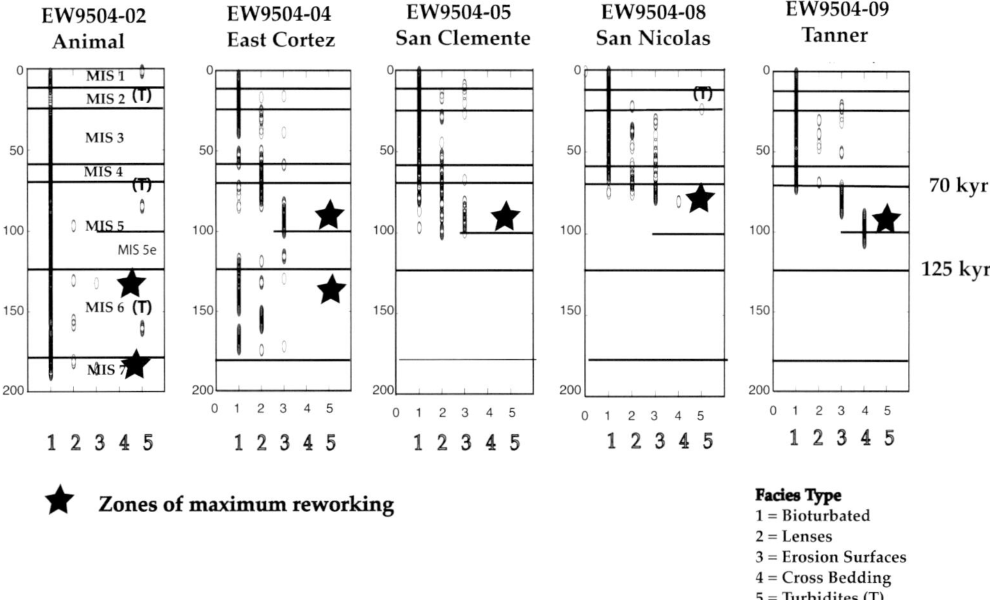

Fig. 6. Facies type distributions v. age in the cores. Data from X-radiographs of the cores. (See Fig. 5 for examples of the structures.)

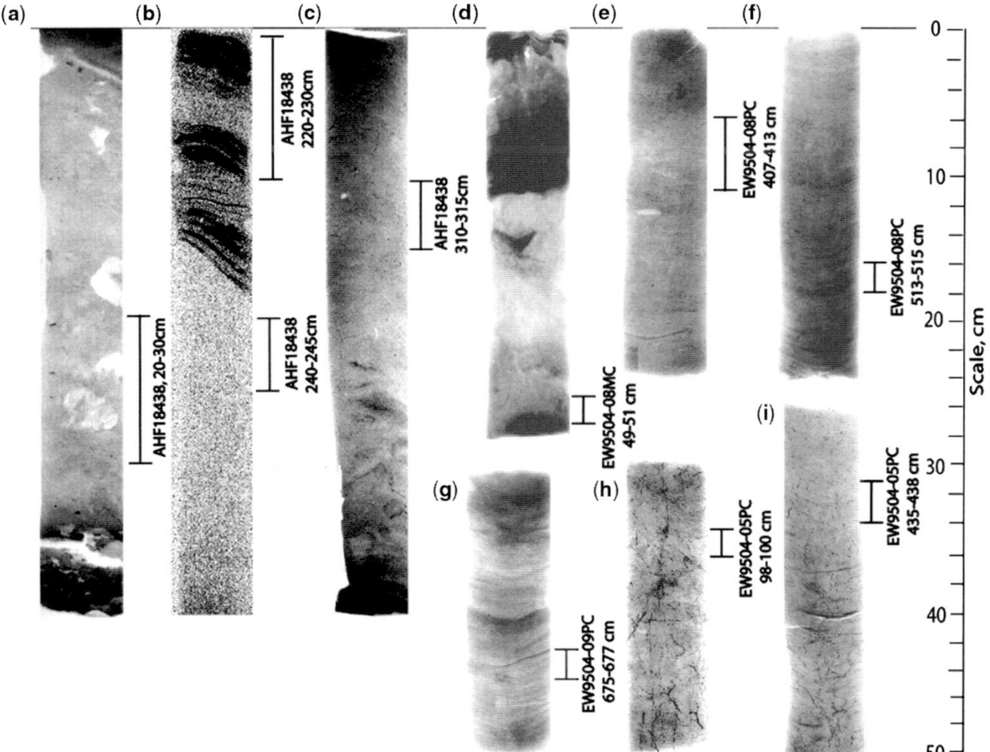

Fig. 7. X-radiographs of the zones from which size analysis samples were taken, (See Fig. 8 for the frequency diagrams of the samples.)

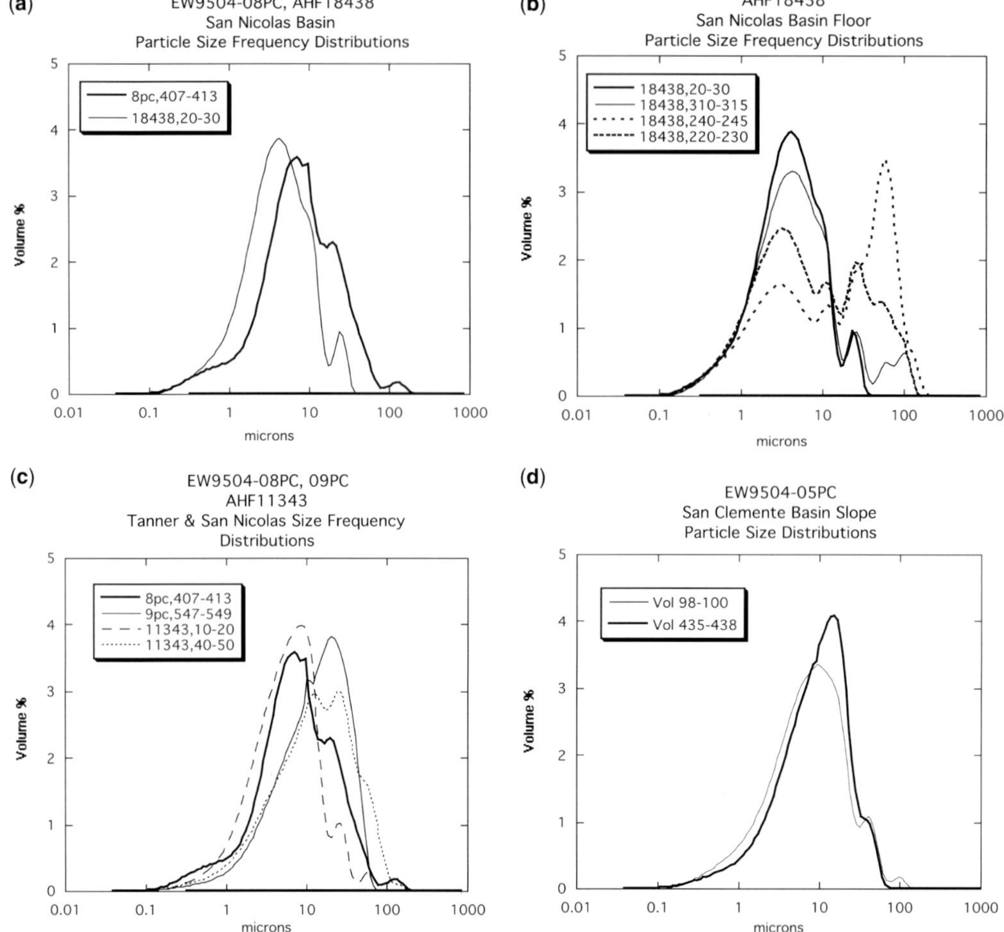

Fig. 8. Representative size–frequency curves illustrating general textural characteristics of the samples. Because the structures are at millimetre scale in thickness and the samples were at centimetre scale, the distributions are typically multimodal, showing the presence of more than one sediment type in the distribution. Preservation of the modes shows that the units were not homogenized. The basin-floor hemipelagic muds have mean diameters of 6–10 μm; the slope hemipelagic muds are coarser, with mean diameters of about 10 μm.

increase in MIS 5, whereas the Holocene rates in that core are only slightly greater than during the glacial times. Similar patterns are repeated in organic carbon MAR.

Organic carbon MAR is shown in Figure 10b. East Cortez and San Clemente show higher values in MIS 1 and 5, but San Nicolas shows a significant high in MIS 4 and 5. Tanner and Animal record essentially uniform organic carbon accumulation v. age, which may reflect the distal and deeper location of Animal south of the major counter current upwelling, and the relatively uniform influence of the California Current in Tanner at all times. The increase in carbonate and organic carbon MAR in San Nicolas may signal the shift of the counter-current upwelling over those basins in MIS 4 and late MIS 5.

Discussion of basin-slope sedimentary structures

Examination of sedimentary structures in X-radiographs of a set of piston cores from the lower slopes of central and outer basins in the California Borderland show the presence of discontinuities in the otherwise continuous but variably bioturbated hemipelagic mud.

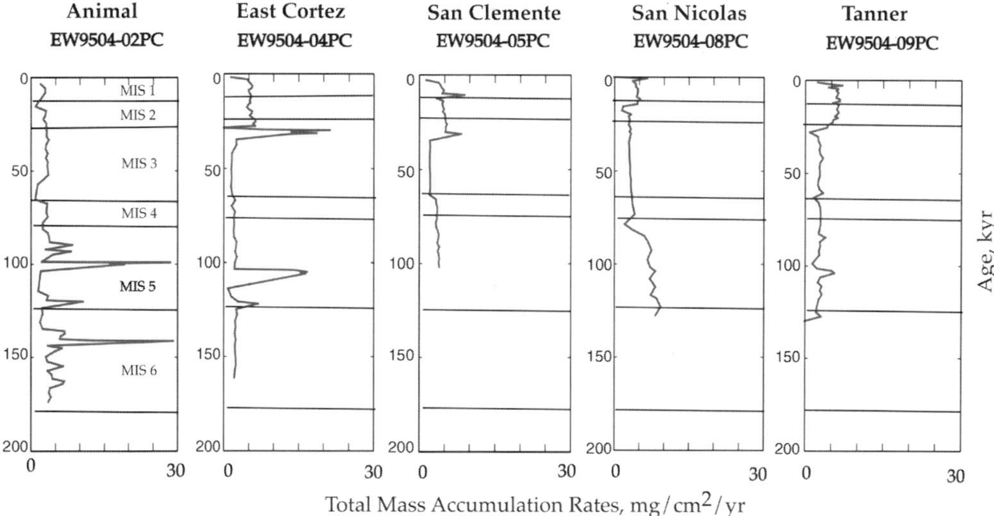

Fig. 9. Total mass accumulation rates v. age for the cores based on age models by Stott *et al.* (2000).

The cores have undergone varying degrees of bioturbation tiering (Savrda *et al.* 1984) and the last overprinting ichnostructures in many of the cores from both slopes and basin floors are two distinct size populations of fine burrows a few tens of microns in diameter that may be formed by bacteria and/or nematodes (Fig. 5a and b). These are most commonly seen in sediments in low-oxygen or anoxic environments (Fossing *et al.* 1995). The concentrations of these micro-burrows vary with time, and they are rare in some cores. These ichnotraces are good indicators of episodes of decreased oxygen during the depositional histories recorded in the cores. Isotopic analyses discussed by Stott *et al.* (2000) also indicate lower oxygenation of the borderland basins in comparison with eastern Pacific cores. Because micro-bioturbation is apparent, the conditions may have been dysaerobic rather than fully anaerobic.

Early diagenetic pyritization of the organic linings of the micro-burrows below depths in cores of about 1 m produces high-density contrast images of these ichno-traces in the radiographs (Fig. 5a and b). Fragments of these pyritized burrows contribute to the coarse fractions of the reworked sediments and are evidence of the reworking. The winnowing structures appear to truncate pre-existing micro-burrows.

As noted above, Figure 6 shows the presence of the various reworking structures v. age (MIS) and indicates the location in the cores of highest flow regime structures. The figure indicates that the highest flow regimes occur mainly in the transition from MIS 5 to 4, a time of relatively rapid sea-level fall (70–125 ka). Cross-bedding structures are seen only in this period, and only in San Nicolas and Tanner. The Animal and East Cortez cores penetrate to MIS 7 and they both record relatively higher flow regimes in that transition from MIS 7 to 6, which is another period of sea-level fall.

In contrast to the intensively reworked large-scale ripple drift structures and thick accumulations seen in such slope locations as the Atlantic margin of the USA, the features described here are interpreted to be the product of winnowing of fine high water content silt–clays by relatively low-velocity currents (most probably $>10\ \mathrm{cm\ s^{-1}}$). As noted above, the resulting sequence at increasing energy levels includes lenses that may be starved isolated silt ripples, thin beds that appear to be erosional surfaces with overlying residual debris, and cross-bedded silts and fine sands (Fig. 5a and b). The primary sediments are fine to medium clay–silts with modal particle sizes of about 10 μm. The lenses and possible starved ripples also attest to the relatively low current velocities, as erosion produces residual fine debris on the erosional surfaces that apparently is adequate to inhibit further erosion.

These structures appear to be very similar to reworked clays and silts reported by Lucchi & Rebesco (2004, 2007) from a part of the Antarctic margin where slow countercurrents impinge on fine sediments. They noted well-defined fine laminations as well as irregular or wispy laminations and lenses that may be very similar to the structures that we report here.

Although we consider bottom-current winnowing to be the most plausible process, other mechanisms

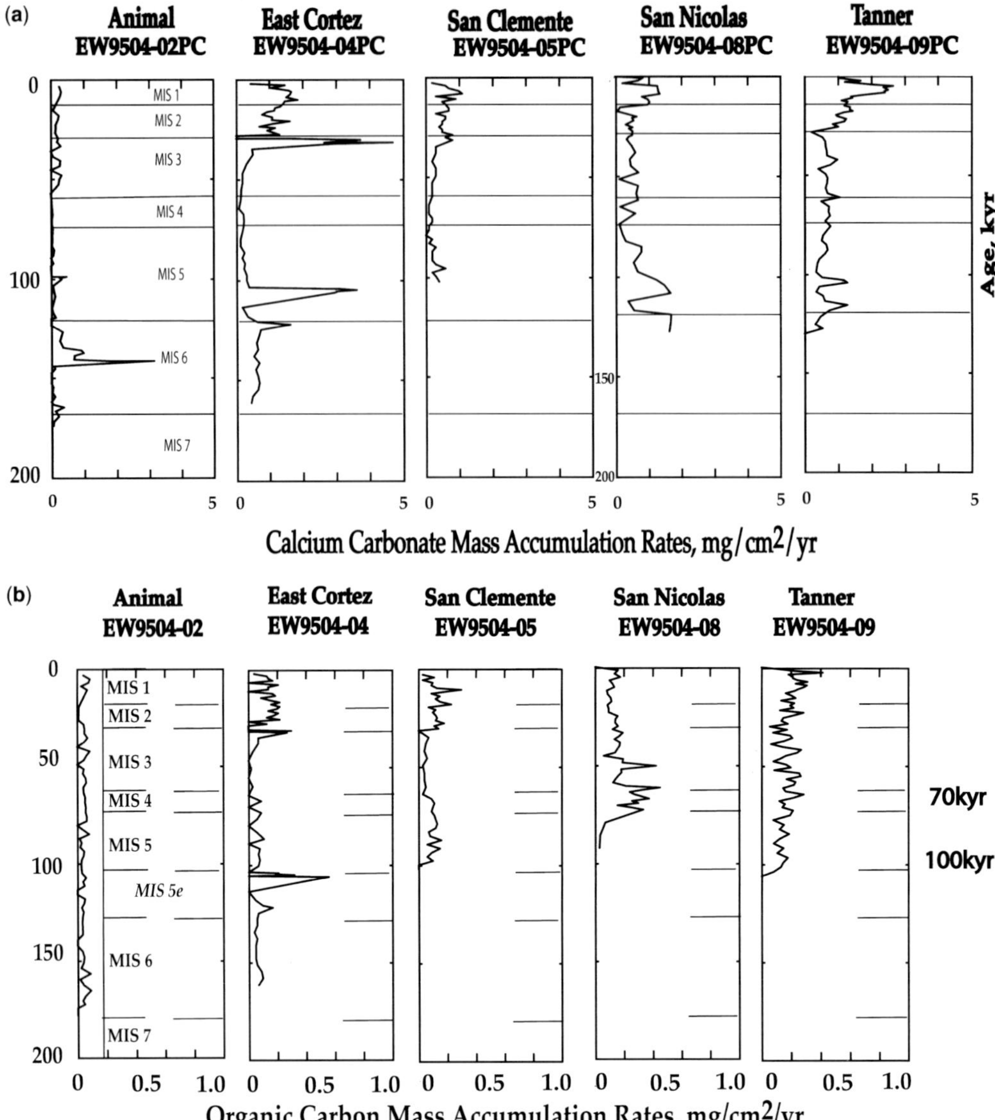

Fig. 10. (**a**) Calcium carbonate mass accumulation rates v. age for the cores based on age models by Stott *et al.* (2000). (**b**) Organic carbon mass accumulation rates v. age based on age models by Stott *et al.* (2000).

might be possible. Well-defined turbidity current deposits are generally absent from these cores although they are characteristic of the adjacent basin-floor cores.

The core sites were selected based on acoustic profiling to be in areas without slope failures and away from submarine canyons. Turbidites in the borderland basins typically originate from the toes of slope slumps and from submarine canyons. The basin locations discussed here are also distant and downcurrent from the major strong upwelling areas north of Point Conception. Thus these locations are not likely to have been affected by bottom currents associated with those strong systems.

The winnowing structures are not present or are rare and weakly defined during the recent 30 ka (MIS 1, 2 and late 3) and begin to increase in the older parts of the cores prior to about 30 ka. They reach a maximum in concentration and in current intensity at about 70–100 ka (MIS 4–5).

Two explanations can be put forth to explain the temporal occurrence of the reworking structures. Bottom-current strength on the basin margins may have varied as a result of the changing physical conditions in the region, or alternatively, bottom-current activity may be a constant feature of these basins but the rates of bioturbation may have varied as a result of changes in bottom-water oxygenation conditions. We will address both options below.

Sedimentologically, the cores can be grouped qualitatively into three sets: (1) Tanner and San Nicolas; (2) San Clemente, East Cortez and Descanso Plain; (3) Animal Basin. Animal Basin is the deepest location, with low mass accumulation rates. The Animal Basin core shows only weakly preserved current features with only minor sedimentological variation over the past 200 ka. It delimits the southern limit of influences of the California Current and of the major mainland sediment sources. The Tanner and San Nicolas basins provide records of relatively intense and nearly continuous reworking, and as such, they represent the other end-member group. San Clemente and East Cortez core records) represent intermediate reworking regimes (see Figs 10–12) Descanso is so similar to the latter two that we have not detailed its characteristics here.

The strongest events, represented by cross-bedded silts and fine sands, are best expressed in the Tanner Basin. The San Nicolas basin-slope core (EW9504-08PC; Fig. 6a) also exhibits cross-bedded silts in the MIS 4–5 interval. These cores are located in basins with restricted passages for bottom-water flow (Fig. 3a and b). Changes in sea level with the onset of glaciation would lead to an intensification of flow in these regions. In the Tanner Basin, a branch of the California Current passing through the San Miguel gap was probably accelerated because of constricted flow associated with a reduced vertical cross-section as well as the emergence of the Channel Island banks to the east during MIS 2 and MIS 5–4. Associated with this condition, the California Undercurrent shifted south and away from the present coast. This effect would deflect its main flow through the San Nicolas and Santa Cruz Basins and north across the outer Santa Cruz Basin sill.

Our sedimentological data show that the two main groups of cores (Tanner–San Nicolas (TSN) and San Clemente–East Cortez–Descanso (SCECD)) show differences in rates of accumulation records. The SCECD group has a general pattern of higher rates in MIS 1 and 5 and very low rates of all components in the intervening colder stages. This suggests that the bathymetric changes associated with lower sea level (Fig. 3a and b) moved the present California Counter Current and upwelling zone, thus markedly decreasing the biogenic contribution to the sediments of those locations.

The changing bathymetric pattern resulting from sea-level changes is probably a major driving factor. The largest changes would occur during shifts from high sea level to low sea level, such as the period from MIS 5 to MIS 4 and perhaps from MIS 7 to MIS 6. Shifts from low to high sea level would decrease overall flow strength as cross-sections increase.

Conclusions

Slope current variations occur in both western and eastern boundary situations, but are more intense, have higher velocities and develop major drifts in the western systems, whereas the eastern low-velocity systems produce small-scale structures of limited stratigraphic thickness. The western boundary systems may also generally be in regions of higher terrigenous contribution. The prevailing low-oxygen environments associated with the eastern boundary upwelling systems tend to preserve these micro-structures from bioturbation and should be present in the black shales of older similar margin settings such as the Monterey (Miocene) formation of California. High-resolution studies of these types of deposits may reveal small-scale current-produced deposits, which would be of great palaeo-environmental significance.

Although the slope-current deposits in the eastern boundary current system regimes do not produce potential reservoir targets, the subtle effects of these reworking processes can affect interpretations of deposits generally assumed to be laid down in environments of continuous deposition.

Funds for the study of the cores used here have come from the National Science Foundation, National Oceanic and Atmospheric Administration, Bureau of Land Management and Office of Naval Research over the past 30 years. L. D. Stott kindly provided the stable isotope and age dating data. The late L. J. Doyle and S. P. Prensky performed the early analysis of the *Velero* cores. Numerous students aided in the collections of the cores on a number of cruises. Reviews by A. Viana and M. Rebesco much improved the paper and its presentation, and provided leads to additional useful references.

References

ATWATER, T. 1989. Plate tectonics history of the northeast Pacific and western North America. *In*: WINTERER, E. L., HUSSONG, D. M. & DECKER, R. W. (eds) *Decade of North American Geology, N. The Eastern Pacific and Hawaii*. Geological Society of America, Boulder, CO, 21–71.

Behl, R. J. 1995. Sedimentary facies and sedimentology: in late Quaternary Santa Barbara Basin, Site 893. *In*: Kennett, J. P., Bauldauf, J. & Lyle, M. (eds) *Proceedings of the Ocean Drilling Program, Scientific Results, 146*. Ocean Drilling Program, College Station, TX, 295–308.

Behl, R. J. & Kennett, J. P. 1996. Brief interstadial events in the Santa Barbara Basin, NE Pacific, during the past 60 kyr. *Nature*, **379**, 243–246.

Berelson, W. M. 1991. The flushing of two deep-sea basins, Southern California Borderland. *Limnology and Oceanography*, **36**, 1150–1167.

Bouma, A. H. & Hollister, C. D. 1963. *Deep ocean sedimentation*. Pacific Section SEPM. The Sedimentary Geology Society, Short Course Notes, **79**.

Christensen, C. J., Gorsline, D. S., Hammond, D. E. & Lund, S. P. 1994. Non-annual laminations and expansion of anoxic basin-floor conditions in Santa Monica Basin, California Borderland, over the past four centuries. *Marine Geology*, **116**, 399–418.

DeDiego, T. & Douglas, R. G. 1999. Oxygen-related sediment microfabrics in modern 'black shales', Gulf of California, Mexico. *Journal of Foraminiferal Research*, **29**, 453–464.

Doyle, L. J. 1973. *Marine Geology of the Baja California Continental Borderland, Mexico*. PhD Dissertation, University of Southern California, Los Angeles.

Doyle, L. J. & Gorsline, D. S. 1977. Marine geology of Baja California Continental Borderland. *AAPG Bulletin*, **61**, 903–917.

Emery, K. O. 1960. *The Sea off Southern California*. Wiley, New York.

Emery, K. O. & Hulsemann, J. 1962. The relationships of sediments, life and water in a marine basin. *Deep-Sea Research*, **8**, 165–180.

Fleischer, P. 1970. Mineralogy and sedimentation history. Santa Barbara Basin, California. *Journal of Sedimentary Petrology*, **42**, 49–58.

Fossing, H., Gallardo, V. A., Jorgensen, B. B., et al. 1995. Concentration and transport of nitrate by the mat-forming sulphur bacterium *Thioploca*. *Nature*, **374**, 713–715.

Gorsline, D. S. & Prensky, S. P. 1975. Paleoclimatic inferences for Late Pleistocene and Holocene from California Continental Borderland Basins. *In*: Suggate, R. P. & Cresswell, M. M. (eds) *Quaternary Studies*. Royal Society of New Zealand, Auckland, 147–154.

Gorsline, D. S. & Teng, L. S.-Y. 1989. The California Continental Borderland. *In*: Winterer, E. L., Hussong, D. M. & Decker, R. W. (eds) *Decade of North American Geology, N. The Eastern Pacific Ocean and Hawaii*. Geological Society of America, Boulder, CO, 471–487.

Gorsline, D. S., Nava-Sanchez, E. M. & Murillo de Nava, J. M. 1996. A review of the occurrences of Holocene and Late Pleistocene laminated sediments in California Continental Borderland basins: products of a variety of depositional processes. *In*: Kemp, A. E. S. (ed.) *Paleoclimatology and Paleoceanography from Laminated Sediments*. Geological Society, London, Special Publications, **116**, 93–110.

Hams, J. E. 1987. *Turbidite sedimentation in a modern basin floor environment: San Nicolas Basin, California Continental Borderland*. MS thesis, California State University, Los Angeles.

Heezen, B. C. & Hollister, C. D. 1964. Evidence of deep-sea bottom currents from abyssal sediments. *Marine Geology*, **1**, 141–174.

Hickey, B. M. 1979. The California Current System—hypotheses and facts. *Progress in Oceanography*. **8**, 191–279.

Hickey, B. M. 1992. Circulation over the Santa Monica–San Pedro Basins and shelf. *Progress in Oceanography*, **30**, 37–116.

Hickey, B. M. 1993. *Physical Oceanography in Ecology of the Southern California Bight*. University of California Press, Berkeley, 19–70.

Hollister, C. D., Johnson, D. A. & Lonsdal, P. 1974. Current-controlled abyssal sedimentation: Samoan Passage, Equatorial west Pacific. *Journal of Geology*, **82**, 275–300.

Keigwin, L. D. & Jones, G. A. 1989. Glacial–Holocene stratigraphy, chronology and paleoceanographic observations on some North Atlantic sediment drifts. *Deep-Sea Research*, **36**, 845–868.

Kennett, J. P. & Ingram, B. L. 1995. Paleoclimatic evolution of Santa Barbara Basin during the last 20 ky: marine evidence from Hole 893A. *In*: Kennett, J. P., Bauldauf, J. G. & Lyle, M. (eds) *Proceedings of the Ocean Drilling Program, Scientific Results, 146*. Ocean Drilling Program, College Station, TX, 309–328.

Kolpack, R. L. & Bell, S. A. 1968. Gasometric determination of carbon in sediments by hydroxide absorption. *Journal of Sedimentary Petrology*, **38**, 617–620.

Lonsdale, P. 1981. Drifts and ponds of reworked pelagic sediment in part of the southwest Pacific. *Marine Geology*, **43**, 153–193.

Lonsdale, P. & Malfait, B. 1974. Abyssal dunes of foraminiferal sand on the Carnegie Ridge. *Geological Society of America Bulletin*, **85**, 1697–1712.

Lucchi, R. G. & Rebesco, M. 2004. Hybrid contourites from the Antarctic Continental Margin. *In*: *32nd International Geological Congress, Abstracts*, 316.

Lucchi, R. G. & Rebesco, M. 2007. Glacial contourites on the Antarctic Peninsula margin: insight for palaeoenvironmental and palaeoclimatic conditions. *In*: Viana, A. R. & Rebesco, M. (eds) *Economic and Palaeoceanographic Significance of Contourites*. Geological Society, London, Special Publications, **276**, 111–127.

Lyle, M., Mix, A., Stott, L., Francis, B. A. & Budyhpramono, S. 1995. *Collaborative Research: Late Pleistocene Climate-induced Fluctuations in the Californa Current*. Paleoceanographic Study and ODP Site Survey Cruise Report No. BSU-CGISS 95-11. Lamont–Doherty Earth Observatory. Palisades, NY.

Middleton, G. V. & Southard, J. B. 1977. *Mechanics of Sediment Movement*. SEPM, The

Sedimentary Geology Society, Short Course Lecture Notes, **3**.

MOLLENHAUER, G., SCHNEIDER, R. R., MULLER, P. J., SPIETZ, V. & WEFER, G. 2002. Glacial/interglacial variability in the Benguela upwelling system: spatial distribution and budgets of organic carbon accumulation. *Global Biogeochemical Cycles*, **16**, 81/1–81/15.

RASMUSSEN, E. S. 1994. The relationship between submarine canyon fill and sea-level change: an example from Middle Miocene offshore Gabon, West Africa. *Sedimentary Geology*, **90**, 61–75.

RASMUSSEN, E. S. 1997. Depositional evolution and sequence stratigraphy of the shelf and slope area off south Gabon, West Africa. *Journal of Sedimentary Research*, **67**, 715–724.

RASMUSSEN, E. S. 2000. Discussion. 'Oligocene to Holocene sediment drifts and bottom-currents on the slope of Gabon continental margin (West Africa). Consequences for sedimentation and southeast Atlantic upwelling. *Sedimentary Geology* **128**, 179–199.' *Sedimentary Geology*, **136**, 157–161.

REID, J. L., JR. 1965. *The Intermediate Waters of the Pacific Ocean*. Johns Hopkins Press, Baltimore, MD.

ROBINSON, B. S. 1997. *Late Pleistocene depositional history of the California Continental Borderland*. MS thesis, University of Southern California, Los Angeles.

SAVRDA, C. E., BOTTJER, D. J. & GORSLINE, D. S. 1984. Development of a comprehensive oxygen-deficient marine biofacies model: evidence from Santa Monica, San Pedro and Santa Barbara Basins, California Continental Borderland. *AAPG Bulletin*, **68**, 1179–1192.

Scripps Institution of Oceanography, 1971. *Bathymetric Atlas of the Northeastern Pacific Ocean*. US Naval Oceanographic Office, Hydrographic Office Publication, **1303**, map 1206N.

SERANNE, M. & ABEIGNE, C.-R. N. 1999. Oligocene to Holocene sediment drifts and bottom currents on the slope of Gabon continental margin (west Africa). Consequences for sedimentation and southeast Atlantic upwelling. *Sedimentary Geology*, **128**, 179–199.

SERANNE, M., ABEIGNE, C.-R. N. & LOPEZ, M. 2000. Reply to 'Oligocene to Holocene sediment drifts and bottom-currents on the slope of Gabon continental margin (West Africa). Consequences for sedimentation and southeast Atlantic upwelling. *Sedimentary Geology*, **128**, 179–199 (1999).' *Sedimentary Geology*, **136**, 163–168.

SHEPARD, F. P. & EMERY, K. O. 1941. *Submarine topography off the California coast; canyons and tectonic interpretations*. Geological Society of America, Special Papers, **31**.

SLOSS, L. L. 1962. Stratigraphic models in exploration. *Journal of Sedimentary Petrology*, **32**, 415–422.

STOTT, L. D., NEUMAN, M. J. & HAMMOND, D. E. 2000. Intermediate water ventilation on the south-eastern Pacific margin during the late Pleistocene inferred from benthic foraminiferal δC^{13}. *Pale oceanography*, **15**, 161–169.

STOW, D. A. V., FAUGÈRES, J.-C., VIANA, A. & GONTHIER, E. 1998. Fossil contourites: a critical review. *Sedimentary Geology*, **115**, 3–31.

VAN GEEN, A., FAIRBANKS, R. G., DARTNELL, P., MCGANN, M., GARDNER, J. V. & KASHGARIAN, M. 1996. Ventilation changes in the northeastern Pacific during the last deglaciation. *Paleoceanography*, **11**, 519–528.

VIANA, A. R., FAUGÈRES, J.-C. & STOW, D. A. V. 1998*a*. Bottom-current-controlled sand deposits—a review of modern shallow- to deep-water environments. *Sedimentary Geology*, **115**, 53–80.

VIANA, A. R., FAUGÈRES, J. C., KOWSMANN, R. O., LIMA, J. A. M., CADDAH, L. F. G. & RIZZO, J. G. 1998*b*. Hydrology, morphology and sedimentology of the Campos continental margin, offshore Brazil. *Sedimentary Geology*, **115**, 133–157.

Santos Drift System: stratigraphic organization and implications for late Cenozoic palaeocirculation in the Santos Basin, SW Atlantic Ocean

CLÁUDIO S. L. DUARTE & ADRIANO R. VIANA

Petrobras S.A./E&P-Exploration, 65, República do Chile Av., 13th Floor, Rio de Janeiro, RJ, 20031-912, Brazil (e-mail: aviana@petrobras.com.br)

Abstract: High-quality 2D and 3D seismic data were analysed to investigate the stratigraphic organization of the sedimentary deposits and the impact of the palaeocirculation of the SW Atlantic Ocean in the construction of the Santos Basin slope from late Palaeogene to Recent time. Seven seismic sequences were identified based on their external geometry, internal seismic pattern and seismic boundaries. A correlation between these sequences, the glacio-eustatic curves and the major climatic–palaeoceanographic events was attempted. The base of the studied sequences is the *c.* 28 Ma intra-Oligocene Rupelian–Chatian unconformity. Six other seismic horizons corresponding to the sequence boundaries were mapped ranging from the late Oligocene to the Pliocene. Variations in the bottom-current intensity were characterized by the seismic pattern of each sequence and by the evolution of margin physiography. A 100 km long channel-like gutter, the Santos Channel, was excavated at the foot of an intra-slope escarpment. It was the locus of major flow circulation until the middle Miocene, when the margin physiography reorganization transferred the main axis of deep current action downslope and excavated the São Paulo Channel at the foot of the present continental slope. Two major contourite drifts were accumulated in response to the different combinations of bottom-current axis position and slope physiography. Those drifts constitute the Santos Drift System. Conversely to the present-day circulation pattern, with southward flow above the slope dominated by the western boundary Brazil Current, both at surface and deep waters, the geological record indicates that the palaeocirculation in the Santos Basin was marked by the opposite sense of circulation of surface waters (southward palaeo Brazil Current) and of the intermediate to deep waters (northward Southern Ocean Current). The study indicates that periods of relative sea-level rise to highstands correspond to increase in drift accumulation whereas during lowstands slope drift sedimentation is reduced.

The Brazilian SE margin comprises a set of sedimentary basins (Espírito Santo, Campos and Santos) that have been a focus of attention of the oil industry in the past 20 years. They hold most of the country's petroleum reserves known to date. The Santos Basin is a large sedimentary basin with a total area in excess of 350 000 km^2. It extends from the present coastline to the outer boundary of the São Paulo Plateau (water depths >3500 m). To the north, it is separated from the Campos Basin by the Cabo Frio High. To the south, the Florianópolis Fracture Zone, which defines the Florianópolis High (Fig. 1), separates it from the Pelotas Basin (Gambôa & Rabinowitz 1981; Dias *et al.* 1994). Despite several earlier studies and the economic interest owing to its potential hydrocarbon reserves (more than 100 exploratory wells have been drilled), its complex geological evolution remains poorly known.

Seismic refraction surveys carried out in the mid-1960s identified the basin (Ewing *et al.* 1969). The first lithostratigraphic classification was proposed by Ojeda & Cesero (1973; see also Ojeda 1982). The continuity of oil-industry investigations has resulted in a growing knowledge on the basin, involving both tectono-structural approaches (Chang & Kowsmann 1987; Macedo 1989, 1990; Chang *et al.* 1992; Demercian *et al.* 1993; Mohriak *et al.* 1995; Cobbold *et al.* 2001; Meisling *et al.* 2001) and sedimentary–stratigraphic approaches (Pereira & Macedo 1990; Pereira 1994; Modica & Brush 2004).

Broadly, its history follows that of the other SE Brazilian basins, the Campos and Espírito Santo Basins, with a rift phase developed during the Hauterivian and Barremian, followed by a transitional period, marked by thick deposition of evaporites during the late Aptian. The present marine drift phase started during the Albian (Pereira & Macedo 1990; Mohriak *et al.* 1995). The continuous deepening of the basin was accompanied in the Late Cretaceous by a huge transfer of continental sediments towards the basin, mostly influenced by the flexural tilting of the margin, which uplifted the coastal ranges (Serra do Mar and Serra da Mantiqueira). The uplift induced the denudation

From: VIANA, A. R. & REBESCO, M. (eds) *Economic and Palaeoceanographic Significance of Contourite Deposits*. Geological Society, London, Special Publications, **276**, 171–198.
0305-8719/07/$15.00

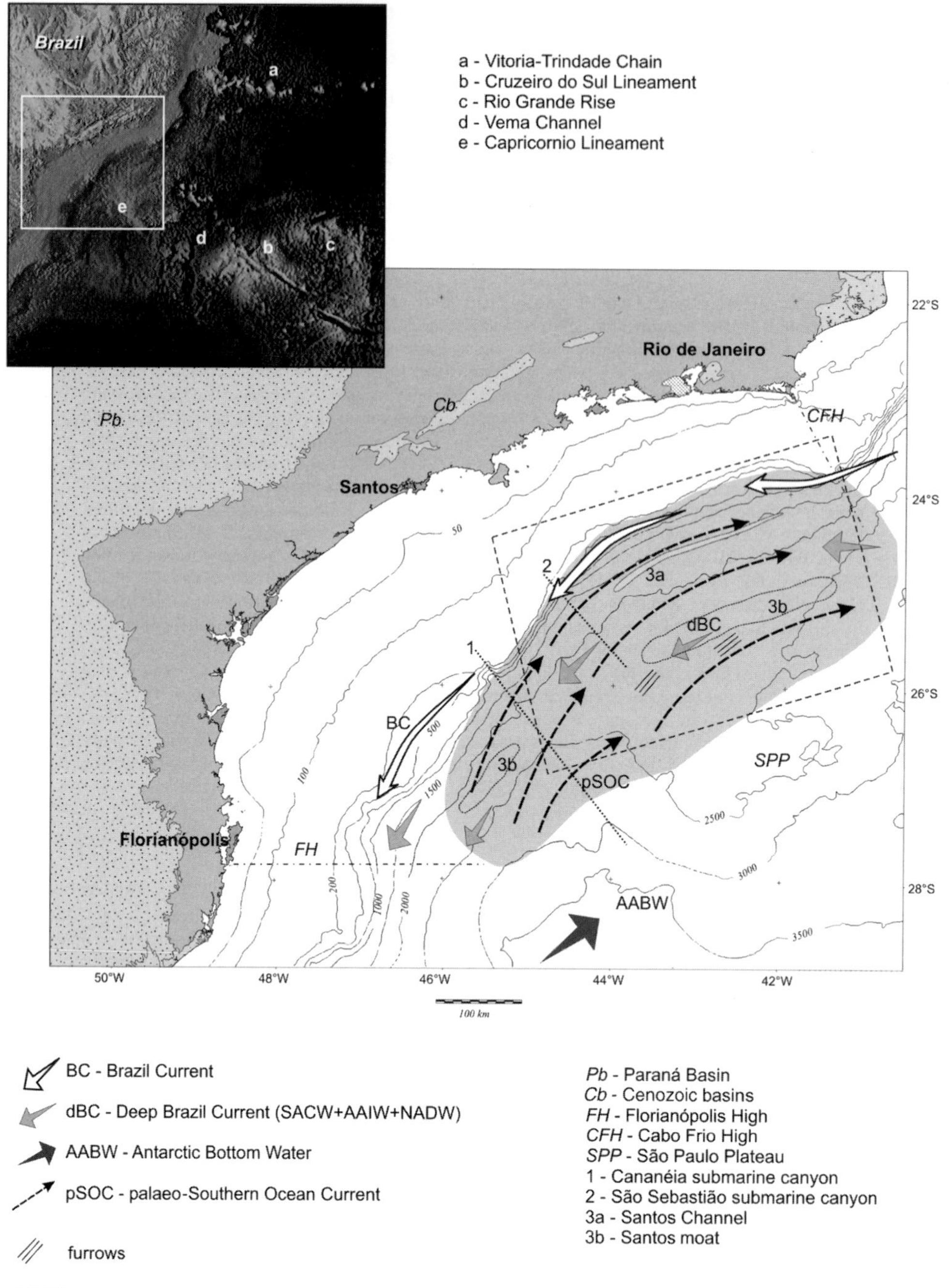

Fig. 1. Location of the study area. Shaded area indicates the location of the Santos Drift. Continuous-line arrows indicate the present-day circulation pattern of the Santos Basin; dashed arrows indicate the path of the palaeo-Southern Ocean Current, which drained the lower basin with high intensity in different climatic conditions, as described in the text. The main physiographic features referred to in the text are shown. Features 3a and 3b represent the position of the axis of the Santos Channel and the Santos moat, respectively. Dashed rectangle indicates the main study area. The inset is a GEBCO topographic chart of the continental and oceanic area of SE Brazil showing the main submarine tectonic features. White rectangle in the inset marks the study area.

of the highlands and provided the sedimentary load that filled the inner portion of the basin (Pereira & Feijó 1994; Almeida & Carneiro 1998; Meisling *et al.* 2001; Modica & Brush 2004).

During the Neogene, after the northward shift of the locus of continental sediment transfer towards the ocean, the Santos Basin was dominated by marine processes, which redistributed the already deposited sediments and those that arrived from the coastal systems during successive high-frequency lowstands.

High-quality industrial 2D and 3D seismic lines were shot along the central–east portion of the Santos Basin, from the present-day shelf edge to water depths in excess of 2000 m. The interpreted seismic horizons were time-correlated through biostratigraphic dating from proprietary well data and provided the basis for a regional seismic stratigraphic framework. The seismic lines show evidence of strong bottom currents influencing the regional sedimentation pattern from late Cretaceous time at least, with a strong increase from late Palaeogene to Recent time, with a broad signature of this process being present throughout the Neogene. The analysis of Deep Sea Drilling Project (DSDP) data from leg 72 (Barker *et al.* 1981) retrieved in the Brazil Basin and the Rio Grande Rise, basinward from the study area, indicated the presence of active systems of bottom currents sweeping the outer margin since the early Eocene, preceding the commonly accepted period of Antarctic Bottom Water onset located at the Eocene–Oligocene boundary. In the study area, a large sediment drift was developed covering the basin from the shelf break to its deep portions, far beyond the 2000 m water depth, and was responsible for defining the present local configuration of the margin slope.

The goal of this paper is to present the general characteristics of this large drift deposit developed mostly during the Neogene and occupying a wide area along the Santos Basin slope, its stratigraphic framework and evolution, and the palaeoceanograpic implications resulting from the recognition of strong bottom currents being active in this portion of the Brazil margin.

Geological context

Tectonic framework

The SE Brazil basins developed over a Precambrian framework of cratonic nuclei that were welded along fold-and-thrust belts during the Brasiliano Cycle (0.93–0.50 Ga). The Precambrian domains are interpreted as part of a transform plate margin where dextral transpressive motion was dominant during the Late Proterozoic (Szatmari *et al.* 1996). The Santos Basin's structural framework followed pre-existing weakness zones and its general SW–NE trend reflects this inheritance. Sediment pathways responsible for transferring and redistributing large amounts of sand into deep-water settings were strongly influenced by pre-existing structural features.

The Santos Basin was part of the East Brazil rift system (Chang *et al.* 1992), formed in the Jurassic–Early Cretaceous, that led to the opening of the Atlantic (Rabinowitz & LaBrecque 1979; Nurnberg & Muller 1991). Rifting initiated in the south, at the southern tip of South Africa, in the Late Triassic–Early Jurassic (220–200 Ma), and propagated northward into the southeastern Brazil margin to the Florianópolis fracture zone in the Late Jurassic–Early Cretaceous (*c.* 140 Ma) (Sibuet *et al.* 1984; Szatmari *et al.* 1985; Conceição *et al.* 1988). The main phase of rifting in the Santos Basin probably began in the Hauterivian–Barremian, possibly following an earlier volcanic phase of rifting (Peate 1997). Important transfer zones caused by extensional stresses were developed during the rift phase: the Rio de Janeiro Transfer Zone, the Curitiba Transfer Zone, and the Florianópolis Transfer Zone. The first is situated in the passage to the Campos Basin at the north, the second divides the Santos Basin into two sectors (north and south), and the third separates it from the Pelotas Basin (Chang & Kowsmann 1984; Macedo 1989; Demercian *et al.* 1993; Mohriak *et al.* 1995; Cobbold *et al.* 2001).

During the drift phase, thin-skinned salt-cored structures overlay a detachment layer of Aptian evaporites. Almost all researchers have attributed these structures to gravitational gliding (Demercian *et al.* 1993; Cobbold *et al.* 1995; Szatmari *et al.* 1996). In the Santos Basin, growth folds were identified on both dip-oriented and strike-oriented seismic lines, providing evidence for radially convergent gliding in variable directions perpendicular to the arcuate coastline. Discrete phases of salt tectonics occurred in the Albian, Campanian, early Cenozoic and Neogene (Demercian *et al.* 1993). Compressive stresses responsible for this deformation style are mostly related to the differential slopes created by the thick siliciclastic-dominated sedimentary column deposited during the late Cretaceous progradation as a result of a continuous and vigorous denudation of the highlands (Szatmari *et al.* 1996). The thick sedimentary prism induced the development of a regional SW–NE-trending antithetic fault, associated with the rafting of Albian carbonates and salt welding during the Senonian (Demercian *et al.* 1993; Mohriak *et al.* 1995; Demercian 1996; Szatmari *et al.* 1996).

A subsidence phase occurred from late Cretaceous to Recent time, and was probably controlled by intensive Mesozoic alkaline magmatism and by the uplift of the Serra do Mar range.

The locus of clastic deposition into the basin shifted during the Late Cretaceous and Cenozoic as a result of onshore block faulting and drainage reorganization. Cretaceous sedimentary rocks were folded, tilted, eroded, and unconformably onlapped above an inferred Neocomian Moho uplift to produce an accentuated nearshore hinge line (Cobbold *et al.* 2001). Neotectonic fault-block tilting has resulted in mountain ranges up to 2700 m high and extensive river capture. Based on fission-track data, the mountains were exhumed in the Cretaceous and Eocene (Cobbold *et al.* 2001; Saenz *et al.* 2003). The continuous northward shift of the locus of clastic deposition resulted in an increased subsidence during the Neogene and a complete abandonment of the basin by large drainage systems, which induced drowning and domination of oceanic processes in sedimentation. Off Cabo Frio, the continental platform salient that separates the Campos and Santos Basins (the Cabo Frio anticline; Mohriak *et al.* 1995) plays an important role in the oceanic circulation in the area. Cabo Frio marks the transition between the tropical environment (to the north) and the subtropical (to the south) (Rocha *et al.* 1975).

Sedimentary and stratigraphic framework

The geological evolution of the Santos Basin comprises three major phases (Pereira & Macedo 1990; Pereira & Feijó 1994): lake (continental rift); gulf (restricted marine); oceanic (open marine) (Fig. 2). The oceanic phase started in the Cenomanian and was marked by an intense progradation, which attained its peak during the Santonian and the Campanian (Juréia Progradation) and reducing in intensity from the Maastrichtian to the Eocene, the last strong tectonic period recorded in the basin.

A thick sequence of coarse-grained siliciclastic sediments prograded basinward during the Late Cretaceous, whereas finer-grained sediments accumulated in the Cenozoic (Pereira & Macedo 1990). These contrasting patterns and phases have been attributed to uplift and exhumation of the Serra do Mar coastal mountains (the Juréia Progradation; Pereira & Feijó 1994; Almeida & Carneiro 1998; Meisling *et al.* 2001). The present coastal plains and the isobaths are parallel to the Serra do Mar range. Apatite fission-track analysis indicated that the Serra do Mar underwent successive episodes of heating followed by linear cooling from the late Cretaceous to the Neogene (Saenz *et al.* 2003), which were expressed in the Santos Basin as regional unconformities and accommodation space creation.

The cooling between 90 and 60 Ma is corroborated by the huge amount of continentally derived sediments transferred to the basin (Juréia Progradation, Santos and Juréia Formations; Almeida & Carneiro 1998). The large volumes of Late Cretaceous and Palaeogene clastic sediments in the central Santos Basin must have been transported by major river systems. The headwaters of two major rivers present in the area, the Paraíba do Sul and the Tietê, lie a few tens of kilometres from the coast. They were probably responsible for draining the Santos Basin during the late Cretaceous before the coastal uplift. During the Palaeogene other features developed such as the important regional unconformities corresponding to the beginning of a late progradation in the western part of the Santos Basin (Meisling *et al.* 2001; Modica & Brush 2004).

On the northeastern flank of the basin, during the Maastrichtian–Eocene, sediment depocentres migrated progressively eastward toward Cabo Frio (Mohriak *et al.* 1995). In the late Eocene, sedimentation rates decreased greatly in the central parts of the Santos Basin but increased on both flanks.

The late tectonic readjustment that occurred from the Eocene onwards induced a shift of the tectonic locus, which migrated northward, thus shutting off sediment supply to the Santos Basin, blocking the main issues of sediment feeder systems to the sea, and diverting them to the Campos Basin, via the present path of the Paraíba do Sul river (Karner & Driscoll 1999). The Paraíba do Sul river is at present responsible for the development of a large wave-dominated delta in the Campos Basin.

From the Palaeocene to the Eocene, the Cenozoic progradation studied by Moreira *et al.* (2001) partially masked the effects of the vigorous bottom currents (see the Discussion for further details).

Morphological and hydrographic context

Morphological context

The Santos Basin extends 800 km in a NE–SW shoreline-parallel trend, and more than 450 km in a NW–SE dip trend (locally this dip trend extends more than 600 km). To the west, the basin is limited by a complex of mountain ranges up to 2200 m high, comprising the Serra do Mar and the Serra da Mantiqueira, that confine the basin to the offshore domain (Milani 2000). A very narrow coastal plain is developed marked by the absence of large rivers and, especially in the central area between Rio de Janeiro and Santos, by a series of large bays and several islands derived from the cropping out of the Precambrian basement. Towards the south and in the extreme NE, the

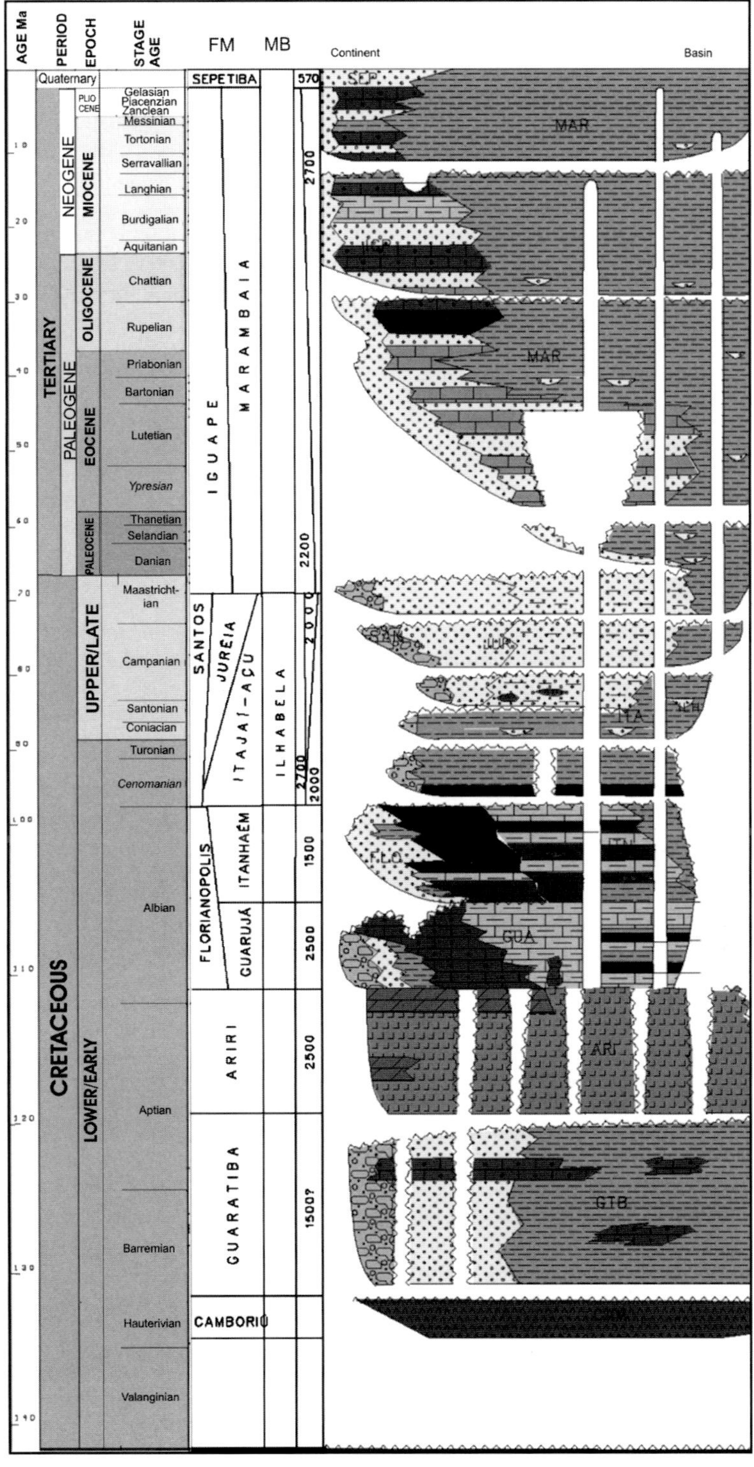

Fig. 2. Santos Basin stratigraphic chart (modified from Pereira & Feijó 1994).

wave-dominated littoral zone develops systems of lagoons and beach ridges.

The present physiography of the margin is a combination of structures inherited from rigid tectonic processes (basement and rift-derived), salt tectonics and sediment reworking by bottom-current processes in deep waters, and by shallow-water processes on the continental shelf (Souza 1991). The continental shelf ranges between 100 km and 200 km width on average (Fig. 1). Locally, close to Cabo Frio, it narrows to about 70 km. The present shelf break is situated near the 200 m isobath and is generally marked by a smooth passage to the slope. Only in the central and the easternmost parts of the basin is the shelf–slope boundary more accentuated, coinciding with the steeper slope of the basin, with a 3° average gradient. In general terms, the slope is smoother in its upper part (from the shelf break down to 1000 m) and steeper downslope. The toe of the slope occurs between 2200 m and 1800 m, being shallower towards the north.

Hydrological context

The water masses and the circulation regime of the South Atlantic have been well studied through a series of international projects and scientific efforts in recent years (Reid 1989, 1996; Stramma 1989; Garfield 1990; Peterson & Stramma 1991; DeMadron & Weatherly 1994; Siedler *et al.* 1996; Silveira *et al.* 2000). The more recent studies indicate that the ocean circulation pattern associated with the intermediate and deep-water masses is more complex than previously assumed (Silveira *et al.* 2000).

The Brazilian southeastern margin is characterized by the stacking of several water masses (Fig. 3a): the Tropical Water (TW), the South Atlantic Central Water (SACW), the Antarctic Intermediate Water (AAIW), the North Atlantic Deep Water (NADW) and the Antarctic Bottom Water (AABW). In the study area, the surface waters are strongly influenced by the atmospheric circulation pattern and comprise two distinct water masses: the TW and the SACW. The surface waters result from the mixing of three water masses: the TW, a hot ($T > 18$ °C) and high-salinity water ($S > 36‰$), the coastal waters and the waters resulting from vertical excursions (upwelling) of the SACW (Signorini 1978; Evans *et al.* 1983; Garfield 1990). The SACW is colder ($T < 18$ °C) and less saline ($S < 36‰$).

The surface waters are driven to the south by the Brazil Current (BC), the western boundary current associated with the anticyclonic South Atlantic Subtropical Gyre. It originates near 10°S, where the southern branch of the South Equatorial Current bifurcates to form also the North Brazil Current (Stramma 1991). The BC flows southward along the Brazilian margin to the Subtropical Convergence zone at *c.* 35°S, where it merges with the northward-flowing Malvinas Current and separates from the coast (Souza 2000). In the Santos Basin, the BC flows along the shelf–slope boundary and carries to the SW the hot and saline STW at water depths between the sea surface and 200 m, and the SACW down to 500–600 m (Miranda & Castro 1998). The BC has its axis located at about the 200 m isobath (Fig. 3b). Near Cabo Frio, its average transport between the shelf break and uppermost slope (300 m water depth) is -2.01 ± 0.98 Sv (Sv $= 10^6$ m^3 s^{-1}), with near-bottom velocities often reaching values around 0.5 m s^{-1} (Souza 2000). Towards the south, the BC broadens to more than 600 m deep and its transport increases to 7.3 Sv (Campos *et al.* 1995). Direct observations and numerical modelling indicate a shear of the BC close to the slope. This facilitates the onshelf penetration of the deeper SACW during the summer (Rezende 2003), inducing coastward bottom circulation reaching 0.2 m s^{-1} and upwelling (Fig. 4a). Satellite images of sea surface temperatures show the presence of mesoscale meanders, cyclonic and anticyclonic gyres, and the onshelf penetration of slope waters (Fig. 4b). Campos *et al.* (1995) attributed the development of the BC eddy activity to the abrupt change in the shelf–slope boundary trend, which starts to the north, offshore the Cabo de São Tomé (22°S, Campos Basin), and is amplified by the strong margin bend at Cabo Frio, which induces the BC to meander cyclonically in response to the potential vorticity forcing. Mahiques *et al.* (2004) suggested that, similarly to what was observed in the Campos Basin (Viana 1998; Viana & Faugères 1998; Viana *et al.* 1998, 2002*a*), the oscillations in the BC are responsible for different sedimentation rates along its path.

Mahiques *et al.* (2004) also suggested that in the northern Santos Basin, areas with modern low sedimentation rates are associated with the main flux of the BC, which would prevent settling. Higher values of organic carbon are observed in zones of higher sedimentation rates, indicating that pelagic processes associated with the northward propagation of southerly cold waters as well as with upwelling processes control the sedimentation.

The Antarctic Intermediate Water (AAIW) occurs immediately below the SACW, between the 500–600 m and 1200 m isobaths. It is characterized by temperatures ranging between 2 and 6 °C, a high dissolved oxygen content and salinity minimum of 34.2‰ (Piola & Gordon 1989; Reid 1989). It flows northward to the Subtropical Convergence Zone (South Atlantic Bight), near 35°S,

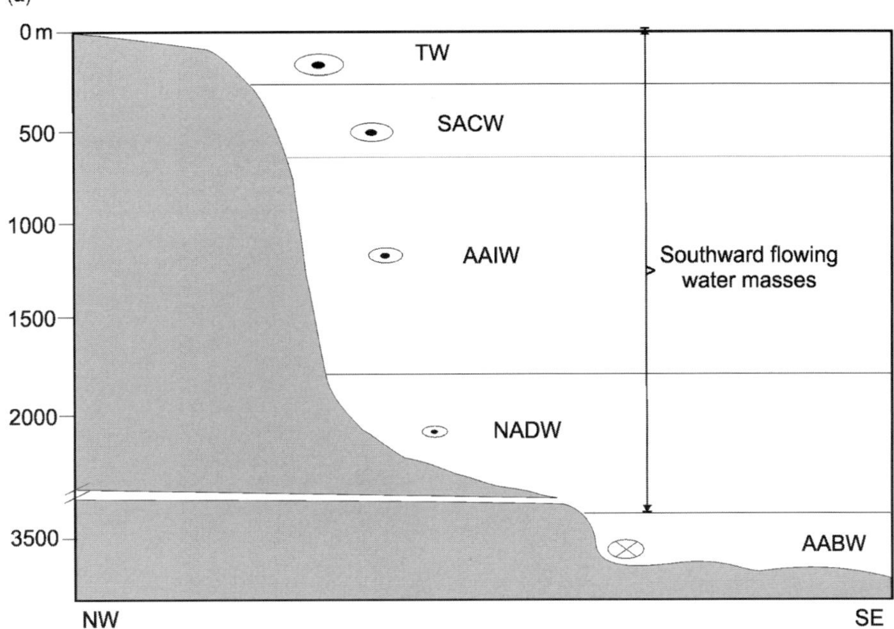

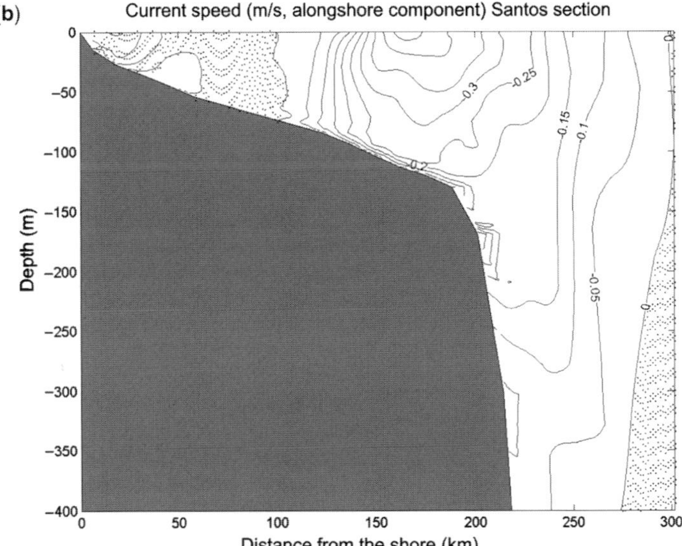

Fig. 3. (**a**) NW–SE cross-section with the vertical distribution of main present-day water masses in the study area and their relative sense of flow. TW, Tropical Water; SACW, South Atlantic Central Water; AAIW, Antarctic Intermediate Water; NADW, North Atlantic Deep Water; AABW, Antarctic Bottom Water. It should be noted that all waters move towards the south, except the deepest AABW, which flows to the north. (**b**) Cross-section of shore-parallel component of current velocity (m s^{-1}), in the central Santos Basin (from Silveira *et al.* 2000).

(a)

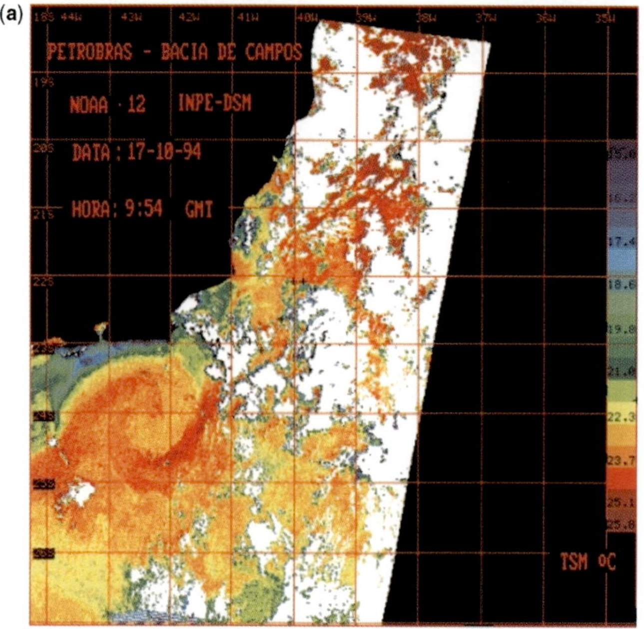

(b)

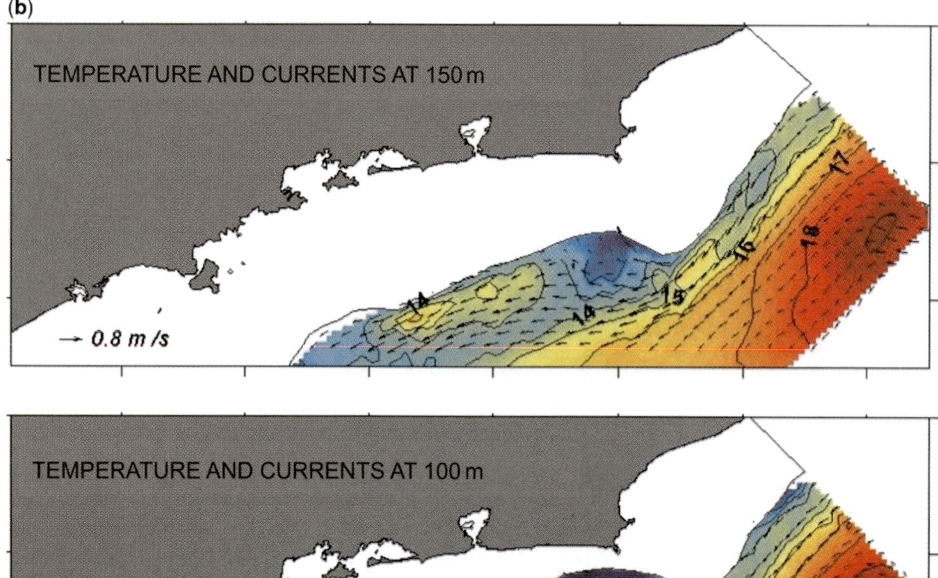

Fig. 4. (**a**) Sea surface temperature satellite image (Landast NOAA-12) illustrating the meandering pattern of the warm Brazil Current and the giant eddy developed at the Cabo Frio salient and penetrating onto the shelf. (**b**) Numerical model plots of Brazil Current behaviour in the Santos Basin based on temperature (°C) and intensity represented by the length of the arrows (m s^{-1}) (from Souza 2000).

where it abandons the western boundary contour pattern and shifts eastward accompanying the southern hemisphere anticyclonic subtropical gyre. On its way back to the west after crossing the Mid-Atlantic Ridge and being incorporated with the Benguela Current along the western Africa margin, the AAIW crosses the northern flank of the Rio Grande Rise and reaches the South America margin near 22°S, at the Cabo Frio salient (Reid 1989; Peterson & Stramma 1991). At this point, the AAIW bifurcates and flows in two opposite senses. One branch flows to the north, along the Campos Basin margin, and the other to the south, being incorporated into the general southward flow of surface and mid-depth waters. This southward-flowing branch of AAIW was recently confirmed by Muller *et al.* (1998).

The North Atlantic Deep Water (NADW), occurring below the AAIW, is characterized by temperatures ranging between 3 and 4 °C and salinities between 34.6 and 35‰. It forms an organized southward flow following the western boundary of the margin to 32°S, where part of the flows turns back to the north (Reid 1989). Beyond 25°S, TW, SACW, AAIW and NADW all flow southward (Figs 1 and 3a), introducing some questions about the actual thickness of the Brazil Current that are still unsolved (Silveira *et al.* 2000).

Beyond the São Paulo Plateau, in water depths greater than 3500 m, the Antarctic Bottom Water continues toward the northern hemisphere, along the contours of the Plateau–continental rise border. It does not occur in the study area.

The complex circulation pattern of the modern Santos Basin complicates the task of reconstructing the palaeocirculation in the study area. Seismic evidence suggests that sedimentation from Neogene to Recent time was dominated by oceanic circulation redistributing the sediments transferred to the basin during both relative sea-level highstands and lowstands, and gives some indication of the path and relative intensity of the bottom currents that passed through the Santos Basin in different climatic–oceanographic conditions. Seismic evidence and our interpretation of how those currents influenced the basin infill will be given below.

Results

Physiography of the study area

The integrated analysis of 2D and 3D seismic and satellite data help to identify physiographic features and their interplay with dynamic oceanic and sedimentary processes.

Continental slope physiography. Alternating salients and embayments form the general features of the shelf–slope transition. Embayments occur in areas where the upper slope is gentler. Association of margin projections and re-entrants with a coupled control by the pre-existing tectonic structures and the oscillations of the slope boundary currents is proposed here but remains to be proven. Sea surface temperature satellite images show the meandering and gyre detachment of the surface slope western boundary Brazil Current (BC) in the Santos Basin (Figs 3 and 4). These images and oceanographic measurements indicate frequent onshelf penetration of the BC, which could be implicated in the physiographic configuration of the upper margin and development of the large-scale sinusoidal configuration (wavelength of around 200 km) that characterizes the shelf–slope transition (Fig. 1). The shelf break is characterized by a 100 m high, 5° escarpment, passing to a relatively smooth and gently dipping slope. The passage from middle to lower slope, at about 1500 m water depth, is marked by the sea-floor expression of the Cabo Frio Fault, defining a parallel-to-slope channel and marking the inner limit of the region in which salt movements reach the modern sea floor. Such features are observed from that depth downslope, creating an intricate pattern of mini-basins and topographic highs. The base of the slope is marked by a conspicuous NE–SW escarpment corresponding to the inner flank of a second slope-parallel channel.

Slope-parallel channels. Two distinct SW–NE-trending along-isobath features were identified in the continental slope. The upper one is a narrow channel herein called the Santos Channel (SC) (Figs 1 and 5), about 2 km wide and 100 m deep, situated in the transition from the middle to the lower slope (*c.* 1500 m water depth). It is a conspicuous feature, at least 200 km long, associated with the present sea-floor expression of the antithetic Cabo Frio Fault.

The downslope channel, herein called the São Paulo Channel (SPC), is a wide, asymmetric submarine valley, 15–20 km wide and 300 m deep, with steeper walls on its inner flank. The SPC is parallel to the Santos Channel and marks the passage from the continental slope to the São Paulo Plateau. It locally has strong sea-floor erosion with the cropping out of sediments as old as Eocene age.

Submarine canyons. Two huge NW–SE incisions occur in the central Santos Basin slope, in the western portion of the study area. The first is 10 km wide, with steep walls, and is located around 26°S and 45°W. The other, 100 km to the NE, is narrower (5 km wide) and with gentler walls, both probably related to voluminous mass

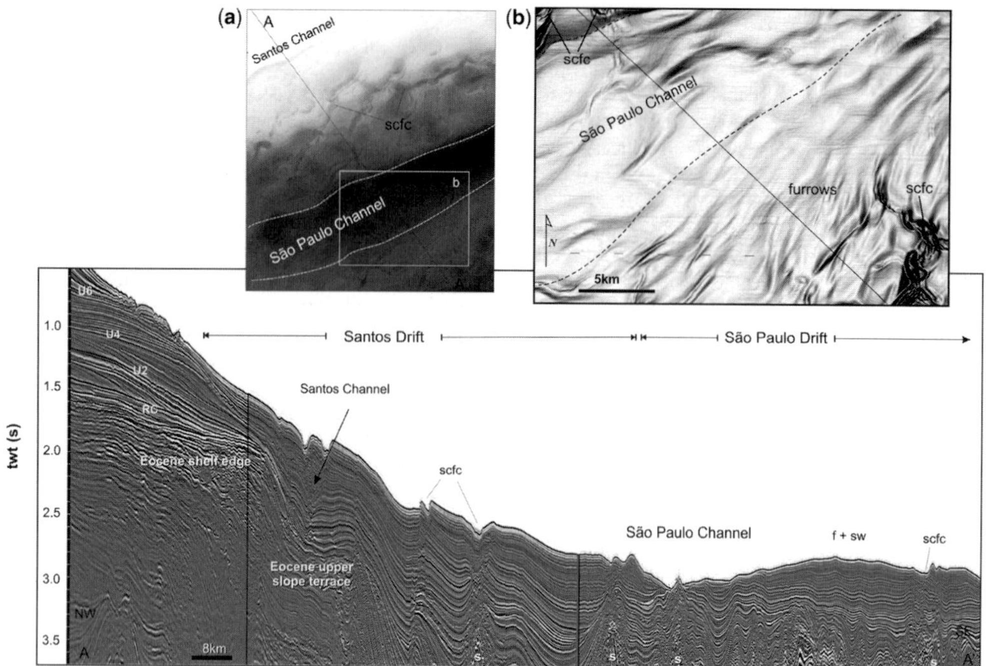

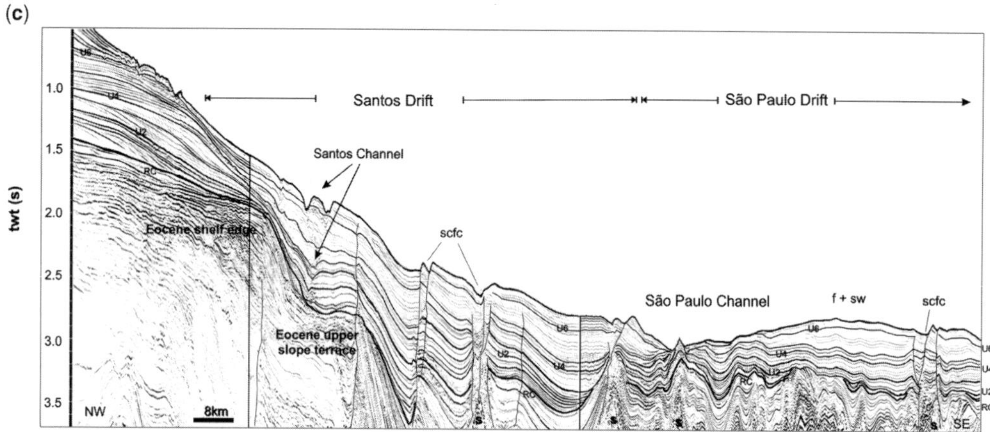

Fig. 5. (**a**) Grey-scale bathymetric chart representing the passage from the slope to the São Paulo Channel, highlighting the Santos Channel and the Santos moat features. (**b**) Coherence map of the sea floor, indicating detailed relief perturbations, where furrows (f), salt crestal fault channels (scfc) and the moat axis are well defined. Continuous line indicates the position of the seismic profile. (**c**) Regional dip seismic line and line drawing of the the Santos Drift System main elements. sw, sediment waves; twt, two-way travel time.

movement episodes. Their deposits form aprons tens of kilometres wide that disturb the São Paulo Plateau relief. The wider submarine incision is herein called the Cananéia canyon; it breaches the shelf and seems apparent in the 50 m isobath. Downslope, on the São Paulo Plateau, it is connected to a large submarine valley that collects the surrounding submarine drainage network and continues towards the basin with a NW–SE trend. This submarine drainage coincides with a crustal lineament, the Capricornio lineament, recently defined by Bueno *et al.* (2004). The other incision, herein called the São Sebastião canyon, is connected to a subsidiary drainage system, developed above the sea-floor expression of salt crestal grabens, which feeds downslope to the main gathering system described above.

The São Paulo Plateau. The São Paulo Plateau (SPP) occurs from the base of the slope down to the limit of the oceanic crust, at about 3.5 km water depth (Fig. 1). Its inner part corresponds to the São Paulo Channel axis. The SPP is characterized by a smooth gradient (less than 0.5°) and by the wide and gentle mounded geometry enhanced by the São Paulo Drift. The SPP is also marked by the sea-floor expression of salt movements. Crestal faults of diapirs and salt walls often reach the sea floor, modifying the local relief and creating obstacles or passageways for dense, sediment-load flows, or clear water oceanic currents. Patches of furrow-like features were observed in 3D seismic coherence maps, especially in the SPP sector (Figs 1 and 5).

Mass-flow deposits form a positive relief, especially close to the Cananéia and São Sebastião canyons, and disturb the SPP landscape by introducing a huge amount of chaotic deposits. These deposits locally interrupt the continuity of the Santos and São Paulo channels.

The suppression of sediment supply to the Santos Basin after the Eocene occurred along with a strong marine transgression, which drowned the lower Cenozoic continental shelf and created the space for the rearranged ocean circulation, and became the dominant sedimentary agent. The depositional geometries, the large-scale discontinuities and the reflection pattern in the interior of depositional bodies identified in the studied data support the idea that bottom currents eroded, transported and deposited a large amount of sediments on the Santos Basin slope and the adjacent São Paulo Plateau. Such a process was responsible for the development of several contourite deposits that constitute the Santos Drift System (SDS). Below we will discuss the genesis and provide seismic evidence for the characterization of different contourite deposits that form this system and their relationship to the action of long-lived oceanic bottom currents that alternate their flow direction and intensity in response to climatic-induced global oceanographic changes from the late Palaeogene to the Recent.

Deposit characteristics and development of the Santos Drift System (SDS)

Seismic-stratigraphic interpretation of the Late Palaeogene to Recent section of the northern Santos Basin indicates that during that period a large sedimentary wedge was deposited from the upper slope to the distal portion of the São Paulo Plateau, in water depths ranging between 200 m and >3000 m. Seven third-order sequences were distinguished and several contourite drift deposits were identified, with different geometries responding to shifts in the locus and intensity of current action (Fig. 6). These shifts resulted from accommodation space modifications caused by glacio- and tectono-eustatic changes.

The present configuration of the continental slope corresponds to a very large drift system, more than 300 km long, 200 km wide and 1 km thick. Slope-plastered sediment drifts, separated drifts, sediment waves, and alongslope channel erosion occur throughout the system evolution and their combination gives rise to the Santos Drift System (SDS). At present, the SDS initiates at the upper–middle slope transition (*c.* 600 m water depth), and extends more than 100 km downslope to water depths in excess of 2500 m. The SDS consists of two major drift complexes, the Santos Drift and the São Paulo Drift.

The Santos Drift (SD) is an elongated slope-plastered drift with a wedged to mounded external geometry up to 1 km thick. Its upper margin portion is characterized by a decrease in the slope gradient, which corresponds to the area where the strong slope currents hinder sediment deposition and start sediment drift downcurrent and downstream. An upslope migration of the system is marked by the progressive onlap of the seismic reflections above a basal erosive surface (Fig. 6). The reflectors diverge from the erosional axis and develop an external geometry of gently mounded accumulation with continuous parallel to slightly undulated reflections. In its central portion the SD is interrupted by the Santos Channel (SC; Figs 5–8), the most prominent feature of this drift, marked by periods of dominant erosion and dominant deposition. The axis of the SC remains almost stacked throughout its history and corresponds to the shallower expression of the Cabo Frio Fault (Figs 5–8). A basal erosional surface is clearly defined above which the SD stacks or slightly migrates upslope. At the foot of the slope, the SD thins and is dominated by erosional processes responsible for the development of the contour-parallel São Paulo Channel (SPC).

Basinward from the SPC (water depths >2000 m) is observed the development of the São Paulo Drift (SPD), a thick (>700 m) elongated depositional moat–drift system with a mounded external geometry, more than 250 km long and 100 km wide (Fig. 5), genetically connected to the SPC. The SPD presents continuous aggrading parallel seismic reflections, diverging from the SPC outer flank (Fig. 5). Standing sediment waves and furrows prevail above the present sea floor of the SP (Fig. 5).

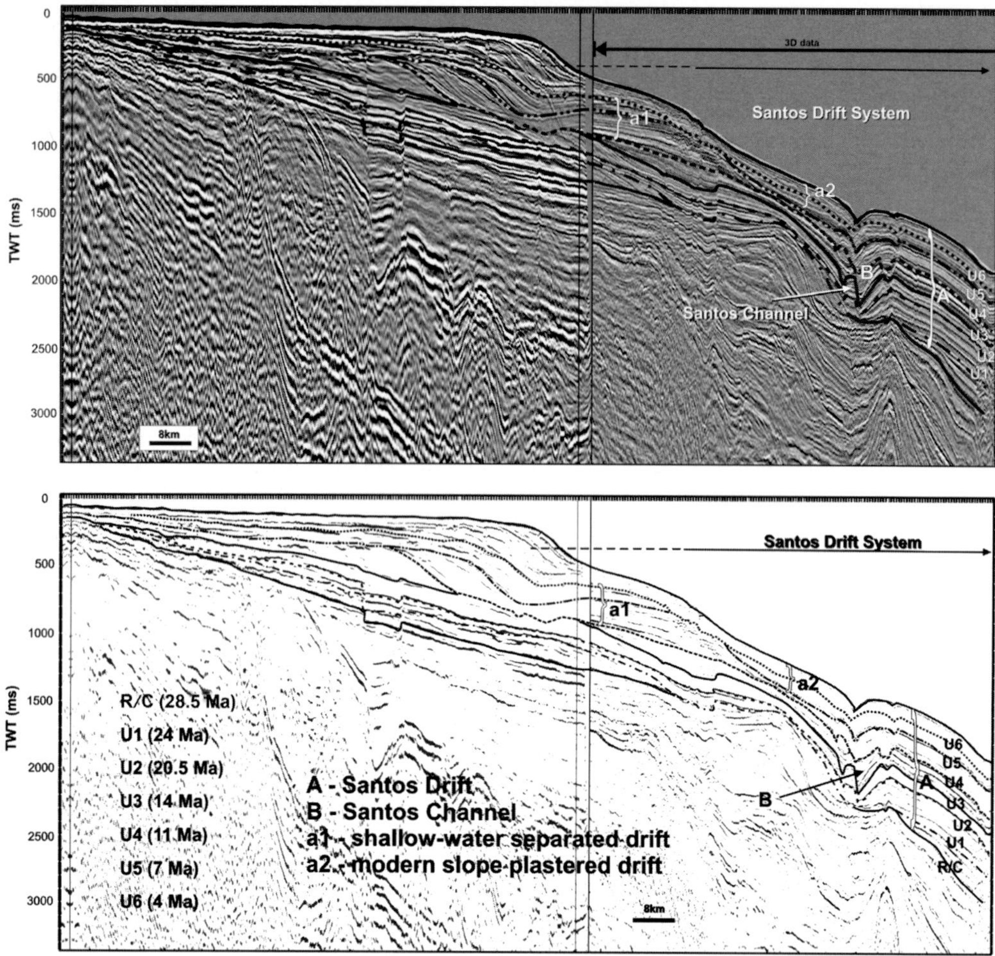

Fig. 6. A 2D–3D composite seismic line across the Santos Basin outer shelf and slope, indicating the main horizons identified in this study. Zones of current-induced slope erosion and the palaeo-Santos Channel developed above an inherited Eocene upper to mid-slope terrace are indicated.

The longer axis of the Santos Drift System (Santos Drift + São Paulo Drift) is aligned NE–SW, parallel to the slope trend. Seismic data analysis indicates a two-direction development trend for the SDS, an upper slope SW trend, which shifts to a NE migration trend from the mid- to lower slope section. This alternation suggests two main current regimes acting and reworking sediments in the area. The shallower one flows to the south or SW and is responsible for the linear features developed at the outer shelf–upper slope; the other involves mid- and deep waters and flows to the north or NE. These features resemble those found in a similar environment in the Campos Basin by Viana (2001), and Viana *et al.* (2002*a*, *b*), and related by those workers to the action of the surface slope boundary Brazil Current.

In its earlier stages, the SD had a different depositional geometry, influenced by the inherited Eocene slope configuration. The Palaeogene sequence of the northern area of the Santos Basin is marked by an intense progradation phase, which attained its maximum expression during the early–middle Eocene, resulting in large sets of prograding shelf-edge deltas and their related gravitational deposits (Moreira *et al.* 2001; Moreira & Carminatti 2004).

Clinoform shelf-to-slope geometry and associated mud successions are generally interpreted as evidence for sediment bypass across the shelf into deeper-water, lower-energy depositional locations (Lu & Fulthorpe 2004). However, this process alone cannot explain the well-developed drift-like sedimentation that occurs on many seismic profiles through the Santos prograding shelf–slope complex

(Fig. 5). The steep slope–terrace geometry of the Eocene upper slope resulted in an important physiographic feature, which induced an augmentation of the western boundary current action against the slope. The SD was developed above the Eocene clinoforms, suggesting a change in the sedimentation style probably associated with the prevention of terrigenous discharge as a result of the tectonic shift of the main denudation area towards the north and the dominant role played by oceanic processes in sedimentation. The drowning of the Eocene upper margin provided space for the major drift accumulation that is the subject of this study. Several mud-dominated third-order sequences were deposited above a regional unconformity that is correlated with the drowning surface of the Rupelian transgression (early Oligocene).

Following the northward shift of both the denudation area and the river drainage system in the Eocene, the Santos Basin was drowned and underwent a starvation period when sediment reworking by bottom currents prevailed. Stratigraphic correlations suggest that the transgression probably occurred from the late Eocene to the early Oligocene, and was responsible for the complete drowning of the Eocene shelf. The early Oligocene period was marked by a very low sedimentation rate and coincides with a regional highstand, which was responsible for the deposition in the Campos Basin of the Blue Marker, a nannofossil ooze that drapes all deep-water deposits (Gambôa *et al.* 1986; Shimabukuro 1994). In the Santos Basin this period was characterized by the development of the thick and extensive SDS, which occupies the entire slope and São Paulo Plateau. In the present study we will deal with data for its northern expression, but its continuity towards the south has been observed in unpublished, industrial seismic lines.

By the end of the Eocene the face of the clinoforms of shelf-edge deltas defined a steep upper slope escarpment, up to 400 m high, and the flat area at their toe gave rise to an upper slope terrace (Fig. 5). The face of the clinoforms acted as a palaeotopographic restriction for mid-slope currents flowing against the slope. The flow restriction was locally enhanced by adjacent salt-related topographic highs (Figs 5 and 6). The flow constriction induced a major increase in bottom-current speed and conspicuous erosion of the sea floor. This initial activity of a strong bottom-current regime was partially masked by the intense sediment progradation that prevailed during this period and formed the substratum over which the SDS was later developed. The evolution of this system to a moat–drift occurred after the drowning of the Eocene margin during the late Eocene–early Oligocene. The SC results from the combination of this palaeomorphological configuration and intense bottom-current activity. The sediment eroded and transported by these currents drifted downstream and downslope, to form the SD (Figs 5 and 6).

During the middle Miocene, an intensification of the transgression phase probably accompanied and/or induced by a tectonic tilt of the basin modified the accommodation space and caused retreat of the depositional systems towards the continent. Large progradation sets were established over the drowned and inclined shelf, producing a new topographic profile (Fig. 6). This ramp-like shelf permitted the incursion of northeasterly surface slope boundary currents onto the shelf. These currents eroded the face and the toe of the 200 m high clinoforms, promoting the development of a shallow-water separated drift, which maintained this profile until the early Pliocene (Figs 5 and 6). The steep slope established in this phase acted as bypass zone for the gravity-derived sediment flows as well as a topographic barrier for deeper-water currents. Thus, the deeper northward currents flowing in an opposite sense to the surface currents eroded the steeper face of the upper slope and developed a plastered drift on the middle slope that was connected downslope to the SC (Figs 5 and 6). From Pliocene to Recent time, the reduction of the accommodation space was accompanied by an intense progradation system and by the suppression of the shallow-water drift as a result of the direct connection of delta clinoforms to the upper slope (Fig. 6).

Stratigraphic sequences of the Santos Drift

The application of the classic concepts of sequence stratigraphy (Mitchum *et al.* 1977; Vail *et al.* 1977*a*, *b*; Vail 1987) was fundamental in the identification of the main episodes of slope construction and to better constrain the periods of bottom-current activity. The use of this technique was facilitated by the tectonic evolution (slow thermal subsidence) and predominant glacio-eustatic relative sea-level variations of this late Palaeogene to Quaternary section. The seismic-stratigraphic analysis of this section of the northern Santos Basin permitted the recognition of major reflections associated with basin-scale unconformities developed across the entire continental slope and locally recognizable on the shelf. The extent of this sea-floor erosion and its association with sedimentary drifts suggests that besides the possible triggering by eustatic changes it could be amplified by the action of vigorous bottom currents.

In the Santos Drift area (slope setting) seven third-order sequences were characterized (Figs 6 and 8), which were tentatively correlated with the global relative sea-level fluctuations curve (Fig. 9). Only the two oldest unconformities (R–C unconformity and U1) could be stratigraphically tied to nannofossil biozones identified in unpublished, confidential exploratory wells drilled in the study area; the uppermost section was not sampled in those wells.

The identification of the seven major depositional sequences took into account the seismic character of the sedimentary packages resulting from both high current activity (moat-carving and widespread slope erosion) and low current activity (standing sediment waves and channel aggradation). The external geometry and internal reflection configuration of the mounded drift were locally modified by strong salt activity, which prevented clear recognition of the system (Fig. 7). Salt movements affected the modification of the accommodation space, providing topographic restrictions for both gravity and bottom currents, and modifying their intensity and locus of action.

Sequence 1 (S1)

The lowest seismic sequence studied was deposited during the late Oligocene (28.5–24 Ma). This sequence onlaps the horizon R. This horizon corresponds to a strongly erosive unconformity, which erodes the underlying section especially in the shelf break and upper slope settings. This erosion defines a new bathymetric profile, which is smoother that the preceding physiography in which the steep clinoforms and adjacent terrace marked the upper slope (Figs 6 and 8). Horizon R–C is tied to the Rupelian–Chatian unconformity (Fig. 9). From the base to top of sequence S1, the internal seismic configuration changes from divergent to parallel reflections. The reflectors show an onlap succession progressively migrating upslope above R–C. Its external geometry resembles a flat elongated mound characterizing a slope-plastered drift. A coeval sequence developed on the shelf is characterized by low-angle, prograding clinoforms, abruptly interrupted on the outer shelf (Fig. 6). The physical discontinuity observed between the shallow- and deep-water portions of this sequence is attributed to the highly active surface slope boundary currents, which must have swept the shelf edge and the uppermost slope from NE to SW, similarly to the modern Brazil Current (Figs 3 and 4), hindering deposition in that setting. The upper sequence boundary (U1) corresponds to a strong conformable reflection in the lower slope, slightly undulating near the position of a strong middle slope gradient change, corresponding to the position of the future axis of the Santos Channel. From that point upslope the slope is strongly eroded by the action of strong bottom currents. On the shelf U1 truncates the underlying shelf-edge deltas. Seismic data suggest that U1 was a diachronous surface, produced during the strong sea-level fall that immediately preceded the Oligocene–Miocene boundary, 24 Ma, and the rapid and high-amplitude subsequent sea-level rise (Figs 6, 8 and 9). This surface is related to the D3 seismic horizon mapped by Gambôa *et al.* (1983) at DSDP site 515, located on the northern side of the Rio Grande Rise, in the deep Brazil Basin at 4250 m water depth, in the same latitude as the study area (Fig. 1).

Sequence 2 (S2)

S2 was deposited during a period of high relative sea level that occurred during the earliest Miocene (Aquitanian) and succeeded the strong relative sea-level fall responsible for the development of U1 in the shelf and uppermost slope areas. This sequence is marked by the initial incision of the Santos Channel (SC). A moat–drift configuration is initiated, with erosional features along the channel axis and diverging reflections on the downslope flank of the channel (Figs 6 and 8). The WSW–ENE-trending SC is up to 2 km wide and 300 m deep (Fig. 7). Downslope, the drifted sediments develop a continuous parallel reflection pattern that smoothly offlaps downslope (Fig. 8). The SC geometry and the erosion–non-deposition characteristics of that time are illustrated in the structural and isopach maps (Figs 7 and 8). S2 records a period of extreme activity of the bottom currents on the Santos slope as suggested by the development of diverging reflectors downslope from the SC (Fig. 7, lines 3 and 4). The sea-level rise that occurred during that period (earliest Miocene) induced a continuous upslope migration of the northward-flowing current core. The encounter of that current with the steep slope inherited from the Eocene progradation induced the acceleration of the bottom currents and the increase of their erosion potential. This period coincides with the initiation of formation of the SC, above the mid-slope terrace developed at the foot of the upper slope escarpment inherited from the Eocene clinoforms. Movements of the Aptian salt modified the accommodation space, locally enhancing the current activity, especially to the NE of the study area (Fig. 7). During the deposition of S2 the shelf was occupied by a prograding, mixed siliciclastic–carbonate platform. Siliciclastic deposits are more abundant to the ENE, where a shingled pattern of internal reflections dominates. To the WSW, carbonates organized in a prograding pattern with small, high-amplitude clinoforms are dominant. The top of S2 is marked by a strong conformable reflection within the drift deposit

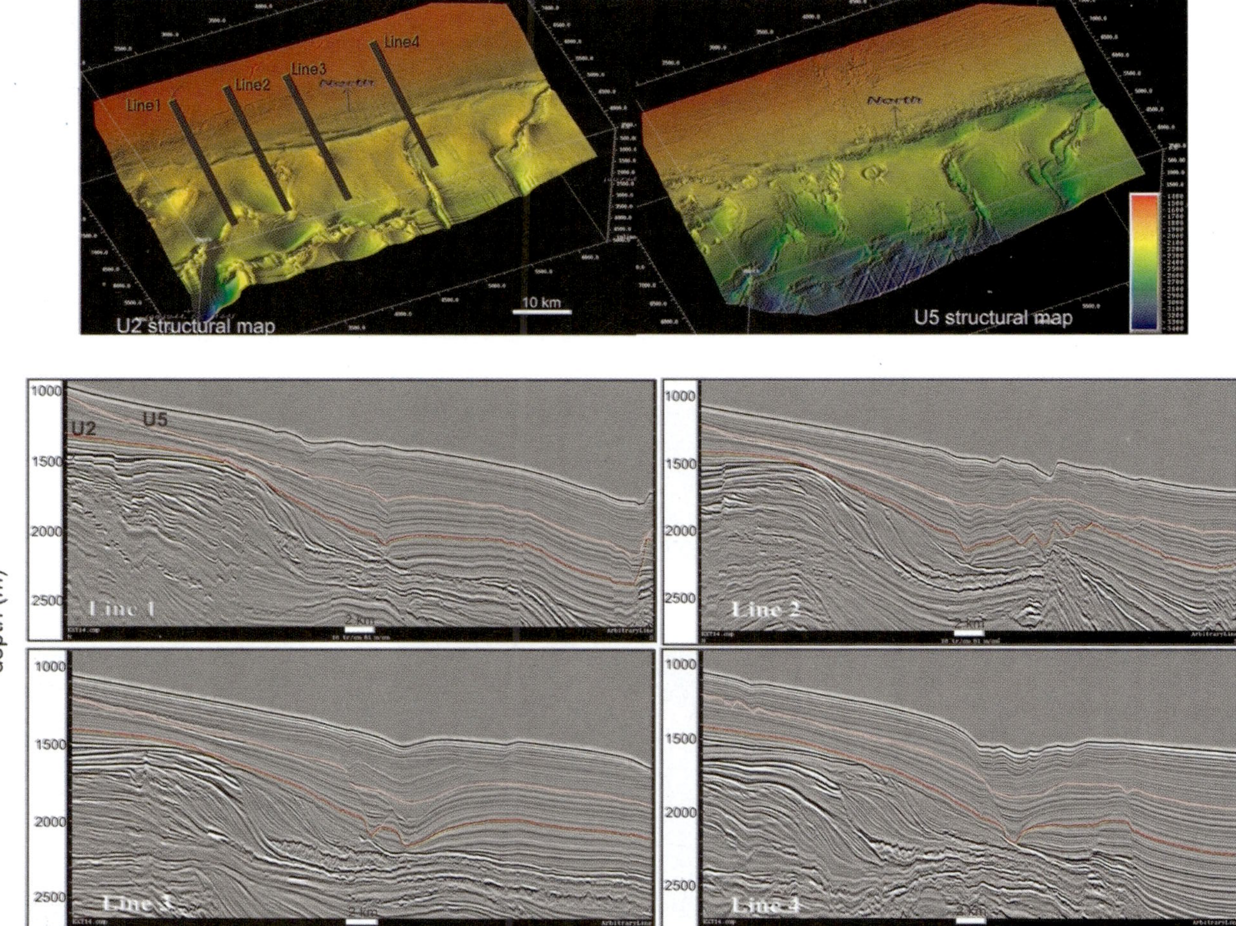

Fig. 7. A 3D perspective of structural maps of U2 and U5 horizons with location of 3D dip seismic lines crossing the Santos Channel. (Note the channel stacking pattern and the influence of salt movements in changing the accommodation space.)

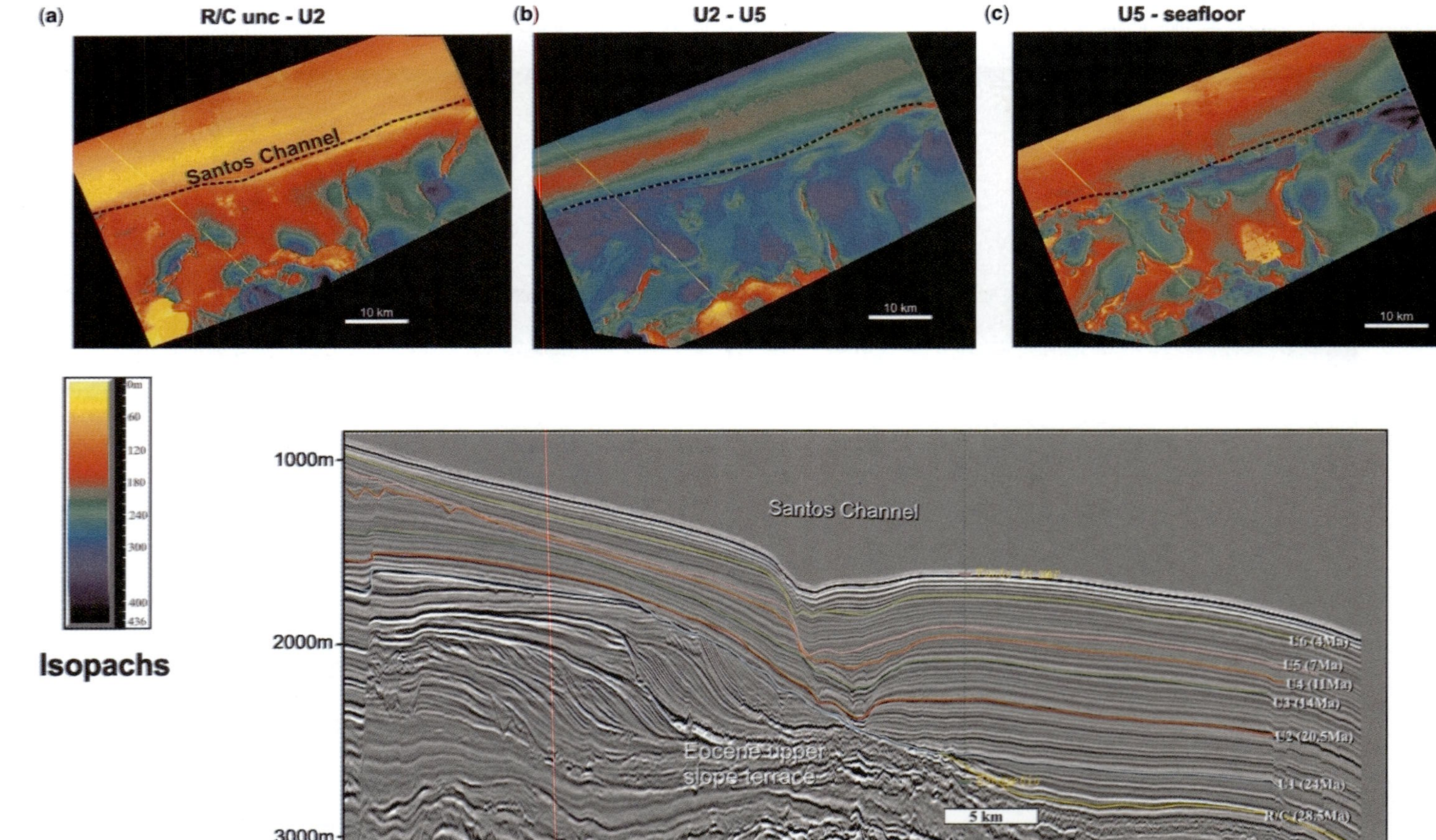

Fig. 8. Isopach maps of three distinct intervals of Santos Drift construction: (**a**) Rupelian–Chatian unconformity (R/C) to early Miocene unconformity (U2) isopachs suggest contour drift deposits infilling the salt mini-basins; (**b**) early Miocene unconformity (U2) to late Miocene unconformity (U5) isopachs indicate the upslope migration of the drift core (thickest isopachs); (**c**) U5 to sea-floor isopachs indicate that the Santos Channel is the main depocentre, with smoothing of the slope relief. Yellow line indicates the position of the seismic line.

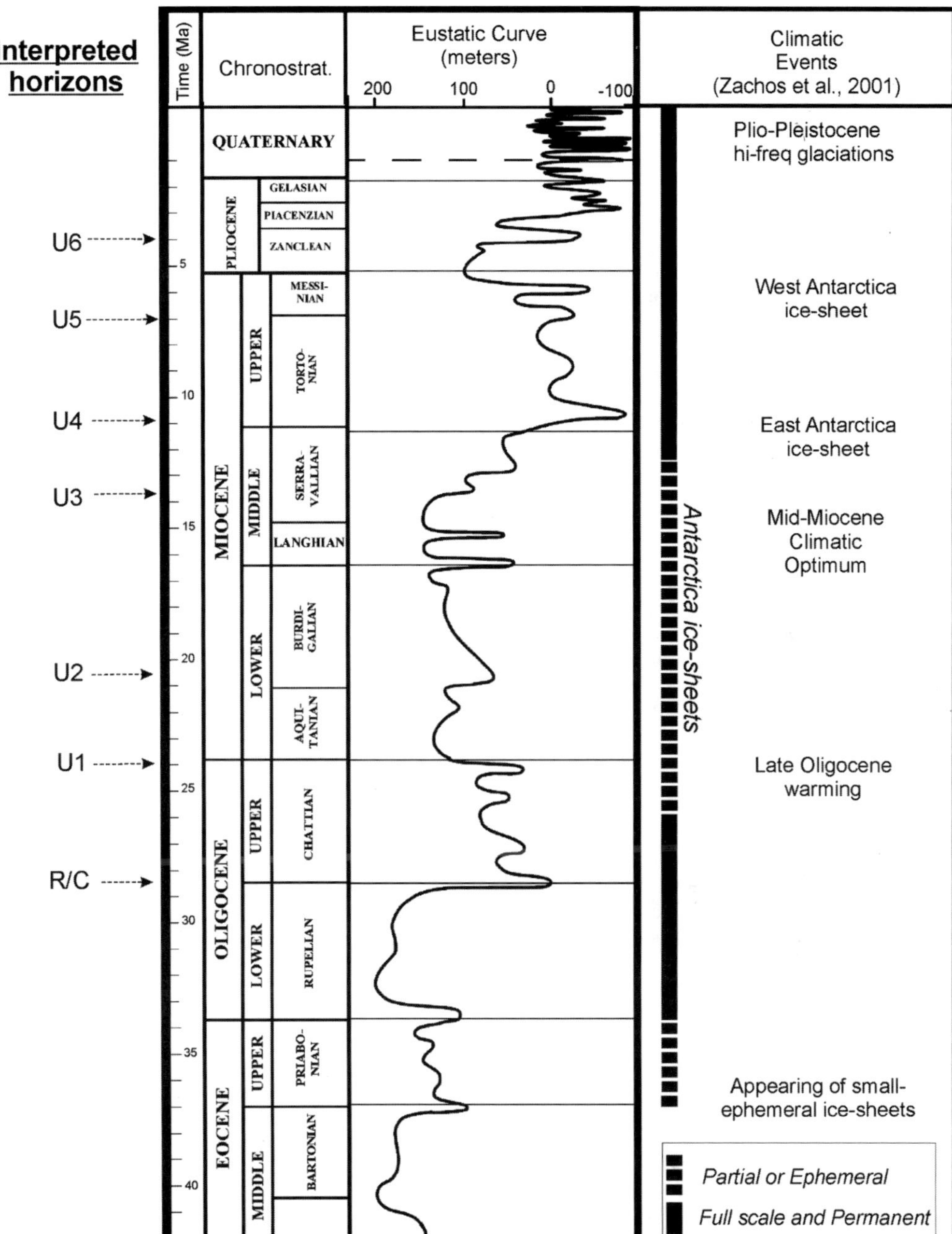

Fig. 9. Correlation of the studied seismic horizons with a chronostratigraphic table and geological ages (from the International Stratigraphic Chart, IUGS 2004), with eustatic sea-level curves from Haq *et al.* (1987), and with climatic events as presented by Zachos *et al.* (2001).

(U2). On the shelf, U2 is a downlap surface and corresponds to the drowning of an Oligocene–Miocene carbonate platform. Biostratigraphic data from exploratory wells date U2 to *c.* 20.5 Ma (early Miocene unconformity).

Sequence 3 (S3)

S3 was probably deposited from early to middle Miocene time (20.5–14 Ma). The internal reflection pattern in the Santos Drift is characterized by continuous, low-frequency, high-amplitude reflectors gently offlapping basinward from the study area (Fig. 5). The moat–drift geometry is preserved and enhanced by the augmentation in the drift thickness. The erosion or non-deposition pattern of the SC is marked by alternating phases of erosion and deposition along its axis, suggesting oscillations in bottom-current activity (Figs 6 and 7). Upslope, this sequence corresponds to the installation of a prograding shelf marked by a rapid slopeward migration of the shelf break and an increase in the shelf-edge escarpment declivity (Fig. 6). The S3 upper boundary U3 is related to the strong middle Miocene sea-level fall (early Serravallian, *c.* 14 Ma) and corresponds to an erosive seismic horizon that truncates the previously constructed shelf-edge deltas. The strong landward shift of the depositional systems of the overlying sequences (>50 km) suggests a major accommodation space creation (tectonic subsidence) at the end of S3 (Fig. 6).

Sequence 4 (S4)

Regionally, this sequence marks a strong shift of the depositional systems towards the continent and a rapid increase in basin subsidence, culminating with a strong progradation of the shelf. A complete margin reorganization is required related to this newly accommodation space. The shelf is marked by the development of large prograding clinoform sets, indicating the reoccupation of the depositional space by transitional to shallow-water systems involving sediment transfer towards the basin, probably induced by successive phases of large-amplitude relative sea-level oscillations. The reinstallation of a steep shelf break at the face of the clinoforms promotes the constriction of the surface slope boundary currents, which erode and prevent deposition at the foot of the shelf-edge escarpment. An erosive terrace is then installed at the foot of the shelf-edge delta clinoforms and sediment is drifted along- and downslope. On the upper to middle slope, sediments are deposited as the result of the interaction between the gravity processes related to the slopeward progradation of shelf clinoforms and the along-isobath action of the surface slope boundary currents. In the SD, low-frequency, high-amplitude reflectors prevail in the base of the sequence, with higher frequency and lower amplitude reflectors towards the top. Mass-flow deposits are frequent because of the destabilization of the uppermost slope and the erosion of the inner flank of the SC. The resultant chaotic reflectors are intercalated with the high-amplitude parallel reflections at the base of the sequence. The SC is well marked and currents flowing along its axis remove fine-grained sediments delivered by gravity processes. During this phase, the change in the slope physiography induces the development of a second and deeper site of current action, along the São Paulo Channel (SPC) axis. Eroded sediments are drifted and give rise to initiation of the São Paulo Drift (SPD; Fig. 5). Some of the transfer conduits connecting the shelf edge to the slope allow sediment to bypass the SC and communicate with lower slope conduits, such as salt wall crestal grabens and mid-slope canyons. The map of r.m.s. amplitudes extracted between U2 and U5 (Fig. 10) indicates the capture of some of the upper slope conduits by the SC. Also suggested in these seismic images are sediment bypass, mass-flow deposits and slope conduits related to salt crestal grabens. The S4 upper boundary (U4) corresponds to a highly erosive unconformity at the shelf edge and on the uppermost slope and to a conformable horizon downslope separating the high-amplitude, low- to moderate-frequency reflectors of S4 from the high-frequency, low-amplitude S5. U4 is proposed to correspond to the D4 horizon of Gambôa *et al.* (1983) and is dated as a late Miocene unconformity (Serravallian–Tortonian, *c.* 11 Ma), related to the final expansion of the East Antarctic Ice Cap (Savin *et al.* 1975; Wise *et al.* 1985; Miller *et al.* 1987).

Sequence 5 (S5)

S5 was probably deposited during a period of low relative sea level, punctuated by low-amplitude oscillations spanning the whole Tortonian stage (late Miocene, *c.* 11 to *c.* 7 Ma). The strong erosion related to basal unconformity U4 is responsible for the removal of sediments from the outer shelf and upper slope from both S4 and part of S3. The erosional terrace developed during S4 at the toe of the shelf-edge prograding clinoforms, at the very top of the slope, attains its greatest extension (>20 km wide downslope, Fig. 6) and remains the site of strong incision by the surface slope boundary currents. During S5 this terrace is marked by an onlapping succession of upslope-migrating, high-amplitude, low-frequency reflectors prevailing in the basal portion, which is gradually covered by low-amplitude, high-frequency reflectors. Along-slope erosional features include

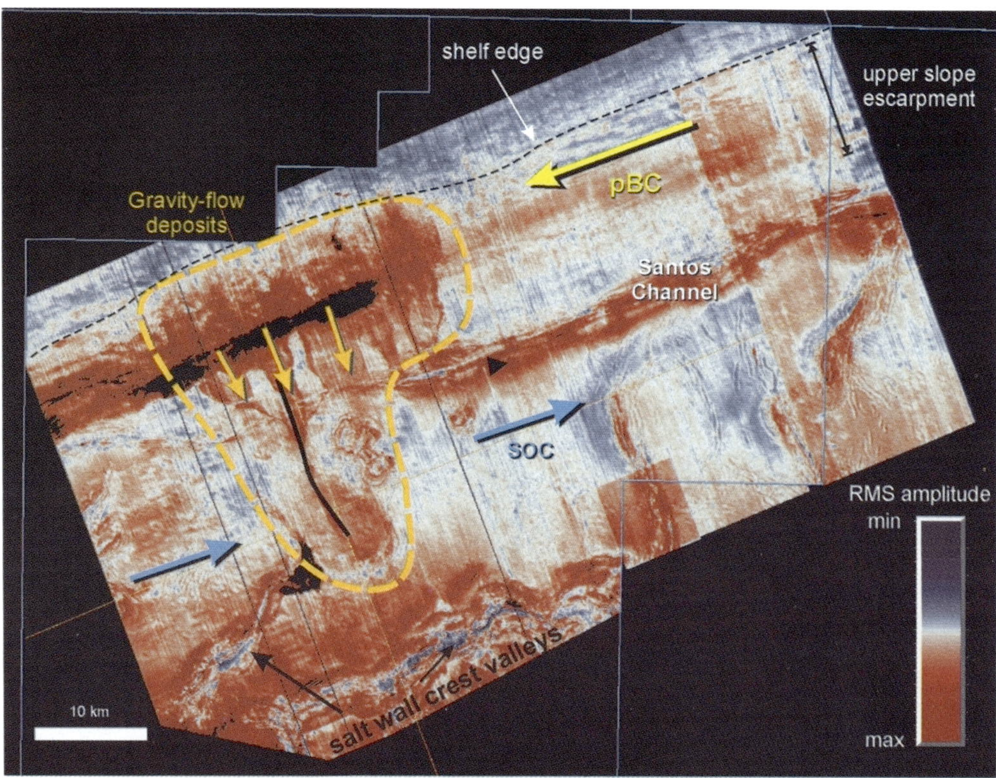

Fig. 10. A RMS amplitude map extracted between U2 and U5, showing the areal coexistence of gravitational deposits and contour-current processes. Surface slope boundary current (pBC, palaeo-Brazil Current) develops linear features at the shelf edge and erodes the top of the upper slope, inducing gravity flows. Shelf-edge sediments transferred downslope by gravity flows are partially subjected to the action of deeper contour currents (SOC, Southern Ocean Current) and locally bypass the Santos Channel.

narrow channels and erosional ridges and furrows. The constraining of shallow slope currents against the foot of the clinoforms focuses and enhances the erosional activity of such currents and drifts sediments, which accumulate above the terrace in a mounded geometry similar to a shallow-water separated drift, connected downslope to the plastered portion of the SD (Fig. 6). The basal high-amplitude set is probably related to coarse-grained material eroded and drifted from the top of the previous sequences by the slope surface currents. Such a configuration is similar to the shallow-water contourites observed in the Quaternary and late Miocene of the Campos Basin (Viana & Faugères 1998; Viana *et al.* 2002*a*), in the Quaternary of Antarctica (Rodríguez & Anderson 2004) and in the Aptian–Albian of the Vocontian Basin, SE France (Viana 1998). No significant record of S5 is observed on the shelf probably because of absence of accommodation space creation, implied by the bypass of the shallow-water systems to the basin, the slopeward migration of the shelf edge, and the strong erosion related to the upper unconformity U5 (Fig. 6). Shingled high-amplitude, high-frequency reflectors are observed at the lower part of the shelf edge clinoforms and seem to correspond to prodeltaic gravity-flow deposits. Downslope, the steep middle slope acts as a bypass zone for gravity-driven sediment flows and is submitted to intense along-slope erosion (Figs 6 and 8). Mass flow episodes are recurrent, resulting in intermingled chaotic and parallel reflections. The increase in the mid-slope declivity during S4 and though S5 creates a new area of augmented intensity of bottom currents. Progressive upslope migration of the SD occurs, with its upper (landward) flank developing a flat-lying plastered drift (Figs 6 and 8). In the SD sector the seismic pattern is marked by high-frequency, low-amplitude reflections. The bottom-current activity along the SC axis is confined to its shelfward (inner) flank and an aggradation pattern dominates in its valley and the adjacent drifted deposits. The upper sequence boundary (U5) is tied

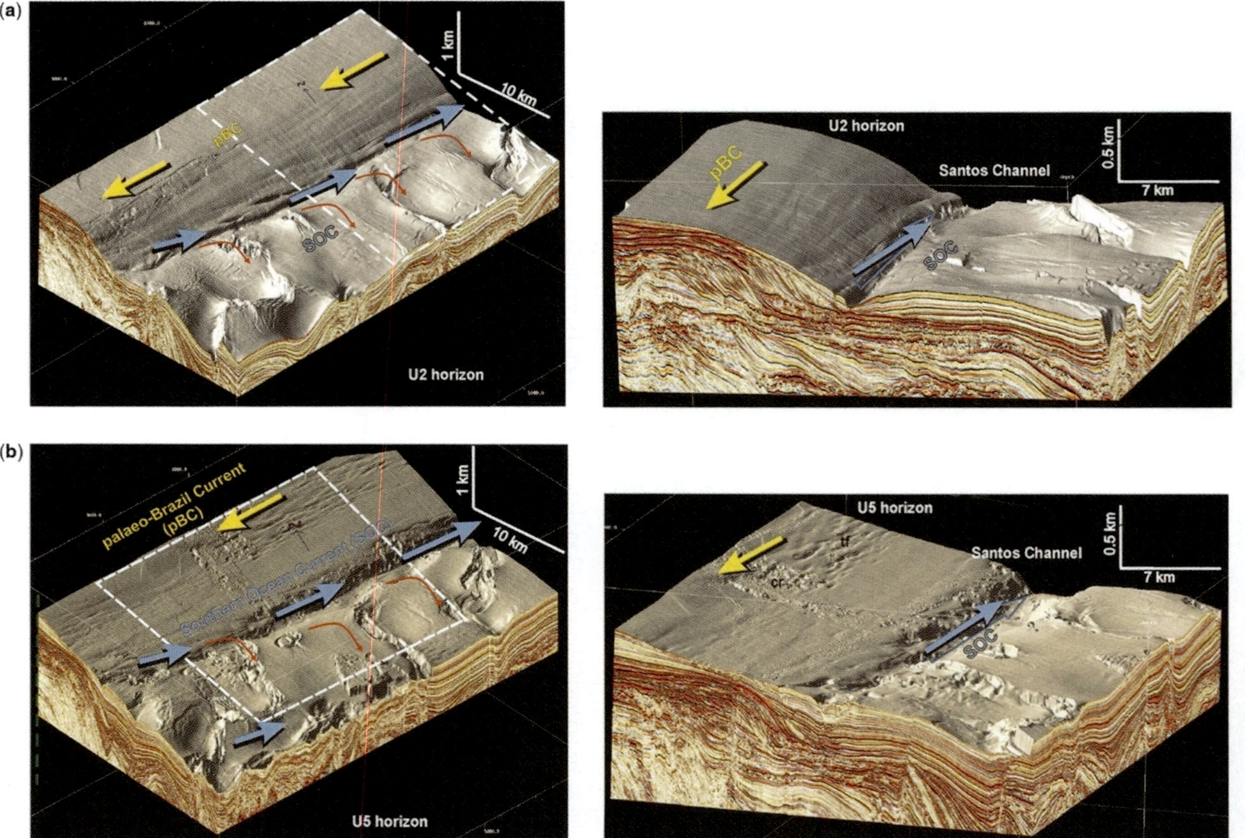

Fig. 11. Seismic block diagrams representing the action of bottom currents in constructing the Santos Drift at stages U2 (early Miocene) and U5 (late Miocene). The white dashed line indicates the area of detailed view. The salt walls influence the sea floor in both stages, with well-marked crestal valleys being potential sediment fairways in connection with upper slope canyons. (**a**) At stage U2, strong activity of the Southern Ocean Current (SOC) deeply eroded the Santos Channel and developed the adjacent drift; upslope, the southward-flowing palaeo-Brazil Current (pBC) locally eroded the steeper portion of the slope, forming lineations and locally destabilizing the slope. (**b**) At stage U5, the drift developed downslope in the Santos Channel starts to weld with the upper slope deposits, reducing the slope gradient. The locus of the SOC shifts eastward, marked by the stronger Santos Channel incision towards that direction. Mass-flow deposits probably related to dismantling of the toe of the prograding clinoforms, and affected by along-slope erosion by the pBC, may be noted (cr, chaotic relief; tf, shallow-seated traction faults).

to the late Miocene unconformity that marks the passage from Tortonian to Messinian (*c.* 7 Ma). U5 presents a moderate to locally strong erosive capacity (Fig. 11). It erodes the foot of the shelf-edge escarpment, the steeper face of the middle slope and the inner flank of the SC (Figs 6 and 8). Downslope, it corresponds to a conformable, continuous, high-amplitude seismic reflector.

Sequence 6 (S6)

S6 is marked by the greatest thickening of the contourite levees of the Santos Drift and São Paulo Drift and by an intense upslope migration of the Santos Drift. Aggradation and parallel reflections are the dominant pattern in the slope setting of this sequence. The upper slope terrace constructed in the underlying sequence is preserved with minor erosive features, suggesting the presence of a less intense surface slope boundary current at the toe of the shelf-edge clinoforms. Lower amplitudes of the reflections above the S6 terrace suggest a drop in the current activity at that site (Fig. 6). Mass-flow deposits are still recurrent, probably related to the upper to mid-slope erosion by the bottom currents. Aggradation occurs also in the interior of the SC, suggesting a general decrease in current activity. Erosion and non-deposition is focused in two main regions, one at about 500 m palaeo-water depth in the slope between the upper slope terrace and the SC, and the other at about 2000 m palaeo-water depth along the SPC, preventing deposition along its axis during S6 (Figs 5–7). Sediment eroded from the axis of the SPC is accumulated on the SPD, which assumes a mounded geometry. Its top is eroded by the upper boundary U6, especially close to the São Paulo Channel, where bottom current activity seems to have attained its maximum energy. The erosive processes related to U6 widen the SC and the SPC axis. Salt-related growing faults are still active and accentuate the physiographic perturbations of the mid- and lower slope (Fig. 5). The shelf presents seismic evidence of progressive drowning with the continuous excursion of the shallow-water depositional systems towards the continent. In the base of S6 a sigmoidal set is developed onlapping the shelf-edge escarpment. During the final period, a downlapping set covers the upper slope terrace and is truncated at the shelf by U6. The upper sequence boundary U6 is related to the late Zanclean (mid-Pliocene) sea-level drop that occurred *c.* 4.0 Ma (Fig. 9).

Sequence 7 (S7)

The uppermost sequence S7 presents predominant prograding characteristics of shallow-water systems with the development of steep progradation sets (Fig. 6). The upper slope terrace is progressively buried by the shelf-edge prograding clinoforms. The sediments from clinoform toes are transferred downslope with minor reworking by the currents above the terrace. The upper slope angle increases and the uppermost part of the upslope-migrating SD is situated at the declivity change from the upper to the mid-slope, at about 700 m water depth (Figs 5 and 6). Chaotic reflections associated with mass-flow movements are frequent in the upper to mid-slope and close to the SC axis. The SC is progressively filled by downslope divergent reflections, which show narrowing of its axis (Fig. 8). Erosion is focused along its inner flank and parallel draping reflectors dominate the internal characteristics of the Santos Drift. The constant thickness from its upslope to lower slope boundary and the absence of marked current-related erosional features suggest that the SD was swept by less intense bottom currents during S7. In contrast, downslope the SPC widens to more than 20 km and truncates the subjacent reflectors. Migrating sediment waves characterize the SPD, which is locally disturbed by salt-related faults reaching the sea floor (Fig. 5). The reduced thickness of SPD during S7 compared with S6 suggests that erosion and transport prevailed during this phase in deep waters, whereas along the slope (SD) depositional processes were dominant. The final stage of S7 (late Quaternary) is marked by the dominant activity of the Brazil Current (BC), which sweeps the uppermost slope characterized by the truncation of reflectors against the sea floor (Fig. 12). Such a high-energy shelf edge–upper slope, where both the BC and the mesoscale eddies are very intense (Fig. 4a), is responsible for transferring sediment downslope, feeding the SD. The upper boundary of S7 is the modern sea floor.

Discussion

The application of sequence-stratigraphic concepts to current drift deposits has always been controversial (Faugères *et al.* 1999; Stow *et al.* 2002). The lack of reliable time correlation, mostly because of the diachronous character of the current regime record, has prevented their wider utilization. The sequences studied here were identified based on their seismic geometry and facies, which, as in several other ocean margin basins, reflect different combinations of controls on the development of their architectural styles. Biostratigraphic tying with exploratory wells, when data were available, shows a good correlation with oxygen isotope records and with relative sea-level curves (Haq *et al.* 1987; Miller *et al.* 1987, 1991). The Santos Drift System sequence boundaries show a

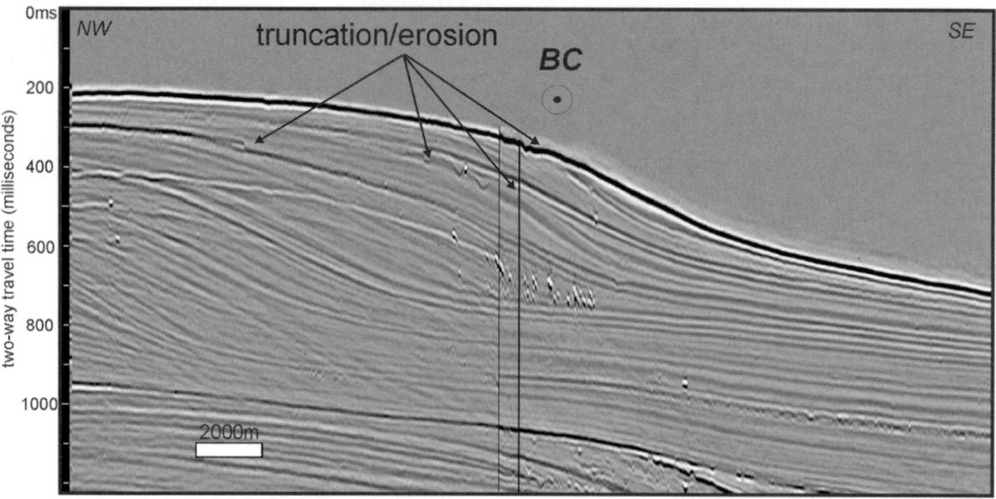

Fig. 12. Seismic line showing the truncation of reflectors and the erosion of the sea floor at the shelf edge–upper slope under the action of the Brazil Current (BC), at different stages of S7.

good correlation with the $\delta^{18}O$ isotope curves for the Atlantic Ocean (Abreu & Anderson 1998), suggesting that the major global climatic oscillations had affected drift development (Fig. 9).

Pre-Santos Drift circulation regime

The initiation of the strong bottom circulation regime that is responsible for the construction of the slope-plastered SD and the SC was initially recorded at the late Palaeocene–early Oligocene. Seismic amplitude maps for that age interval produced by Bulhões and Nunes and presented by Viana *et al.* (2007) permitted the identification of two distinct current regimes. In deep waters, in a lower slope setting, the seismic maps recorded the presence of SW–NE-trending furrows related to NE-flowing contour currents. The same circulation pattern was advocated by Viana *et al.* to be responsible for the development of a large field of submarine barchan dunes and an along-isobath channel–moat system at the toe of the Eocene shelf-edge clinoforms described by Moreira *et al.* (2001). The uppermost slope circulation trend was seismically characterized by shelf-edge and along-slope erosional features and lineaments (Fig. 10). These features were related to a SW-trending surface slope boundary current, similar to the present-day Brazil Current, and to the features described in the Campos Basin related to the same current (Viana *et al.* 2002*a*). Although the proposed existence of a Palaeocene circumpolar flow as the origin of erosive intermediate to deep waters penetrating northward into the Atlantic Ocean is controversial (Wise *et al.* 1985), seismic evidence in the Santos Basin favours the hypothesis of strong deep-water circulation sweeping the basin from at least the Palaeocene. This period immediately preceded the installation of the SDS.

Circulation pattern during and mechanisms of Santos Drift System deposition

The SDS results from the action of two main current systems. One, occurring in intermediate to deep-water settings, flowed northeastward. The other, a surface slope boundary current regime, flowed from south to north. The deeper currents flowed in an opposite sense to the present-day circulation and are related to the Southern Ocean Current (SOC) proposed by Viana *et al.* (2002*a*) to have existed in the Campos Basin as a relatively strong mid- to deep-water current occurring from the transition from the last glacial maximum to early deglaciation. That current occupied the whole middle slope to basin setting following the retreat of a sluggish Glacial North Atlantic Intermediate Water (GNAIW) at the end of the glacial period. This circulation system seems to have prevailed since the late Palaeogene, and, although it lacks isotopic confirmation, seismic evidence seems to corroborate this hypothesis. These observations match the assumption of Wright *et al.* (1991, 1992), who suggested that on a large scale the hydrodynamics of the South Atlantic have not significantly changed from the late Oligocene–early Miocene to Quaternary, and that the oceanic processes observed in the Quaternary can be extrapolated to that period. In the Santos Basin, the bottom currents were constrained against the slope and migrated upslope (northward) under the influence of the Coriolis effect and the locus of their deposition migrated northeastward consistent with the current direction. The uppermost slope was swept by the south-flowing surface current (palaeo-Brazil Current?) and the locus of its deposition migrated southward accompanying its flow (Figs 10 and 11).

Variations in the locus and intensity of the slope boundary currents with time strongly influenced internal and external characteristics of the different aspects of the system, such as the drift isopachs and the distribution and amplitude of erosional features (Figs 5–8). A balance between downslope and along-slope processes is observed through the coeval development of prograding shelf-edge clinoform sets and the upper slope terrace. Similarly to the Canterbury Drift in New Zealand (Carter *et al.* 2004; Lu & Fulthorpe 2004), the clinoform geometry of the SD is influenced by the surface slope boundary current intensity. Partial erosion and the development of the terrace at the clinoform toes were observed in periods of high current intensity (Fig. 12), whereas in periods of reduced current activity burial of the terrace occurred. Sea-floor sweeping by mesoscale eddies could also have helped to transfer fine-grained sediments towards the slope.

At any time, the active drifts were concentrated upon the slope, seaward of the contemporary shelf edge in the earlier stages of the drift development. The upper slope terrace separated the drifts from the shelf-edge clinoforms. The continuous upslope migration of the drift system induced the welding of its upper flank to the upper slope, the thickening of the drift deposits above the slope, and the development of a new slope profile. This new profile, with a conspicuous high-angle slope foot, promoted the installation of a deeper core of the SOC, which was responsible for the development of the São Paulo Drift (Fig. 5).

Stratigraphic–palaeoceanographic control on sequence deposition

The current-dominated slope sequences S1–S7 coincide with third-order relative sea-level variations and provide a basis to better constrain the periods of higher along-slope current activity and those of predominant gravitational-dominated shelf–basin transfer processes.

The transition from Eocene to Oligocene (*c.* 34 Ma) corresponds to a period when the continuous Cenozoic global cooling, which attained a 10 °C drop in the average global temperature (Prothero & Dott 1994), permitted the installation of a vigorous thermohaline circulation, usually correlated with the onset of the AABW (Savin *et al.* 1975; Shackleton & Kennett 1975; Kennett 1977; Barker *et al.* 1983; Johnson 1985). At the end of this period there was a climatic change from a greenhouse period to an icehouse period. The Rupelian–Chatian unconformity (horizon R–C), which defines the base of the study section, was developed in a global context in which successive glaciations trapped large quantities of seawater, inducing the greatest relative sea-level fall during the Cenozoic. Warm-water planktonic forms were extinct and cold-water microplankton proliferated. The Oligocene–Miocene transition was marked by high $\delta^{18}O$ values (icehouse conditions), and a great thermal variation between tropical and high-latitude waters, inducing the installation of vigorous thermohaline circulation. This period would coincide with the establishment of S1.

The abrupt increase in the bottom circulation dynamics during the late Oligocene, accompanying the onset of the AABW, was probably at the origin of the erosion observed in the Santos Basin slope (horizon R–C), where a sparse quantity of upper Eocene to lower Oligocene sediments is preserved. Along with the increase in bottom currents, a major transfer of shallow-water sediments towards the basin occurred during that period of relative low sea-level stand (Fig. 9). The combination of those two processes is responsible for the development of S1.

The intensity of the bottom currents attained its peak especially during the deposition of S2 and S3, after a remarkable rise of relative sea level that occurred during the early and middle Miocene (Fig. 9), induced by a large-scale global warming. That period, marked by low $\delta^{18}O$ values (Miller *et al.* 1987; Abreu & Anderson 1998), was accompanied by a rise in the temperature of Atlantic waters and by the retreat of the Antarctic polar front. Those conditions allowed the development of a mixed carbonate ramp on the Santos Basin shelf edge, whereas the SD experienced a period of intense erosion along the SC and thick accumulation of sediments above the plastered drift, suggesting high acceleration of the bottom currents from the south. Significant accumulation of drifted sediments is noticed in deeper waters with the initiation of the SPD construction, and strong erosion occurred along the axis of the incipient SPC.

Two main events, one tectonic and the other climatic, coincide to give rise to the S3 upper boundary U3. At that time a major regional tectonic tilt, as suggested by the seismic line of Figures 5 and 6, produced a shift of the marine systems towards the continent, followed by an intense shelf progradation and an increase in the slope declivity characteristic of S4. This period coincides with the beginning of the global cooling that occurred between 14 and 11 Ma, coeval with the formation of an ice cap on East Antarctic following the opening of the Drake Passage and the onset of the Antarctic Circumpolar Current (Savin *et al.* 1975; Shackleton & Kennett 1975; Cieselski *et al.* 1982). The re-establishment of a major ice sheet on East Antarctic was reflected in the Santos Drift by the development of S4, characterized by a relatively low sediment accumulation, a predominant aggradation pattern and rare erosional features, indicating a relative drop in the bottom-current intensity during this period.

The temperature drop in deep waters during the late Miocene is time-correlated with several strong erosional events in the Southern Ocean (Wise *et al.* 1985). High-amplitude sea-level falls related to the high-frequency glaciations induced a seaward shift of depositional systems, expressed in the study area by the large prograding clinoforms of the proximal parts of S5. In the SD, this period coincides initially with an aggradation pattern infill of the SC followed by an upslope migration of the drift and a low accumulation rate in both the SD and SPD (Figs 5 and 6).

The Pliocene warm period, which prevailed during S6 deposition, would have partially deglaciated the West Antarctic Ice Sheet (Ciesielski *et al.* 1982) and is related to the drowning of the continental shelf in the study area. Barker *et al.* (1983), analysing data from Ocean Drilling Program (ODP) Leg 72, suggested that AABW would have extended to substantially shallower areas and then could be related to the continuous upslope migration of the SD, the intense erosion along the SPC, and the sediment accumulation on both the SD and SPD.

High-frequency, high-amplitude glacial oscillations occurred from the Pliocene to the Holocene (S7), following the increase of ice volume in Antarctica and the beginning of the Northern Hemisphere glacial expansion (Ciesielski & Wise 1977; Wise *et al.* 1985), and are expressed in the study area by a strong progradation of the shelf edge, intense erosion along the axis of the SC and SPC, and minor drift thickening. These characteristics suggest a high erosion potential of deep waters during this period.

Conclusions

The Santos Drift System (SDS) consists of a complex sedimentary system that developed from Oligocene to Recent time in the Santos Basin, SE Brazil, under the action of strong slope currents. It involves a large variety of contourite drift geometries within which mass flows locally intervene. Two main contourite drifts are recognized, the Santos Drift (SD) and the São Paulo Drift (SPD). The SD is developed in intermediate depths and consists of a gently mounded slope-plastered drift of 1 km in thickness, and is genetically related to the narrow, slope-parallel Santos Channel (SC). The SPD is developed in close association with the wide moat-like São Paulo Channel (SPC), and advances downslope to water depths in excess of 2400 m.

The present study showed that deposition of the SDS was controlled by three major factors: (1) climate, which controls the oceanic and atmospheric circulation regime, determining the intensity of the bottom currents; (2) the nature and volume of available sediments; (3) the morphotectonic context of the basin, which controls the accommodation space.

Our analysis of seismic reflection data suggests that the deposition of the SDS occurred in concert with the northward flow of a mid- to deep-water Southern Ocean Current (SOC), similar to that described by Viana *et al.* (2002*b*) for the northern Campos Basin as flowing during the latest Quaternary in the transition between the maximum lowstand and early sea-level rise, and proposed by those workers to prevail at least since the early Miocene. The presence of a southward-flowing surface slope boundary current similar to the present Brazil Current was responsible for the shelf-edge and upper slope erosion and distribution of sediments delivered by the shelf.

A record of the relationship between intermediate-depth water flow, drift sedimentation and climate change at different time scales was obtained through the use of sequence-stratigraphic tools. Periods of bottom-current increase and more intense sequence development correspond to periods of rising relative sea level or highstand. We faced inherent difficulties in this approach, such as differentiating the origin of regional slope unconformities, which may be related to the basinward transfer of depositional systems predominant during a relative sea-level fall or to along-slope current-related erosive processes predominant in the opposite scenario.

Thus, based on the correlation with the global sea-level curve and Cenozoic oxygen isotopic curves, our study suggests that cold climate periods with low relative sea level were accompanied by a relative decrease in bottom-current intensity expressed by minor sediment accumulation on the drifts. Conversely, periods in which glaciations were less intense, with a relative warming of the ocean waters and sea-level rise, corresponded to an increase in bottom-current intensity marked by local sea-floor erosion and drift aggradation.

The pre-existing margin physiography played an important role as well. The Eocene upper slope escarpment was fundamental in the constriction of the southerly flows and the development of the drift, as much as the sea-floor expression of salt movements, modifying the depositional space and generating current restrictions and obstacles that caused current acceleration or deviation. The upslope drift migration modified the slope declivity and merged with the prograding clinoforms to form the modern margin configuration of the Santos Basin.

This paper is part of the MSc dissertation of the first author. The authors are grateful to Petrobras for granting permission to publish this paper. PGS kindly gave permission to publish 3D seismic lines. We also acknowledge the importance of the comments by the reviewers. Our colleagues M. C. V. Nunez and E. B. Mattos were fundamental in identifying the seismic record of bottom currents in the pre-Santos Drift phase. Discussions with L. A. P. Gambôa, G. V. Bueno and P. Szatmari provided a wider comprehension of the structural evolution of the São Paulo Plateau and of the role played by salt tectonics. Thanks are also extended to A. Grassi, A. Scofano and J. A. M. Lima, who provided valuable information on the past (Grassi) and present-day circulation (Scofano and Lima) along the SE Brazil margin. The authors are very grateful to J.-C. Faugères, whose comments were fundamental for the improvement of the original manuscript.

References

Abreu, V. S. & Anderson, J. B. 1998. Glacial eustasy during the Cenozoic: sequence stratigraphic implications. *AAPG Bulletin*, **82**(7), 1385–1400.

Almeida, F. F. M. & Carneiro, C. D. R. 1998. Origem e evolução da Serra do Mar. *Revista Brasileira de Geociências*, **28**, 135–150.

Barker, P. F., Carlson, R. L., Johnson, D. A., et al. 1981. Deep Sea Drilling Project Leg 72: Southwest Atlantic paleocirculation and Rio Grande rise tectonics. *Geological Society of America Bulletin*, **92**(5), 294–309.

Bueno, G. V., Machado, D. L. Jr, Oliveira, J. A. B. & Marques, E. J. J. 2004. A influência do Lineamento Capricórnio na evolução tectono-sedimentar da Bacia de Santos. *In*: *Congresso Brasileiro de Geologia, 52, Araxá. Annals, Sociedade Brasileira de Geologia, Simposium 28—Petróleo: Geologia e Exploração*, T 773.

Campos, E. J. D., Gonçalves, J. E. & Ikeda, Y. 1995. Water mass structure and geostrophic circulation in the South Brazil Bight—Summer of 1991. *Journal of Geophysical Research*, **100**(C9), 18537–18550.

Carter, R. M., Fulthorpe, C. S. & Lu, H. 2004. Canterbury Drifts at Ocean Drilling Program Site 1119, New Zealand: climatic modulation of southwest Pacific intermediate water flows since 3.9 Ma. *Geology*, **32**, 1005–1008.

Chang, H. K. & Kowsmann, R. O. 1987. Interpretação genética das seqüências estratigráficas das bacias da margem continental brasileira. *Revista Brasileira de Geociências*, **17**, 74–80.

Chang, H. K., Kowsmann, R. O., Figueiredo, A. M. F. & Bender, A. A. 1992. Tectonics and stratigraphy of the East Brazil rift system: an overview. *Tectonophysics*, **213**, 97–138.

Ciesielski, P. F. & Wise, S. W. 1977. Geologic history of the Maurice Ewing Bank of the Falkland Plateau (Southwest Atlantic sector of the Southern Ocean) based on piston and drill cores. *Marine Geology*, **25**, 175–207.

Ciesielski, P. F., Ledbetter, M. T. & Ellwood, B. B. 1982. The development of Antarctic glaciation and the Neogene paleoenvironment of the Maurice Ewing Bank. *Marine Geology*, **46**, 1–51.

Cobbold, P. R., Szatmari, P., Demercian, L. S., Coelho, D. & Rossello, E. 1995. Seismic and experimental evidence for thin-skinned horizontal shortening by convergent radial gliding on evaporites, deep-water Santos Basin, Brazil. *In*: Jackson, M. P. A., Roberts, D. G. & Snelson, S. (eds) *Salt Tectonics: a Global Perspective*. AAPG, Memoirs, **65**, 305–321.

Cobbold, P. R., Meisling, K. E. & Mount, V. S. 2001. Segmentation of an obliquely rifted margin, Campos and Santos basins, southeastern Brazil. *AAPG Bulletin*, **85**, 1925–1944.

Conceição, J. C. J., Zalan, P. V. & Wolff, S. 1988. Mecanismo, evolução e cronologia do rift sul-atlantico. *Boletim de Geociencias da Petrobras*, **2**, 255–265.

DeMadron, X. D. & Weatherly, G. 1994. Circulation, transport and bottom boundary layers of the deep currents in the Brazil Basin. *Journal of Marine Research*, **52**, 583–638.

Demercian, L. S. 1996. *A halocinese na evolução do sul da Bacia de Santos do Aptiano ao Cretáceo Superior*. MSc thesis, Universidade Federal do Rio Grande do Sul, Porto Alegre.

Demercian, L. S., Szatmari, P. & Cobbold, P. R. 1993. Style and pattern of salt diapirs due to thin-skinned gravitational gliding, Campos and Santos basins, offshore Brazil. *Tectonophysics*, **228**, 393–433.

Dias, J. L., Sad, A. R. E., Fontana, R. L. & Feijó, F. J. 1994. Bacia de Pelotas. *Boletim de Geociências da Petrobras*, **8**, 235–245.

Evans, D. L., Signorini, S. R. & Miranda, L. B. 1983. A note on the transport of the Brazil Current. *Journal of Physical Oceanography*, **13**, 1732–1738.

Ewing, J., Leyden, R. & Ewing, M. 1969. Refraction shooting with expendable sonobuoys. *AAPG Bulletin*, **53**, 174–181.

Faugères, J.-C., Stow, D. A. V., Imbert, P. & Viana, A. R. 1999. Seismic features diagnostic of contourite drifts. *Marine Geology*, **162**, 1–38.

GAMBÔA, L. A. P. & RABINOWITZ, P. D. 1981. The Rio Grande Rise Fracture Zone in the Western South Atlantic and its tectonic implications. *Earth and Planetary Science Letters*, **52**, 410–418.

GAMBÔA, L. A. P., BUFFLER, R. T. & BARKER, P. F. 1983. Seismic stratigraphy and geologic history of the Rio Grande gap and Southern Brazil basin. *In*: BARKER, P. F., CARLSON, R. L., JOHNSON, D. A., ET AL. (eds) *Initial Reports, Deep Sea Drilling Project Leg 72*. US Government Printing Office, Washington, DC, 481–497.

GAMBÔA, L. A. P., ESTEVES, F. R., CARMINATTI, M., PEREZ, W. E. & SOUZA CRUZ, C. E. 1986. Evidencias de variacoes de nivel do mar durante o Oligoceno e suas implicacoes faciologicas. *34th Congresso Brasileiro de Geologia*, **1**, 8–22.

GARFIELD, N. 1990. *The Brazil Current at subtropical latitudes*. PhD thesis, University of Rhode Island, Narragansett.

HAQ, B. U., HARDENBOL, J. & VAIL, P. R. 1987. Chronology of fluctuating sea levels since the Triassic. *Science*, **235**, 1156–1187.

IUGS 2004. *International Stratigraphic Chart*. International Union of Geological Sciences: International Commission on Stratigraphy, Paris.

JOHNSON, D. A. 1985. Abyssal teleconnections II. Initiation of Antarctic Bottom Water flow in the southwestern Atlantic. *In*: HSÜ, K. J. & WEISSERT, H. J. (eds) *South Atlantic Paleoceanography*. Cambridge University Press, Cambridge, 243–282.

KARNER, G. & DRISCOLL, N. W. 1999. Tectonic and stratigraphic development of the west African and eastern Brazilian margins: insights from quantitative basin modeling. *In*: CAMERON, N. R., BATE, R. H. & CLURE, V. S. (eds) *The Oil and Gas Habitats of the South Atlantic*. Geological Society, London, Special Publications, **153**, 11–40.

KENNETT, J. P. 1977. Cenozoic evolution of Antarctic glaciation, the circum-Antarctic Ocean, and their impact on global paleoceanography. *Journal of Geophysical Research*, **82**, 3843–3860.

LU, H. & FULTHORPE, C. S. 2004. Controls on sequence stratigraphy of a middle Miocene–Holocene, current-swept, passive margin: offshore Canterbury Basin, New Zealand. *Geological Society of America Bulletin*, **116**, 1345–1366.

MACEDO, J. M. 1989. Evolução tectônica da Bacia de Santos e áreas continentais adjacentes. *Boletim de Geociências da Petrobras*, **3**, 159–173.

MACEDO, J. M. 1990. Evolução tectônica da Bacia de Santos e áreas continentais adjacentes. *In*: GABAGLIA, G. P. R. & MILANI, E. J. (eds) *Origem e Evolução de Bacias Sedimentares*. Petrobras, Rio de Janeiro, 361–376.

MAHIQUES, M. M., TESSLER, M. G., CIOTTI, A. M., ET AL. 2004. Hydrodynamically driven patterns of recent sedimentation in the shelf and upper slope off Southeast Brazil. *Continental Shelf Research*, **24**, 1685–1697.

MEISLING, K. E., COBBOLD, P. R. & MOUNT, V. S. 2001. Segmentation of an obliquely rifted margin, Campos and Santos basins, southeastern Brazil. *AAPG Bulletin*, **85**, 1903–1924.

MILANI, E. J. 2000. Cratonic interior. *In*: CORDANI, U. G., MILANI, E. J., THOMAS FO, A. & CAMPOS, D. A. (eds) *Tectonic Evolution of South America. XXXI International Geological Congress, Rio de Janeiro*. IUGS Special Publication, 392–449.

MILLER, K. G., FAIRBANKS, R. G. & MOUNTAIN, G. S. 1987. Tertiary oxygen isotope synthesis, sea level history, and continental margin erosion. *Paleoceanography*, **2**, 1–19.

MILLER, K. G., WRIGHT, J. D. & FAIRBANKS, R. G. 1991. Unlocking the icehouse: Oligocene–Miocene oxygen isotopes, eustasy, and margin erosion. *Journal of Geophysical Research*, **96**, 6829–6848.

MIRANDA, L. B. & CASTRO, B. M. 1998. Physical oceanography of the western Atlantic continental shelf located between 4°N and 34°S: coastal segment (4°W). *In*: ROBINSON, A. R. & BRINK, K. H. (eds) *The Sea – The Global Coastal Ocean-Regional Studies and Synthesis*. John Wiley & Sons, New York, **11**, 209–251.

MITCHUM, R. M. JR, VAIL, P. R. & THOMPSON, S. 1977. The depositional sequence as a basic unit for stratigraphic analysis. *In*: PAYTON, C. E. (ed.) *Seismic Stratigraphy—Applications to Hydrocarbon Exploration*. American Association of Petroleum Geologists, Memoirs, **26**, 53–62.

MODICA, C. J. & BRUSH, E. R. 2004. Postrift sequence stratigraphy, paleogeography, and fill history of the deep-water Santos Basin, offshore southeast Brazil. *AAPG Bulletin*, **88**, 923–945.

MOHRIAK, W. U., MACEDO, J. M., CASTELLANI, R. T., ET AL. 1995. Salt tectonics and structural styles in the deep-water province of the Cabo Frio region, Rio de Janeiro, Brazil. *In*: JACKSON, M. P. A., ROBERTS, D. G. & SNELSON, S. (eds) *Salt Tectonics: a Global Perspective*. AAPG, Memoirs, **65**, 273–304.

MOREIRA, J. L. P. & CARMINATTI, M. 2004. Sistemas deposicionais de talude e de bacia no Eoceno da Bacia de Santos. *Boletim de Geociências da Petrobras*, **12**(1), 73–87.

MOREIRA, J. L. P., NALPAS, T., JOSEPH, P. & GUILLOCHEAU, F. 2001. Stratigraphie sismique de la marge éocène du bassin de Santos (Brésil): relations plate-forme/systèmes turbiditiques; distorsion des séquences de depot. *Comptes Rendus de l'Académie des Sciences, Science de la Tèrre et des Planètes*, **332**, 491–498.

MULLER, T. J., IKEDA, Y., ZANGENBERG, N. & NONATO, L. V. 1998. Direct measurements of the western boundary currents between 20°S and 28°S. *Journal of Geophysical Research*, **103**(C3), 5429–5437.

NURNBERG, D. & MULLER, R. D. 1991. The tectonic evolution of the South Atlantic from Late Jurassic to present. *Tectonophysics*, **191**, 27–53.

OJEDA, A. H. 1982. Structural framework, stratigraphy, and evolution of Brazilian marginal basins. *AAPG Bulletin*, **66**, 732–749.

OJEDA, H. A. O. & CESERO, P. 1973. *Bacia de Santos e Pelotas: Geologia e perspectivas petroliferas*. PETROBRAS Internal Report.

PEATE, D. W. 1997. The Paraná–Etendeka province. *In*: MAHONEY, J. J. & COFFIN, M. F. (eds) *Large Igneous Provinces*. Geophysical Monograph, American Geophysical Union, **100**, 217–245.

PEREIRA, M. J. 1994. *Sequências deposicionais de 2a/3a ordens (50 a 2Ma) e tectono-estratigrafia no Cretáceo de cinco bacias marginais do Brasil: comparações com outras áreas do globo e implicações geodinâmicas*. PhD thesis, Universidade Federal do Rio Grande do Sul, Porto Alegre.

PEREIRA, M. J. & FEIJÓ, F. J. 1994. *Bacia de Santos. Boletim de Geociencias da Petrobras*, **8**, 219–234.

PEREIRA, M. J. & MACEDO, J. M. 1990. A Bacia de Santos: perspectivas de uma nova província petrolífera na plataforma continental sudeste brasileira. *Boletim de Geociencias da Petrobras*, **4**, 3–11.

PETERSON, R. G. & STRAMMA, L. 1991. Upper-level circulation in the South Atlantic Ocean. *Progress in Oceanography*, **26**, 1–73.

PIOLA, A. R. & GORDON, A. L. 1989. Intermediate waters in the southwest South Atlantic. *Deep-Sea Research*, **36**(1), 1–16.

PROTHERO, D. R. & DOTT, R. H. JR. 1994. *Evolution of the Earth*, 7th edn. McGraw-Hill, New York, 450–505.

RABINOWITZ, P. D. & LABRECQUE, J. 1979. The Mesozoic South Atlantic Ocean and evolution of its continental margins. *Journal of Geophysical Research*, **84**, 5973–6002.

REID, J. L. 1989. On the total geostrophic circulation of the South Atlantic Ocean: flow patterns, tracers, and transports. *Progress in Oceanography*, **23**, 149–244.

REID, J. L. 1996. On the circulation of the South Atlantic Ocean. *In*: WEFER, G., BERGER, W. H., SIEDLER, G. & WEBB, D. J. (eds) *The South Atlantic, Present and Past Circulation*. Springer, Berlin, 13–44.

REZENDE, J. H. M. 2003. *Intrusões da Água Central do Atlântico Sul na Plataforma Continental Sudeste durante o verão*. M.Sc dissertation, Universidade São Paulo.

ROCHA, J. M., MILLIMAN, J. D., SANTANA, C. I. & VICALVI, M. A. 1975. Southern Brazil. *In*: MILLIMAN, J. D. & SUMMERHAYES, C. (eds) *Upper continental margin off Brazil*. Contributions to Sedimentology, **4**, 117–150.

RODRÍGUEZ, A. B. & ANDERSON, J. B. 2004. Contourite origin for shelf and upper slope sand sheet, offshore Antarctica. *Sedimentology*, **51**, 699–711.

SAENZ, C. A. T., HACKSPACHER, P. C., HADLER NETO, J. C., IUNES, P. J., GUEDES, S., RIBEIRO, L. F. B. & PAULO, S. R. 2003. Recognition of Cretaceous, Paleocene, and Neogene tectonic reactivation through apatite fission-track analysis in Precambrian areas of southeast Brazil: association with the opening of the South Atlantic Ocean. *Journal of South American Earth Sciences*, **15**, 765–774.

SAVIN, S. M., DOUGLAS, R. G. & STEHLI, F. G. 1975. Tertiary marine paleotemperatures. *Geological Society of America Bulletin*, **86**, 1499–1510.

SHACKLETON, N. J. & KENNETT, J. P. 1975. Late Cenozoic oxygen and carbon isotope changes at DSDP site 284: implications for glacial history of the northern hemisphere and Antarctica. *In*: KENNETT, J. P. & HOUTE, R. E. (eds) *Initial Reports Deep Sea Drilling Project Leg 29*. US Government Printing Office, Washington, DC, 801–807.

SHIMABUKURO, S. 1994. *Braarudosphaera chalk: investigações sobre a gênese de um marco estratigráfico*. MSc dissertation, Universidade Federal Rio Grande do Sul, Proto Alegre.

SIBUET, J. C., HAY, W. W., PRUNIER, A., MONTADERT, L., HINZ, K. & FRITSCH, J. 1984. The early evolution of the South Atlantic Ocean: role of the rifting episode. *In*: AMIDEI, R. (ed.) *Initial Reports of the Deep Sea Drilling Project Leg 75*. US Government Printing Office, Washington, DC, 469–481.

SIEDLER, G., MÜLLER, T. J., ONKEN, R., ARHAN, M., MERCIER, H., KING, B. A. & SAUNDERS, P. M. 1996. The Zonal WOCE Sections in the South Atlantic. *In*: WEFER, G., BERGER, W. H., SIEDLER, G. & WEBB, D. J. (eds) *The South Atlantic, Present and Past Circulation*. Springer, Berlin, 83–104.

SIGNORINI, S. R. 1978. On the circulation and volume transport of the Brazil Current between the Cape of São Tome and Guanabara Bay. *Deep-Sea Research*, **25**, 481–490.

SILVEIRA, I. C. A., SCHMIDT, A. C. K., CAMPOS, E. J. D., GODOI, S. S. & IKEDA, Y. 2000. A Corrente do Brasil ao largo da costa leste brasileira. *Revista Brasileira de Oceanografia*, **48**, 171–183.

SOUZA, K. G. 1991. *La marge continentale bresilienne sub-orientale et les domaines océaniques adjacents: structure et évolution*. PhD dissertation, Université Pierre et Marie Curie.

SOUZA, M. C. A. 2000. *A Corrente do Brasil ao largo de Santos: medições diretas*. MSc thesis, Universidade São Paulo.

STOW, D. A. V., FAUGÈRES, J.-C., HOWE, J. A., PUDSEY, C. J. & VIANA, A. R. 2002. Bottom currents, contourites and deep-sea sediment drifts: current state-of-the-art. *In*: STOW, D. A. V., PUDSEY, C. J., HOWE, J. A., FAUGÈRES, J.-C. & VIANA, A. R. (eds) *Deep-Water Contourites: Modern Drifts and Ancient Series, Seismic and Sedimentary Characteristics*. Geological Society, London, Memoirs, **22**, 7–20.

STRAMMA, L. 1989. The Brazil Current transport south of 23°S. *Deep-Sea Research*, **36**(4A), 639–646.

STRAMMA, L. 1991. Geostrophic transport of the South Equatorial Current in the Atlantic. *Journal of Marine Research*, **49**, 281–294.

SZATMARI, P., MILANI, E., LANA, M., CONCEICAO, J. & LOBO, A. 1985. How South Atlantic rifting affects Brazilian oil reserves. *Oil & Gas Journal*, **83**, 107–113.

SZATMARI, P., GUERRA, M. C. M. & PEQUENO, M. A. 1996. Genesis of large counter-regional faults by flow of Cretaceous salt in the South Atlantic Santos Basin, Brazil. *In*: ALSOP, G. I., BLUNDELL, D. J. & DAVISON, I. (eds) *Salt Tectonics*. Geological Society, London, Special Publications, **100**, 259–264.

VAIL, P. R. 1987. Seismic stratigraphy interpretation using sequence stratigraphy. Part 1: seismic stratigraphy interpretation procedure. *In*: BALLY, A. W. (ed.) *Atlas of Seismic Stratigraphy*. AAPG Studies in Geology, **27**, 1–10.

VAIL, P. R., MITCHUM, R. M., TODD, R. G., *ET AL.* 1977*a*. Seismic stratigraphy and global changes in sea level. *In*: PAYTON, C. E. (ed.) *Seismic Stratigraphy—Applications to Hydrocarbon Exploration*. AAPG Memoirs, **26**, 49–212.

VAIL, P. R., TODD, R. G. & SANGREE, J. B. 1977*b*. Seismic stratigraphy and global changes of sea level, part 5: chronostratigraphic significance of seismic reflections. *In*: PAYTON, C. E. (ed.) *Seismic Stratigraphy—Applications to Hydrocarbon Exploration*. AAPG Memoirs, **26**, 99–116.

VIANA, A. R. 1998. *Le rôle et l'enregistrement des courants océaniques dans les dépôts de marges continentales: la marge du bassin Sud-Est brésilien*. Doctoral thesis, Université de Bordeaux 1.

VIANA, A. R. 2001. Seismic expression of shallow- to deep-water contourites along the south-eastern Brazilian margin. *Marine Geophysical Researches*, **22**, 509–521.

VIANA, A. R. & FAUGÈRES, J.-C. 1998. Upper slope sand deposits: the example of Campos Basin, a latest Pleistocene/Holocene record of the interaction between along-slope and downslope currents. *In*: STOKER, M. S., EVANS, D. & CRAMP, A. (eds) *Geological Processes on Continental Margins: Sedimentation, Mass-Wasting and Stability*. Geological Society, London, Special Publications, **29**, 287–316.

VIANA, A. R., FAUGÈRES, J.-C., KOWSMANN, R. O., LIMA, J. A. M., CADDAH, L. F. G. & RIZZO, J. G. 1998. Hydrology, morphology and sedimentology of the Campos Continental Margin, Offshore Brazil. *Sedimentary Geology*, **115**, 133–158.

VIANA, A. R., ALMEIDA, W. JR. & ALMEIDA, C. F. W. 2002*a*. Upper slope sands—the late Quaternary shallow-water sandy contourites of Campos Basin, SW Atlantic margin. *In*: STOW, D. A. V., FAUGÈRES, J.-C., HOWE, J. A., PUDSEY, C. J. & VIANA, A. R. (eds) *Deep-Water Contourite Systems: Modern Drifts and Ancient Series, Seismic and Sedimentary Characteristics*. Geological Society, London, Memoirs, **22**, 261–270.

VIANA, A. R., HERCOS, C. M., ALMEIDA, W. JR, MAGALHÃES, J. L. C. & ANDRADE, S. B. 2002*b*. Evidence of bottom current influence on the Neogene to Quaternary sedimentation along the northern Campos slope, SW Atlantic margin. *In*: STOW, D. A. V., FAUGÈRES, J.-C., HOWE, J. A., PUDSEY, C. J. & VIANA, A. R. (eds) *Deep-Water Contourite Systems: Modern Drifts and Ancient Series, Seismic and Sedimentary Characteristics*. Geological Society, London, Memoirs, **22**, 249–259.

VIANA, A. R., ALMEIDA, W. JR, NUNES, M. C. V. & BULHÕES, E. M. 2007. The economic importance of contourites. *In*: VIANA, A. R. & REBESCO, M. (eds) *Economic and Palaeoceanographic Significance of Contourite Deposits*. Geological Society, London, Special Publications, **276**, 1–24.

WISE, S. W., GOMBOS, A. M. & MUZA, J. P. 1985. Cenozoic evolution of polar water masses, southwest Atlantic Ocean. *In*: HSÜ, K. J. & WEISSERT, H. J. (eds) *South Atlantic Paleoceanography*. Cambridge University Press, Cambridge, 283–324.

WRIGHT, J. D., MILLER, K. G. & FAIRBANKS, R. G. 1991. Evolution of modern deep water circulation: evidence from the late Miocene Southern Ocean. *Paleoceanography*, **6**, 275–290.

WRIGHT, J. D., MILLER, K. G. & FAIRBANKS, R. G. 1992. Early and middle Miocene stable isotopes: implications for deep water circulation and climate. *Paleoceanography*, **7**, 357–390.

ZACHOS, J., PAGANI, M., SLOAN, L., THOMAS, E. & BILLUPS, K. 2001. Trends, rhythms, and aberrations in global climate 65 Ma to Present. *Science*, **292**, 686–693.

Mediterranean bottom-current deposits: an example from the Southwestern Adriatic Margin

GIUSEPPE VERDICCHIO[1,2], FABIO TRINCARDI[1] & ALESSANDRA ASIOLI[3]

[1]*ISMAR (CNR), via Gobetti 101, 40129 Bologna, Italy*
(e-mail: giuseppe.verdicchio@bo.ismar.cnr.it)

[2]*Università di Bologna, Dipartimento di Scienze Geologico Ambientali, Piazza di Porta San Donato 1, 40126 Bologna, Italy*

[3]*IGG (CNR), C.so Garibaldi, 37, 35137 Padova, Italy*

Abstract: The identification of bottom-current deposits is a key to understanding the long-term deep-sea circulation and its changes through geological times. The Southwestern Adriatic Margin (SAM) is a small Mediterranean sub-basin that represents a key site to study bottom-current deposits in a Mediterranean context and hence to improve our knowledge of changes in Mediterranean deep-water circulation during the recent geological past. The SAM is characterized by complex stratification and circulation related to an interaction between two south-flowing bottom water masses: the cold North Adriatic Dense Water (NAdDW), formed in the shallow northern Adriatic through cold wind forcing and winter heat loss, and the highly saline Levantine Intermediate Water (LIW), generated in the Eastern Mediterranean through intense evaporation and flowing along the slope in a depth range of 200–600 m. Chirp-sonar profiles, TOBI mosaics and sediment cores acquired along the SAM reveal distinctive sediment drift types (elongated, plastered and isolated drifts) and extensive fields of sediment waves. Non-depositional and erosional features related to bottom-current activity include moats between drifts and the steep slope, widespread upper-slope erosional areas and extensive furrowed areas, which are particularly developed where change in slope orientation blocks the current circulation. The distribution, morphology and size of bottom-current features along the SAM result from an interaction between current regime and slope morphology, characterized by structural highs perpendicular to the slope contour (e.g. Dauno Seamount), multiple slope incisions (e.g. Bari Canyon and slump scars) and extensive block-slide deposits. Morphobathymetric and seismic stratigraphic data on the SAM show that bottom-current deposits are best developed where the regional slope flattens seaward of a very steep, often erosional, upper slope. The roughness of the lower slope, in particular, seems to correlate with the complexity and decreasing size of the bottom-current deposits. Like other land-locked basins, the Adriatic underwent dramatic palaeogeographical and palaeoceanographic rearrangements during the Late Quaternary sea-level oscillations. Indeed, during the Last Glacial Maximum (LGM), most of the areas where NAdDW is formed today were subaerially exposed. Concurrently, during glacial times the LIW production was probably reduced compared with the present-day conditions. The SAM slope is a key site to study the impact of changing current regime on late Quaternary slope deposits. Other Mediterranean late Quaternary contourite deposits are either in water depths compatible with the LIW, particularly in the case of shallow sill basins (e.g. Sicily, Corsica Channel), or at the slope base reflecting the flow of Mediterranean deep waters. The SAM bottom-current deposits, instead, seems to record the changing interaction between two distinctive bottom-hugging currents along the same pathway.

Contourite deposits shaped by deep currents of thermohaline origin have been extensively described in most oceanic basins. In these settings, the activity of bottom currents throughout the Cenozoic produced extensive, and in some cases huge, sediment drifts, spatially associated with deep-sea hiatuses and erosional surfaces (Stow & Holbrook 1984; Faugères *et al.* 1999; Stow *et al.* 2002).

The Mediterranean is a geologically young basin with intense tectonic and volcanic activity, and extensive mass-wasting processes (Rothwell *et al.* 1998; Canals *et al.* 2004). These factors may mask most sedimentological and stratigraphic evidence of bottom-current activity on Mediterranean margins (Reeder *et al.* 2002). Increasing evidence of bottom-current deposits and their laterally equivalent erosional features in the Mediterranean comes from restricted sub-basins and particularly from passageways where local topography induces flow restriction and bottom-current acceleration (Marani *et al.* 1993; Roveri

From: VIANA, A. R. & REBESCO, M. (eds) *Economic and Palaeoceanographic Significance of Contourite Deposits.* Geological Society, London, Special Publications, **276**, 199–224.
0305-8719/07/$15.00

2002). Bottom-current deposits recognized on the Mediterranean sea floor have external shapes and internal geometries that are very similar to those of some of the sediment drifts of the Atlantic Ocean (e.g. Faugères *et al.* 1993; Howe *et al.* 1994; Stoker 1998; Viana 2002), but are smaller in size and represent a more reduced time interval encompassing, at most, the last few hundred thousand years instead of several million years.

This paper describes sedimentological evidence of the impact of bottom currents on the Southwestern Adriatic Margin (SAM; Figs 1 and 2). This relatively small sub-basin, located in a key position of the central Mediterranean, is affected by the formation of cold deep water in the North Adriatic. The area is also affected by the southward flow of intermediate water entering the Adriatic from the SE (Orlic *et al.* 1992). This complex circulation pattern may have undergone changes in intensity,

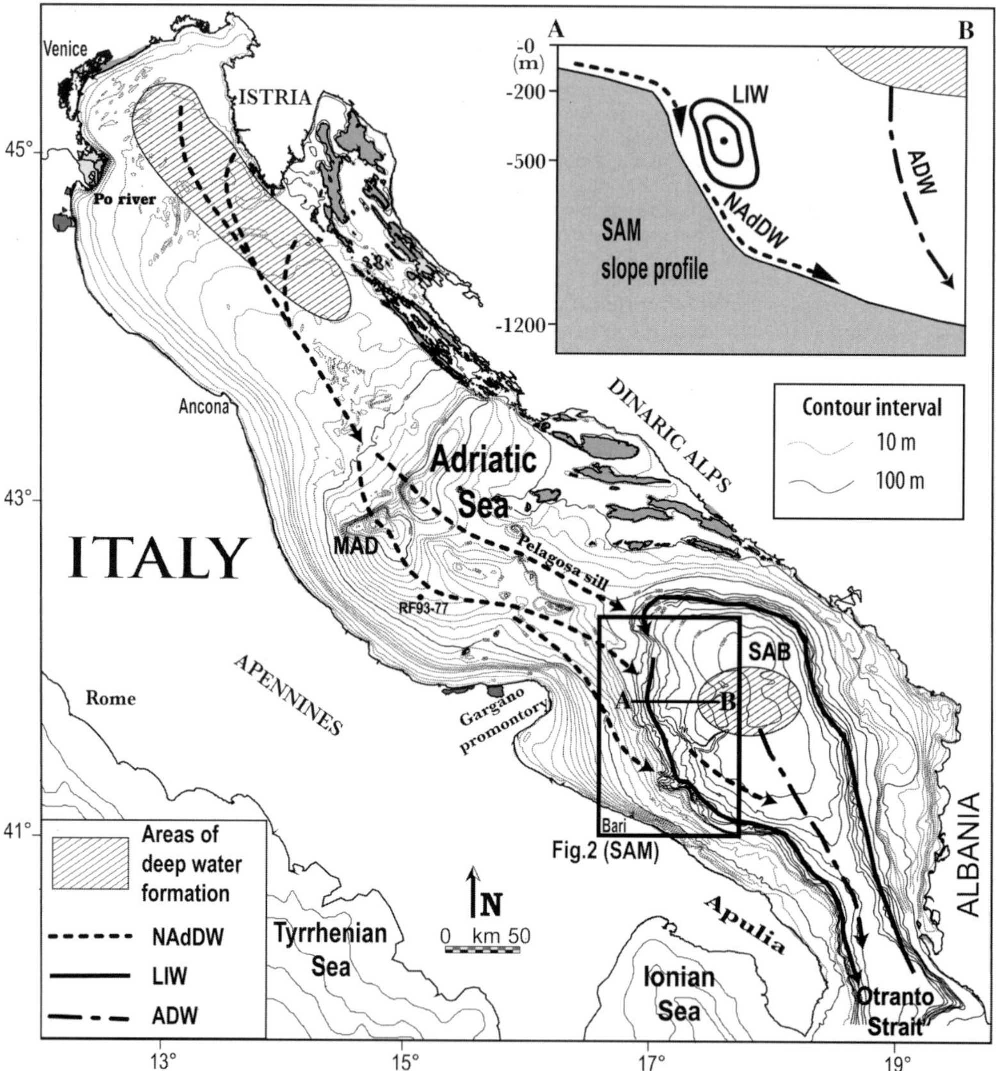

Fig. 1. Schematic reconstruction of regional intermediate- and bottom-water circulation pattern in the Adriatic basin, and location of the study area. Inset shows schematic reconstruction of the intermediate–deep-water circulation on the Southwestern Adriatic Margin (SAM) slope. NAdDW, North Adriatic Dense Water; LIW, Levantine Intermediate Water; ADW, Adriatic Dense Water; MAD, Mid Adriatic Deep; SAB, South Adriatic Basin.

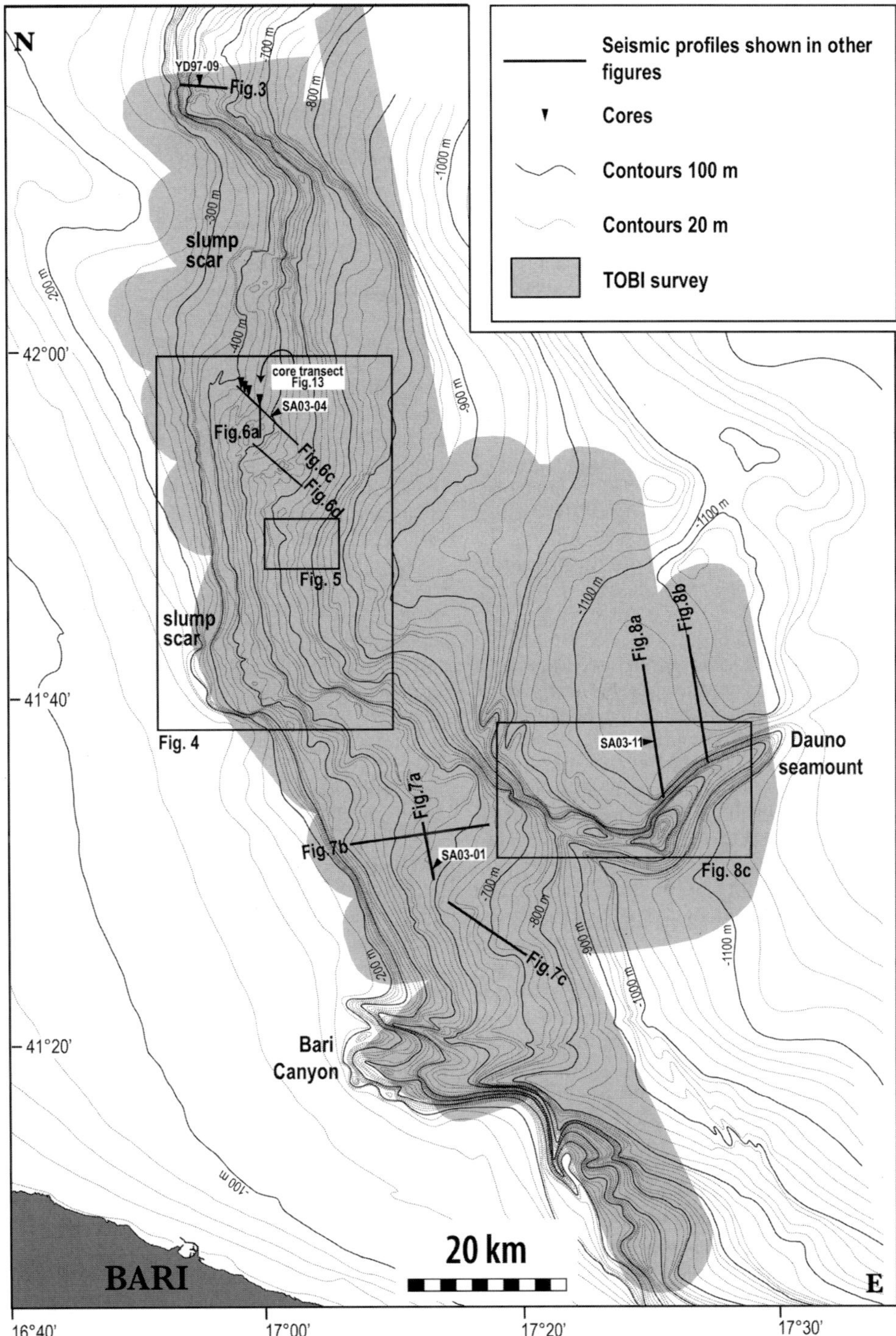

Fig. 2. Detailed bathymetry of the Southwestern Adriatic Margin (SAM) with 20 m contour interval and location of Chirp-sonar profiles and TOBI mosaic (grey pattern).

or there may have even been complete rearrangements of the overall Mediterranean circulation driven by climate and sea-level changes in the late Quaternary (Myers *et al.* 1998). The main aims of this paper are to: (1) describe the characteristics of both depositional and erosional features that created by bottom currents along the SAM, and their relationship with the complex margin morphology; (2) understand the genetic relation between these deposits and the bottom-current regime, with particular reference to the present-day circulation; (3) investigate if the phases of growth and quiescence of bottom-current deposits are affected by Quaternary climatic changes, which affected the bottom circulation.

Methods

A new bathymetric map of the SAM (Fig. 2), with a 20 m contour interval, was compiled by interpolating Global Positioning System (GPS)-located single-beam data, acquired with a high-frequency echosounder (DESO 20) during two oceanographic cruises onboard R.V. *Urania* (cruises SAGA03 and STRATA04) carried out by ISMAR-CNR Bologna in the summers of 2003 and 2004.

High-resolution seismic profiles were gathered using a 16-transducer hull-mounted chirp-sonar profiler with a sweep modulated outgoing signal equivalent to a 3.5 kHz profiler. Chirp-sonar profiles allow observation of subtle variations in seismic facies with vertical resolution of 0.5 m or less. Knowledge of the sea-floor morphology and backscatter comes from a 30 kHz TOBI sidescan sonar mosaic acquired during cruise SAGA03 across the entire SAM slope north and seaward of the Bari Canyon. Differential GPS located the ship's is position every 5 s; TOBI was positioned relative to the ship by calculating its distance accurately with Hydroacoustic Position References (HPR). Post-cruise processing was carried out at Southampton Oceanographic Centre, following the methods of Le Bas *et al.* (1995).

Sediment cores were collected using a piston corer with variable barrel length (5–20 m), reaching a maximum core recovery of *c.* 16 m. Several core sites were resampled using a lightweight coring device (SW104) with a 1.35 m barrel that allows recovery of the undisturbed core top and a detailed study of the uppermost part of the sedimentary section (Magagnoli & Mengoli 1995).

Geological and oceanographic setting

Geological setting

The western Adriatic basin is the Oligo-Miocene to early Pleistocene foreland domain of the Apennine chain (Ricci Lucchi 1986; Ori *et al.* 1986; Argnani *et al.* 1991). This foreland shows contrasting tectonic deformation trends north and south of the Gargano Promontory, possibly reflecting deep crustal discontinuity and variations in lithosphere thickness of the westward-dipping Adriatic plate (Royden *et al.* 1987; de Alteriis & Aiello 1993; Doglioni *et al.* 1994). The southern Adriatic margin, surrounding the Gargano Promontory and Apulia region, was affected by Quaternary uplift and deformation (Bertotti *et al.* 1999). Seismo-stratigraphic data and exploration wells indicate that the western margin of the south Adriatic basin underwent extension since the Oligocene (de Dominicis & Mazzoldi 1987). During the last half million years, the SAM has been built through the deposition of a composite stack of regressive depositional sequences, separated by shelf-wide erosional unconformities (Ridente & Trincardi 2002*a*). These sequences record high-frequency glacio-eustatic cycles, characterized by 100–120 ka duration (Trincardi & Correggiari 2000). On a regional scale, seismic data show that the four late Pleistocene sequences form a composite wedge with an overall back-stepping architecture in the central Adriatic and a forestepping stacking pattern south of the Gargano Promontory. In this area the four sequences show an overall seaward shift of the landward pinch-out with time, indicating the progressive seaward tilt of the margin in response to the long-term uplift documented on land (Ridente & Trincardi 2002*a*).

The Gondola fault (Finetti 1984; Finetti *et al.* 1987) is the main deformation system in the SAM. Deformational features in superficial deposits indicate that the system has been active up to recent times (Tramontana *et al.* 1995; Ridente & Trincardi 2002*a*). This observation is consistent with the evidence of recent deformation through syndepositional gentle folding on the continental shelf and particularly along the east–west-trending Gondola deformation belt (Ridente & Trincardi 2002*b*).

Oceanographic setting

The large-scale circulation pattern of the Mediterranean Sea has been well known since the pioneering studies of Wüst (1961). The present-day Mediterranean thermohaline circulation, both as a whole and at the scale of its sub-basins, is anti-estuarine with surface inflow of light waters and deep outflow of relatively dense water (Myers *et al.* 1998).

The Adriatic Sea (Fig. 1) was identified as the most important source of the bottom water of the Eastern Mediterranean (Wüst 1961;

Malanotte-Rizzoli *et al.* 1997; Manca *et al.* 2002), with a thermohaline circulation characterized by a combination of both estuarine and anti-estuarine components (Cushman-Roisin *et al.* 2001).

Estuarine circulation is triggered when accumulation of lighter waters from river runoff and solar heating creates a higher sea level in the Adriatic relative to the adjacent Ionian Sea, further south. Anti-estuarine circulation in the Adriatic is instead created by internal accumulation of waters of greater density relative to those of the Ionian basin. This denser water mass tends to flows southward, through the Otranto Strait, forming the Dense Water Outflow Current (Artegiani *et al.* 1989, 1997), which contributes to the formation of the Eastern Mediterranean Deep Water (Bignami *et al.* 1990).

The water masses classically identified in the southern Adriatic are as follows. In the surface layer (0–200 m), the colder and low-salinity Adriatic Surface Water (ASW) flows out from the Adriatic and the highly saline Ionian Surface Water (ISW) enters from the southern side of the basin (Manca & Scarazzato 2001). In the intermediate layer, between 200 and 700 m, the Levantine Intermediate Water (LIW) is the dominant and the saltiest water mass, which forms in the Eastern Mediterranean (Levantine basin) through intense evaporation. LIW, with a density excess of 29.0 kg m^{-3}, enters the south Adriatic on the eastern side of the Otranto Strait and flows along the SAM, following the perimeter of the south Adriatic basin (Orlic *et al.* 1992). In the bottom layer, a very dense, cold but less saline water mass, identified as the North Adriatic Dense Water (NAdDW), flows southward across the SAM. The NAdDW represents the densest water of the entire Mediterranean, with a potential density excess reaching values of about 29.8 kg m^{-3}, and forms in the shallow northern Adriatic shelf because of cold wind forcing, related to Bora events, and winter heat loss (Cushman-Roisin *et al.* 2001). In the deep basin, NAdDW combines with the Adriatic Dense Water (ADW), which forms during winter by deep convection in the southern Adriatic, and leaves the basin through the Otranto Strait (Vilibic & Orlic 2002; Mantziafou & Lascaratos 2004).

Therefore, the SAM is probably affected by two south-flowing and interacting bottom-water masses: the LIW and the cascading NAdDW (Fig. 1). Oceanographic data collected on the SAM slope confirm this current pattern and reveal high current velocities, exceeding 60 cm s^{-1} in 700 m depth north of the Bari Canyon during events of dense water flow (S. Miserocchi, pers. comm.).

Bathymetry

The SAM is a complex continental margin extending roughly north–south for 150 km and reaching 1100 m below sea level (Fig. 2). The narrow Pelagosa Sill (about 180 m deep at present), shown in Figure 1, represents the only connection between the central and the southern Adriatic during the Last Glacial Maximum (LGM) sea-level lowstand (Trincardi *et al.* 1996; Asioli *et al.* 2001).

In the SAM the continental shelf reaches its widest extent (*c.* 70–80 km) south of the Gargano Promontory in the Manfredonia Gulf and becomes narrower to the north and south of this area. Shelf gradient varies between 0.1° and 0.3°, and slope gradient ranges between 1° and 8°. Slope depth typically ranges between 200 and 900 m, with a steeper upper slope and a gentler, but morphologically more irregular, lower slope. In the northern part of the SAM the slope trend changes locally from north–south to east–west.

Several large incisions, including slump scars, gullies and canyons, dissect the slope, commonly breaching the shelf edge. The largest incision is the Bari Canyon, a markedly asymmetric, east–west-trending erosional feature that has multiple heads and two main branches, almost parallel to each other, reaching the basin floor at 1000 m water depth. The southernmost wall of these branches is particularly steep (>30°) and is likely to represent a morphological barrier for the southward-flowing bottom waters.

The Dauno Seamount is a structural high in the middle of the SAM area, extending into the basin. The seamount stretches for more than 20 km to the east in a water depth range of 700–1100 m. The Dauno Seamount, possibly related to the tectonic activity of the Gondola fault system (Tramontana *et al.* 1995), is characterized by a relief of about 400 m above the surrounding basin floor and steeper north flank, which is likely to represent an important hydrological barrier.

Seismic stratigraphy

The analysis of high-resolution Chirp-sonar profiles and TOBI sidescan sonar mosaics reveals a very complex morphology of the SAM slope and shows several sedimentary features that reflect the interaction of multiple processes. Large-scale morphology of the SAM is characterized by changes in slope trend and by the presence of topographic highs (e.g. Dauno Seamount) and large incisions (Bari Canyon). Moreover, the SAM continental slope presents clear evidence of sea-floor instability, including extensive slide scars on the upper slope, large slide blocks and debris on the sea floor. Thick acoustically transparent deposits are buried at the base of the slope and beneath most of the basin floor (Trincardi *et al.* 2004). Widespread erosional surfaces are common in the study area and are usually characterized by high backscatter on TOBI images and by high-amplitude reflectors on seismic profiles. Slope erosion is particularly evident on the upper slope (200–350 m depth), where the sea floor is steeper than the bedding planes of the underlying strata. This margin configuration, at places, results in rather complex sea-floor morphology where erosional remnants define a stepped sea floor.

Well-stratified (conformable) deposits accumulate on the slope with a patchy distribution. These stratified deposits are organized in broad mounds with internal convex-upward seismic-reflector configurations, and show great variability in thickness, morphology and orientation with respect to the regional slope. Reflectors converging towards the sea floor in areas of dominant erosion, short distance change in thickness of reflector packages and occurrence of undulating reflectors on seismic profiles are consistent with the activity of bottom currents impinging on the sea floor. By comparison with similar sedimentary deposits ascribed to bottom-current activity on other continental margins, these mounded deposits are interpreted as sediment drifts and sediment waves from bottom currents.

Unfortunately, the complex morphology of the study area and the extreme variability in sediment thickness hinder precise seismic correlation throughout the margin. The main characteristics of the superficial deposits along the SAM are described in detail below, for three major subareas: the northern slope, the southern slope and the basin floor.

The northern slope area

In the northern part of the SAM a sediment drift is located in a depth range of 500–600 m just downslope of the area where the continental slope is steepest (Fig. 3). Drift accretion generated maximum relief of 50 m above the sea floor with a drift axis oriented parallel to the regional contour. In cross-section, the drift is asymmetric

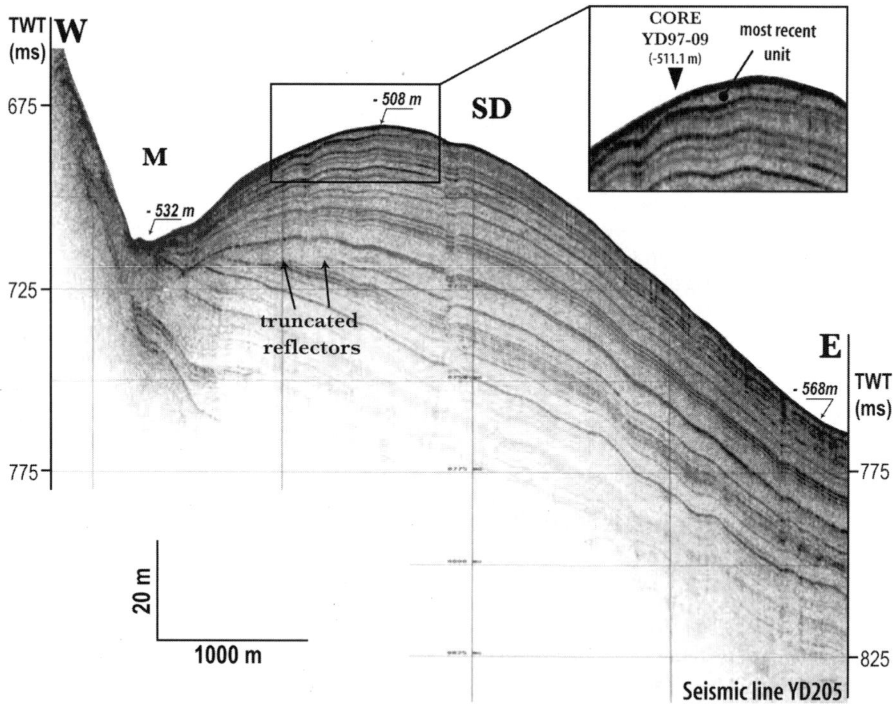

Fig. 3. Chirp-sonar profile across the northern sector of the study area showing a separated drift rimmed by a well-defined moat on its upslope side. Inset shows an enlargement of the drift crest, with the most recent unit developed only on the upslope side of the drift. Profile and core YD97-09 locations are shown in Figure 2. SD, separated drift; M, moat; TWT, two-way travel time.

with a short, steep flank facing up the slope towards a contour-parallel moat and a gentler flank dipping down the slope. Layered, continuous, convex-upwards subparallel reflectors characterize the internal stratigraphy of this sediment drift and converge toward the moat. Using the classification scheme proposed by Faugères *et al.* (1999), this deposit is a separated drift, similar in shape to several North Atlantic drifts (e.g. the Flemish Drift described by Kennard *et al.* 1990). High-resolution seismic profiles also show sediment waves on the downslope gentler flank of the sediment drift. The waves occupy a small region and have wavelengths of about 1 km and typical heights of 20 m. Internal seismic reflectors tend to converge, suggesting preferential erosion or reduced deposition on the downslope flank. Piston core YD97-C09 from the drift crest shows 5.8 m of intensely bioturbated muddy silt, with localized enrichments in bioclastic debris and organic matter.

To the south, an extensive field of sediment waves covers an area of about 280 km^2 where the eastward-dipping slope is markedly irregular as a result of previous erosion. Wave height is up to 40 m and wavelength between 1 and 2 km. On 30 kHz TOBI sidescan sonar images the sediment waves appear as a series of alternating stripes with high, intermediate and low backscatter corresponding to erosional troughs, downslope-facing, steeper flanks and areas of maximum sediment thickness, respectively (Fig. 4). Wave crests range from straight to sinuous and in some cases develop a barchan-like form (Fig. 5a). Backscatter striations correspond to the erosional remnants below the sediment waves and are the expression of outcropping internal layers of different lithology, as suggested by the truncated reflectors on Chirp profiles (Fig. 5b). This is particularly evident in areas where the sediment waves lose continuity and appear as isolated barchan-like features (Fig. 5). The dominant orientation of these features is NW–SE but varies from subparallel to the slope (i.e. north–south), typically in shallower areas, to perpendicular to the regional slope (i.e. east–west) in deeper water; this change in orientation does not appear to be random, but occurs consistently along the slope from north to south. Concurrently with this change in orientation, the sediment waves tend to become smaller and more isolated (Fig. 4 inset).

Seismic profiles collected in the sediment wave field reveal a complex internal structure with distinctive seismic facies and reflector geometries on the two flanks of each wave (Fig. 6). Sediment waves display variable degrees of asymmetry resulting from the presence of thicker deposits on the NW flank (upcurrent) and thinner deposits or subtle erosional surfaces on the southern flank. Typically, only few reflectors can be traced across an entire sediment wave, from flank to flank. The migration of the wave crests is approximately northwestward (Fig. 6). Therefore, these sediment waves migrate upcurrent and upslope, similarly to several examples of bottom-current sediment waves described in oceanic contexts (e.g. Manley & Flood 1993; Howe 1996; Masson *et al.* 2002).

In the sediment wave field, seismic profiles allow us to distinguish distinct seismic units based on their stratigraphic position and internal geometry: B, basal; L, L′, L″, lateral aggradation; D, drape; U, prevalent upward aggradation (Fig. 6). At least in some cases, these units are separated by widespread erosional surfaces (E1 and E2; Fig. 6). Unit B is characterized by continuous, parallel and slightly undulating reflectors, and represents the lowest unit visible on seismic profiles. Units L, L′ and L″ are more expanded on the northwestern (upcurrent) flank of the sediment waves and show continuous to discontinuous, slightly convex-upward, reflectors, which onlap or downlap on the downcurrent flanks. Unit D is characterized by continuous and parallel reflectors, which drape the underlying morphological relief. Unit U is similar to unit L, but presents a marked upward aggradation and more continuous internal reflectors.

Toward the south the sediment waves evolve into isolated barchan-like bedforms (Fig. 5) and then into a zone characterized by longitudinal stripes of alternating low and high backscatter oriented NW–SE and roughly parallel to the regional bottom-current pathway. The low-backscatter stripes are 1–11 km long and 20–100 m wide and are typically separated by 100–200 m wide areas of high backscatter. Comparison with similar sedimentary features described on other continental margins (e.g. Hollister *et al.* 1974; Flood 1994; Kuijpers *et al.* 2003) suggests that these features can be interpreted as abyssal furrows parallel to the main flow of the bottom currents (Flood 1983). TOBI mosaics indicate that abyssal furrows are also superimposed on the sediment wave field (Fig. 5). In this case the furrows are shorter and occur mainly on the downslope flank of the sediment waves or in the erosional trough between them. Further to the south, instead, an area adjacent to the main furrowed area is characterized by the occurrence of large slide blocks at the sea floor. In this area sediment accumulation is confined preferentially on the north or NW side of the slide blocks and defines comet-like elongated deposits that are subparallel to the furrows. Seismic profiles indicate that the high-backscatter areas on TOBI mosaics (Fig. 4) are slide blocks, whereas low-backscatter areas indicate preferential

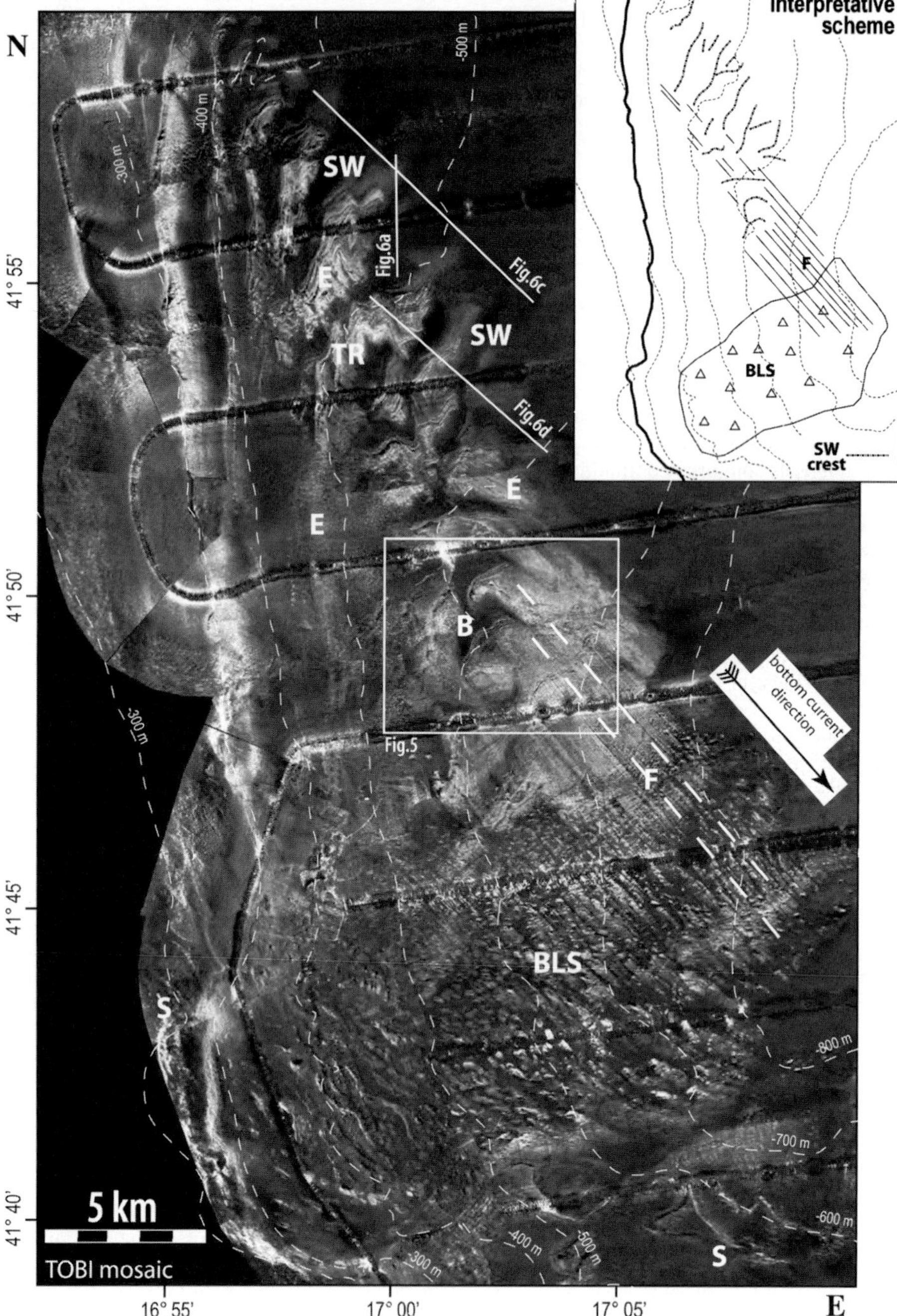

Fig. 4. TOBI sidescan sonar mosaic, in the northern sector of the SAM, showing the field of sediment waves and the large blocky slide. The areas of high backscatter (light tones) denote prevailing erosion whereas those of low backscatter (dark tones) indicate prevailing deposition. The different forms and orientations of the sediment waves from north to south, and the NW–SE lineations (furrows) superimposed on the sediment waves and on the slide blocks, should be noted. Inset shows an interpretative scheme of the TOBI image. Mosaic location is shown in Figure 2. BLS, Blocky slide deposits; S, slide scar; F, furrows; E, erosion; SW, sediment wave deposit; B, barchan-like sediment waves; TR, sediment wave trough.

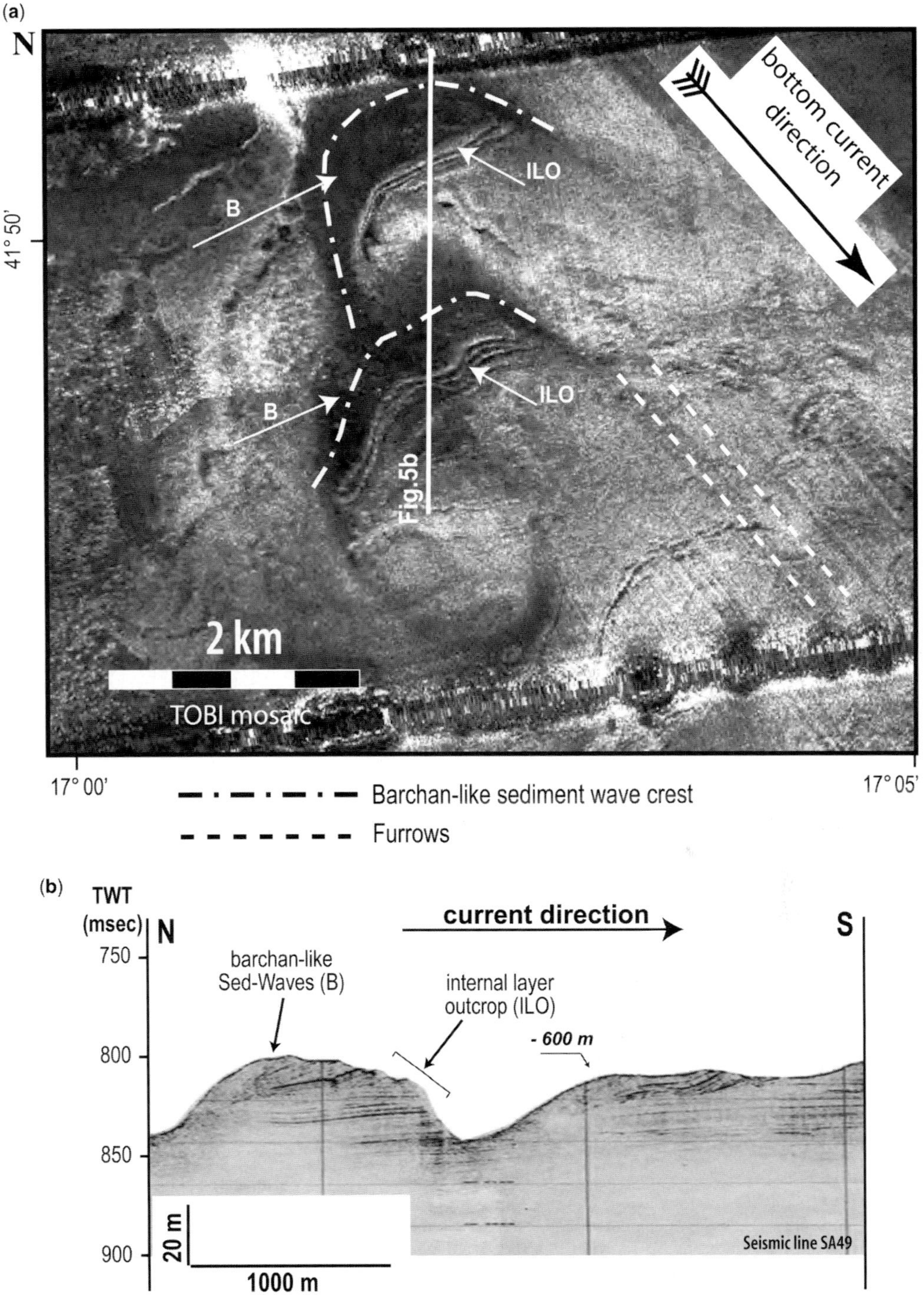

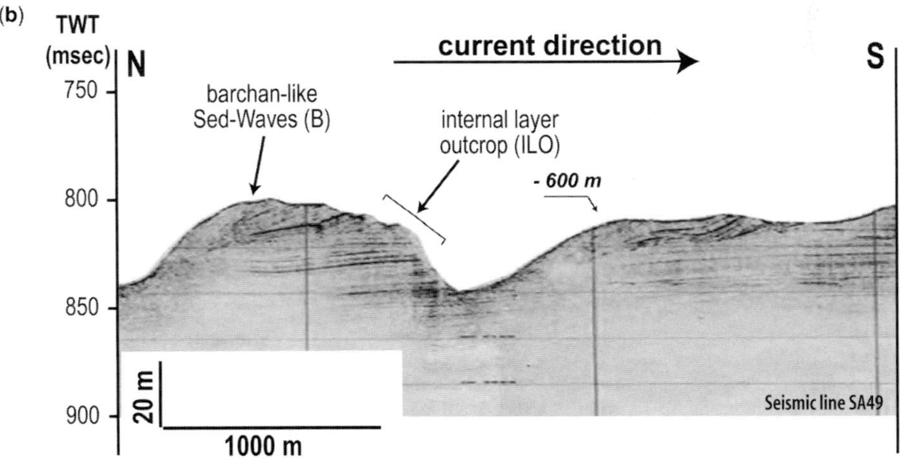

Fig. 5. TOBI sidescan sonar image (**a**) and Chirp-sonar profile (**b**) of large, isolated barchan-like sediment waves adjacent to an area of erosional furrows in the northern part of the SAM. Backscatter striations parallel to the wave crest are the expression of internal layer outcrop (ILO) on the erosional remnants below the sediment waves.

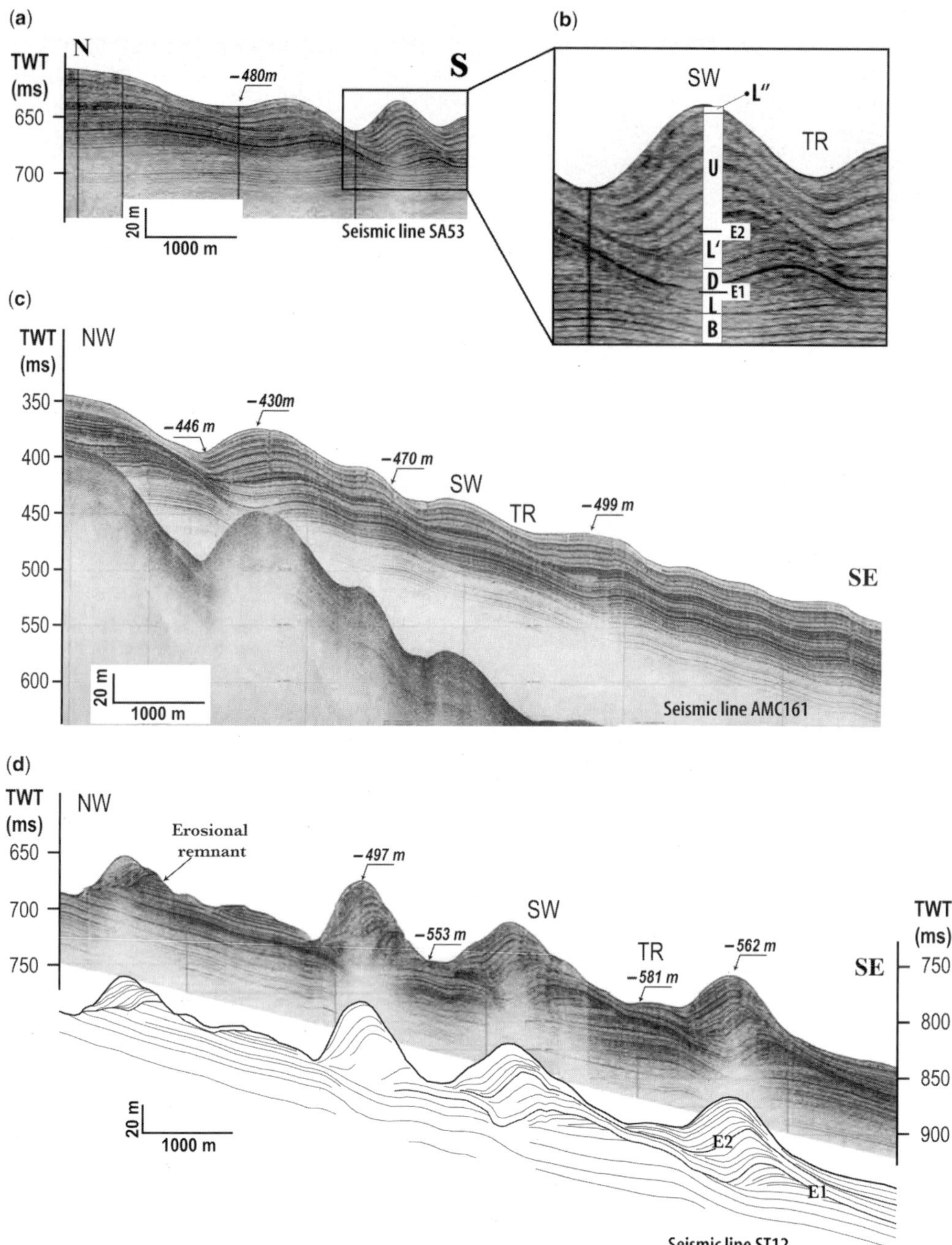

Fig. 6. Chirp-sonar profiles across the large sediment waves in the northern sector of the SAM. Classification of seismic units illustrated in the text is shown in (**b**). (For location see Fig. 2.) SW, Sediment wave deposit; TR, sediment wave trough.

accumulation of fine-grained sediment interpreted as current deposits.

The southern slope area

In the southern sector of the SAM, between the large blocky slide and the Bari Canyon area, the slope displays a marked change in gradient: the upper slope (between 200 and 350 m) is steeper (6°) and the lower slope is gentler. Widespread erosional areas characterize the upper slope. Extensive mass-wasting deposits with irregular tops occur at the sea floor or are buried and are common on the lower slope. The lower slope is also dissected by several deep incisions perpendicular or slightly oblique to the regional slope. The resulting slope morphology is very complex, with numerous mounded deposits having a patchy distribution along the slope. In cross-section, these deposits show a well-layered acoustic sequence with divergent or convergent internal reflectors and occasional evidence of subtle and localized truncations. Two main types of sedimentary bodies are identified: plastered drifts, where the slope is smoother, and isolated mounds, where the slope has irregularities (Fig. 7). The plastered drifts are up to 15–20 m in thickness and are similar in shape to the larger plastered drifts described by Wood & Davy (1994) on the eastern New Zealand margin. The mounds reach more than 40 m in thickness, are typically isolated, and have a large variety of morphology and orientations. Both types of deposits display a slight component of up-slope (northwestward) migration.

A long piston core (SA03-C01) collected from the sediment drift (Fig. 7a) recovered 14 m of bioturbated muddy silt.

The basin-floor area

An elongated drift occurs at the base of the steep flank of the Dauno Seamount, in a depth range of 1000–1100 m, and displays an asymmetric mounded profile. Chirp profiles and TOBI mosaics (Fig. 8) allow us to trace the drift axis along the base of the Dauno Seamount for about 20 km. The drift has a width of more than 3 km and its relief above the sea floor is rather subdued (a few metres) on its northern end but appears more abrupt on its southward limb, plunging into a well-defined erosional moat (>20 m deep). On Chirp profiles (Fig. 8a and b) it is possible to observe a basal unit characterized by a continuous parallel, slightly undulating locally, reflector, and an upper unit with a layered structure in which the reflectors range from parallel to convergent or divergent and result in an overall stratigraphic expansion defining the mound crest and pinching-out towards the moat.

Between the two units described above, there is a lenticular and acoustically transparent layer with erosional base and irregular top. This layer is interpreted as a mud-flow deposit, in which the lack of internal stratification indicates mechanical remoulding and the good penetration of seismic signal is consistent with a dominant fine-grained lithology (Trincardi *et al.* 2004).

On seismic profiles, the character of the erosional moat changes, commonly over a short distance, from a deep and well-defined depression to a broader and shallower depression. Within the moat, TOBI mosaics suggest the occasional presence of furrows aligned parallel to the moat axis (Fig. 8c). We interpret this pattern, together with the convergence of downlapping deposits toward the moat axis, as a result of an intensified bottom current preventing deposition or favouring sea-floor erosion and sediment winnowing. Both seismic profile and backscatter pattern on TOBI images indicate that fine-grained sediments are removed from the axis of the moat. The distribution of bottom-current features along the SAM is summarized in Figure 9.

Sediments

The sediment cores taken in the upper 15 m of the sediment drifts and sediment waves of the SAM encountered essentially muddy successions; this evidence, together with the great penetration of high-frequency acoustic signal, is consistent with an overall dominance of muddy sediments not only at the sea floor but also at deeper stratigraphic levels. The most common lithology recovered from the drifts is Foraminifera-rich, intensely bioturbated mud or silt-rich mud (Fig. 10). Primary sedimentary structures related to traction processes have not been observed. Thin coarser-grained beds are rare and mainly bioclastic (Foraminifera and mollusc fragments) in the coarse fraction (up to fine sand). Several centimetre-scale tephra layers (under study) are commonly graded; additional micro-tephra can be detected by microscope only. The study of these tephra layers will further constrain core correlation and age assignment.

Stratigraphic correlation

The planktonic Foraminifera ecobiostratigraphy adopted in this study refers to the published literature on the Adriatic for the last 15 ka interval (Jorissen *et al.* 1993; Rohling *et al.* 1997; Asioli *et al.* 1999, 2001; Capotondi *et al.* 1999; Hayes *et al.* 1999; Ariztegui *et al.* 2000). For the interval between 15 and 125 ka the Foraminifera stratigraphy published on the area is poor

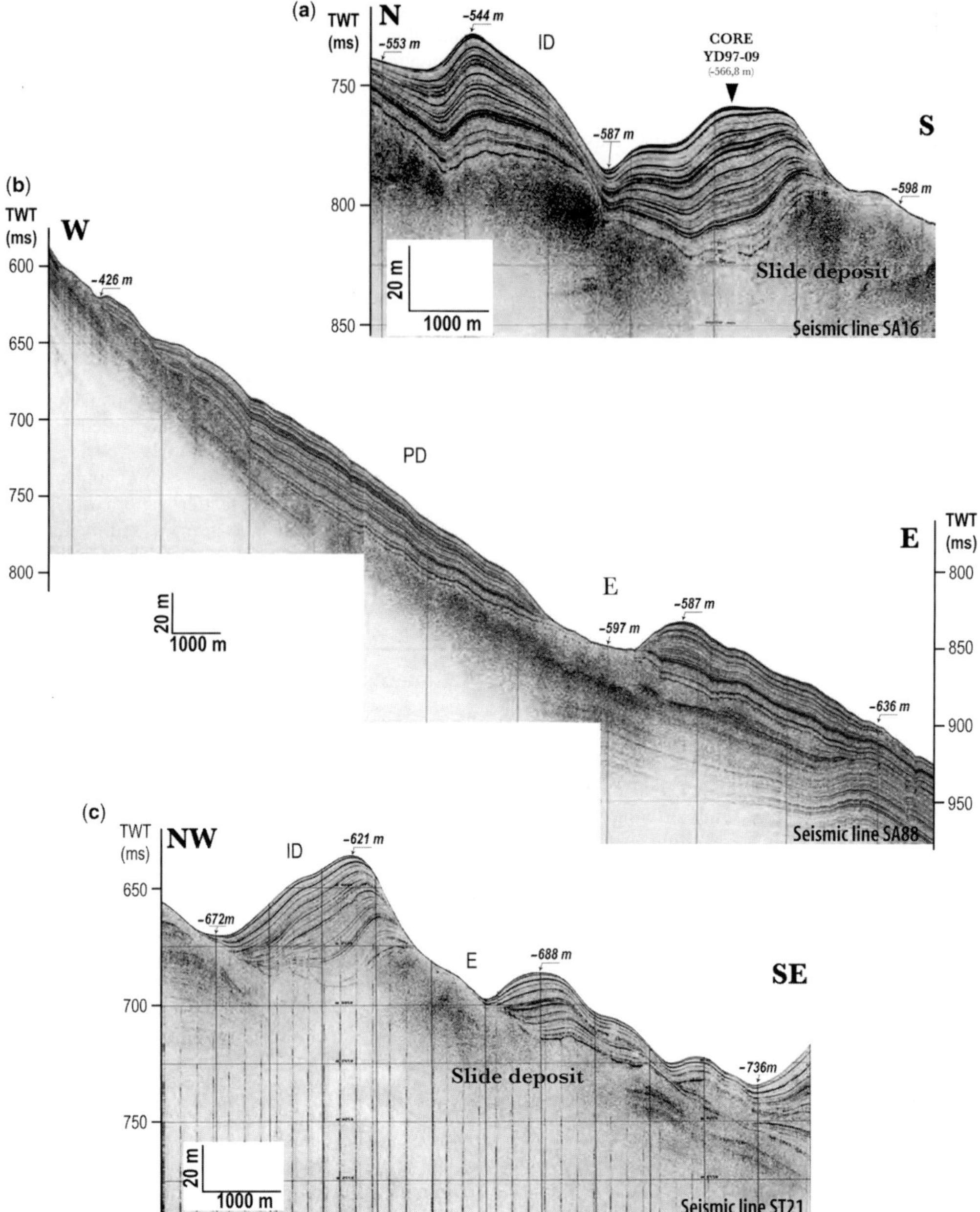

Fig. 7. Chirp-sonar profiles showing different types of bottom-current deposits in the southern sector of the SAM. (For location see Fig. 2.) ID, Isolated drift; PD, plastered drift; E, erosional area.

(Borsetti *et al.* 1995; Asioli 1996). For this interval, the most detailed stratigraphic information comes from core RF93-77, from the Mid Adriatic Deep (MAD; Figs 1 and 11). This core is constrained by oxygen stable isotope stratigraphy, accelerator mass spectrometry ^{14}C datings, tephrochronology, pollen record, and planktic and benthic Foraminifera assemblages in a well-defined seismic stratigraphic context (Ariztegui *et al.* 1996; Asioli 1996; Trincardi *et al.* 1996; Calanchi *et al.* 1998).

The stratigraphic analysis of each core resulting in a sequence of ecozones is illustrated in Figure 11 together with a synthesis of the ecozone

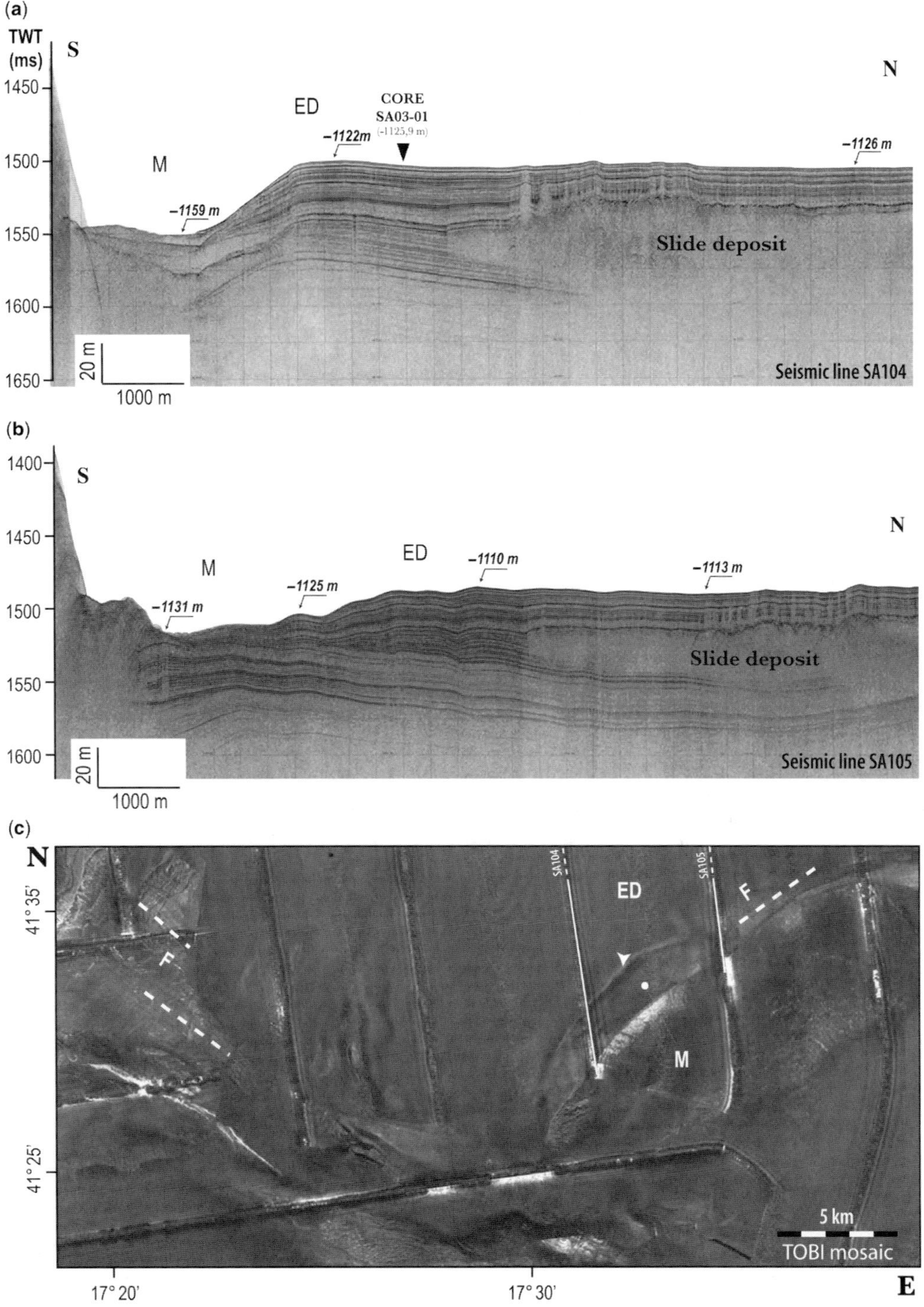

Fig. 8. Elongated drift and broad erosional moat north of the steep flank of the Dauno Seamount. (**a**) and (**b**) north–south Chirp-sonar profiles showing eastward (downcurrent) decrease in drift relief and increase in moat width. (**c**) TOBI mosaic showing the spatial relation between drift crest and moat. Furrows are locally superimposed within the moat. ED, Elongated drift; M, moat; F, furrows.

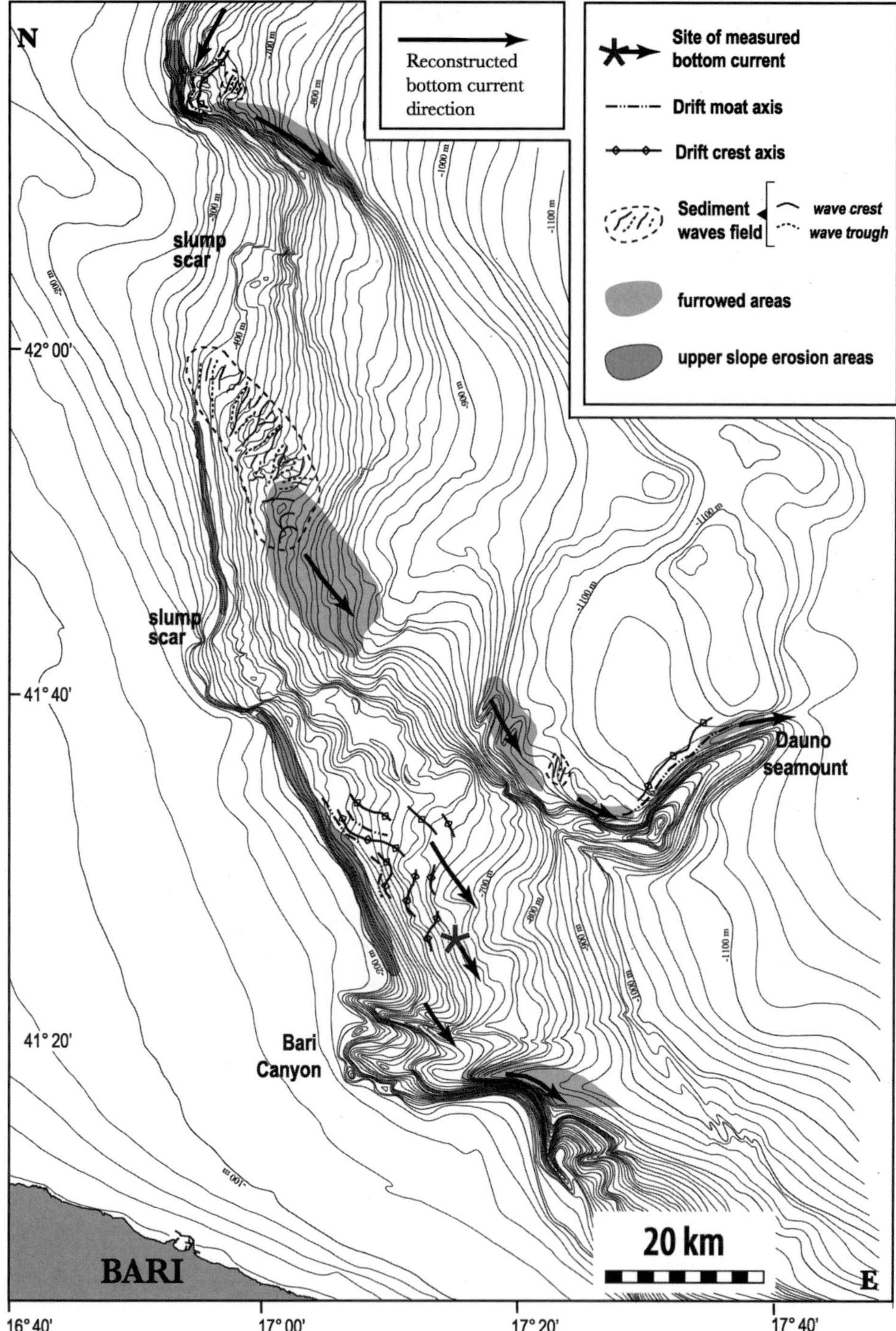

Fig. 9. Distribution of the main morphological and depositional elements in the SAM, based on TOBI mosaics and Chirp-sonar profiles. Arrows represent the bottom-current direction based on oceanographic data and on furrow alignment (see text for details).

Fig. 10. Photograph of core YD97-09. Core location is shown in Figure 2.

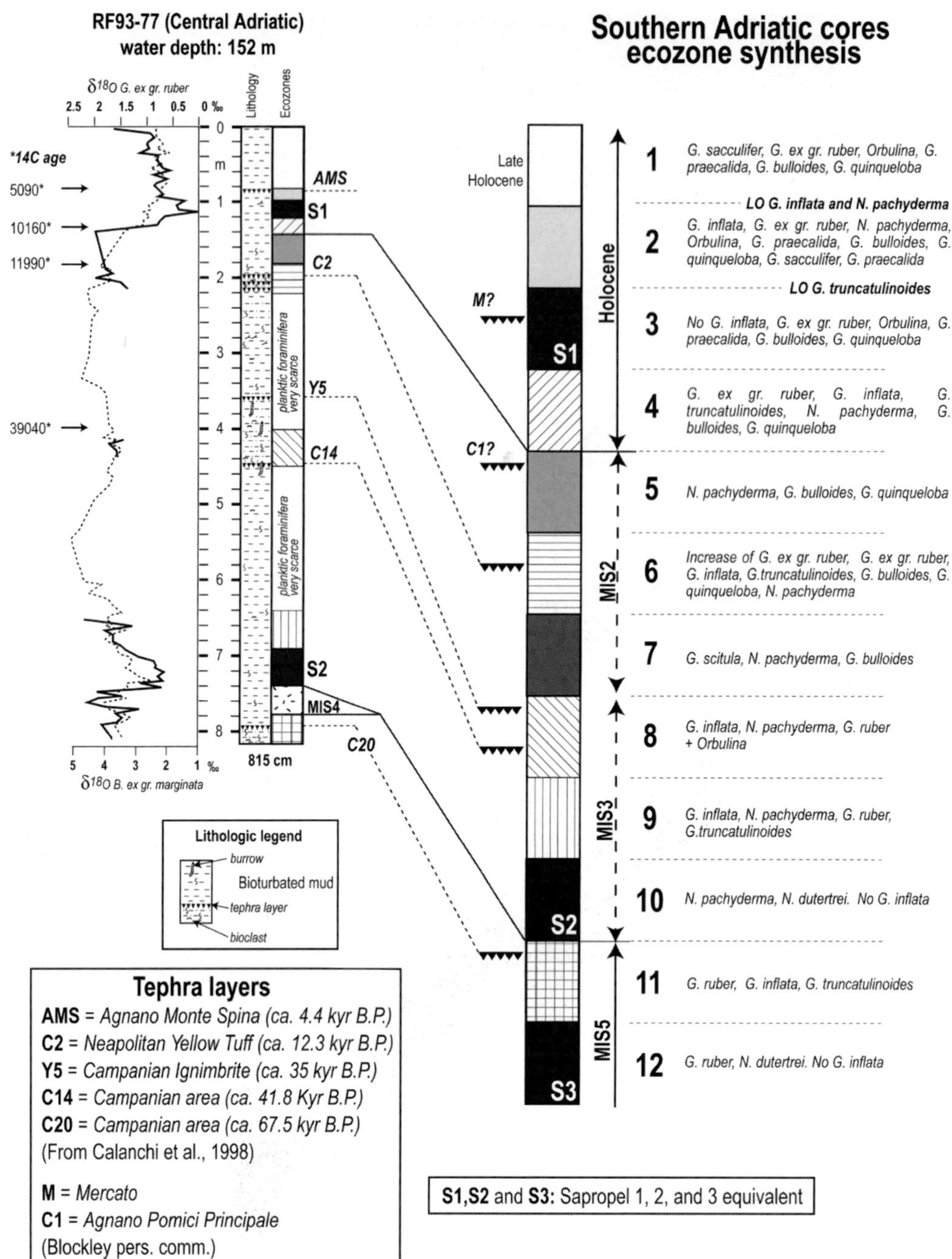

Fig. 11. On the left is the ecobiostratigraphy from reference core RF93-77 (see Fig. 1 for location) plotted against the lithology, the identified tephra layers and the $\delta^{18}O$ curve on the planktonic Foraminifera *Globigerinoides* ex gr. *ruber* (continuous line) and the benthic Foraminifera *Bulimina* ex gr. *marginata* (dashed line). On the right is a synthesis of the ecozones recognized in the South Adriatic cores discussed in this paper. The position of some key tephra layers is proposed based on their possible correlation to the RF93-77 ash layers (dashed lines); two additional tephras (Mercato and Pomice Principale) were recognized by S. Blockley (pers. comm.) and appear consistent with the age of the ecobiozone in which they occur.

sequence for the Southern Adriatic with reference to core RF93-77. During interglacial periods, the entry of intermediate (*Globorotalia inflata, Neogloboquadrina pachyderma*) and, occasionally, deep-water dwellers (*Globorotalia truncatulinoides*) in the MAD allows precise ecobiostratigraphic correlation to the study area. During glacial sea-level lowstands the connection between the two areas was restricted to the Pelagosa sill (Fig. 1), and planktic Foraminifera in the MAD were very scarce and dominated by surface dwellers, such as *Globigerina quinqueloba*. In the southern basin (>1200 m deep) deep dwellers, such as *Globorotalia scitula*, were instead abundant during sea-level lowstands. In summary, the planktonic Foraminifera ecozones defined for the central Adriatic for the last 15 ka (Asioli *et al.* 1999, 2001) correlate with the southern Adriatic, whereas the ecozones older than 15 ka defined for the southern Adriatic show a good similarity to the ecozones described in reference core RF93-77, except for glacial maxima.

On the basis of the planktic Foraminifera assemblages, the studied cores reached Sapropel 3 deposited during Marine Oxygen Isotope Stage (MIS) 5a and consequently span the last 75 ka (Emeis *et al.* 1998). No planktic Foraminifera assemblage that could be referred to MIS 4 was found in core SA03-3, indicating nondeposition or an erosional hiatus (Fig. 11). On the basis of literature no clear bioevent or ecozone defines the boundary between MIS 2 and MIS 3, and it is therefore represented as a dashed line in Figure 11.

Stratigraphic correlation between three long piston-cores (YD97-09; SA03-04; SA03-01) collected along the SAM slope indicates a substantial increase in sedimentation rates from north to south since the Last Glacial Maximum (LGM) (Fig. 12). Beside the study of micro-Foraminifera assemblages this correlation relies on whole-core

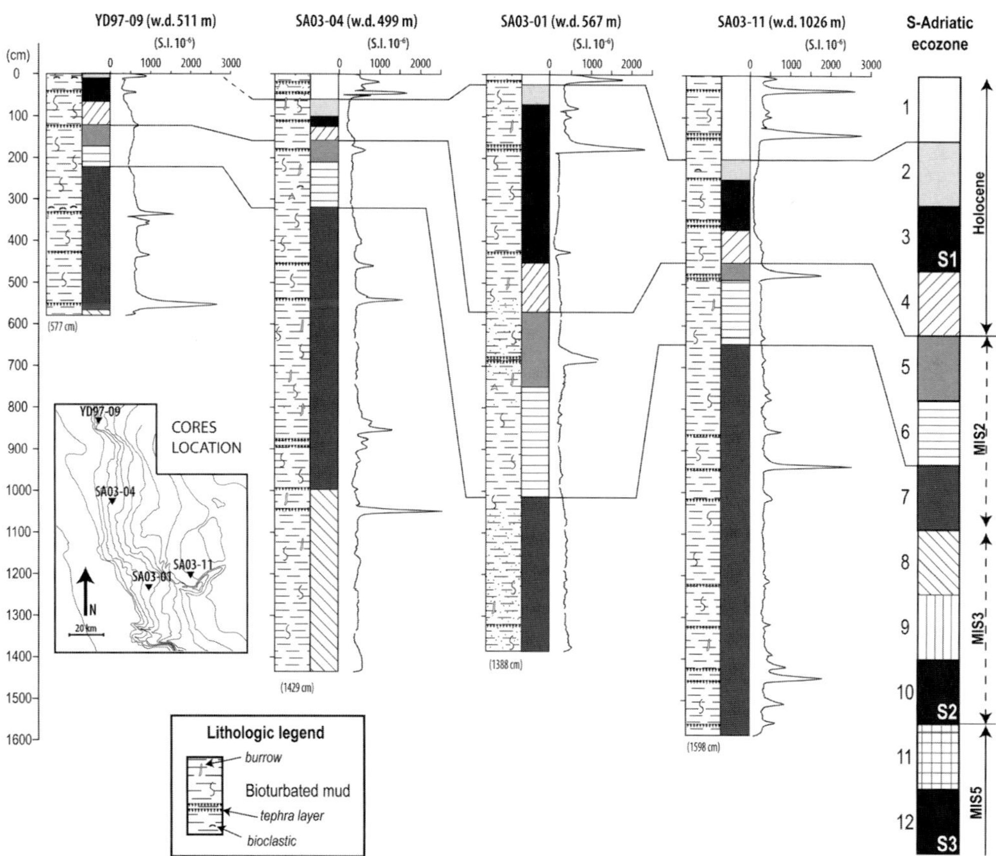

Fig. 12. Stratigraphic correlation between three piston cores (YD97-09; SA03-04; SA03-01) collected on the slope and a basinal core (SA03-11). The stratigraphic expansion of the late glacial–interglacial deposits toward the south should be noted, w.d., water depth.

magnetic susceptibility curves. Peaks in magnetic susceptibility commonly indicate the presence of tephra layers. Ages are ascribed to biozone boundaries based on the available literature and should be taken as indicative values.

The larger variability in sediment accumulation rates occurs during the last glacial–interglacial transition: in the northern area after the late glacial period sedimentation rates are of the order of 25 cm ka^{-1}, whereas in the southern area sediment accumulation rates are up to about 185 cm ka^{-1}. The southward increase in sediment accumulation rates during the LGM reflects at least two concurrent factors: (1) a higher current intensity in the north, and/or (2) an increased off-shelf sediment transport proceeding to the south. Along the SAM slope, deposition seems to be reduced during the last 6 ka, with sedimentation rates varying from 2 cm ka^{-1} (in core YD97-09) to 10 cm ka^{-1} (in core SA03-04). In the deep basin (core SA03-11), the difference between the sedimentation rates during the last glacial–interglacial phase and those of the last 6 ka is less dramatic, with approximate sedimentation rates of about 90 cm ka^{-1} and 40 cm ka^{-1} respectively. This evidence suggests bypass of sediment across the slope.

At a more local scale, a core-transect on the sediment wave field (Fig. 13) denotes stratigraphic expansion on the wave crests and condensation on the downcurrent flanks and in the troughs of the waves. Core SA03-03, 10.5 m long, on the erosional–condensed downslope limb, has no deposit younger than the LGM. Nevertheless, this core reaches the deepest stratigraphic unit of the dataset: the sapropel layer S3 within the MIS 5a interval.

These marked changes in thickness of the cored deposits are consistent with geometric observations from seismic reflection profiles and reinforce our interpretation of these features as actively growing bottom-current deposits.

Discussion

The SAM offers a good example of interaction between bottom currents and slope configuration in controlling distribution and morphology of bottom-current deposits. Seismo-stratigraphic data allow us to discuss possible mechanisms of formation of bottom-current deposits and define the impact of sea-level and climatic change.

Bottom-current deposits and margin morphology

Size, morphology and distribution of bottom-current deposits along the SAM vary as a function of the interaction between bottom currents, regional slope orientation and local sea-floor topography. The widespread occurrence of furrows along the SAM may have resulted from a dominantly along-slope current flow, similar to the helicoidal flows in the benthic boundary layer described by Hollister *et al.* (1976) and Viekman *et al.* (1989). Furrows are, by definition, parallel to the bottom currents. Several studies have described these features developing in cohesive sediment (Flood 1981, 1982) in areas where currents reach velocities higher than 30 cm s^{-1} (Hollister *et al.* 1974; Flood 1983, 1994). Fields of furrows are extensive and common on the SAM, where oceanographic data indicate south-flowing currents with velocities exceeding 60 cm s^{-1} (S. Miserocchi, pers. comm.). These furrows can therefore be used as indicators of the modern bottom-current flow pattern. It is worth noting that in the SAM furrows are common also in areas characterized by the presence of sediment waves, as in the Bahama Outer Ridge and described by Tucholke (1979) and Flood & Giosan (2002). In the case of the SAM, erosion takes place mainly in the troughs between sediment waves and/or on the downcurrent flank of the waves. Moreover, some of the furrowed areas (see Fig. 9 for the location) are upcurrent of major sea-floor morphological barriers that are likely to be able to alter bottom flows. The map in Figure 9 shows a reconstruction of the bottom-current pathways by combining oceanographic data with orientation of the furrows.

The role of morphological barriers and regional topography. Morphological barriers play an important role in constraining and possibly accelerating bottom flows and, consequently, in redistributing sediments along the margin. Where constrained, the flow intensifies, leading to the development of an erosional moat. Away from the moat axis, flow velocities decrease, favouring preferential deposition and sediment drift construction (Faugères *et al.* 1999). Morphological barriers were recognized in several oceanic contexts as influencing bottom-current flows and being conducive to sediment drift development. As an example, the Wyville Thomson Ridge deflects to the west the south-flowing Norwegian Sea Overflow Water (Kuijpers *et al.* 1998), and the topographic relief of the North Scotia Ridge influences the flow of the Circumpolar Deep Water across the western Falkland Trough (Howe *et al.* 1997). Within the Mediterranean basin, other occurrences of sediment drifts have been attributed to the presence of sea-floor sills or slope bulges confining bottom-current flows (Millot & Monaco 1984; Marani *et al.* 1993; Roveri 2002).

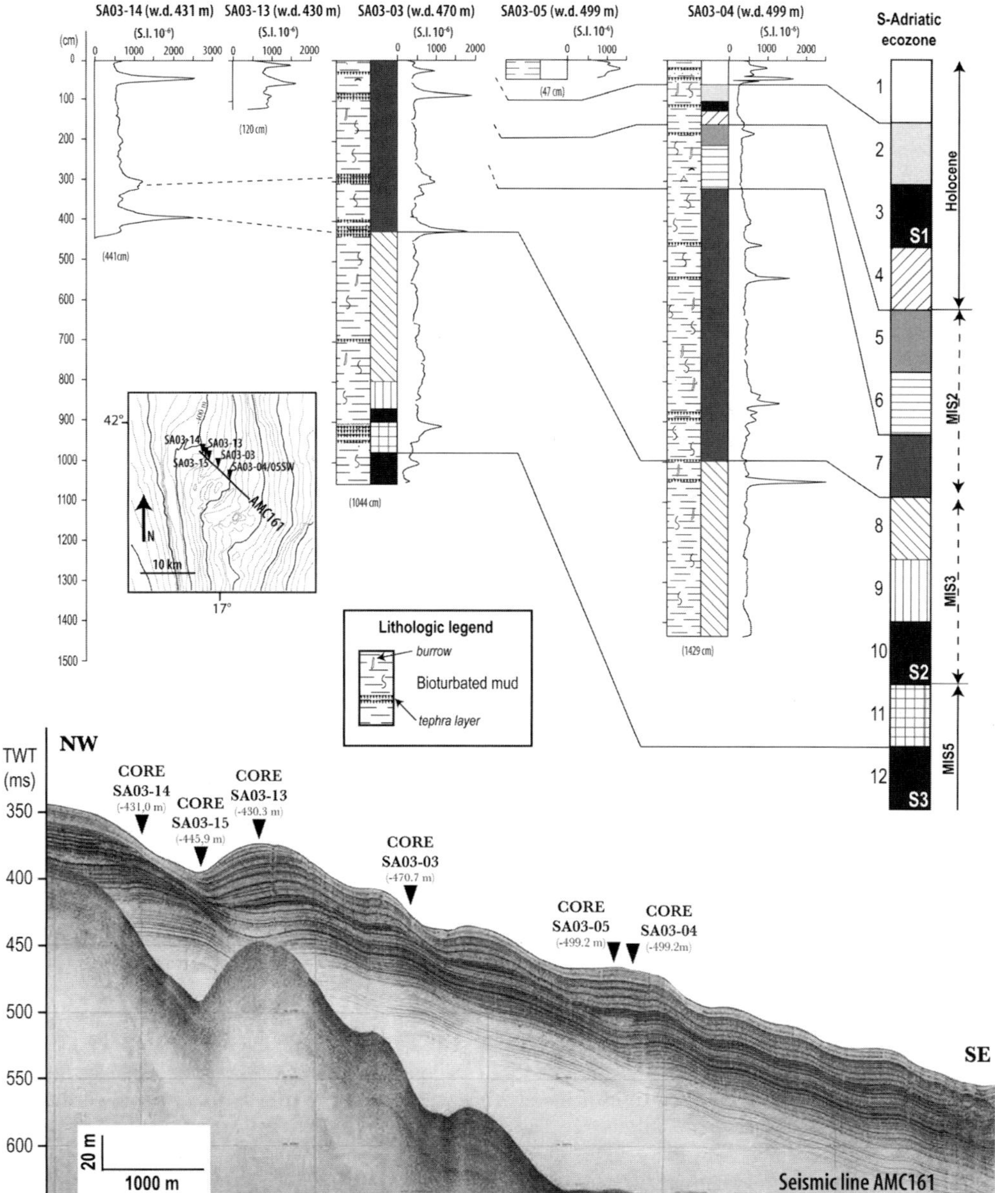

Fig. 13. Core transect along the sediment wave field in the northern part of the SAM. (Note the significant variation in sediment thickness from bedform crests to the troughs.)

In the SAM, structural control results in the presence of steep slope areas affecting the southward-flowing current, such as (1) the NW–SE slope in the northernmost sector of the study area, (2) the southern flank of the east–west-trending Bari Canyon, and (3) the east–west elongated Dauno Seamount, extending to the basin area. In all these cases, extensive fields of furrows develop in the area where current velocity is likely to be the greatest. Down-flow of these large-scale obstacles, the furrows appear to diverge and gradually disappear, indicating deceleration of flows and consequent sediment deposition. Where the slope changes direction, bottom currents are constrained and erode the upper slope. In many areas a well-developed simple moat separates an elongated mounded drift from the slope. This is particularly marked on seismic profiles in the northernmost sector of the SAM (Fig. 3) and on the TOBI mosaic on the northern flank of the Dauno Seamount, where NW–SE oriented furrows within the moat indicate that currents impinge on the basin floor (Fig. 8c).

The role of slope gradient reductions and small-scale topography. The SAM is an area of varied morphology with slope gradient reductions and different types of topographic obstacles such as erosional remnants and complex slide topography (e.g. slide blocks on the sea floor; Fig. 4). This complex morphology interacting with the south-flowing bottom-water masses promotes the development of varied drift types (separated drift, plastered drift) and sediment waves. In addition, several small-scale patched drifts occur at various locations on the margin, commonly separated from broad erosional areas.

Where the slope is smooth and gradient change is abrupt, separated drifts develop along the side of an erosional moat, as in the northernmost part of the SAM at 500 m depth (Fig. 3). As in oceanic contexts, this type of separated drift indicates a lateral gradient of decreasing velocity away from the core of the current located in the moat (e.g. Stoker *et al.* 1998).

In the southern part of the study area, slope gradients decrease more gradually and the lower slope is characterized by irregular topography. The coexistence of plastered drifts, patched mound drifts and widespread erosional areas hints at a more complex interaction between current regime and sea-floor morphology. These features may form in response to a splitting of the current flow into several strands, perhaps influenced both by sea-floor irregularities and by developing drift topography.

Fine-grained deep-water sediment waves, described in many basinal contexts, form fields of hundreds to thousands of square kilometres and have wavelengths and wave heights ranging from 1 to 10 km and from 15 to 50 m, respectively (Wynn & Stow 2002). These bedforms can be produced by either turbidity currents (e.g. Migeon *et al.* 2001; Normark *et al.* 2002) or contour currents (e.g. Masson *et al.* 2002; Habgood *et al.* 2003). If not from the regional setting in which they occur, these two mechanisms are difficult to distinguish, although bottom-current sediment waves are usually characterized by greater wave crest regularity (Howe 1996). Bottom-current sediment waves are often associated with large contourite sedimentary bodies. Their crests are usually straight or slightly sinuous, and appear oblique (typically 10–50°) to the main bottom flow (Wynn & Stow 2002). Assuming that the mean bottom-current flow is parallel to the slope, most of these oblique waves are seen to migrate upcurrent and therefore upslope (e.g. Manley & Flood 1993; Masson *et al.* 2002). To explain the migration of fine-grained bottom-current sediment waves over a pre-existing wavy topography, Flood (1988) proposed the lee-wave model (refined by Hopfauf *et al.* 2001). This model also predicts that wave migration occurs under flow velocities ranging between 9 and 30 cm s^{-1}, and that if the flow velocity is greater than about 17 cm s^{-1} deposition occurs only on the upstream flank. No model however, has comprehensively explained the processes by which sediment waves initiate (von Lom-Keil *et al.* 2002), although some pre-existing sea-floor irregularity seems to be required (Blumsack & Weatherly 1989; Howe *et al.* 1998; Cattaneo *et al.* 2004). On the SAM a downcurrent change in orientation of the wave crests is likely to reflect local adjustments to the irregular sea-floor morphology and/or to an increase in bottom-current intensity. A southward increase in bottom-current intensity is suggested by a gradual change in wave morphology, from continuous straight-crested to isolated barchan-like (Figs 4 and 5). Previous studies have revealed that submarine barchans form under a broad range of unidirectional bottom currents (Kenyon & Belderson 1973; Lonsdale & Malfait 1974; Kenyon 1986). The scheme in Figure 14 shows a relation between the slope morphology and scale and the form of the bottom-current deposits.

Bottom-current features v. regional circulation pattern

The distribution of bottom-current deposits along the SAM indicates that the present deep-water circulation has been active for prolonged geological times. Bottom currents in water depths shallower than 600 m reflect an interaction between the south-flowing LIW and the NAdDW coming from the north and cascading from the shallower water depths. Depositional and erosional features in water depths greater than 600 m reflect, instead, only downslope flows of the NAdDW (Fig. 1).

Cores collected on the SAM suggest that sediment drifts began to grow before MIS 2. Cores collected in the sediment wave field, in the northern slope area, show that these bedforms have actively grown and migrated, at least since MIS 5. Therefore, the growth of depositional features was particularly active during the last glacial–interglacial transition.

Palaeoceanographic records suggest that, as in other oceanic basins, Mediterranean circulation during glacial times was different from that today.

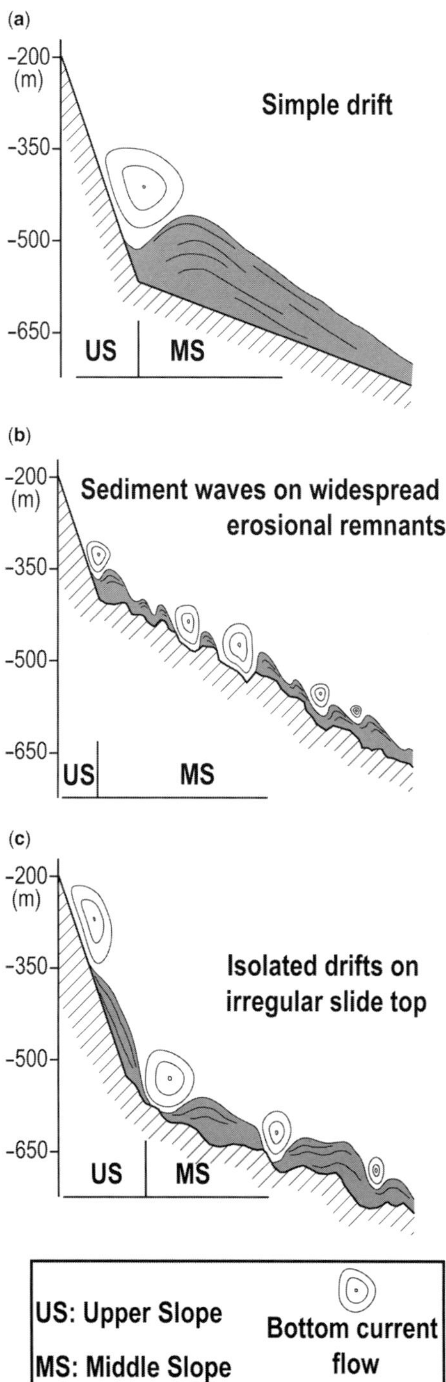

Fig. 14. Schematic representation of the relation between bottom currents, change in slope gradient and sea-floor roughness owing to pre-existing erosional surfaces or mass transport deposits. Current flow is seaward (right) and oblique to the three sections.

During the LGM (MIS 2), *c.* 18 ka ago, global sea level was *c.* 125 m lower than today (Fairbanks 1989). Like other land-locked basins, the Adriatic underwent dramatic palaeogeographical and palaeoceanographic rearrangements during Quaternary climatic change and related variations of sea level.

The northern shallow Adriatic, where the NAdDW forms today, was subaerially exposed during the LGM and consequently NAdDW was strongly reduced. Concurrently, in the Levantine basin, higher LGM salinities allowed wintertime cooling to increase the density of surface water sufficiently for it to form bottom water, called Levantine Deep Water (LDW; Myers *et al.* 1998). In this configuration the circulation in the Adriatic during the LGM was completely reversed from the present pattern. Salty and dense LDW flowed up over the Otranto Strait, filling the deepest part of the Adriatic basin and stabilizing water mass stratification (Myers *et al.* 1998). The combined reduction of NAdDW and LIW formation appears consistent with a phase of more drape-like deposition in the sediment wave area and more uniform thickness of LGM deposits within the cores.

However, in the sediment wave field, seismic profiles (Fig. 6) show that cyclic variations in growth and in migration are present. Figure 15 schematically represents the relation between the sediment wave growth and the current velocity change at constant sediment supply. Increase in bottom currents causes the formation of small sediment waves with a slightly upslope migration (Fig. 15a). Progressive increase in bottom-current energy leads to the formation of a widespread erosional surface (Fig. 15b); as bottom-current energy decreases, a drape is formed above the surface (Fig. 15c). A new increase in bottom-current energy triggers a new phase of upslope migration of sediment waves, culminating in a new phase of erosion (Fig. 15d). Core SA03-03 indicates that this erosional phase, marked by truncation of high-amplitude reflectors, corresponds to the upper part of MIS 5, the last interglacial period. Above this surface, the upslope migration of the sediment waves is less pronounced. The enhanced thickness of the surface layer on the upslope side of the waves (Fig. 15e) indicates an increase of bottom-current flows and that the sediment waves are actively migrating at present. Moreover, the presence of furrows close to and superimposed on the sediment waves confirms the impact of fast-flowing bottom currents on the sea floor. Evidence of high-energy bottom currents during the present-day interglacial condition and during the last interglacial period suggests that bottom currents reach maximum energy during sea-level highstands.

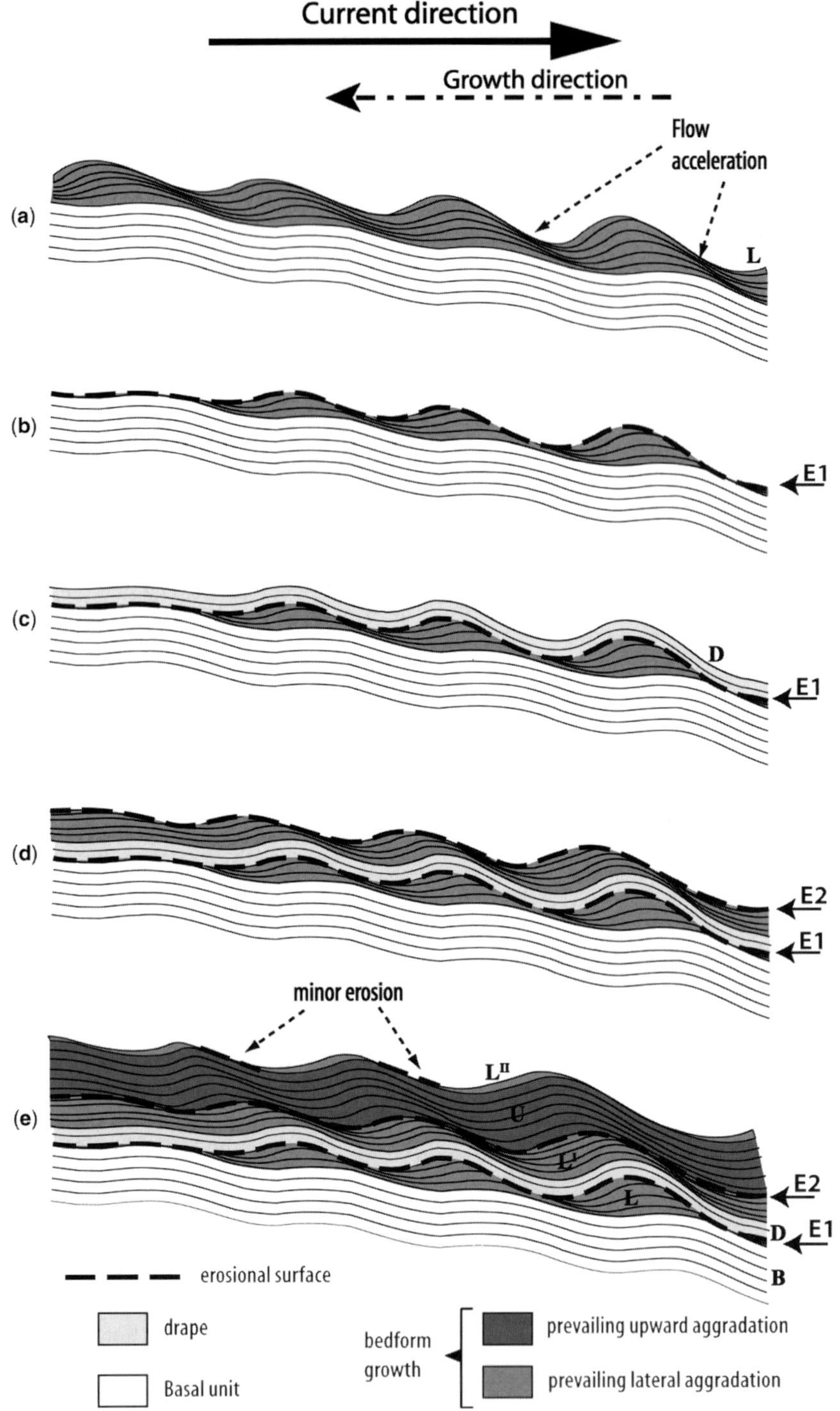

Fig. 15. Schematic representation of the growth and upslope migration of the sediment waves on the SAM. Units L, L′ and L″ denote phases of most persistent lateral aggradation. U, instead, is characterized by significant upward aggradation. E1 and E2 are the main erosional surfaces separating lateral aggradational units (below) from draped or upward aggradational units (above). Minor erosional surfaces are also present, locally, in the troughs.

Conclusion

The Southwestern Adriatic Margin (SAM) is characterized by a complex bottom-circulation pattern related to the interaction between two distinct, southward-flowing bottom water masses: the cold and deep NAdDW, forming in the northern Adriatic, and the highly saline LIW, generated in the Eastern Mediterranean. Along the SAM slope, bottom-current sedimentation is dominant, leading to the development of sediment drifts (elongated drifts, plastered drifts and isolated drifts) and sediment waves. Non-depositional and erosional features related to bottom-current activity were also identified and include moats between drifts and the steep upper slope, areas of extensive erosion on the upper slope, and extended furrowed areas.

The interaction between bottom-current circulation and complex margin morphology controls the distribution, geometry and scale of these deposits: (1) elongate-plastered drifts occur where the slope is smooth and the change in gradient is abrupt; (2) plastered or isolated drifts develop where slope topography is irregular and the slope gradient changes more gradually; (3) sediment waves occur on gentler slopes characterized by irregular topography. Moreover, erosional areas, moats and furrows are mainly located upcurrent of major sea-floor morphological barriers to bottom-current flows.

Sediment cores that collected bottom-current deposits along the SAM slope reveal an overall southward increase in sediment accumulation rates during the last glacial–interglacial transition. This evidence seems to reflect at least two possibly related factors: (1) higher current intensity in the north preventing deposition, and/or (2) increased off-shelf sediment transport proceeding to the south.

The combination of data from sediment cores and Chirp-sonar profiles across the large sediment wave field in the northern part of the SAM shows the relationship between Late Quaternary climatic variations, sea-level change and bottom-current sedimentation. The observations reveal that during glacial periods (and sea-level lowstand) bottom currents were less intense than during interglacial times.

This study was supported by the European project EURO-STRATAFORM (EC contract EVK3-CT-2002-00079) and the European Access to Seafloor Survey Systems 'EASSS III-TOBI sidescan sonar' (HPRICT199900047). We thank the captains and crews of CNR R.V. *Urania*, and all participants of campaigns SAGA03 and STRATA04. The TOBI team at NOC, Southampton, is also acknowledged for its invaluable work on cruise SAGA03. Special thanks go to T. Le Bas for helping and guiding the senior author in the processing of TOBI data, and to V. Ferrante and D. Minisini for contributing to the TOBI data processing. A. Akhmetzhanov and A. Camerlenghi helped improve the text and gave valuable advice on figure quality. This is ISMAR-Bologna (CNR) contribution 1456.

References

ARGNANI, A., ARTONI, A., ORI, G. G. & ROVERI, M. 1991. L'avanfossa centro-Adriatica: stili strutturali e sedimentazione. *Studi Geologici Camerti*, **1**, 371–381.

ARIZTEGUI, D., CHONDROGIANNI, C., WOLFF, G., ASIOLI, A., TERANES, J., BERNASCONI, S. M. & MCKENZIE, J. A. 1996. Paleotemperature and paleosalinity history of the Meso Adriatic depression (MAD) during the Late Quaternary: a stable isotopes and alkenones study. *Memorie dell'Istituto Italiano di Idrobiologia*, **55**, 219–230.

ARIZTEGUI, D., ASIOLI, A., LOWE, J. J., *ET AL.* 2000. Palaeoclimate and the formation of sapropel S1; inferences from late Quaternary lacustrine and marine sequences in the central Mediterranean region. *Palaeogeography, Palaeoclimatology, Palaeoecology*, **158**, 215–240.

ARTEGIANI, A., AZZOLINI, R. & SALUSTI, E. 1989. On dense water in the Adriatic Sea. *Oceanologica Acta*, **12**, 151–160.

ARTEGIANI, A., BREGANT, D., PASCHINI, D., PINARDI, E., RAICICH, N. & RUSSO, F. 1997. The Adriatic General Circulation. Part I. Air–sea interaction and water mass structure. *Journal of Physical Oceanography*, **27**, 1492–1514.

ASIOLI, A. 1996. High resolution Foraminifera biostratigraphy in the central Adriatic basin during the last deglaciation: a contribution to the PALICLAS project. *Memorie dell'Istituto Italiano d'Idrobiologia*, **55**, 197–217.

ASIOLI, A., TRINCARDI, F., LOWE, J. J. & OLDFIELD, F. 1999. Short-term climate change during the last Glacial–Holocene transition: comparison between Mediterranean records and GRIP event stratigraphy. *Journal of Quaternary Science*, **14**, 373–381.

ASIOLI, A., TRINCARDI, F., LOWE, J. J., ARIZTEGUI, D., LANGONE, L. & OLDFIELD, F. 2001. Sub-millennial climatic oscillations in the Central Adriatic during the last deglaciation: paleoceanographic implications. *Quaternary Science Reviews*, **20**, 33–53.

BERTOTTI, G., CASOLARI, E. & PICOTTI, V. 1999. The Gargano Promontory: a Neogene contractional belt within the Adriatic plate. *Terra Nova*, **11**, 168–173.

BIGNAMI, F., SALUSTI, E. & SCHIARINI, S. 1990. Observations on a bottom vein of dense water in the Southern Adriatic and Ionian Seas. *Journal of Geophysical Research*, **95**, 7249–7259.

BLUMSACK, S. L. & WEATHERLY, G. L. 1989. Observations and growth mechanisms for mudwaves. *Deep-Sea Research*, **36**, 1327–1339.

BORSETTI, A. M., CAPOTONDI, L., CATI, F., ET AL. 1995. Biostratigraphic events and Late Quaternary tectonics in the Dosso Gallignani (central–southern Adriatic Sea). *Giornale di Geologia*, **57**, 41–58.

CALANCHI, N., CATTANEO, A., DINELLI, E., GASPAROTTO, G. & LUCCHINI, F. 1998. Tephra layer in Late Quaternary sediments of the Central Adriatic Sea. *Marine Geology*, **149**, 191–209.

CANALS, M., LASTRAS, G., URGELES, R., ET AL. 2004. Slope failure dynamics and impacts from seafloor and shallow sub-seafloor geophysical data: case studies from the COSTA project. *Marine Geology*, **213**, 9–72.

CAPOTONDI, L., BORSETTI, A. M. & MORIGI, C. 1999. Foraminiferal ecozone, a high resolution proxy for the late Quaternary biochronology in the central Mediterranean Sea. *Marine Geology*, **153**, 253–274.

CATTANEO, A., CORREGGIARI, A., MARSSET, T., THOMAS, Y., MARSSET, B. & TRINCARDI, F. 2004. Seafloor undulation pattern on the Adriatic shelf and comparison to deep-water sediment waves. *Marine Geology*, **213**, 121–148.

CUSHMAN-ROISIN, B., GACIC, M., POULAIN, P. M. & ARTEGIANI, A. (eds) 2001. *Physical Oceanography of the Adriatic Sea: Past, Present and Future*. Kluwer Academic Publishers, Dordrecht.

DE ALTERIIS, G. & AIELLO, G. 1993. Stratigraphy and tectonics offshore of Puglia (Italy, Southern Adriatic Sea). *Marine Geology*, **113**, 233–253.

DE DOMINICIS, A. & MAZZOLDI, G. 1987. Interpretazione geologico-strutturale del margine orientale della piattaforma apula. *Memorie della Società Geologica Italiana*, **38**, 163–176.

DOGLIONI, C., MONGELLI, F. & PIERI, P. 1994. The Puglia uplift (SE Italy): an anomaly in the foreland of the Apennine subduction due to buckling of a thick continental lithosphere. *Tectonics*, **13**, 1309–1321.

EMEIS, K. C., SCHULZ, H.-M., STRUCK, U., ET AL. 1998. Stable isotope and alkenone temperature records of sapropels from sites 964 and 967: constraining the physical environment of sapropel formation in the Eastern Mediterranean Sea. *In: Proceedings of the Ocean Drilling Program, Scientific Results, 160*. Ocean Drilling Program College Station, TX, 309–331.

FAIRBANKS, R. G. 1989. A 17,000 year glacio-eustatic sea level record: influence of glacial melting rates on the Younger Dryas event and deep-ocean circulation. *Nature*, **342**, 637–642.

FAUGÈRES, J. C., MÉZERAIS, M. L. & STOW, D. A. V. 1993. Contourite drift types and their distribution in the North and South Atlantic Ocean Basins. *Sedimentary Geology*, **82**, 189–203.

FAUGÈRES, J. C., STOW, D. A. V., IMBERT, P. & VIANA, A. 1999. Seismic features diagnostic of contourite drifts. *Marine Geology*, **162**, 1–38.

FINETTI, I. 1984. Struttura ed evoluzione della microplacca Adriatica. *Bollettino di Oceanologia Teorica e Applicata*, **II**, 115–123.

FINETTI, I., BRICCHI, G., DEL BEN, A., PIPAN, A. & XUAN, Z. 1987. Geophysical study of the Adria plate. *Memorie della Società Geologica Italiana*, **40**, 335–344.

FLOOD, R. D. 1981. Distribution, morphology and origin of sedimentary furrows in cohesive sediments, Southampton Water. *Sedimentology*, **28**, 511–529.

FLOOD, R. D. 1982. Observations, classification, and dynamics of furrows in cohesive sediments. *Bulletin de l'Institut de Géologie du Bassin d'Aquitaine*, **31–32**, 167–179.

FLOOD, R. D. 1983. Classification of sedimentary furrows and a model for furrow initiation and evolution. *Geological Society of America Bulletin*, **94**, 630–639.

FLOOD, R. D. 1988. A lee-wave model for deep-sea mudwave activity. *Deep-Sea Research*, **35**, 973–983.

FLOOD, R. D. 1994. Abyssal bedforms as indicators of changing bottom current flow: examples from the US east-coast continental rise. *Paleoceanography*, **9**, 1049–1060.

FLOOD, R. D. & GIOSAN, L. 2002. Recognition and interpretation of deep-water sediment waves; implications for palaeoceanography, hydrocarbon exploration and flow process interpretation. *Marine Geology*, **192**, 259–273.

HABGOOD, E. L., KENYON, N. H., MASSON, D. G., AKHMETZHANOV, A., WEAVER, P. P. E., GARDNER, J. & MULDER, T. 2003. Deep-water sediment wave fields, bottom current sand channels and gravity flow channel-lobe systems: Gulf of Cadiz, NE Atlantic. *Sedimentology*, **50**, 483–510.

HAYES, A., ROHLING, E. J., DE RIJK, S., KROON, D. & ZACHARIASSE, W. J. 1999. Mediterranean planktonic foraminiferal faunas during the last glacial cycle. *Marine Geology*, **153**, 239–252.

HOLLISTER, C. D., FLOOD, R. D., JOHNSON, D. A., LONSDALE, P. & SOUTHARD, J. B. 1974. Abyssal furrows and hyperbolic echo traces on the Bahama Outer Ridge. *Geology*, **12**, 395–400.

HOLLISTER, C. D., SOUTHARD, J. B., FLOOD, R. D. & LONSDALE, P. 1976. Flow phenomena in the benthic boundary layer and bed forms beneath deep-current system. *In*: MCCAVE, I. N. (ed.) *The Benthic Boundary Layer*. Plenum, New York, 183–204.

HOPFAUF, V., SPIEß, V. & GEOWISSENSCHAFTEN, F. 2001. A three-dimensional theory for the development and migration of deep sea sedimentary waves. *Deep-Sea Research I*, **48**, 2497–2519.

HOWE, J. A. 1996. Turbidite and contourite sediment waves in the northern Rockall Trough, North Atlantic Ocean. *Sedimentology*, **43**, 219–234.

HOWE, J., STOKER, M. S. & STOW, D. A. V. 1994. Late Cenozoic sediment drift complex, Northeast Rockall Trough. *Paleoceanography*, **9**, 989–999.

HOWE, J. A., PUDSEY, C. J. & CUNNINGHAM, A. P. 1997. Pliocene-Holocene contourite deposition under the Antarctic Circumpolar Current, western

Falkland Trough, South Atlantic Ocean. *Marine Geology*, **138**, 27–50.

HOWE, J. A., LIVERMORE, R. A. & MALDONADO, A. 1998. Mudwave activity and current-controlled sedimentation in Powell Basin, northern Weddell Sea, Antarctica. *Marine Geology*, **149**, 229–241.

JORISSEN, F. J., ASIOLI, A., BORSETTI, A. M., *ET AL.* 1993. Late Quaternary central Mediterranean biochronology. *Marine Micropalaeontology*, **21**, 169–189.

KENNARD, L., SCAFER, C. & CARTER, L. 1990. Late Cenozoic evolution of Sackeville Spur: a sediment drift on the Newfoundland continental slope. *Canadian Journal of Earth Science*, **27**, 863–878.

KENYON, N. H. 1986. Evidence from bedforms for a strong poleward current along the upper continental slope of NW Europe. *Marine Geology*, **72**, 187–189.

KENYON, N. H. & BELDERSON, R. H. 1973. Bedforms of the Mediterranean undercurrent observed with side-scan sonar. *Sedimentary Geology*, **9**, 77–100.

KUIJPERS, A., ANDERSEN, M. S., KENYON, N. H., KUNZENDORF, H. & VAN WEERING, T. C. E. 1998. Quaternary sedimentation and Norwegian Sea overflow pathways around Bill Bailey Bank, northeastern Atlantic. *Marine Geology*, **152**, 101–127.

KUIJPERS, A., TROELSTRA, S. R., PRINS, M. A., *ET AL.* 2003. Late Quaternary sedimentary processes and ocean circulation changes at the Southeast Greenland margin. *Marine Geology*, **195**, 109–129.

LE BAS, T. P., MASON, D. C. & MILLARD, N. W. 1995. TOBI image processing—the state of the art. *IEEE Journal of Oceanic Engineering*, **20**, 85–93.

LONSDALE, P. & MALFAIT, B. 1974. Abyssal dunes of foraminiferal sand in Carnegie Ridge. *Geological Society of America Bulletin*, **83**, 289–316.

MAGAGNOLI, A. & MENGOLI, M. 1995. *Carotiere a gravità SW-104*. Rapporto Tecnico CNR, **27**.

MALANOTTE-RIZZOLI, P., MANCA, B., RIBERA D'ALCALA, M., *ET AL.* 1997. A synthesis of the Ionian Sea hydrography, circulation and water mass pathways during POEM Phase I. *Progress in Oceanography*, **39**, 153–204.

MANCA, B. B. & SCARAZZATO, P. 2001. The two regimes of the intermediate/deep circulation in the Ionian–Adriatic seas. *Archivio di Oceanografia e Limnologia*, **22**, 15–26.

MANCA, B. B., KOVACEVIC, V., GACIC, M. & VIEZZOLI, D. 2002. Dense water formation in the Southern Adriatic Sea and interaction with the Ionian Sea in the period 1997–1999. *Journal of Marine Systems*, **33–34**, 133–154.

MANLEY, P. L. & FLOOD, R. D. 1993. Paleoflow history determined from mudwave migration; Argentine Basin. *Deep-Sea Research II*, **40**, 1033–1055.

MANTZIAFOU, A. & LASCARATOS, A. 2004. An eddy resolving numerical study of the general circulation and deep-water formation in the Adriatic Sea. *Deep-Sea Research I*, **51**, 921–952.

MARANI, M., ARGNANI, A., ROVERI, M. & TRINCARDI, F. 1993. Sediment drifts and erosional surfaces in the central Mediterranean: seismic evidence of bottom-current activity. *Sedimentary Geology*, **82**, 207–220.

MASSON, D. G., HOWE, J. A. & STOKER, M. S. 2002. Bottom-current sediment waves, sediment drifts and contourites in the northern Rockall Trough. *Marine Geology*, **192**, 215–237.

MIGEON, S., SAVOYE, B., ZANELLA, E., MULDER, T., FAUGÈRES, J. C. & WEBER, O. 2001. Detailed seismic-reflection and sedimentary study of turbidite sediment waves on the Var Sedimentary Ridge (SE France): significance for sediment transport and deposition and for the mechanisms of sediment wave construction. *Marine and Petroleum Geology*, **18**, 179–208.

MILLOT, C. & MONACO, A. 1984. Deep strong currents and sediment transport in the northwestern Mediterranean Sea. *Geo-Marine Letters*, **4**, 13–17.

MYERS, P. G., HAINES, K. & ROHLING, E. J. 1998. Modeling the paleocirculation of the Mediterranean: the last glacial maximum and the Holocene with emphasis on the formation of Sapropel S1. *Paleoceanography*, **13**(6), 586–606.

NORMARK, W. R., PIPER, D. J. W., POSAMENTIER, H., PIRMEZ, C. & MIGNON, S. 2002. Variability in form and growth of sediment waves on turbidite channel levees. *Marine Geology*, **192**, 23–58.

ORI, G. G., ROVERI, M. & VANNONI, F. 1986. Plio-Pleistocene sedimentation in the Apenninic–Adriatic foredeep (Central Adriatic Sea, Italy). *In*: ALLEN, P. A. & HOMEWOOD, P. (eds) *Foreland Basins*. International Association of sedimentologists, Special Publications, **8**, 183–198.

ORLIC, M., GICIC, M. & LA VIOLETTE, P. E. 1992. The currents and circulation of the Adriatic Sea. *Oceanologica Acta*, **15**, 109–122.

REEDER, M. S., ROTHWELL, G. & STOW, D. A. V. 2002. The Sicilian gateway: anatomy of deep-water connection between East and west Mediterranean basins. *In*: STOW, D. A. V., PUDSEY, C. J., HOWE, J. A., FAUGÈRES, J. C. & VIANA, A. R. (eds) *Deep-Water Contourite Systems: Modern Drift and Ancient Series, Seismic and Sedimentary Characteristics*. Geological Society, London, Memoirs, **22**, 171–189.

RICCI LUCCHI, F. 1986. The Oligocene to Recent foreland basin of the Northern Apennines. *In*: ALLEN, P. & HOMEWOOD, P. (eds) *Foreland Basins. International Association of Sedimentologists, Special Publications*, **8**, 105–139.

RIDENTE, D. & TRINCARDI, F. 2002*a*. Eustatic and tectonic control on deposition and lateral variability of Quaternary regressive sequences in the Adriatic basin. *Marine Geology*, **184**, 273–293.

RIDENTE, D. & TRINCARDI, F. 2002*b*. Late Pleistocene depositional cycles and syn-sedimentary tectonics on the central and south Adriatic shelf. *Memorie della Società Geologica Italiana*, **57**, 516–526.

ROHLING, E. J., JORISSEN, F. J. & DE STIGTER, H. C. 1997. 200 year interruption of Holocene sapropel formation in the Adriatic Sea. *Journal of Micropalaeontology*, **16**, 97–108.

ROTHWELL, R. G., THOMSON, J. & KAHLER. 1998. Low-sea-level emplacement of a very large

late Pleistocene 'megaturbidite' in the western Mediterranean Sea. *Nature*, **392**, 377–380.
ROVERI, M. 2002. Sediment drifts of the Corsica Channel, northern Tyrrenian Sea. *In*: STOW, D. A. V., PUDSEY, C. J., HOWE, J. A., FAUGÈRES, J. C. & VIANA, A. (eds) *Deep-water Contourite Systems: Modern Drifts and Ancient Series, Seismic and Sedimentary Characteristics*. Geological Society, London, Memoirs, **22**, 191–208.
ROYDEN, L. E., PATACCA, E. & SCANDONE, P. 1987. Segmentation and configuration of subducted lithosphere in Italy: an important control on thrust-belt and foredeep-basin evolution. *Geology*, **15**, 714–717.
STOKER, M. S. 1998. Sediment drift development on the Rockall continental margin, off NW Britain. *In*: STOKER, M. S., EVANS, D. & CRAMP, A. (eds) *Geological Processes on Continental Margins: Sedimentation, Mass-Wasting and Stability*. Geological Society, London, Special Publications, **129**, 229–254.
STOKER, M. S., AKHURST, C., HOWE, J. A. & STOW, D. A. V. 1998. Sediment drifts and contourites on the continental margin, off Northwest Britain. *Sedimentary Geology*, **115**, 33–52.
STOW, D. A. V. & HOLBROOK, J. A. 1984. North Atlantic contourite: an overview. *In*: STOW, D. A. V. & PIPER, D. J. W. (eds) *Fine-Grained Sediments: Deep-Water Processes and Facies*. Geological Society, London, Special Publications, **15**, 245–256.
STOW, D. A. V., FAUGÈRES, J. C., PUDSEY, C. J. & VIANA, A. R. 2002. Bottom currents, contourites and deep-sea sediment drifts; current state-of-the-art. *In*: STOW, D. A. V., PUDSEY, C. J., HOWE, J. A., FAUGÈRES, J. C. & VIANA, A. (eds) *Deep-water Contourite Systems: Modern Drifts and Ancient Series, Seismic and Sedimentary Characteristics*. Geological Society, London, Memoirs, **22**, 191–208.
TRAMONTANA, M., MORELLI, D. & COLANTONI, P. 1995. Tettonica plio-quaternaria del sistema sud-garganico (settore orientale) nel quadro evolutivo dell'Adriatico centro meridionale. *Studi Geologici Camerti*, **2**, 467–473.
TRINCARDI, F. & CORREGGIARI, A. 2000. Quaternary forced-regression deposits in the Adriatic basin and the record of composite sea-level cycles. *In*: HUNT, D. & GAWTHORPE, R. (eds) *Depositional Response to Forced Regression*. Geological Society, London, Special Publications, **172**, 245–269.
TRINCARDI, F., ASIOLI, A., CATTANEO, A., CORREGGIARI, A. & LANGONE, L. 1996. Stratigraphy of the late-Quaternary deposits in the Central Adriatic basin and the record of short-term climatic events. *Memorie dell'Istituto Italiano di Idrobiologia*, **55**, 39–70.
TRINCARDI, F., CATTANEO, A., CORREGGIARI, A. & RIDENTE, D. 2004. Evidence of soft-sediment deformation, fluid escape, sediment failure and regional weak layers within the late-Quaternary mud deposits of the Adriatic Sea. *Marine Geology*, **213**, 91–119.
TUCHOLKE, B. E. 1979. Furrows and focused echoes on the Blake Outer Ridge. *Marine Geology*, **31**, 13–20.
VIANA, A. 2002. Seismic expression of shallow- to deep-water contourites along the south-eastern Brazilian margin. *Marine Geophysical Researches*, **22**, 509–521.
VIEKMAN, B. E., WIMBUSH, M. & VAN LEER, J. C. 1989. Secondary circulation in the bottom boundary layer over sedimentary furrows. *Journal of Geophysical Research*, **94**, 9721–9730.
VILIBIC, I. & ORLIC, M. 2002. Adriatic water masses, their rates of formation and transport through the Otranto Strait. *Deep-Sea Research I*, **49**, 1321–1340.
VON LOM-KEIL, H., SPIEß, V. & HOPFAUF, V. 2002. Fine-grained sediment waves on the western flank of the Zapiola Drift, Argentine Basin: evidence for variations in Late Quaternary bottom flow activity. *Marine Geology*, **192**, 239–258.
WOOD, R. & DAVY, B. 1994. The Hikurangi Plateau. *Marine Geology*, **118**, 153–173.
WÜST, G. 1961. On the vertical circulation of the Mediterranean Sea. *Journal of Geophysical Research*, **66**, 321–327.
WYNN, R. B. & STOW, D. A. V. 2002. Classification and characterisation of deep-water sediment waves. *Marine Geology*, **192**, 7–22.

Small mounded contourite drifts associated with deep-water coral banks, Porcupine Seabight, NE Atlantic Ocean

D. VAN ROOIJ[1], D. BLAMART[2], M. KOZACHENKO[3] & J.-P. HENRIET[1]

[1]*Renard Centre of Marine Geology, Ghent University, Krijgslaan 281 S8, B-9000 Gent, Belgium (e-mail: David.VanRooij@UGent.be)*

[2]*Laboratoire des Sciences de Climat et de l'Environnement, Laboratoire mixte CNRS/CEA, Bâtiment 12, 4 avenue de la Terrasse, F-91198 Gif-sur-Yvette, France*

[3]*Coastal and Marine Resources Centre, University College Cork, Cork, Ireland*

Abstract: Numerous studies on sediment drifts have demonstrated a close interaction between sea-bed morphology, palaeoceanography, sediment supply and climate. Contourites have been reported in areas along continental margins directly influenced by the effect of intensive deep-water currents from the global conveyor belt. In this paper, we report the occurrence of a small-scale confined contourite drift from Porcupine Seabight, SW of Ireland, and its association with a province of coral banks. The Porcupine Basin is a relatively shallow, semi-enclosed basin characterized by the presence of cold-water coral bank provinces. These coral banks are often associated to a strong northward-flowing bottom current, created and steered by a complex interaction of the water mass characteristics, tidal influences and sea-bed morphology. Very high-resolution seismic stratigraphy allowed the identification of a small mounded drift, located between a depression created by (1) an irregular palaeotopography caused by a vigorous Late Pliocene erosion event and (2) a north–south alignment of coral banks. Core MD99-2327, taken on the flank of this drift mound, shows the variability of the bottom currents. Sortable silt data show several periods of bottom-current enhancement, which may be linked with warmer periods and an inferred influx of Mediterranean Outflow Water. The glacial part of the core has been interpreted as a muddy contourite with a high content of ice-rafted debris. The lower part of the core is a deep-water massive contourite sand resembling the present-day sea-floor sediments.

The Porcupine Seabight (PSB) is a shallow to deep-water, amphitheatre-shaped basin SW of Ireland, forming a small bight along the North Atlantic margin (Fig. 1). It is the surface expression of the underlying deep sedimentary Porcupine Basin, which is a failed rift of the proto-North Atlantic Ocean, filled with a 10 km thick series of Mesozoic and Cenozoic sediments (Moore & Shannon 1991). The PSB has gained interest because of the presence of special deep-water habitats (Henriet *et al.* 1998; De Mol *et al.* 2002; Huvenne *et al.* 2003), consisting of a province of coral banks, discovered on the eastern slope and described as the Belgica mound province (BMP). Earlier studies (De Mol *et al.* 2002; Huvenne *et al.* 2002; Van Rooij *et al.* 2003) suggested that the geometry of the most recent deposits in this province seemed to be under the influence of bottom currents of variable intensity.

This paper will illustrate the importance of such a relatively small basin as the Porcupine Seabight for the study of bottom-current deposits. Until now, only the large sediment drift systems in the Northern Atlantic Ocean have been extensively studied. High-resolution seismic surveys have allowed description of the details of smaller depositional systems, which were ignored or not observed in earlier studies. In this way, smaller-scale varieties of current-driven deposits were recognized, adding to the complexity of the contourite paradigm (Faugères *et al.* 1999; Rebesco & Stow 2001). Moreover, the local hydrodynamic environment of these deposits may be very complex. Because they mostly are located out of reach of the typical thermohaline deep-sea bottom-current circulation, the driving current process needs to be better understood.

Here, we present and discuss high-resolution seismic and core data. More specifically, this paper highlights the special present and past hydrodynamic environments of the BMP. The role of palaeoceanographic turnovers is discussed in view of the creation of a specific hydrodynamic, morphological and sedimentary environment, ready to host the development of a contourite drift. In addition, the presence of the deep-water coral bank and its interaction with the sedimentary environment is discussed as a controlling key factor. Ultimately, a link between the sedimentary record and past climate changes is proposed.

From: VIANA, A. R. & REBESCO, M. (eds) *Economic and Palaeoceanographic Significance of Contourite Deposits*. Geological Society, London, Special Publications, **276**, 225–244.
0305-8719/07/$15.00

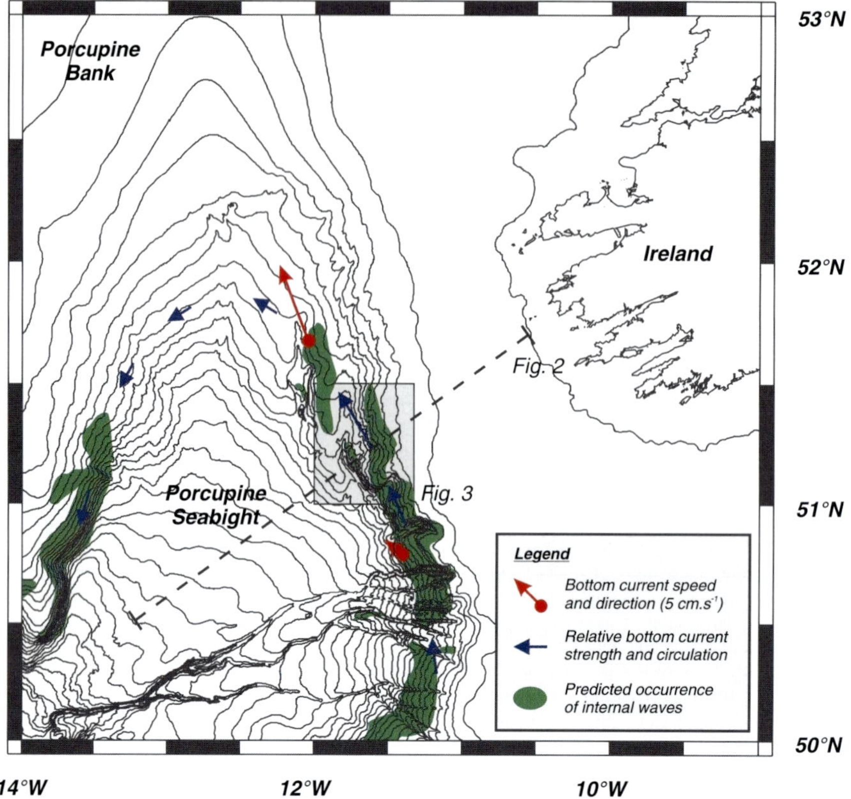

Fig. 1. Location map of the Porcupine Seabight, illustrating the main morphological elements and hydrographic features. The red arrows indicate known general near-bottom current measurements after Pingree & Le Cann (1989); the blue arrows show the assumed general circulation after White (2006). The green areas delineate the predicted occurrence of bottom-current enhancement as a result of internal tides (Rice *et al.* 1990). The GEBCO bathymetry contour intervals are drawn at 100 m intervals.

Material and methods

Very high-resolution seismic profiles

From 1997 to 2003, about 1500 km of single-channel seismic reflection data were collected over the eastern flank of the PSB. These data were acquired with an SIG surface sparker source and recorded with a single-channel surface streamer (vertical resolution between 0.4 and 1 m). These data were recorded digitally using the ELICS Delph 2 system. The basic data processing involved a first, broad-window Butterworth bandpass filter (200 Hz with a 24 dB per octave slope and 2000 Hz with a 36 dB per octave slope), minimum phase predictive deconvolution (with a 20 ms window), followed by a second Butterworth bandpass filter (250 Hz with a 24 dB per octave slope and 700 Hz with a 36 dB per octave slope). Additionally, a true amplitude recovery or an automatic gain control was applied. The obtained network of profiles is organized in a general N60° orientation, running perpendicular to the slope. The distance between adjacent profiles was about 3 km. Only a limited number of long, along-slope (N155°) profiles have been acquired.

Core analyses

Two cores, MD99-2327 and MD01-2449, were analysed for gamma density (g cm^{-3}) and magnetic susceptibility (SI). Grain-size analysis were performed every 5 cm using a Coulter LS130, to provide mean grain size (μm) and sorting (μm). Only core MD99-2327 was analysed for its sortable silt content, at a sample interval of 10 cm. This silt-sized fraction, between 10 and 63 μm, is considered as an indicator of relative bottom-current strength, as bottom currents sort coarse silt during events of resuspension and ensuing deposition.

Stronger currents yield a coarser mean size of the non-cohesive silt fraction, acting through both selective deposition and winnowing (McCave *et al.* 1995).

Core MD99-2327 was also sampled every 10 cm to obtain the coarse sand fraction with a grain size >150 μm and the relative abundance of the planktonic foraminifer *N. pachyderma (s.)* (NPS). A $\delta^{18}O$ stratigraphy for this core was obtained from 10–20 tests of *N. pachyderma (s.)*. The analyses were performed at the LSCE (Gif-sur-Yvette, France) using a VG Optima mass spectrometer equipped with a 'Kiel device' for automatic acidification of individual samples. The isotopic values are reported as per mil deviation with respect to the international V-PDB standard. The uncertainties on the isotope measurements are 0.08‰. Additionally, accelerator mass spectrometry ^{14}C dates were obtained on two monospecific samples of *N. pachyderma (s.)* at 90 and 440 cm downcore. These analyses were also performed at the LSCE and have been corrected for a mean reservoir age of 400 years (Bard 1998).

Hydrography

The present-day water masses of the Porcupine Seabight are in two broad categories: upper water from 0 to 1200 m and deeper water below (Fig. 2). The upper water has sources in the western Atlantic Ocean and the Mediterranean Sea. A permanent thermocline is found from 600 to 1400 m water depth; in this interval the temperature decreases from 10 °C to 4 °C. A seasonal thermocline is formed at about 50 m depth. Within the upper 700 m, the North Atlantic Central Water (NACW) is considered as a saline, winter mode water, formed by strong cooling of water masses NW of Spain (Pollard *et al.* 1996; Van Aken 2000). This winter cooling makes the Eastern North Atlantic Water (ENAW) denser and more saline than NACW. The Mediterranean Outflow Water (MOW) is observed between 700 and 1200 m (White 2006). According to Van Aken (2000), New *et al.* (2001) and White *et al.* (2005), the northward spreading of the MOW occurs as a reasonably steady boundary undercurrent that flows from the Gulf of Cadiz, at least as far as the PSB. Its continuation northward is considered as less certain. The deeper water is characterized by the presence of water masses derived from the Labrador Sea Water (LSW) from 1200 to 1800 m, and North East Atlantic Deep Water (NEADW) with contributions of Norwegian Sea Water (NSW), and is weakly influenced by Antarctic Bottom Water (AABW) (Hargreaves 1984; Rice *et al.* 1991).

A Shelf Edge Current (SEC) is known to be present along the eastern North Atlantic slope, from the Iberian margin to the Norwegian Sea (Pingree & Le Cann 1989, 1990; Rice *et al.* 1991; New *et al.* 2001), carrying warm and saline upper-layer ENAW over the mid-slope in the top 400–500 m (Fig. 2). Within the PSB, a mean northward current is less readily observed, except by near-bottom current meters on the eastern flank. Here, currents are predicted to be strongest at the mid-slope at about 500–600 m (Fig. 1). These bottom currents, with a strong semidiurnal to diurnal tidal

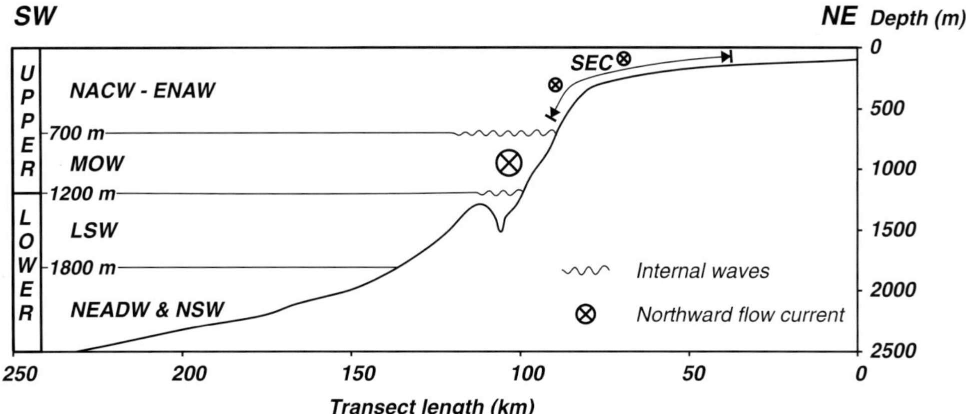

Fig. 2. Compilation of the water stratification along a NE–SW bathymetric transect along the Porcupine Seabight, with indication of the major hydrodynamic processes, after Hargreaves (1984), Rice *et al.* (1990, 1991) and White (2006). The location of this transect is shown in Figure 1. SEC, shelf edge current; NACW, North Atlantic Central Water; ENAW, Eastern North Atlantic Water; MOW, Mediterranean Outflow Water; LSW, Labrador Sea Water; NEADW, North East Atlantic Deep Water; NSW, Norwegian Sea Water.

variation, are strongly steered by the bottom topography (Pingree & Le Cann 1989, 1990; Rice *et al.* 1991). Mean flows of this predominant cyclonic circulation vary between 2 and 5 cm s^{-1} and may reach 10 cm s^{-1}. At the northern end of the PSB, the currents are relatively weaker, flowing towards an anticlockwise pattern around the northern slope (White 2006).

Moreover, Rice *et al.* (1990) and De Mol *et al.* (2002) have suggested that internal tides at a depth between 500 and 1000 m may be responsible for further enhancement of the bottom currents. These are internal waves with tidal periods, generated by the interaction of tidal currents, water stratification and bottom topography. The greatest enhancement is to be expected near the upper boundaries of the interacting water masses, where mixing will occur (Pingree & Le Cann 1989). This occurs predominantly on the eastern flank of the PSB (Fig. 1), where one can expect near-bottom currents estimated at 15–20 cm s^{-1}.

On the eastern slope of the PSB, Kenyon *et al.* (1998) and Wheeler *et al.* (2000) identified current-induced bedforms such as barchan sandwaves, sand ribbons and obstacle marks on high-resolution side-scan sonar imagery. Some of these bedforms indicate that the peak current can reach a considerable speed up to 100 cm s^{-1}. Foubert *et al.* (2005) found visual evidence of this strong hydrodynamic regime using remotely operated vehicle (ROV) submersible investigations. The presence of gravels and boulders within the vicinity of the BMP is very common and some of them are colonized by *Bathylasma* sp. barnacles, indicating strong currents. Sandwaves and superimposed ripples observed on the ROV microbathymetry indicate strong northward currents up to 65 cm s^{-1}. Also, long north–south lineated features with a length up to 250 m can be caused by strong currents of about 150 cm s^{-1}. This is confirmed by direct current measurements at nearby moorings (White *et al.* 2005).

Seismic stratigraphic units

The locations of the seismic lines and cores presented in this study are shown in Figure 3. The sediment drift is located in the central part of the Belgica mound province (Fig. 4). Between the individual coral banks, SW-oriented gullies can be observed. Upslope, another elongated mounded feature is observed, with a relief of *c.* 50–60 m above the sea floor, of 12 km length and maximum 4 km width. At its eastern side, this sediment body is limited by a (north–south elongated) moat channel, which seems to be, in turn, flanked by large south–north elongated scarp. The following section will discuss the internal structure of this sediment body and provide more insight into its construction.

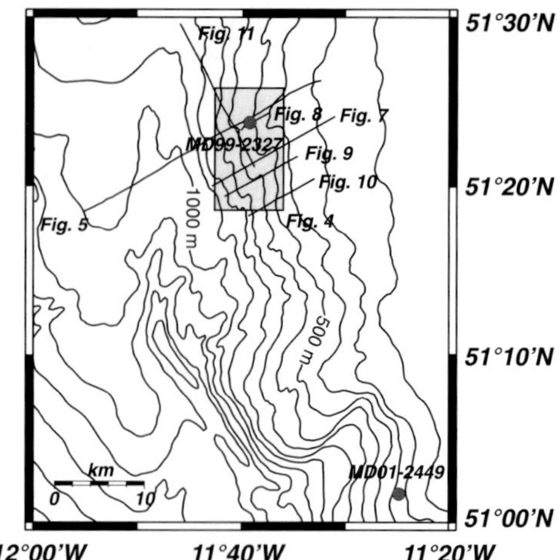

Fig. 3. Map showing the location of the seismic lines, cores (circles) and detailed shaded relief multibeam bathymetry. The GEBCO bathymetry contour intervals are drawn at 100 m intervals.

Pre-contourite drift units (U2–U4)

The stratigraphic record of this area is composed of four units (U4–U1) (Van Rooij *et al.* 2003). The lower boundary of the Quaternary unit U1 is very irregular and is characterized by incisions into underlying units, steep scarps and the presence of coral banks (Fig. 5). These palaeomorphological elements prior to the deposition of U1 are represented in Figure 6.

The acoustically nearly transparent unit U2 has been very heavily incised (Figs 6 and 7). This unit has a limited lateral extent, shows an irregular distribution and is present in two large zones, separated by a large north–south erosional channel (Figs 5 and 6). Generally it disappears towards the NNW and the SW, whereas it extends beyond the data coverage in the northeastern and southeastern part of the study area, where it reaches its maximum thickness of over 200 ms two-way travel time (TWT) (Fig. 6). Several zones have been identified where the thickness of U2 dramatically drops from *c.* 100 ms TWT to 0 ms TWT over about 350 m, creating relatively steep (*c.* 15°) flanks (Fig. 7). Several profiles show the presence of unit U2 at the western side of the erosional channel, still with the same dip as on the eastern side (Fig. 5). The boundary between units U2 and U3 can easily be correlated from one side of the channel to the other. These features strongly

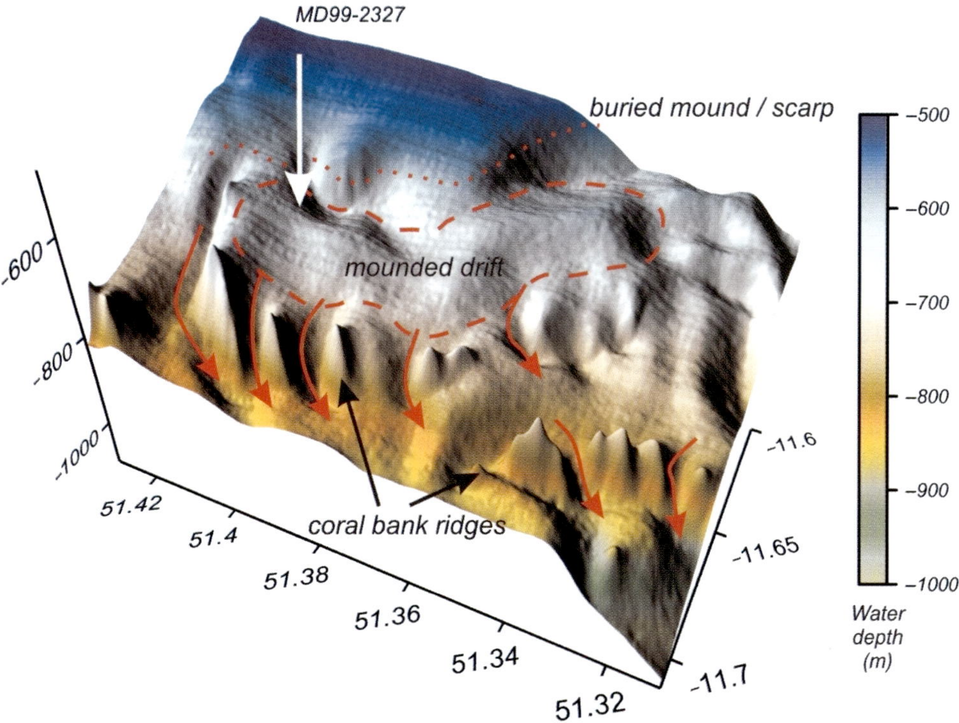

Fig. 4. Surface 3D multibeam bathymetry of the small mounded drift (red dashed lines) with the position of core MD99-2327. The downslope running gullies flanking the coral banks are indicated by the red arrows. The position of the (buried) U2 scarp is indicated by a red dotted line.

suggest that the sediments of unit U2 were deposited over the entire slope and later eroded. Although we have no lithological information on this acoustically transparent unit, its distribution and the seismic facies hint that it might contain very homogeneous sediments.

After the deposition of unit U2, a major change in oceanographic conditions was responsible for a large-scale erosive event along the entire slope of the PSB, leaving a very irregular terrace-like palaeotopography. These deep incisions into unit U2, as well as some U3 strata, created large roughly south–north-oriented ridges with steep flanks (Fig. 6), still evident in the present-day topography (Fig. 4). The timing for this regional hiatus, named RD1, is estimated to be Late Pliocene. It is interpreted in terms of the reintroduction of MOW in the NE Atlantic and the effect of glacial–interglacial events on deep-water circulation (Stow 1982; Pearson & Jenkins 1986).

The Belgica mounds occur in a very narrow bathymetric interval between 700 and 1000 m below sea level, aligned as along-slope-oriented ridges. Most of the observations confirm that these coral banks occupy an elevated position in the palaeobathymetry and are rooted on a scarp or a topographic irregularity corresponding to the RD1 unconformity (Figs 8–10). To the north, the coral banks tend to step off the eroded U2 unit (Figs 6 and 11). In this case, they are resting on a scarp of U3 deposits. Freiwald *et al.* (1999) and De Mol *et al.* (2002) have pointed out that the builders of these coral banks, most often *Lophelia pertusa* and *Madrepora occulata*, prefer to settle on a hard substratum in an elevated position. They need strong bottom currents to obtain sufficient nutrients, and to keep them free from sediment burial. The mere presence of the coral banks may therefore be an indicator of (strong) bottom currents after the Late Pliocene erosion. Moreover, exactly underneath the coral banks, a zone of sigmoidal reflectors observed in unit U3 can be interpreted as sediment waves. Ediger *et al.* (2002), for example, described a similar set of sediment waves in the Mediterranean Cilician basin, migrating upslope in an opposite direction to the present water circulation pattern. This implies the presence of a presumably north- to NW-flowing bottom current on this part

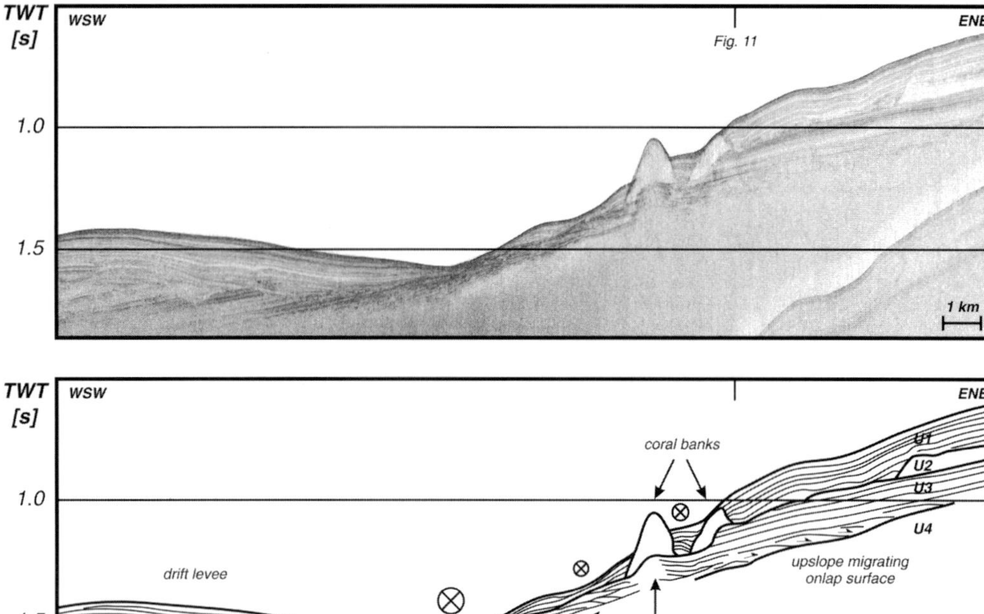

Fig. 5. Profile P980521, illustrating the general seismic stratigraphic setting and showing two coral banks and general sediment drift properties. The several erosional steps along the lower boundary of unit U1 and the incision into unit U3 at the WSW flank of the channel should also be noted.

of the slope during the Early Miocene. If this is the case, this slope section may have been under the influence of enhanced bottom currents since the beginning of the Neogene.

Small mounded contourite drift (U1)

Between the coral banks and the irregular palaeotopography created by the RD1 erosion, reflectors of unit U1 show a mounded feature with a moderate ENE progradation (a; Figs 7 and 9). All the seismic profiles in the study area display a very typical vertical stacking of seismic facies, mainly characterized by changing amplitudes (Figs 7, 10 and 11). The internal geometry of unit U1 seems to be controlled by the length of the passage between the steep flanks of the coral bank and the U2 scarp. In Figure 5, the distance of this passage exceeds 5 km and an undulating–sheeted geometry is observed within unit U1. Only minor irregularities such as downlap features are found against the lower boundary. Figures 7, 9 and 10 on the other hand, show a relatively narrow passage (2 km or less) and the overall geometry of unit U1 shows a mounded appearance with moats on both sides (a). The oldest sediments of unit U1 show a concave geometry downlapping towards the WSW and the ENE. There, unit U1 is deposited within a narrow passage between the steep flank of a coral bank, and a U2 scarp. In particular on the side of this scarp, large moat-like features (width up to 500 m) are observed (Figs 8–10). In some cases, they are filled with flat-topped acoustically transparent deposits (Figs 7 and 8), suggesting that these deposits could be turbidite deposits. The thickness of this mounded sediment body varies from 150 to 250 ms TWT. On the southwestern side of the coral bank, the thickness and reflector terminations are less pronounced, although subtle moats can be inferred.

The geometry and the reflectors of this sediment body suggest that the coral banks were already present before the deposition of unit U1, and that the coral banks were large and steep enough to intensify currents between their foot and the U2 scarp since the start of the deposition of unit U1. The oldest deposits of U1 are always observed in the centre of this passage from where they progressively prograded through time towards the flanks, creating a mounded geometry.

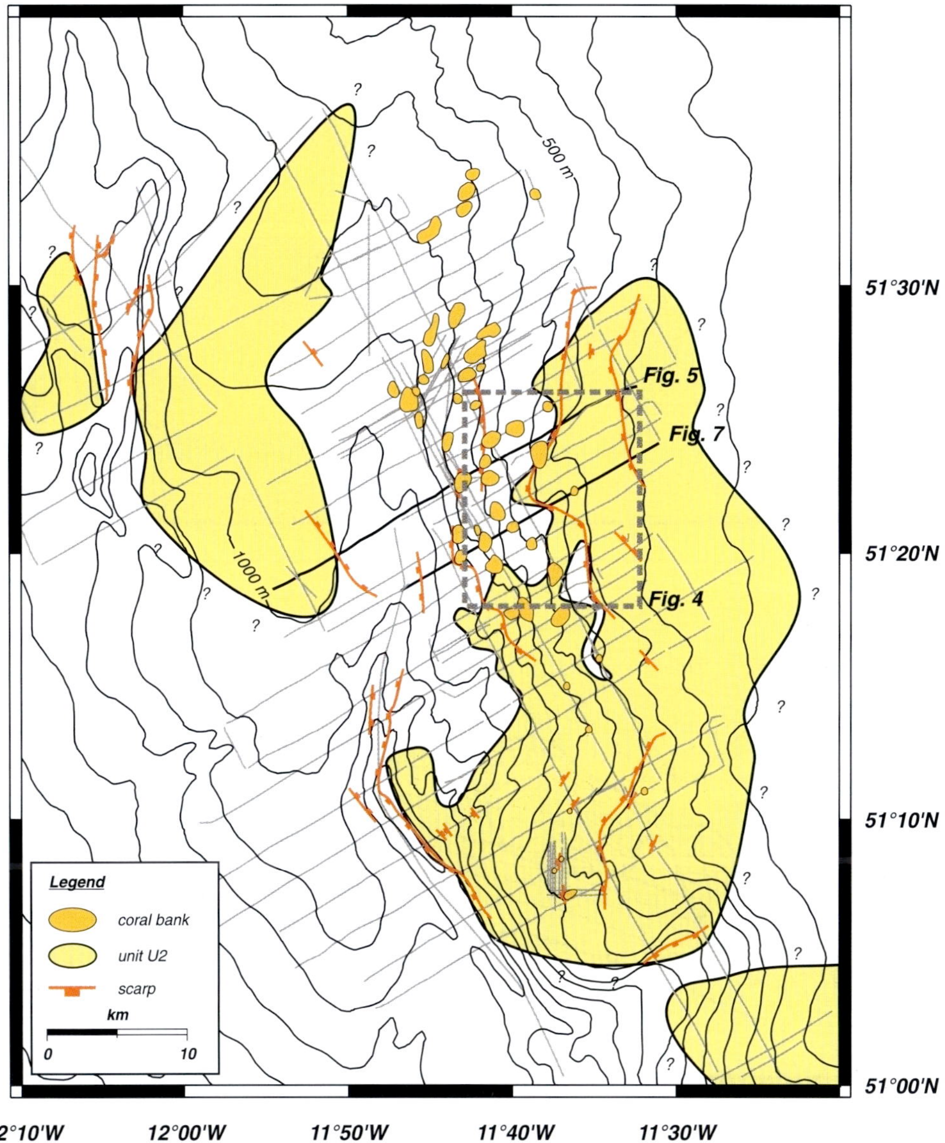

Fig. 6. Mapped distribution of seismic unit U2, the coral banks and the steep scarps created by the RD1 erosion event. All available high-resolution profiles are shown by grey lines.

After the passage became filled the reflectors started to onlap, but currents (still active, although possibly reduced in strength) created moats on both sides. This change is also accompanied by discontinuities on other seismic profiles. In a way, this drift shows similarities to mounded confined drifts such as the Louisville drift (Carter & McCave 1994), the Sicily channel drift (Marani *et al.* 1993) and the Sumba drift (Reed *et al.* 1987) as defined by Faugères *et al.* (1999). Confined drifts appear similar to elongate drifts, but are deposited in passages between tectonic or volcanic highs and are confined by boundary channels on both sides. In the BMP, however, this mounded drift is relatively small and it does not occur in a morpho-tectonically peculiar area such as the Louisville drift (Carter & McCave 1994). In this case, the narrow passage is constructed by an erosion event creating a scarp and by coral banks. Nevertheless, confined drifts remain very rare and this example

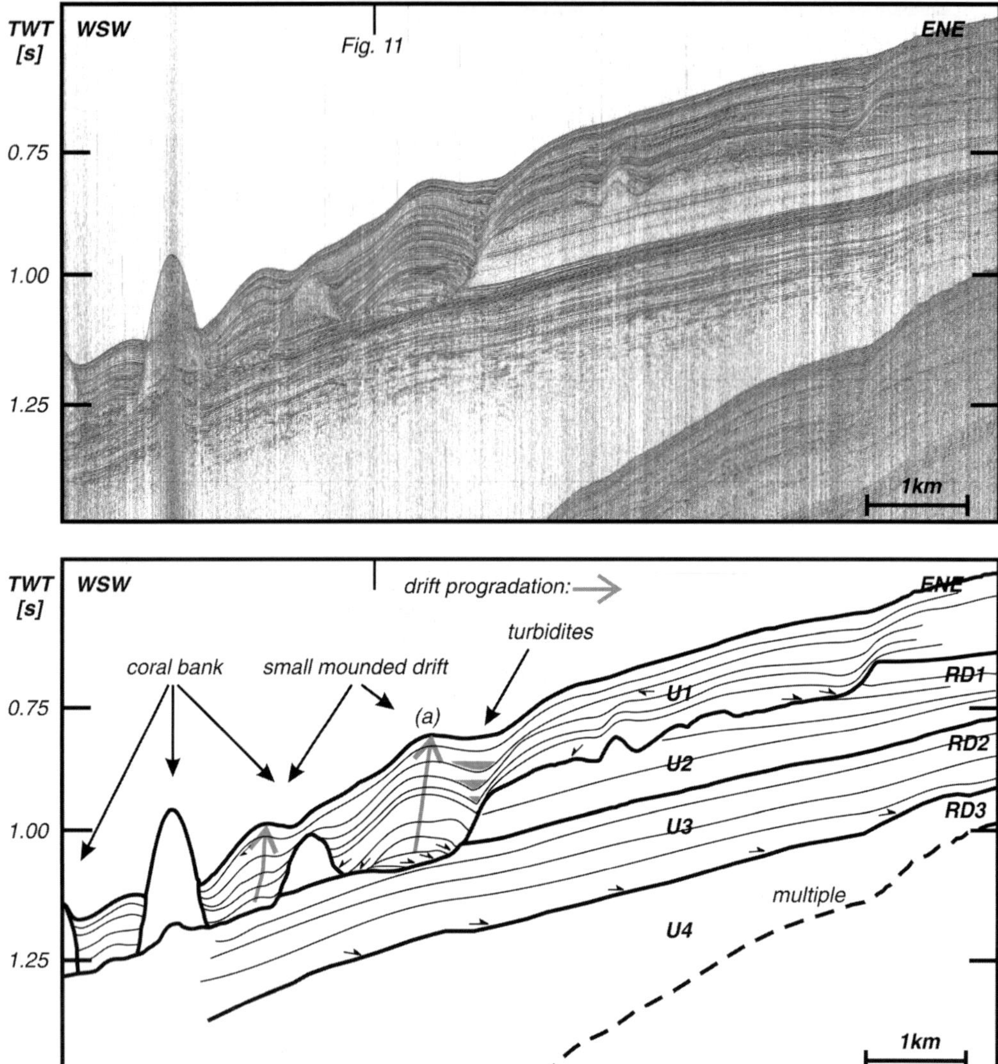

Fig. 7. Profile P010521, illustrating the general setting of the small mounded drift (a). West of the drift, an unconformity surface and a buried coral bank are rooted on the RD1 erosional surface. At its eastern edge, a steep scarp of unit U2 can be observed. The grey arrows indicate drift progradation. The gradual change in acoustic facies along the ENE side of unit U2 should be noted.

on one hand confirms their occurrence in peculiar areas, but on the other hand demonstrates that they do not need to be controlled in a pure morpho-tectonic way.

Core analyses

To further elucidate the true nature of the sediment drift deposits, two cores have been studied (Fig. 3). Core MD99-2327 (2625 cm long) was taken in a water depth of 651 m on the eastern flank of the sediment drift (Fig. 8) to provide an insight into the variability of the hydrodynamic environment. The site of core MD01-2449 (2215 cm long) is located in a water depth of 435 m, about 30 km SE of the sediment drift. Because of its location, it is out of range of the region of high bottom-current velocities. This rules out the possibility that the detrital and biogenic components would have been subject to major reworking through bottom currents.

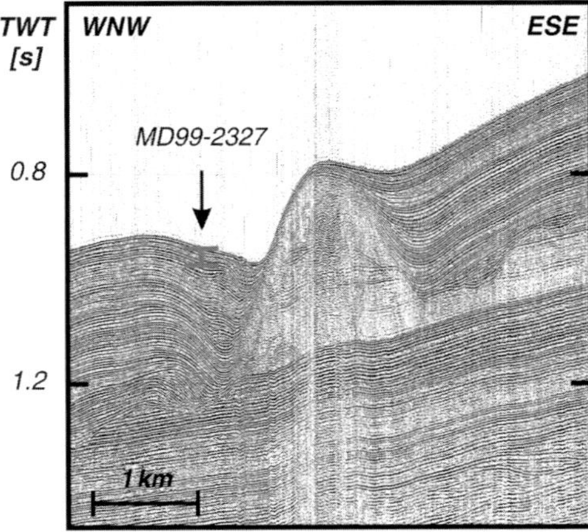

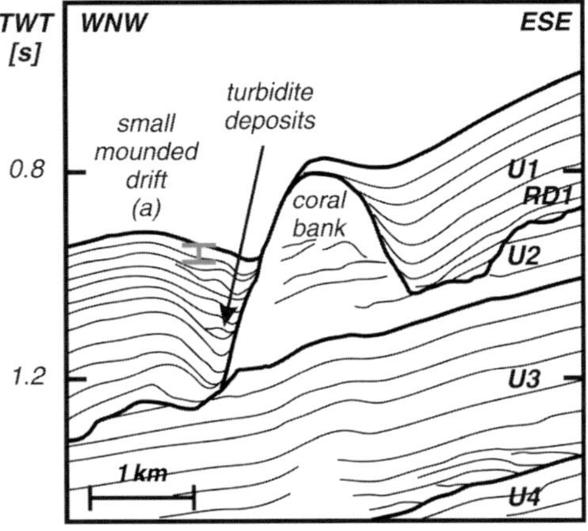

Fig. 8. Profile P010517, showing further detail of the MD99-2327 coring site, at the eastern flank of the small mounded drift.

This core should thus provide a continuous record of the regional palaeoclimatological signals.

Core MD99-2327 stratigraphy

Figure 12 illustrates the main characteristics of core MD99-2327. Although the sedimentary record is rather difficult to interpret in terms of chronostratigraphy, several features might characterize some key palaeoclimatological episodes. In particular, the magnetic susceptibility and the fraction >150 μm clearly show a significant change at about 1400 cm, subdividing the core into an upper and lower unit. A more detailed discussion concerning the atypical stratigraphy of this core has been given by Van Rooij *et al.* (2006).

From 0 to 1400 cm, the core contains homogeneous, olive grey silty clay with some sulphide strikes. At the base of this upper unit (1233–1425 cm), alternating layers of olive grey silty clay and muddy fine sand occur. The sediments are generally very poorly sorted, although sorting increases towards the base of this unit (Fig. 13). X-ray imagery reveals extensive bioturbation (mottling, planolites, chondrites, mycelia) throughout, varying from faint structures to very heavy burrowing (Van Rooij 2004). The upper unit is interpreted to record the last glacial period (Van Rooij *et al.* 2006). An indication of the presence of the cold MIS 4 is observed between 1000 and 1170 cm, with rather high values of NPS and $\delta^{18}O$ and by a rather low amount of coarse sand (Fig. 12). Two anomalies in NPS at 90 and 440 cm were dated respectively at 15.19 ± 0.13 ka BP and 17.38 ± 0.14 ka BP. These dates clearly reflect a gradual cooling with an abrupt culmination in NPS, $\delta^{18}O$ and density. X-ray imagery of these sections clearly shows higher concentrations of ice-rafted debris (IRD) in a heavily burrowed environment (Van Rooij 2004).

The lower unit, from 1500 to 2625 cm, is a very poorly to poorly sorted olive grey foraminiferal fine to medium sand (Fig. 13). Within these sediments occasional shell fragments, lithic grains and dark minerals are observed. Here, a higher sand percentage is present, and the magnetic susceptibility is lower (Fig. 12). Moreover, the $\delta^{18}O$ values also are lower and suggest, together with NPS, an interglacial environment (MIS 5) (Van Rooij *et al.* 2006). The exceptionally high amount of sand >150 μm cannot be interpreted in terms of ice-rafting events, but in terms of hydrodynamic activity.

Core MD01-2449 stratigraphy

The top part of this 2215 cm long core (0–26 cm) consists of olive clayey very fine quartz-foram sands. This grades downcore into olive grey to grey silty clays. Nannofossils and fine sand pockets were commonly observed in varying quantities. Two coarser-grained intervals were also observed; between 600 and 900 cm and between 1200 and 1400 cm. These two intervals are also characterized by elevated plateaux of dry bulk density (Fig. 14). A complete flow-in occurs below 2215 cm. As a consequence, only the first 22 m of this core were described. The variations in mainly dry bulk density, mean grain size, sand content (up to 20%) and magnetic susceptibility are similar to those of MD99-2327.

Common record of the two cores

The data presented here provide a first offshore record of the deglaciation history of the British–Irish Ice Sheet (BIIS). The characteristics and

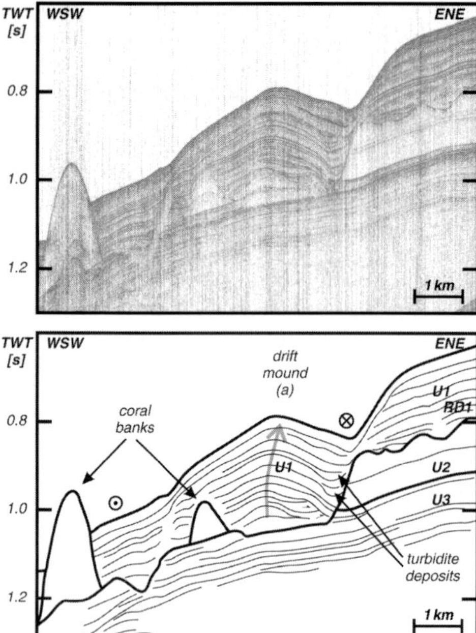

Fig. 9. Profile P980509, which makes a transect across the small mounded drift (a) in the central Belgica mound province. The mounded drift morphology occurs between the foot of a buried coral bank and a scarp in the flank of unit U2. The grey arrow indicates an upslope progradation of the crest of the drift mound. The moat indicating the pathway of a northward flowing current is filled with turbidite deposits.

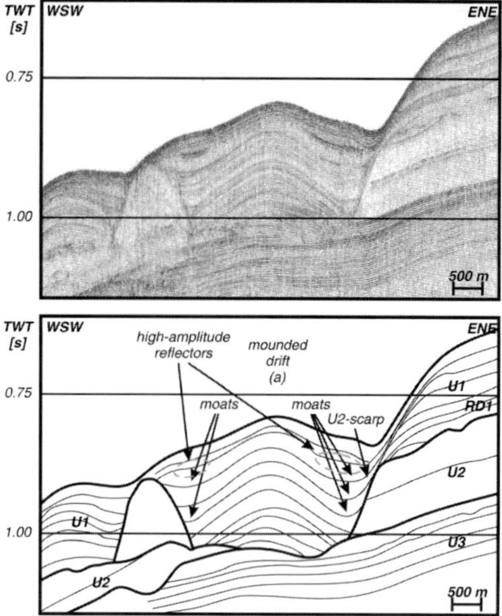

Fig. 10. Profile P970539, showing a typical cross-section of the drift mound within the Belgica mound province. On both sides, the moats show high-amplitude reflectors and probable turbidite deposits. The drift mound occurs between a large coral bank and the scarp in the U2 unit.

variations in dry bulk density, mean grain size and magnetic susceptibility of core MD01-2449 correlate with those of core MD99-2327 (Fig. 14). When the glacial records of the two cores are compared, it can be seen that the sedimentation rates at the site of MD01-2449 are significantly higher than those of the sediment drift site. This can be explained by the higher position on the slope of core MD01-2449 and by the sediment input coming from the Irish channel. This positive comparison strengthens and magnifies the sedimentary history of both cores, showing an atypical situation with a very local ice-rafting signature (Van Rooij *et al.* 2006).

The cold MIS 4 contains a relative minimum of IRD, suggesting a BIIS in a constructional stage with a period of falling sea level (Auffret *et al.* 2002). The high-variability record of IRD starts at about 62 ka (Fig. 13). All subsequent influx of terrigenous particles can be attributed to the BIIS destabilization. During MIS 3, the ice-rafting events succeed each other very closely and even grade smoothly into one another (Figs 13 and 14). This suggests that at the same time no other major sediment-supplying processes must have been active but ice-rafting, hemipelagic rain-out and limited lateral supply of bottom-current transported sediment.

However, with the start of MIS 2, an equal amount of IRD is delivered in a shorter time period compared with MIS 3. This episode of drastic increasing sedimentation rates reflects a higher variability of the BIIS destabilization and could be a record of a millennial-scale disintegration of the BIIS. McCabe & Clark (1998) and Knutz *et al.* (2001) considered the high frequency of BIIS-sourced IRD peaks to contribute to the millennial-scale pattern of iceberg discharges in the NE Atlantic. Several sources report two major shelf-edge advances and BIIS collapses from 26 to 18 ka and at 15 ka (Dowling & Coxon 2001; Knutz *et al.* 2002). These ages correspond to the ages of the two dated NPS anomalies. First of all, McCabe & Clark (1998) and Bowen *et al.* (2002) reported a significant deglaciation of the southern BIIS at about 17.4 ka. This could explain the two δ^{18}O meltwater peaks after 17.38 ka BP (at 400 cm). Moreover, the youngest event of 15.19 ka is consistent with the last collapse and also with the age of HE1 (15 ± 0.7 ka) of Elliot *et al.* (1998).

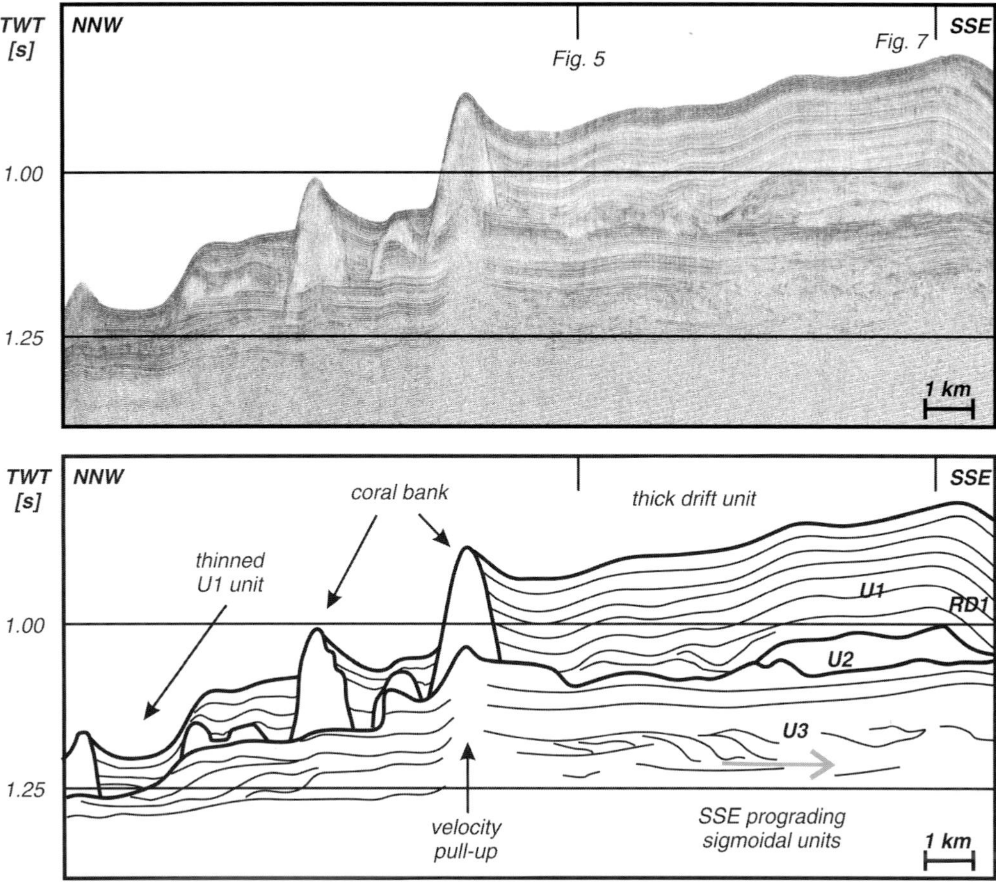

Fig. 11. Profile P980547, the only available alongslope profile crossing the sediment drift. It illustrates the variability of especially units U2 and U3. The marked change in sediment thickness south and north of the coral banks should be noted.

Hydrodynamic interpretation

The description of core MD99-2327 allows the recognition of two intervals, each of which can be interpreted in terms of a changing hydrodynamic environment. The most dominant feature of the glacial unit are the several episodes of ice-rafting events. The inferred IRD abundance is relatively high (5–10%). Particle size analysis shows, in general, that this unit is fine grained with a significantly high percentage of silt (65%), whereas clay tends to make up 8–24% of the sediments and the sand fraction is limited to an average of 18%. The total sample generally is very poorly sorted. The sortable silt index can provide estimates of current strength variations. Starting from the inferred MIS 4, five intervals of peak currents (28–33 μm) and five intervals of low currents (22–26 μm) are observed (Fig. 13). Transitions from minimum to maximum values on the sortable silt curve can be interpreted as accelerations of the bottom current, and vice versa. These peak current events (PCE) obviously behave in a more long-term way compared with the variations in the fraction >150 μm. It does not seem that the five peak current intervals coincide with significant ice-rafting intervals, suggesting another causal event. However, caution should be taken not to completely disconnect changes in sortable silt (thus benthic current variability) and the IRD (ice-rafting), because IRD also includes silt-sized particles. Most of the PCE peak during warm periods. Only PCE1 and partly PCE2 seem to be centred on a cold event. Probably, in this case, current-sorted silt and ice-rafted silt will be interfering factors. Because of the extensive bioturbation and burrowing, seen on X-ray imagery (Van Rooij 2004), it is very difficult to observe primary sedimentary

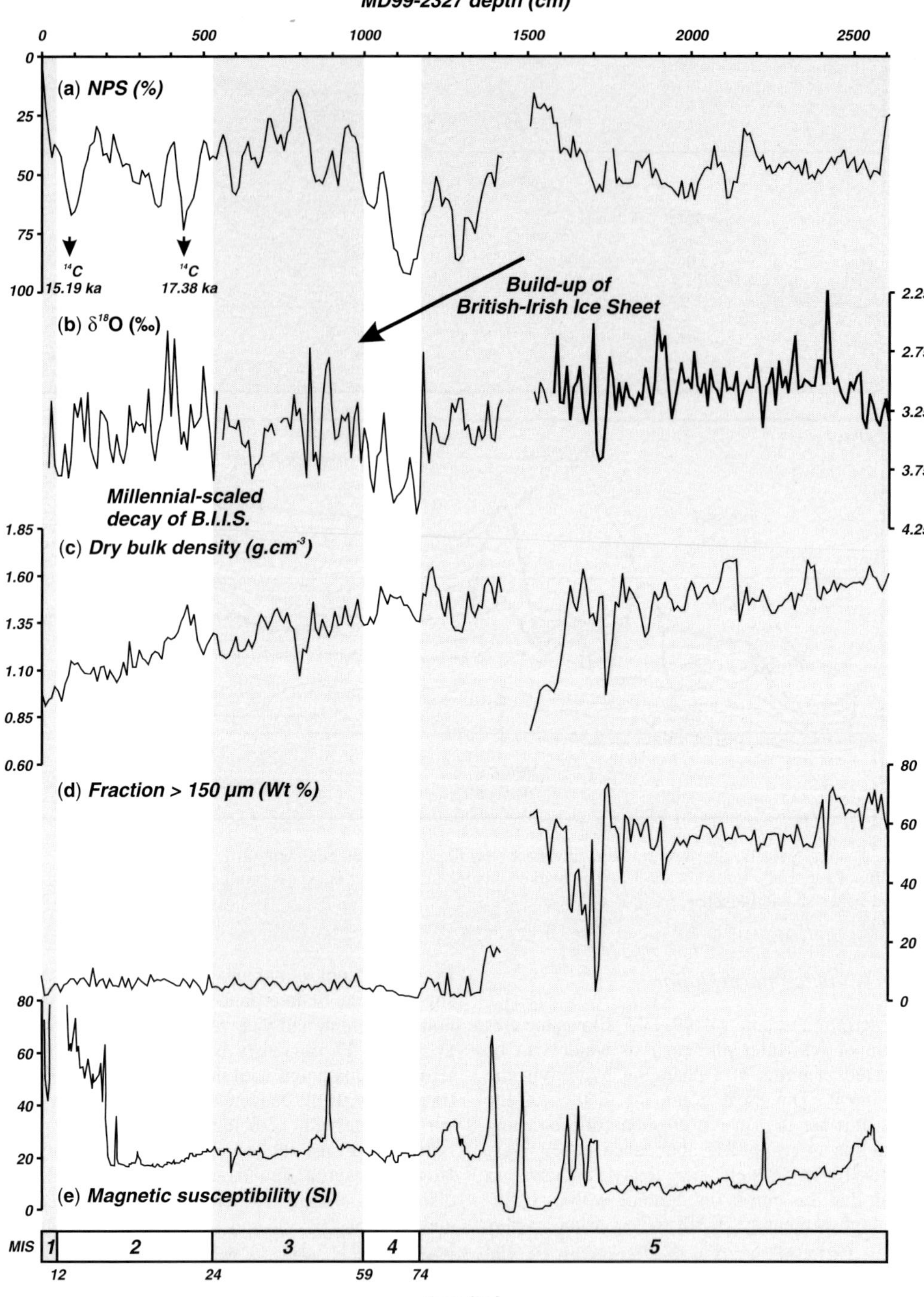

Fig. 12. Physical properties and stratigraphy of core MD99-2327 with (**a**) the relative abundance of the polar planktonic foraminifer *N. pachyderma s.* (%) (NPS (%)), (**b**) $\delta^{18}O$ stable isotopes (‰ v. PDB), (**c**) dry bulk density (g/cm^{-3}), (**d**) fraction >150 µm (wt%) and (**e**) the magnetic susceptibility. The grey-shaded areas highlight warmer (odd) marine isotope stages. The remarkable changes of (**d**) and (**e**) at about 14 m downcore should be noted.

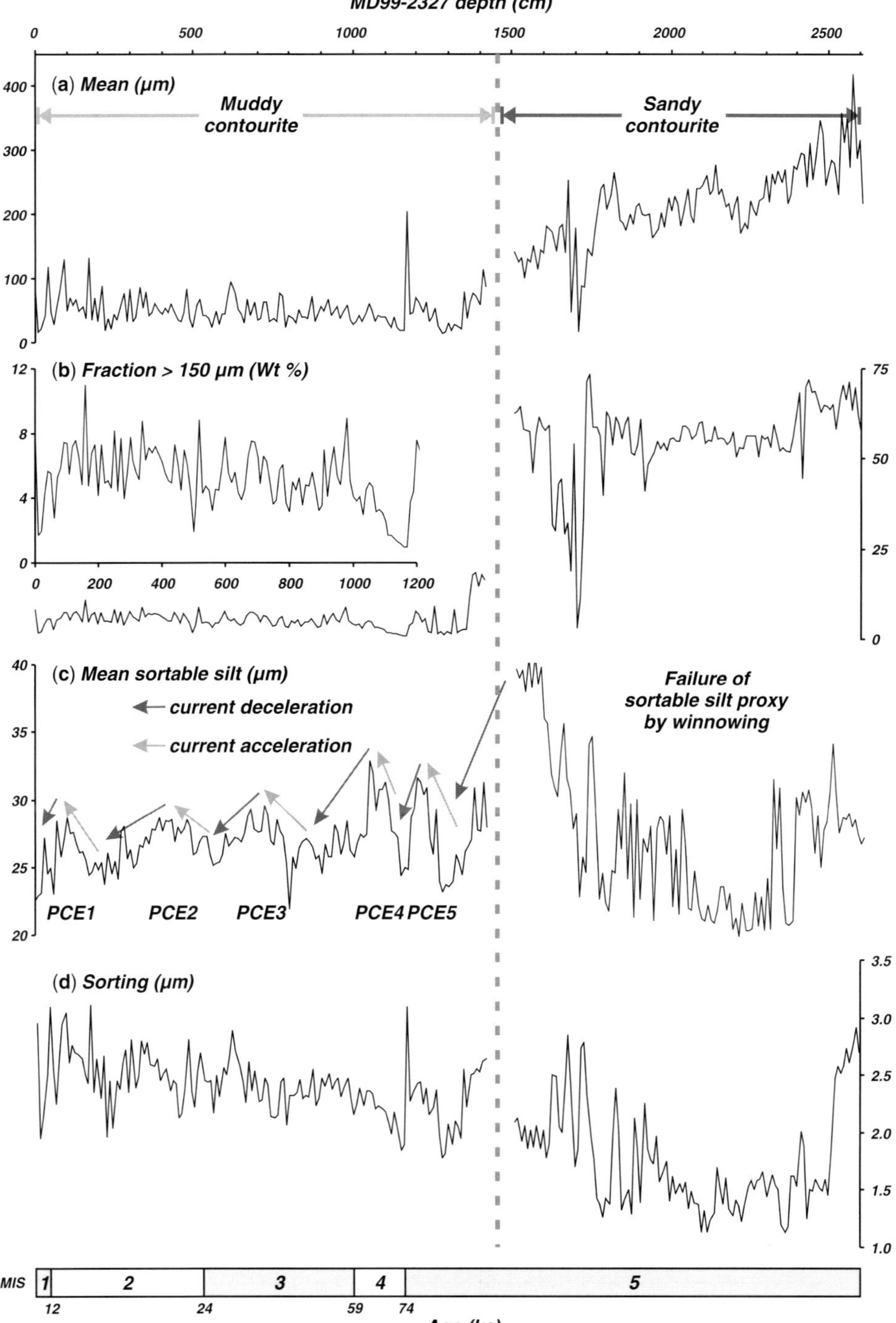

Fig. 13. Sedimentary processes and hydrodynamic interpretation of core MD99-2327 with (**a**) mean (μm), (**b**) fraction >150 μm (wt%), (**c**) mean sortable silt (μm) and (**d**) sorting (μm). Within (**b**), an inset of the glacial detail is provided. The five observed peak current episodes (PCE) are also presented.

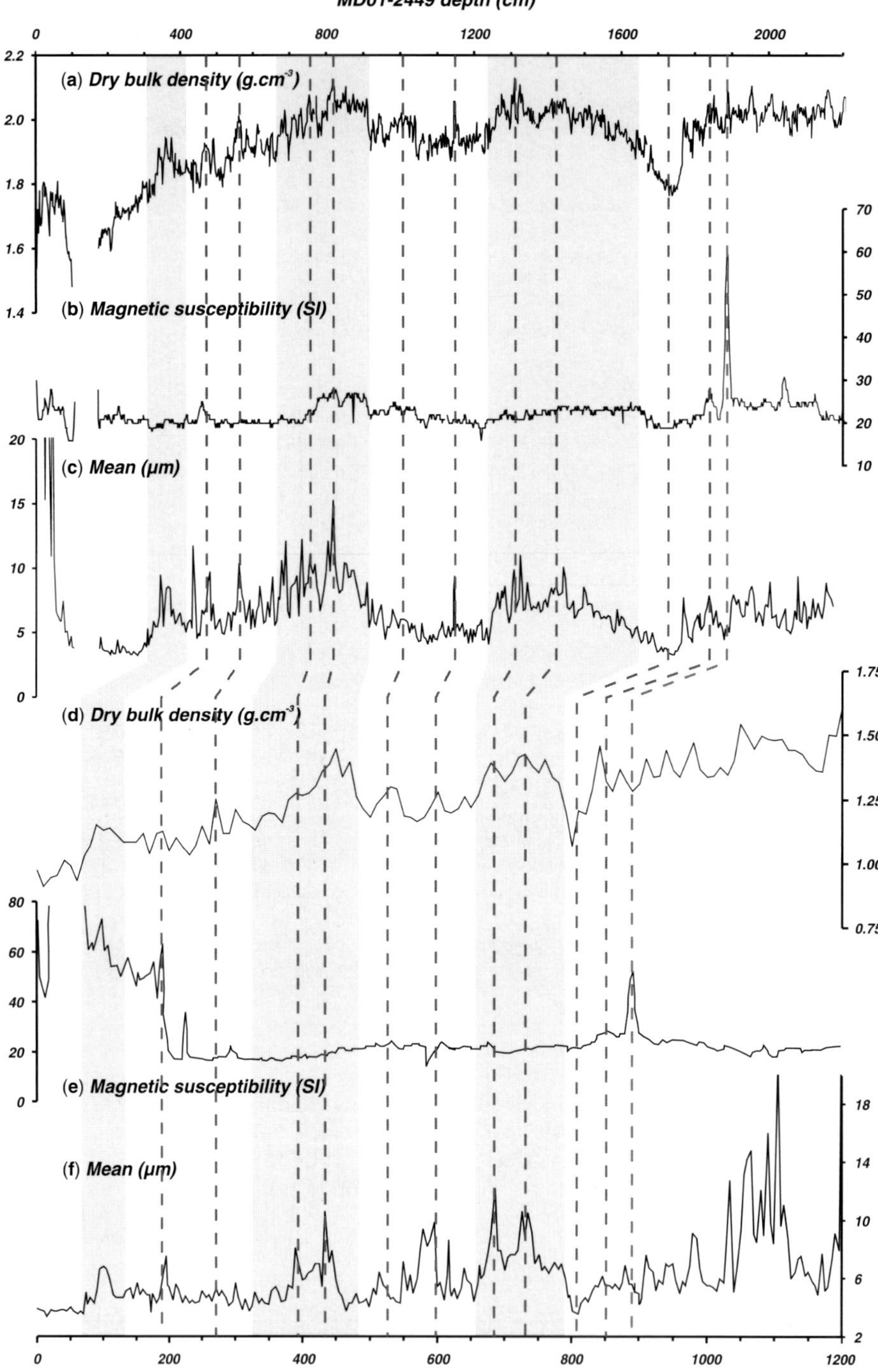

Fig. 14. Correlation of the glacial detail of core MD99-2327 with the higher-resolution data of core MD01-2449, based on respectively (**a**, **d**) dry bulk density (g cm^{-3}), (**b**, **e**) magnetic susceptibility (SI) and (**c**, **f**) mean grain size (μm). Specific correlation areas and points are indicated by grey-shaded areas or dashed lines, respectively.

structures. This bioturbation seems to be episodic and contains variations between mottling, isolated pockets and filaments. All this probably reflects several sediment sources in a biologically and hydrodynamically active area: current-sorted mud, biogenic components and ice-rafted debris.

Conversely, the sand percentage of the lower part of the core is up to 95% (with an average of 84%) and contains mostly quartz grains, detrital carbonate grains, and a high percentage of (reworked) planktonic and benthic foraminifers, reflecting a high biogenic productivity (Figs 12 and 13). This would also explain the large scatter in the oxygen isotope records, which indicates that (some of the) forams were reworked. These sediments are more sorted compared with the overlying units (although they still are poorly sorted), suggesting a significant benthic current influence. However, the sortable silt proxy cannot be applied in this part of the core because of the very low content of silt, which probably has been winnowed by strong currents. An alternative proxy for this benthic current strength can be suggested to be the mean of the 63–150 μm fraction, which encompasses the majority of the sediment, but excludes grains larger than 150 μm, whose origin is more likely to be related to ice-rafting. An increase in this sortable sand index as a result of increasing currents corresponds to a coarsening as the finer components are winnowed. This trend also corresponds to an increase in the degree of sorting, although mostly a fine tail remains present. As such, several fluctuations in current strength can be observed during the interglacial unit, with peak currents within the central part.

Discussion

The Quaternary hydrodynamic environment

Many observations on and around the area of the small mounded drift prove a very active hydrodynamic environment at present. Similar thin surface veneers also were observed on the Hebridean shelf and was interpreted as a contourite sand sheet (Armishaw *et al.* 2000; Akhurst *et al.* 2002). This requires the presence of a relatively strong, semi-permanent benthic current at intermediate depths at a velocity of over 30 cm s^{-1}. The measured and inferred bottom-current velocities within the BMP are consistent with the requirements of Stow *et al.* (2002) for contourites, which involve an average current velocity of 10–20 cm s^{-1} that can be accelerated up to 100 cm s^{-1} or more near steep slopes or narrow passageways. The required sediment supply zone can be variable, with upstream erosion, pirating and winnowing of slope sands (Armishaw *et al.* 2000). In this case, several sources are probable. The sedimentary analyses show that the entire core consists of contourites. The glacial unit meets many of the conditions of the muddy contourite facies as defined by Rebesco & Stow (2001) and Stow *et al.* (2002), although this is combined with a strong ice-rafting component. First, the sediment mainly consists of a siliciclastic fine-grained muddy silt. Indications for a mean grain size have been given by Stow & Piper (1984), who suggested 5–40 μm as a range for muddy contourites. Also, a sand percentage of 10–15% is inferred, but in this case, caution is necessary because a significant part of the sand content originated from ice-rafting. Only sporadically can primary structures be found; most of the core is structureless. The X-ray imagery confirms extensive bioturbation with (sulphide) filaments, planolites, mycelia and chondrites burrows. The nature of the bioturbation, and especially that of the filaments, seems to be variable within the core.

All analyses indicate that the nature and characteristics of the interglacial unit are similar to those of sandy contourites. Generally, they are a mixture of a terrigenous and biogenic content with evidence of abrasion and fragmental bioclasts (Faugères & Stow 1993; Stow *et al.* 2002). A very thick bedded, structureless coarse-grained deposit such as this one is extremely rare in this kind of environment and can, according to Stow & Mayall (2000), be classified as a deep-water massive sand. Mostly these are interpreted as sandy debris flows, although these sands can originate from various processes. In this case, many parameters indicate that the possibility that these interglacial sediments are mass-wasting deposits is rather small. Moreover, Figure 8 does not display features characteristic for a large debris flow or turbidite deposit. The general characteristics of such a deep-water massive sand were defined by Stow & Johansson (2000) and agree with the characteristics of the MIS 5 sediments. They are poorly to moderately sorted and have a high degree of reworking. Examples of sandy contourites as deep-water massive sands are rare and are found in morphologically and hydrodynamically special environments with fluctuating enhanced currents such as near the Gulf of Cadiz (Habgood *et al.* 2003) and the Sicilian gateway (Stow & Johansson 2000). Moreover, the massive (4.4 m) contourite sand deposits described by Habgood *et al.* (2003), have many similarities to the inferred contourite sands in core MD99-2327. In general, the size of a sandy contourite or a deep-water massive sand reported until now is rarely bigger than 1 m (Stow & Johansson 2000). Within this morphological and sedimentary context the presence of a 10 m thick sandy contourite is unique and requires further research.

Although no simple global response to climate can be found in bottom-current activity, it is very likely that the switch from a sandy to a muddy contouritic environment is associated with a particular palaeoclimatological change. Faugères & Stow (1993) linked a glacial dampened bottom-current regime to the presence of sea ice, whereas climate instability means enhanced bottom-current activity. Akhurst *et al.* (2002) have recently describe the presence of sandy contourites on the Hebrides slope during interglacials and interstadials and related them to enhanced current activity. They also recognized the presence of bottom-current action during glacial times, but less intense. This probably also is the case in our study area. Because the presence of enhanced currents in the BMP is highly dependent on the interaction with the MOW (New *et al.* 2001; De Mol *et al.* 2002; White 2006), it is plausible that glacial times seriously weaken the current regime to a muddy contourite sedimentation. The reduced outflow of MOW was then restricted to the Gulf of Cadiz and did not penetrate any further in the Atlantic Ocean (Schönfeld & Zahn 2000), so the conditions for enhanced currents were not met. The variability within the glacial contourite, however, is less clear. The inferred PCE are coeval with presumed warmer periods. During a climatic warmer period, the sea level could be more elevated, especially after a pan-Atlantic ice-rafting event. This could encourage an enhanced MOW production, which also could reach further into the Atlantic Ocean and create weak pulses of enhanced currents within the PSB.

A PSB contourite

The sedimentological record of this site demonstrates that the sediment body located between the coral banks and the steep flank of seismic unit U2 can be classified as a confined drift. The sedimentary facies suggests that the entire unit is influenced by fluctuating current intensities and thus the entire unit can be called a contourite drift at mid-water range (Stow *et al.* 2002). The sediment body shows many similarities to well-known contourite drift systems, such as a downcurrent elongation, subregional discontinuities and subparallel moderate- to low-amplitude reflectors with gradual change in seismic facies (Faugères *et al.* 1999; Rebesco & Stow 2001). Compared with the dimensions of published contourite drifts, the confined drift in the Porcupine Seabight is one of the smaller ones (*c.* 50 km^2). The few other examples of this type of drift are known within small basins (Stow *et al.* 2002), but are much larger. The best comparison can be made with the Sumba drift (Sunda Arc, Indonesia); a smooth asymmetric mound with boundary channels and sandy contourites (15 km elongation) (Reed *et al.* 1987). As a result of lateral current velocity gradients within the Sumba drift, muddy contourites were deposited on the central part and sandy contourites in boundary channels. However, the Sumba confined drift is 15 km wide, whereas our equivalent is only 4 km wide. It could thus be possible that lateral facies changes are less pronounced in the PSB small mounded drift and such changes can only be observed in depth (or time). The nature of the bathymetric restrictions that are responsible for the acceleration of deep currents, however, are of a completely different nature and are at the base of the smaller dimension. Whereas the Sumba drift and other confined drifts have a more tectonically controlled background, the interaction between water-mass mixing and bathymetric interaction is steered in this case by a combination of a turbulent sedimentary history with several erosion episodes and current-controlled biogenic build-ups. Within such a dynamic and irregular environment, it is expected that, besides contourites, turbiditic and other mass-wasting deposits can also be inferred.

The characteristics of the lower drift strata suggest a significant period of non-sedimentation between the onset of coral bank growth in the Late Pliocene and the onset of drift sedimentation. This period of non-sedimentation gives enough time for the 'start-up' phase of the coral bank, required by De Mol *et al.* (2002) for corals to settle on the hard substratum provided by the bottom-current swept and eroded U2 unit. From a palaeoceanographic point of view, we could imagine that the contouritic sedimentation was well under way during a major climatological change within the Pleistocene. The Mid-Pleistocene revolution (MPR, *c.* 940–640 ka) marks the change towards an increasing mean global ice volume and increasing amplitude of 100 ka climatological cycles (Raymo *et al.* 1997; Hernandez-Molina *et al.* 2002). This interval was also characterized by 'weaker' NADW formation, relative to the early and late Pleistocene (Raymo *et al.* 1997). On the other hand, after the MPR the pulsations between glacial and interglacial periods became more pronounced. They could be the cause of the start of the muddy–sandy contourite deposition in the BMP, as well as the acoustic amplitude variations within seismic unit U1.

Within the BMP, the presence of this small mounded drift is a common feature everywhere that the palaeotopography and the presence of the coral banks allow a similar setting. Sidescan sonar imagery of the entire province proves that in a similar depth-range and within the vicinity of the coral banks, enhanced currents are almost always inferred; thus, a similar build-up of hydrodynamic variability through time can be expected. However, as the presence of the current enhancement is strictly bound to the water mass–topography interaction in this region on the slope, we conclude that the presence of this confined contourite drift is a very local feature, bound to several geological, climatological, biological and hydrodynamic variables.

Conclusion

The Belgica mound province, located on the eastern slope of the Porcupine Seabight, is considered a unique environment within the North Atlantic domain because of its particular hydrographic, geological and morphological settings associated with the presence of deep-water coral banks. This paper demonstrates that, since the Neogene, strong northward-flowing currents on this part of the margin were responsible for the deposition of bottom-current controlled deposits (Fig. 15a).

Seismic interpretation of 1500 km of high-resolution seismic profiles across this coral bank province shows that after the deposition of the acoustically transparent unit U2 with as yet unknown lithology, the regional unconformity RD1 (recording a major change in oceanographic conditions) was responsible for the removal of a large part of this unit in the Late Pliocene (Fig. 15b). This unconformity also marks the start of glacial–interglacial cycles and their effects on the deep-water circulation. Subsequent to the RD1 event, corals began to settle on topographic irregularities in the palaeobathymetry. The coral banks were built spectacularly fast in a period when the adjacent areas experienced non-deposition. They are located in a zone within the influence of a complex system of enhanced currents, which is believed to be the main driving force of the controls on their development.

Towards a later phase of the Pleistocene, when the glacial and interglacial periods began to shift towards a 100 ka frequency, a vigorous bottom-current regime was installed and deposited small contouritic deposits closely related to almost full-grown coral banks (Fig. 15c). The special setting of this sediment body, the morphology of which has been influenced by the presence of the coral banks and the underlying palaeotopography, makes this drift probably one of the smallest known confined contourite drifts (Fig. 15d). Side-scan sonar imagery has already proved the presence of a sandy contourite sheet with sand waves over a large part of the confined contourite drift. This Holocene sandy contourite shares many characteristics with the interglacial part of core MD99-2327, located between 1500 and 2625 cm. During

(a) *Late Pliocene*

Pliocene (U2)
Middle Miocene sediment waves (U3a)
Early Miocene buried drift (U3c)

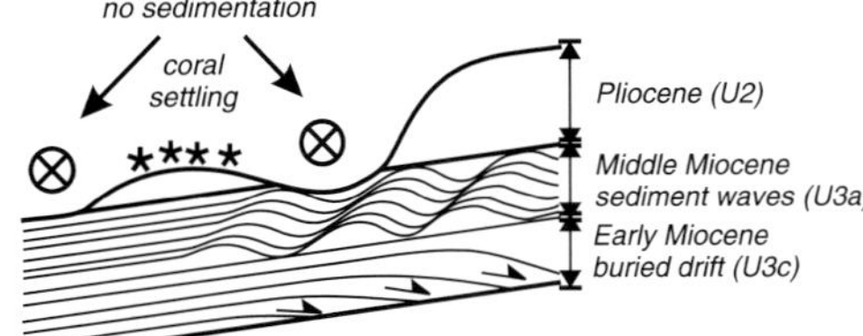

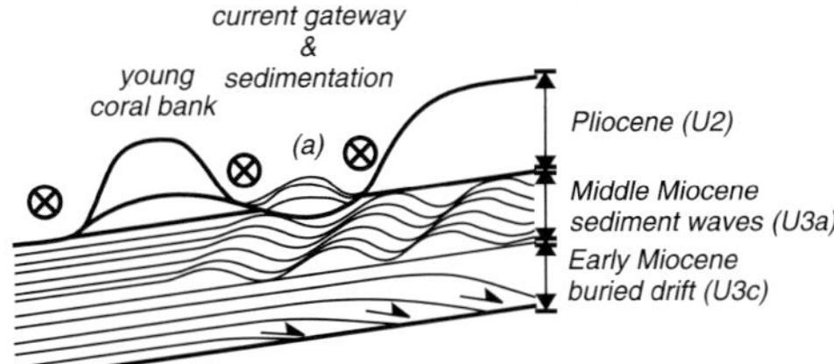

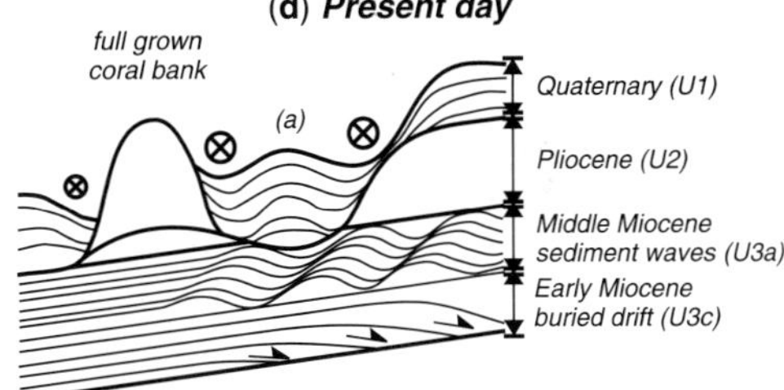

Fig. 15. Reconstruction of the depositional evolution, in an idealized SW–NE section, of the small confined contourite drift (a) in relation to the growth of a coral bank. The crossed circles illustrate the presence of a northward-flowing current.

this period, we can infer an environmental situation similar to that at present. This entire unit can be considered as a sandy contourite and also one of the largest known deep-water massive sands.

During the transition from isotope stage 5 to stage 4, the accumulating ice volumes of the pan-Atlantic ice sheets were responsible for a global sea-level drop and the distribution of the MOW in the Eastern Atlantic was seriously hampered. Because the presence of the MOW was vital to the vigorous BMP hydrodynamic environment, the activity of the glacial bottom currents was reduced. However, evidence of current reactivation, coupled with warmer periods, suggests that enough MOW sporadically entered the Porcupine Seabight to briefly enhance the bottom-current production. This glacial part of the core is a typical muddy contourite with a significant glacio-marine contribution.

Within this relatively small embayment in the NE Atlantic margin, typical features diagnostic of sediment drifts are encountered. The key factors allowing the construction of this contourite system are unique because they result from the interaction between a complex hydrographic system confined by obstacles in the palaeobathymetry, located in a relatively narrow zone of the slope. Evidently, this also calls for a more genetic nomenclature, indicating the causal mechanisms and depth ranges of the contourite deposits. Additionally, this example also proves that the location of this drift seems to be ideal to monitor the changes of the palaeoclimate and palaeoceanography record off Ireland.

This study of deep-sea sediment dynamics within the Porcupine Seabight was carried out within the framework of the EC FP5 programmes ECOMOUND and GEOMOUND and the Ghent University project 'GOA Porcupine–Belgica'. Core MD99-2327 has been acquired under the IMAGES programme. Support is gratefully acknowledged from the Ghent University Palaeontology Department (S. Louwye, S. Van Cauwenberghe) and the Marine Biology Section (M. Vinckx, D. Van Gansbeke and D. Schram) for laboratory use. We would also like to thank the captains and crews of R. V. *Belgica* and R. V. *Marion Dufresne*. The constructive suggestions of S. Ceramicola and B. De Mol significantly improved the manuscript. D.V.R. is a post-doctoral fellow funded by the FWO Flanders.

References

AKHURST, M. C., STOW, D. A. V. & STOKER, M. S. 2002. Late Quaternary glacigenic contourite, debris flow and turbidite process interaction in the Faroe–Shetland Channel, NW European Continental Margin. *In*: STOW, D. A. V., PUDSEY, C. J., HOWE, J. A., FAUGÈRES, J.-C. & VIANA, A. R. (eds) *Deep-Water Contourite Systems: Modern Drifts and Ancient Series, Seismic and Sedimentary Characteristics*. Geological Society, London, Memoirs, **22**, 73–84.

ARMISHAW, J. E., HOLMES, R. W. & STOW, D. A. V. 2000. The Barra Fan: a bottom-current reworked, glacially-fed submarine fans system. *Marine and Petroleum Geology*, **17**, 219–238.

AUFFRET, G. A., ZARAGOSI, S., DENNIELOU, B., *ET AL.* 2002. Terrigenous fluxes at the Celtic margin during the last glacial cycle. *Marine Geology*, **188**, 79–108.

BARD, E. 1998. Geochemical and geophysical implications of the radiocarbon calibration. *Geochimica et Cosmochimica Acta*, **62**, 2025–2038.

BOWEN, D. Q., PHILLIPS, F. M., MCCABE, A. M., KNUTZ, P. C. & SYKES, G. A. 2002. New data for the Last Glacial Maximum in Great Britain and Ireland. *Quaternary Science Reviews*, **21**, 89–101.

CARTER, L. & MCCAVE, I. N. 1994. Late Quaternary sediment pathways through the deep ocean, east of New Zealand. *Paleoceanography*, **9**, 1061–1085.

DE MOL, B., VAN RENSBERGEN, P., PILLEN, S., *ET AL.* 2002. Large deep-water coral banks in the Porcupine Basin, southwest of Ireland. *Marine Geology*, **188**, 193–231.

DOWLING, L. A. & COXON, P. 2001. Current understanding of Pleistocene temperate stages in Ireland. *Quaternary Science Reviews*, **20**, 1631–1642.

EDIGER, V., VELEGRAKIS, A. F. & EVANS, G. 2002. Upper slope sediment waves in the Cilician Basin, northeastern Mediterranean. *Marine Geology*, **192**, 321–333.

ELLIOT, M., LABEYRIE, L. D., BOND, G., CORTIJO, E., TURON, J.-L., TISNERAT, N. & DUPLESSY, J. C. 1998. Millennial-scale iceberg discharges in the Irminger Basin during the last glacial period: relationship with the Heinrich events and environmental settings. *Paleoceanography*, **13**, 433–446.

FAUGÈRES, J.-C. & STOW, D. A. V. 1993. Bottom-current-controlled sedimentation: a synthesis of the contourite problem. *Sedimentary Geology*, **82**, 287–297.

FAUGÈRES, J.-C., STOW, D. A. V., IMBERT, P. & VIANA, A. R. 1999. Seismic features diagnostic of contourite drifts. *Marine Geology*, **162**, 1–38.

FOUBERT, A., BECK, T., WHEELER, A. J., *ET AL.* 2005. New view of the Belgica Mounds, Porcupine Seabight, NE Atlantic: preliminary results from the *Polarstern* ARK-XIX/3a ROV cruise. *In*: FREIWALD, A. & ROBERTS, J. M. (eds) *Deep-water Corals and Ecosystems*. Springer, Heidelberg, 403–415.

FREIWALD, A., WILSON, J. B. & HENRICH, R. 1999. Grounding Pleistocene icebergs shape recent deep-water coral reefs. *Sedimentary Geology*, **125**, 1–8.

HABGOOD, E., KENYON, N. H., MASSON, D. G., AKHMETZHANOV, A. M., WEAVER, P. P. E., GARDNER, J. & MULDER, T. 2003. Deep-water sediment wave fields, bottom current sand channels and gravity flow channel-lobe systems: Gulf of Cadiz, NE Atlantic. *Sedimentology*, **50**, 483–510.

HARGREAVES, P. M. 1984. The distribution of Decapoda (Crustacea) in the open ocean and near-bottom over an adjacent slope in the northern North-East Atlantic Ocean during Autumn 1979. *Journal of the Marine Biological Association of the United Kingdom*, **64**, 829–857.

HENRIET, J.-P., DE MOL, B., PILLEN, S., *ET AL.* 1998. Gas hydrate crystals may help build reefs. *Nature*, **391**, 648–649.

HERNANDEZ-MOLINA, F. J., SOMOZA, L., VAZQUEZ, J. T., LOBO, F., FERNANDEZ-PUGA, M. C., LLAVE, E. & DIAZ-DEL RIO, V. 2002. Quaternary stratigraphic stacking patterns on the continental shelves of the southern Iberian Peninsula: their relationship with global climate and palaeoceanographic changes. *Quaternary International*, **92**, 5–23.

HUVENNE, V., BLONDEL, P. & HENRIET, J.-P. 2002. Textural analyses of sidescan sonar imagery from two mound provinces in the Porcupine Seabight. *Marine Geology*, **189**, 323–341.

HUVENNE, V. A. I., DE MOL, B. & HENRIET, J.-P. 2003. A 3D seismic study of the morphology and spatial distribution of buried coral banks in the Porcupine Basin, SW of Ireland. *Marine Geology*, **198**, 5–25.

KENYON, N. H., IVANOV, M. K., AKHMETZHANOV, A. M. & NEW, A. L. 1998. The current swept continental slope and giant carbonate mounds to the West of Ireland. *In*: DE MOL, B. (ed.) *Geosphere–Biosphere Coupling: Carbonate Mud Mounds and Cold Water Reefs*. IOC Workshop Report, **143**, 24.

KNUTZ, P. C., AUSTIN, W. E. N. & JONES, E. J. W. 2001. Millennial-scaled depositional cycles related to British Ice Sheet variability and North Atlantic paleocirculation since 45 kyr B.P., Barra Fan, U.K. margin. *Paleoceanography*, **16**, 53–64.

KNUTZ, P. C., HALL, I. R., ZAHN, R., RASMUSSEN, T., KUIJPERS, A., MOROS, M. & SHACKLETON, N. J. 2002. Multidecadal ocean variability and NW European ice sheet surges during the last deglaciation. *Geochemistry, Geophysics, Geosystems*, **3**, 1077.

MARANI, M., ARGNANI, A., ROVERI, M. & TRINCARDI, F. 1993. Sediment drifts and erosional surfaces in the central Mediterranean: seismic evidence of bottom-current activity. *Sedimentary Geology*, **82**, 207–220.

MCCABE, A. M. & CLARK, P. U. 1998. Ice-sheet variability around the North Atlantic Ocean during the last deglaciation. *Nature*, **392**, 373–377.

MCCAVE, I. N., MANIGHETTI, B. & ROBINSON, S. G. 1995. Sortable silt and fine sediment size/composition slicing: parameters for palaeocurrent speed and palaeoceanography. *Paleoceanography*, **10**, 593–610.

MOORE, J. G. & SHANNON, P. M. 1991. Slump structures in the Late Tertiary of the Porcupine Basin, offshore Ireland. *Marine and Petroleum Geology*, **8**, 184–197.

NEW, A. L., BARNARD, S., HERRMANN, P. & MOLINES, J.-M. 2001. On the origin and pathway of the saline inflow to the Nordic Seas: insights from models. *Progress in Oceanography*, **48**, 255–287.

PEARSON, I. & JENKINS, D. G. 1986. Unconformities in the Cenozoic of the North-East Atlantic. *In*: SUMMERHAYES, C. P. & SHACKLETON, N. J. (eds) *North Atlantic Palaeoceanography*. Geological Society, London, Special Publications, **21**, 79–86.

PINGREE, R. D. & LE CANN, B. 1989. Celtic and Armorican slope and shelf residual currents. *Progress in Oceanography*, **23**, 303–338.

PINGREE, R. D. & LE CANN, B. 1990. Structure, strength and seasonality of the slope currents in the Bay of Biscay region. *Journal of the Marine Biological Association of the United Kingdom*, **70**, 857–885.

POLLARD, R. T., GRIFFITHS, M. J., CUNNINGHAM, S. A., READ, J. F., PÉREZ, F. F. & RIOS, A. F. 1996. Vivaldi 1991—a study of the formation, circulation and ventilation of Eastern North Atlantic Central Water. *Progress in Oceanography*, **37**, 167–172.

RAYMO, M. E., OPPO, D. & CURRY, W. B. 1997. The mid-Pleistocene climate transition: a deep sea carbon isotopic perspective. *Paleoceanography*, **12**, 546–559.

REBESCO, M. & STOW, D. A. V. 2001. Seismic expression of contourites and related deposits: a preface. *Marine Geophysical Researches*, **22**, 303–308.

REED, D. L., MEYER, A. W., SILVER, E. A. & PRASETYO, H. 1987. Contourite sedimentation in an intraoceanic forearc system: eastern Sunda Arc, Indonesia. *Marine Geology*, **76**, 223–242.

RICE, A. L., THURSTON, M. H. & NEW, A. L. 1990. Dense aggregations of a hexactinellid sponge, *Pheromena carpenteri*, in the Porcupine Seabight (northeast Atlantic Ocean), and possible causes. *Progress in Oceanography*, **24**, 179–196.

RICE, A. L., BILLET, D. S. M., THURSTON, M. H. & LAMPITT, R. S. 1991. The Institute of Oceanographic Sciences Biology programme in the Porcupine Seabight: background and general introduction. *Journal of the Marine Biological Association of the United Kingdom*, **71**, 281–310.

SCHÖNFELD, J. & ZAHN, R. 2000. Late Glacial to Holocene history of the Mediterranean Outflow. Evidence from benthic foraminiferal assemblages and stable isotopes at the Portuguese margin. *Palaeogeography, Palaeoclimatology, Palaeoecology*, **159**, 85–111.

STOW, D. A. V. 1982. Bottom currents and contourites in the North Atlantic. *Bulletin de l'Institut de Géologie du Bassin d'Aquitaine*, **31**, 151–166.

STOW, D. A. V. & JOHANSSON, M. 2000. Deep-water massive sands: nature, origin and hydrocarbon implications. *Marine and Petroleum Geology*, **17**, 145–174.

STOW, D. A. V. & MAYALL, M. 2000. Deep-water sedimentary systems: new models for the 21st century. *Marine and Petroleum Geology*, **17**, 125–135.

STOW, D. A. V. & PIPER, D. J. W. 1984. Deep-water fine-grained sediments: facies models. *In*: STOW, D. A. V. & PIPER, D. J. W. (eds) *Fine-Grained Sediments, Deep-Water Processes and Facies*. Geological Society, London, Special Publications, **15**, 611–646.

STOW, D. A. V., FAUGÈRES, J.-C., HOWE, J. A., PUDSEY, C. J. & VIANA, A. R. 2002. Bottom currents, contourites and deep-sea sediment drifts: current state-of-the-art. *In*: STOW, D. A. V., PUDSEY, C. J., HOWE, J. A., FAUGÈRES, J.-C. & VIANA, A. R. (eds) *Deep-Water Contourite Systems: Modern Drifts and Ancient Series,*

Seismic and Sedimentary Characteristics. Geological Society, London, Memoirs, **22**, 7–20.

Van Aken, H. M. 2000. The hydrography of the mid-latitude northeast Atlantic Ocean I. *Deep-Sea Research I*, **47**, 757–788.

Van Rooij, D. 2004. *An integrated study of Quaternary sedimentary processes on the eastern slope of the Porcupine Seabight, SW of Ireland*. PhD thesis, Ghent University.

Van Rooij, D., De Mol, B., Huvenne, V., Ivanov, M. K. & Henriet, J.-P. 2003. Seismic evidence of current-controlled sedimentation in the Belgica mound province, upper Porcupine slope, southwest of Ireland. *Marine Geology*, **195**, 31–53.

Van Rooij, D., Blamart, D., Richter, T., Wheeler, A., Kozachenko, M. & Henriet, J.-P. 2006. Quaternary sediment dynamics in the Belgica mound province, Porcupine Seabight: ice rafting events and contour current processes. *International Journal of Earth Sciences*, doi: 10.1007/s00531-006-0068-6.

Wheeler, A. J., Bett, B. J., Billet, D. S. M. & Masson, D. G. 2000. Very high resolution side-scan mapping of deep-water coral mounds: surface morphology and processes affecting growth. *EOS Transactions, American Geophysical Union*, **81**, 48.

White, M. 2006. The hydrographic setting for the carbonate mounds of the Porcupine Bank and Sea Bight. *International Journal of Earth Sciences*, doi: 10.1007/s00531-006-0099-1.

White, M., Mohn, C., de Stigter, H. & Mottram, G. 2005. Deep-water coral development as a function of hydrodynamics and surface productivity around the submarine banks of the Rockall Trough, NE Atlantic. *In*: Freiwald, A. & Roberts, J. M. (eds) *Deep-water Corals and Ecosystems*. Springer, Heidelberg, 503–514.

The Eirik Drift: a long-term barometer of North Atlantic deepwater flux south of Cape Farewell, Greenland

S. E. HUNTER, D. WILKINSON, J. STANFORD, D. A. V. STOW, S. BACON, A. M. AKHMETZHANOV & N. H. KENYON

National Oceanography Centre Southampton, University of Southampton Waterfront Campus, European Way, Southampton SO14 3ZH, UK (e-mail: sallyh@noc.soton.ac.uk)

Abstract: The Eirik Drift lies on the slope and rise off the southern tip of the Greenland margin where it formed under the influence of the North Atlantic deep western boundary current. The drift contains a semi-continuous and often expanded sedimentary record ranging from Early Eocene to Holocene and so contains a record of bottom current strengths over decadal to millennial time scales. These variations in current strength can be related to changes in thermohaline circulation and climate. The drift body is composed of four seismic sequences, with a number of internal discontinuities, reflecting a variety of palaeoceanographic events. Three secondary ridges are observed trending to the NW from the main ridge crest. The presence of these ridges, which have been active since the Early Pliocene, suggests that the deep current separates into three strands as it crosses the Eirik Drift, with each strand depositing a separate ridge. Variation in the degree of lateral migration within the Early to Late Pliocene sequence between ridges reflects local variation in the angle of slope on which the ridges formed. Cyclicity of reflector amplitude within the Late Pliocene to Pleistocene sequence could reflect changes in carbonate accumulation and deep current strength linked to glacial–interglacial variations.

The Eirik Drift is an elongate, mounded contourite drift lying on the slope and rise off the southern tip of the Greenland margin to the south of Cape Farewell (Fig. 1). The drift formed under the influence of the North Atlantic deep western boundary current (DWBC) mainly during the Pliocene and Pleistocene (e.g. Arthur *et al.* 1989). This current forms the main part of the deep, southward-flowing section of the North Atlantic thermohaline circulation system (THC). It is widely accepted that the shallow, northward-flowing limb of this circulation system is responsible for the relatively warm modern Northern European climate. THC is known to have varied on a number of time scales associated with different climatic events triggered by different forcings. On longer time scales, THC changes are thought to amplify the orbitally forced glacial–interglacial cycles. Increasing high-latitude sea-ice cover during cold periods can restrict deep-water formation, weakening or repositioning the THC. The resultant reduction in northward heat transport leads to regional cooling. This may in turn increase ice cover and planetary albedo, leading to global cooling. THC changes can also be triggered by more localized events such as the massive release of glacial meltwaters in areas of deep-water formation. Climate changes triggered by THC shifts can be exceptionally fast, typically less than 100 years. Examples of such 'rapid' climate changes include cooling events such as the Younger Dryas and 8.2 ka event (Broeker 2000; Clark *et al.* 2002; Rahmstorf 2002).

Contourite drifts contain millennial-scale records, but with a resolution high enough to resolve decadal-scale events. The Eirik Drift contains a semi-continuous sedimentary record ranging from the Early Eocene to Holocene and, as such, provides an excellent opportunity to study past changes in North Atlantic THC. A considerable volume of literature exists regarding various aspects of the evolution, sedimentology, biostratigraphy and isotope geochemistry of the drift (e.g. Chough & Hesse 1985; Arthur *et al.* 1989) including a number of significant contributions resulting from Ocean Drilling Program (ODP) Leg 105 to the Labrador Sea and Baffin Bay (Srivastava *et al.* 1989*a*). The first aim of this paper is to provide a review of the existing literature regarding the Eirik Drift, focusing on large-scale drift evolution and associated palaeoceanographic events. Following this, new seismic data are presented and correlated with the existing seismic database for the area. Finally, contour mapping of the major seismic stratigraphic sequences is used to identify morphological variations throughout the evolution of the drift. Analysis of the seismic character and morphological variations within the seismic sequences allows improved understanding of drift development processes and the identification of long-term cycles of current variability within the Late Pliocene–Pleistocene sequence.

From: VIANA, A. R. & REBESCO, M. (eds) *Economic and Palaeoceanographic Significance of Contourite Deposits*. Geological Society, London, Special Publications, **276**, 245–263.
0305-8719/07/$15.00

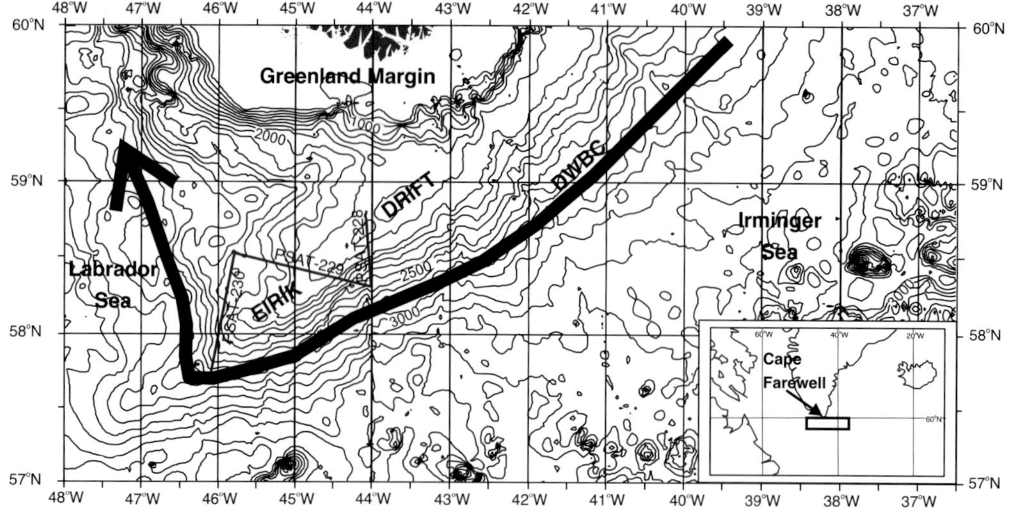

Fig. 1. Regional bathymetric map of the Eirik Drift area (Smith & Sandwell 1997). (See inset for location.) DWBC, Deep Western Boundary Current.

This paper forms part of a continuing study of the sedimentary history of the Eirik Drift and the palaeoceanographic development in the Cape Farewell region, funded as part of the NERC Rapid Climate Change Program, which aims to provide a high-resolution, continuous, calibrated record of North Atlantic DWBC flux through the deglacial to Holocene period.

Tectonic setting

The Eirik Drift lies off the southern tip of Greenland at the junction between the Labrador Sea and the northern North Atlantic. The tectonic evolution of the area is therefore dominated by the opening history of these two basins. The formation of the Labrador Basin began with rifting between Greenland and Labrador during the Cenomanian (early Late Cretaceous) at the same time as sea-floor spreading was beginning in the North Atlantic. Rifting was followed by active sea-floor spreading in the southern Labrador Basin during the Campanian (mid-Late Cretaceous) and continued until the Late Eocene–Early Oligocene (Srivastava & Tapscott 1986).

The dominant structural trends in the study area relate to opening of the Labrador Basin and are composed of a series of NE–SW- and ENE–WSW-oriented fracture zones and associated perpendicular magnetic anomalies (Roest & Srivastava 1989). The Leif Fracture Zone, which lies just to the south of the Eirik Drift, separates older, continental–transitional crust of the southern Greenland margin to the NW from younger, deeply subsided oceanic crust to the SE, forming a relatively steep SE-facing slope (Arthur *et al.* 1989). An older fracture zone, the Farewell Fracture Zone, underlies the Eirik Drift and is associated with a WSW-trending basement high, which is thought to have controlled the initial formation of the drift (Le Pichon *et al.* 1971; Srivastava & Arthur 1989). A series of NW–SE-trending basement highs underlie the NW flank of the drift and run parallel to the magnetic anomalies in this area (Srivastava & Arthur 1989).

Tectonics in the Norwegian–Greenland Sea region, particularly the subsidence history of the Greenland–Scotland Ridge, have also exerted a significant influence on the evolution of the Eirik Drift by forming a structural control on the flow of northern-sourced bottom waters into the North Atlantic. Rifting began in the Norwegian–Greenland Sea area during the mid-Late Cretaceous, but active sea-floor spreading did not begin here until the Early Eocene (Srivastava & Tapscott 1986). The Greenland–Scotland Ridge is a large regional swell in the oceanic crust between Greenland and Scotland associated with the mantle plume that currently underlies Iceland (Wright 1998). The subsidence history of the ridge has yet to be fully resolved but it has been established that the ridge had subsided sufficiently to allow the flow of northern-sourced bottom waters into the northern North Atlantic by the middle Miocene (Wright & Miller 1996; Wright 1998).

Oceanographic setting

The modern DWBC in the region of Cape Farewell is concentrated between the 1900 m and 3000 m isobaths towards the bottom of the continental slope (Clarke 1984). The DWBC transport is commonly accepted to be about 13–14 Sv (1 Sv = $1 \times 10^6\ m^3\ s^{-1}$); for example, Dickson & Brown (1994) quoted 13.3 Sv for the flow below the 27.80 isopycnal. Although this value is often referred to it is largely based on a single dataset collected in 1978 by the R. V. *Hudson* (Clarke 1984). Bacon (1998) calculated a much lower value of 6 Sv from data collected in 1991 by R. R. S. *Charles Darwin*, and Bacon (1998) argued that a comparison of data collected between 1958 and 1997 illustrates the decadal variability of the DWBC, which he attributed to changes in the output from the Nordic Seas.

The DWBC in the vicinity of Cape Farewell is composed of four main water masses (e.g. Dickson & Brown 1994): the Denmark Strait Overflow Water (DSOW), Iceland Scotland Overflow Water (ISOW), Labrador Sea Water (LSW) and modified Antarctic Bottom Water (AABW) (Fig. 2).

DSOW is composed of Nordic Sea intermediate waters that cross the Denmark Strait sill with a maximum depth of about 550 m. After crossing the sill the overflow waters descend rapidly, entraining ambient waters, primarily LSW. The resultant modified DSOW is identifiable as the lower layer of the DWBC off Cape Farewell. The transport of DSOW across the sill and into the DWBC is estimated to be 2.9 Sv (Dickson & Brown 1994; Fig. 2) increasing to around 10 Sv through entrainment on route to Cape Farewell.

Similarly, ISOW is composed of Nordic Sea intermediate waters that cross the Iceland Scotland Ridge to the east of Iceland. Dickson & Brown (1994) estimated the total eastern overflows to be about 2.7 Sv, of which 1.7 Sv flows through the Faeroe Bank Channel, where the maximum sill depth is about 850 m. The remainder overflows via a series of five smaller channels between Iceland and the Faeroes. The density of the ISOW is reduced by entrainment as it travels around the Reykjanes Ridge and into the Irminger Sea via the Charlie Gibbs Fracture Zone (CGFZ), such that it forms the upper layer of the DWBC at a depth of around 2000 m. The contribution of this modified ISOW to the DWBC off Cape Farewell is estimated at between 2 and 3 Sv (Dickson & Brown 1994; Schmitz 1996).

LSW is formed by wintertime deep convection in the Labrador and Irminger Seas. Originally, as the name suggests, it was thought to be formed solely in the Labrador Sea, but more recent work (Bacon *et al.* 2003; Pickart *et al.* 2003) has concluded that a second formation site exists in the Irminger Sea. LSW spreads across the North Atlantic and populates the low-velocity layer between about 700 m and 1500 m off the east coast of Greenland. LSW contributes a significant proportion of the DWBC as a result of entrainment with the two types of overflow water. Dickson & Brown (1994) estimated the contribution of LSW to the DWBC at around 4 Sv off Cape Farewell. However, inverse modelling has produced a value as high as 8 Sv (Alvarez *et al.* 2004).

AABW spreads north from its point of formation in the Antarctic and after modification joins the southward-flowing DWBC at various points in the North Atlantic. Estimates of the component joining off Greenland are in the region of 1–2 Sv (Schmitz & McCartney 1993; Schmitz 1996). A further 2 Sv is thought to be entrained, equally split between sites off Newfoundland and Florida.

In summary, the DWBC off Cape Farewell provides the major input to North Atlantic Deep Water (NADW). NADW is usually considered to have formed by the time the DWBC reaches the Grand Banks of Newfoundland (the deep water transported in the vicinity of Cape Farewell is therefore referred to as Proto North Atlantic Deep Water in Fig. 2) after further addition of LSW, AABW and ISOW in the Labrador Basin, although further modification does occur along its southward path. The transport of the DWBC is typically considered to be between 16 and 18 Sv as it flows across the equator and into the Southern Atlantic (Schmitz & McCartney 1993). As NADW is essentially the lower limb of the North Atlantic THC, the strength of the DWBC off Cape Farewell has a major influence on the THC. However, we still lack detailed knowledge about the drivers and variability of this current. This is an area of active research, and a series of moorings placed on the continental slope off Cape Farewell in summer 2005 as part of the NERC Rapid Climate Change Program is expected to provide valuable new data on the variability of the DWBC.

Database and methods

The seismic database used for this study consists of three new high-resolution, single-channel seismic lines along with a number of older, published single- and multi-channel sections (Arthur *et al.* 1989; Srivastava *et al.* 1989*b*; see Table 1). The new lines were acquired during the Training Through Research-13 (TTR-13) cruise to the northern North Atlantic onboard the R. V. *Professor Logachev* during July 2003 (Kenyon *et al.* 2004). Four seismic sequences (Seismic Sequences

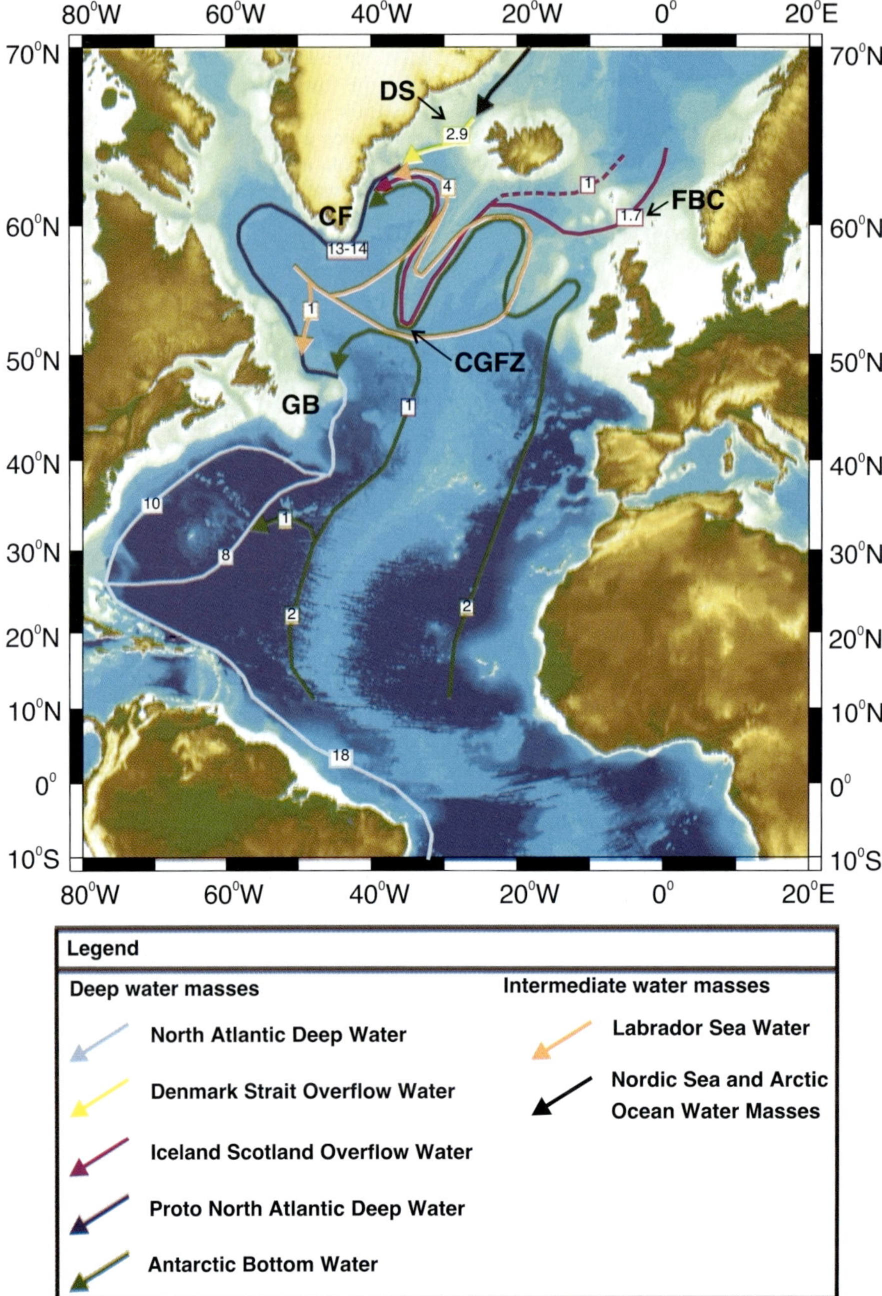

Fig. 2. Map of the North Atlantic region showing the water masses contributing to the formation of the North Atlantic Deep Water (modified from Schmitz 1996; Pickart *et al.* 2003). Boxed numbers refer to the volume flux in Sverdrups of the water masses at the locations indicated (see text for description). DS, Denmark Straits; CF, Cape Farewell; FBC, Faeroe Bank Channel; GB, Grand Banks; CGFZ, Charlie Gibbs Fracture Zone.

Table 1. *Seismic sections used in this study*

Line number	Type	Cruise	Reference
PSAT-228	Single-channel	TTR-13	This study
PSAT-229	Single-channel	TTR-13	This study
PSAT-230	Single-channel	TTR-13	This study
Line 10	Single-channel	HU-84-30	Srivastava *et al.* 1989*b*
Line 11	Single-channel	HU-84-30	Srivastava *et al.* 1989*b*
Line 12	Single-channel	HU-84-30	Srivastava *et al.* 1989*b*
Line 14	Single-channel	HU-84-30	Srivastava *et al.* 1989*b*
Line 15	Single-channel	HU-84-30	Srivastava *et al.* 1989*b*
Line 16	Single-channel	HU-84-30	Srivastava *et al.* 1989*b*
Line 19	Single-channel	HU-84-30	Srivastava *et al.* 1989*b*
Line 20	Single-channel	HU-84-30	Srivastava *et al.* 1989*b*
Line 21	Single-channel	HU-84-30	Srivastava *et al.* 1989*b*
BGR-1	Multi-channel	BGR77	Arthur *et al.* 1989
BGR-2	Multi-channel	BGR77	Arthur *et al.* 1989

1, 2, 3 and 4) have been identified within the published sections (Arthur *et al.* 1989). These range in age from Early Eocene to Pleistocene and are described in following sections. Only the upper two of these sequences can be recognized in TTR-13 lines, because of their much shallower penetration.

Depths (in seconds two-way travel time (TWT)) to the base of the upper two seismic sequences have been mapped and contoured, along with the thickness (in seconds TWT) of Seismic Sequences 1 and 2 and the combined thickness of Seismic Sequences 3 and 4. A map of depth to basement has been constructed from a compilation of published maps (Srivastava & Arthur 1989; Le Pichon *et al.* 1971; Tucholke & Fry 1985) with additional data from the new and old seismic sections.

Bathymetry

A new bathymetric map of the Eirik Drift has been constructed, based on the seismic lines listed in Table 1 along with unpublished bathymetric data from a recent cruise to the area by the R. R. S. *Charles Darwin*, and merged with the bathymetry of Smith & Sandwell (1997) around the margins of the drift. The new bathymetric map, presented in Figure 3, shows a similar pattern to the regional bathymetry of Smith & Sandwell (1997; Fig. 1), but more clearly shows the presence of three NW-trending secondary ridges on the northern drift flank. The new map is significantly different from GEBCO bathymetric charts, with the major difference being that the current dataset shows no evidence of a secondary, SW-trending ridge to the north of the main ridge crest as shown in GEBCO charts.

The Eirik Drift has an elongated, mounded morphology with a length/width ratio of *c.* 2:1 and elongation direction to the SW, oblique to the southern Greenland margin from which the drift extends. The main drift crest descends from around 1500 m adjacent to the Greenland slope to 3500 m 360 km to the SW. The southern flank of the drift, facing the SW-flowing limb of the DWBC, is characterized by a relatively steep and regular slope of around 1.3°. The northern drift flank and drift crest display marked changes in slope, with variation between 0.3 and 1.5°. These variations in slope define three secondary ridges, which extend to the NW from the main drift crest and have relatively steep southwestern flanks facing the DWBC as it flows NW into the Labrador Sea. The secondary ridge crests occur at 2000–2300 m, 2100–2600 m and 3200–3400 m, with the depth of each increasing to the NW, and are numbered Secondary Ridge (SR) 1, 2 and 3, respectively.

Seismic stratigraphy and phases of drift construction: syntheses of the ODP Site 646 results

Seismic sequences

Four seismic sequences have been identified in the Eirik Drift region (Arthur *et al.* 1989; Srivastava *et al.* 1989*b*). These sequences have been described in detail by Arthur *et al.* (1989), with reference to drilling results from ODP Site 646 on the northern flank of the Eirik Drift (location shown in Fig. 3). The following description therefore derives from Arthur *et al.* (1989) unless otherwise stated.

The oldest seismic sequence, Sequence 4, is Early Eocene to Late Miocene in age and is generally acoustically transparent. It overlies basement and is characterized by marked thickness variations

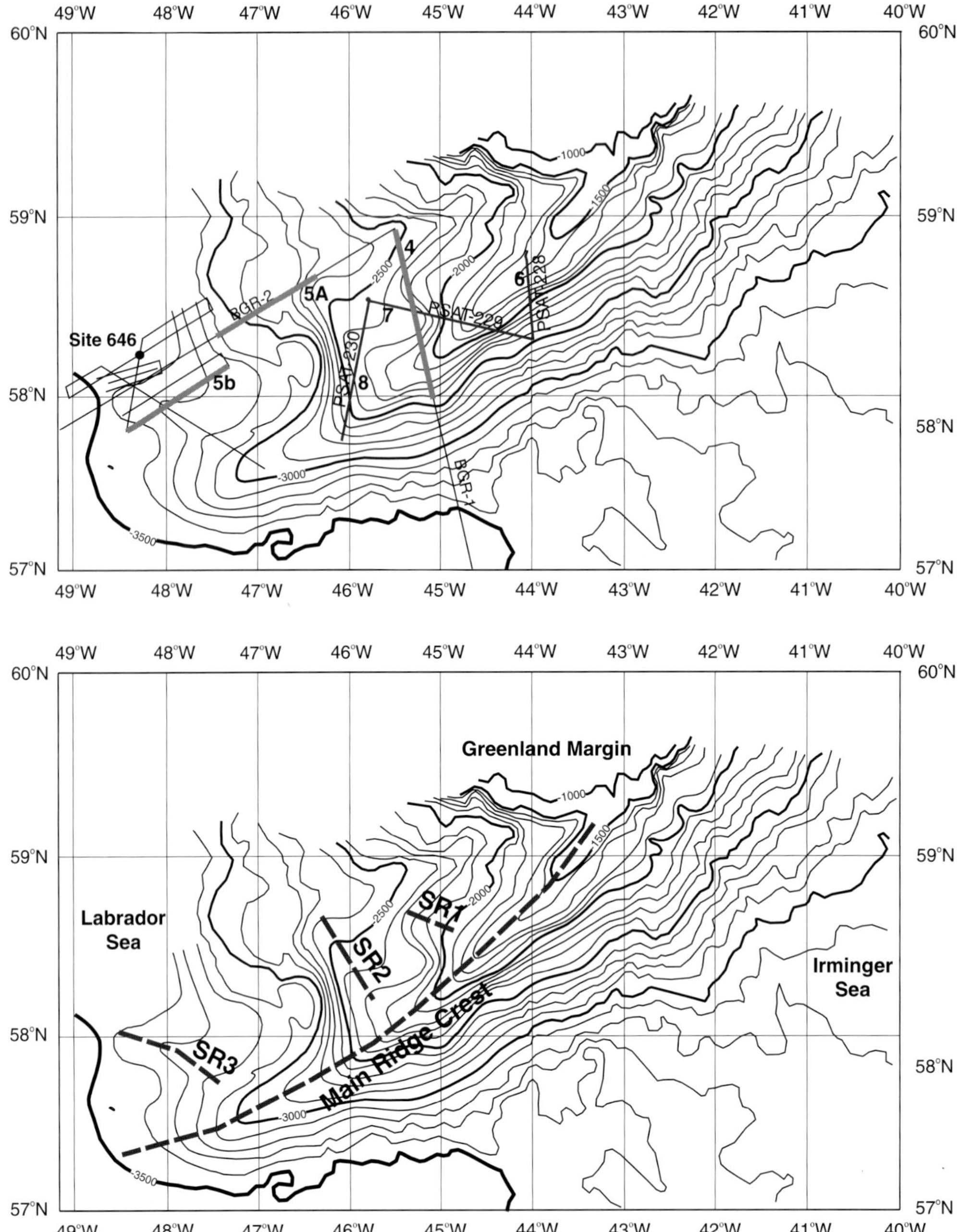

Fig. 3. New bathymetric map of the Eirik Ridge. The upper map shows the location of the seismic sections used in this study and the location of ODP Site 646. The positions of the main ridge crest and three secondary ridge crests are highlighted on the lower map.

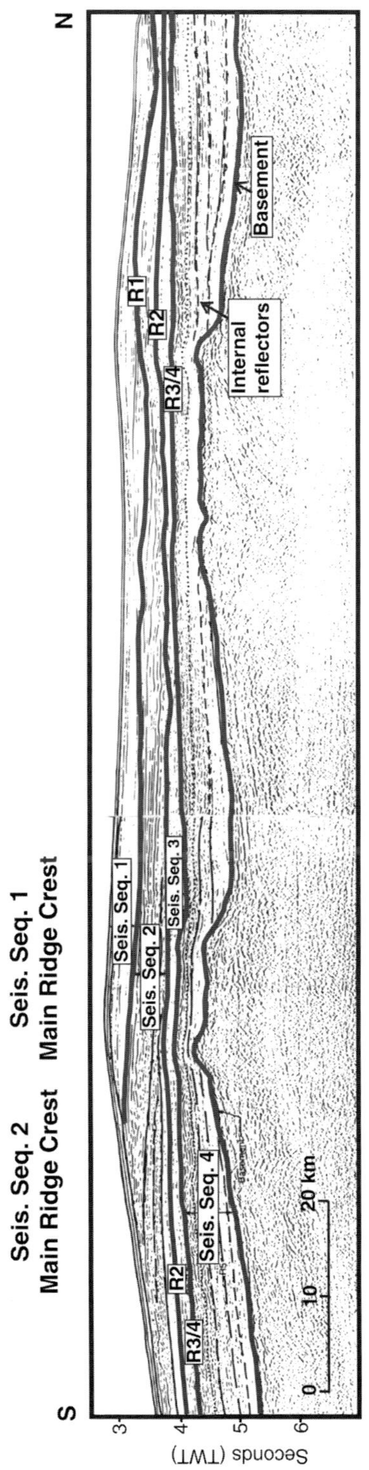

Fig. 4. Section from multi-channel line BGR-1 crossing the main ridge crest. From Arthur *et al.* (1989). (See location in Fig. 3.)

resulting from the infilling of irregular basement topography (Figs 4 and 5a and b). The overlying Sequence 3 is Late Miocene to Early Pliocene in age and consists of low- to moderate-amplitude reflectors forming a package of relatively uniform thickness throughout the area (Fig. 4). The upper part of Sequence 3 contains a prominent reflector (R2), which is dated at around 5.6 Ma, and thought to result from a short-term increase in carbonate preservation. Sequences 3 and 4 are separated by a high-amplitude double reflector (R3–R4) that represents changes in carbonate content and deposit physical properties relating to a short-term decrease in sedimentation rate.

Seismic Sequence 2 (of Early to Late Pliocene age) overlies Sequence 3 with an erosional unconformity across most of the area, which is dated at 4.5 Ma and marks the onset of strong bottom-current activity. The sequence is characterized by the presence of very high-amplitude, parallel to subparallel, northward-dipping reflectors that form four distinct and coeval depositional ridges. The ridges show northward progradation and approximately underlie the ridges seen in the modern bathymetry (Figs 4 and 5a and b). Migrating sediment waves are described within this sequence, although these are difficult to resolve within the published sections (Fig. 5b). Arthur *et al.* (1989) interpreted these characteristics as indicating that this sequence was deposited under the influence of significant bottom-current flow. A prominent reflector at 340 mbsf (metres below sea floor) at Site 646, dated at around 4 Ma, marks a change in the dominant biogenic material within the sediment from calcareous below to bio-siliceous above. This change reflects a significant cooling in surface waters at a time somewhat later than the onset of significant bottom-current activity. Also described at this time interval are erosional modification of the drift and changes in sediment grain-size and sorting parameters.

The uppermost seismic sequence, Sequence 1, is Late Pliocene to Pleistocene in age and consists of moderate- to high-amplitude reflectors that are parallel to subparallel to the sea floor and thought to be a function of local variations in the relative proportion of carbonate and clay in the sediment. The base of Sequence 1 is marked by a prominent reflector that coincides with the onset of ice-rafted sediment deposition in the area. Sequence 1 is generally conformable with Sequence 2, except in the vicinity of Sequence 2 ridge crests, where Sequence 1 onlaps the underlying unit (Fig. 4). Sequence 1 shows pronounced variations in thickness related to the upslope migration of ridge crests from Seismic Sequence 2 to 1. The characteristics of these sequences are summarized in Table 2.

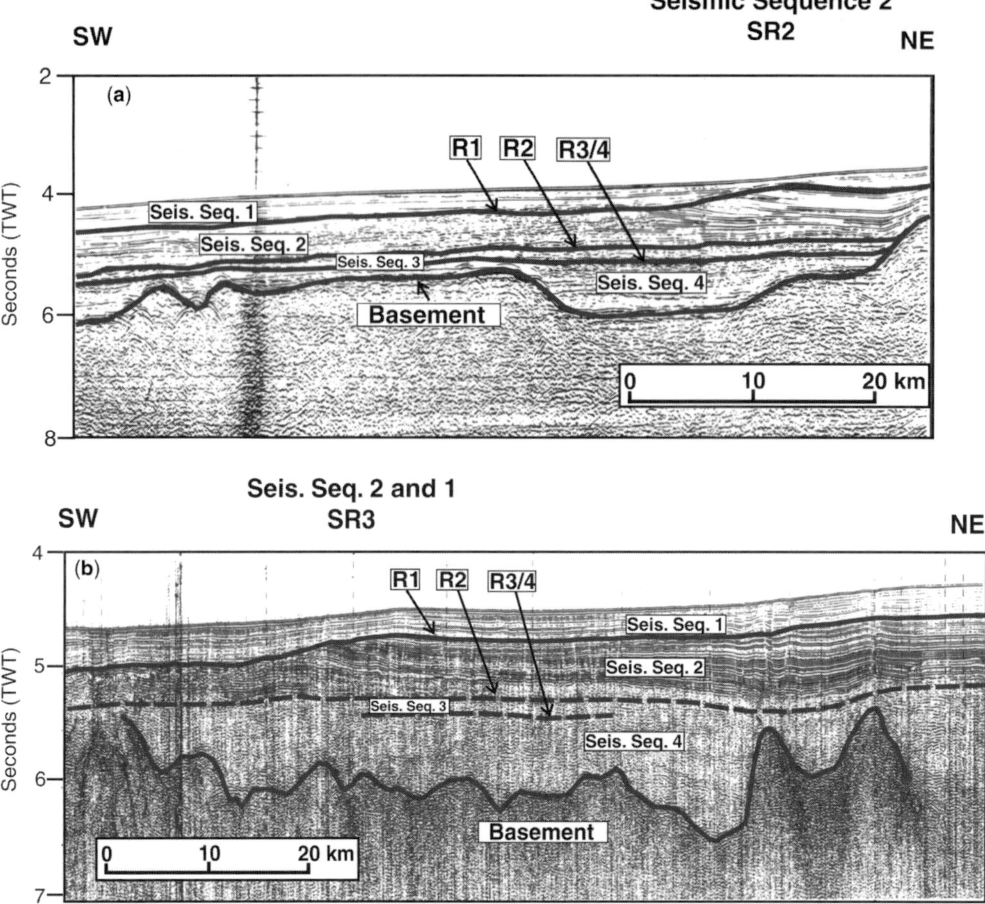

Fig. 5. (**a**) Section from multi-channel line BGR-2 crossing the central secondary ridge crest (SR2); (**b**) single-channel Line 15 crossing the southerly secondary ridge (SR3). From Arthur *et al.* (1989) and Srivastava *et al.* (1989*b*). (See locations in Fig. 3.)

Sedimentological, biostratigraphic and isotopic characteristics

Drilling results from ODP Site 646 show that all four seismic sequences are dominated by silty clays and clayey silts that are generally strongly bioturbated and contain variable proportions of biogenic material (Cremer *et al.* 1989), consistent with earlier core descriptions (Chough & Hesse 1985). Cremer *et al.* (1989) divided the sedimentary section of the Eirik Drift into two lithological sequences, with the upper lithological sequence corresponding to Seismic Sequence 1 and the lower sequence corresponding to Seismic Sequences 2, 3 and 4. These lithological sequences are distinguished on the basis of coarse sediment content, with the upper sequence showing a marked increase in the greater than 63 μm fraction. This coarse fraction in the upper lithological sequence contains some very large clasts and reflects the onset of ice rafting in the Eirik Drift area.

Variations in grain size within the lower lithological sequence were interpreted by Cremer (1989) as reflecting changes in bottom-current intensity during the Late Miocene and Early Pliocene. Cremer noted an increase in median grain size around the R3–R4 reflector (separating Seismic Sequences 3 and 4) resulting from increased bottom-current influence at around 7.5 Ma. This event is followed by generally decreasing grain size, indicating weakening bottom-current influence, until around the depth of the R2 reflector, when grain size increases again, marking increased

Table 2. *Principal characteristics of the seismic sequences in the Eirik Drift area summarized from Arthur et al. (1989) and Srivastava et al. (1989*b*)*

Principal characteristics
Seismic Sequence 1; Late Pliocene to Pleistocene
Moderate- to high-amplitude reflectors, parallel to subparallel to the sea floor
Base (reflector R1) is conformable to unconformable with Sequence 2 with onlap in some areas
R1 is a very high-amplitude reflector closely corresponding to the onset of ice rafting in the area
Thinnest over Sequence 2 ridge crests and thickest on the lee sides of Sequence 2 ridges
Seismic Sequence 2; Early to Late Pliocene
High-amplitude parallel to subparallel, commonly dipping reflectors
Sequence contains a number of pronounced depositional ridges
Migrating sediment waves common
Base conformable to unconformable with Sequence 3, with an erosional contact in some areas
Seismic Sequence 3; Late Miocene to Early Pliocene
Continuous to discontinuous low- to moderate-amplitude reflectors
Contains a moderate- to high-amplitude reflector (R2) in upper part of sequence
Sequence has a generally constant thickness
Base marked by high-amplitude double reflector (R3–R4) and is conformable to unconformable with Sequence 4
Seismic Sequence 4; Early Eocene to Late Miocene
Generally acoustically transparent with the exception of a prominent reflector (R5) near the base of the sequence
The sequence overlies basement and shows pronounced thickness variations, thinning across basement highs and thickening into adjacent troughs

bottom-current activity at around 5.6 Ma. Silt beds first appear in the sequence at around the base of Seismic Sequence 2, indicating the onset of strong bottom-current activity (Cremer 1989) concurrently with the onset of drift construction observed in seismic sections (Arthur *et al.* 1989).

Study of the benthic foraminifers revealed the occurrence of several distinct assemblages characterizing the different seismic sequences, with turnovers in assemblage reflecting changes in water-mass properties (Kaminski *et al.* 1989). The following description of benthic foraminiferal assemblages and associated palaeoenvironmental and palaeoceanographic interpretations is summarized from Kaminski *et al.* (1989).

Seismic Sequence 4 (Early Eocene to Late Miocene) is associated with a benthic foraminiferal assemblage dominated by *Nuttallides umbonifera*, with associated fine agglutinated species. This assemblage represents an environment with low current energy and corrosive bottom waters undersaturated with respect to calcium carbonate. In contrast, Seismic Sequence 3 (Late Miocene to Early Pliocene) is dominated by an assemblage of coarse agglutinated taxa with affinities to Norwegian–Greenland Sea faunas, with subordinate species associated with environments influenced by NADW-type water masses. This assemblage suggests an environment with significant northern-sourced bottom-current flow. This major change in benthic ecology between Seismic Sequences 4 and 3 was interpreted by Kaminski *et al.* (1989) as marking the onset of the flow of Denmark Straits Overflow Water (DSOW) into the Eirik Drift area.

Agglutinated taxa disappear before the deposition of Seismic Sequence 2 (Early to Late Pliocene), suggesting increasing current strength. Seismic Sequence 2 contains a high proportion of calcareous species typical of modern deep-water environments influenced by components of the NADW. This association therefore suggests increased strength of northern-sourced bottom-water currents. Seismic Sequence 1 (Late Pliocene to Pleistocene) contains low-abundance benthic associations typical of glacial environments, indicating the onset of glacial conditions.

Isotopic, planktonic formainiferal, sediment flux and magnetic grain-size studies of the Pleistocene part of the drift sequence reveal strong glacial–interglacial cyclicity. Deep-water areas show generally increased terrigenous and pelagic sediment accumulation rates during interglacial stages and low magnetic grain size and relatively low sediment accumulation rates during glacial stages (Hall *et al.* 1989; Hillaire-Marcel *et al.* 1994). Conversely, sediment accumulation rates at intermediate depths are higher during glacial stages, with relatively condensed interglacial sediments reflecting low sedimentation rates and/or erosion (Hillaire-Marcel *et al.* 1994).

New seismic sections

Seismic sequences and deposit geometry

The new seismic sections are of higher resolution than the published sections and cross both the main SW-trending ridge and the northern and

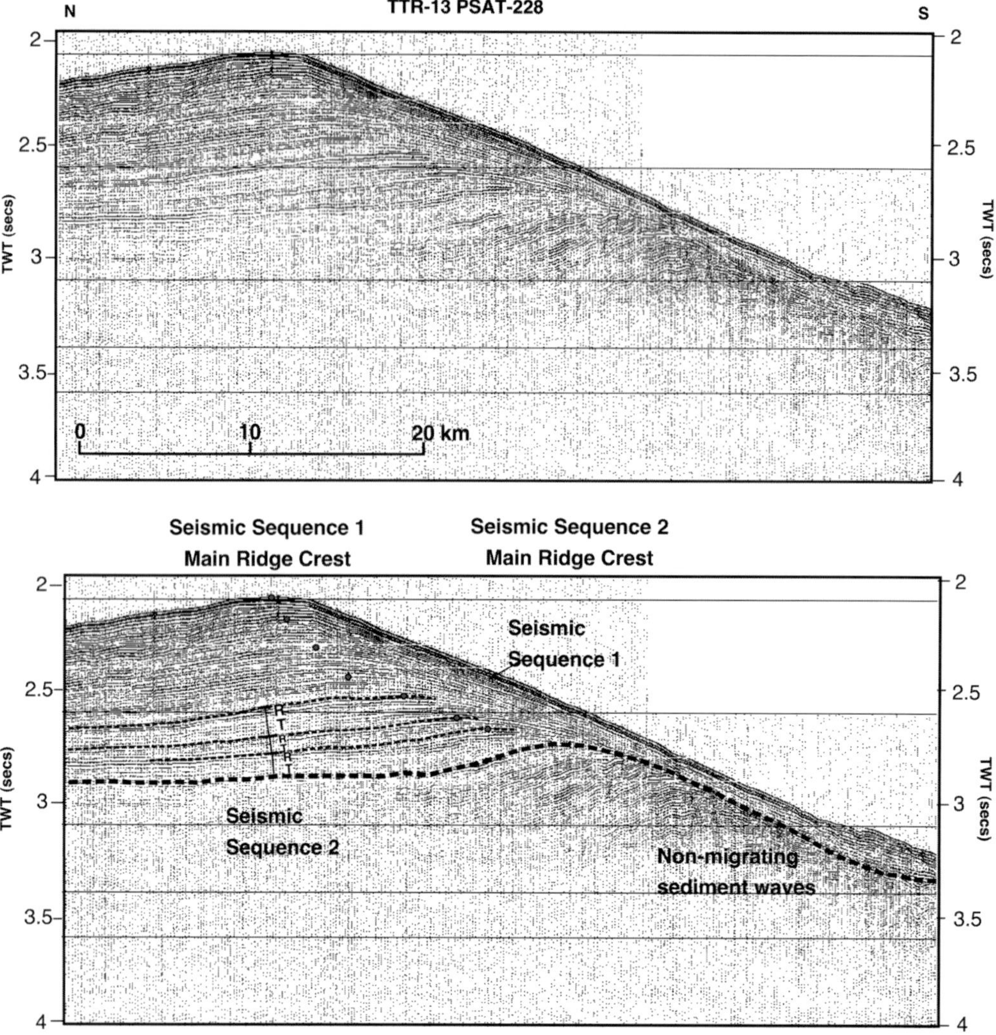

Fig. 6. Line PSAT-228 with uninterpreted section above and interpreted section below. (See Fig. 3 for location.) Within Seismic Sequence 1, T and R indicate alternating transparent and reflective seismic character, and the grey spots mark the position of successive ridge crests to highlight variations in lateral migration.

central secondary NW-trending ridges (locations shown in Fig. 3). PSAT-228 (Fig. 6) crosses the main ridge crest at the most northerly point of any of the seismic sections used in this study. PSAT-229 (Fig. 7) also crosses the main ridge, just to the south of PSAT-228, as well as the southern end of the most northerly secondary ridge (SR1). PSAT-230 (Fig. 8) does not cross the main SW-trending ridge crest, but instead crosses the central secondary ridge (SR2).

Two sequences can be recognized within these high-resolution seismic sections, with the upper and lower sequences correlating with Seismic Sequences 1 and 2 of Srivastava *et al.* (1989*b*), respectively. The upper sequence can therefore be dated as Late Pliocene to Pleistocene in age and the lower sequence as Early to Late Pliocene.

Seismic Sequence 2: Early to Late Pliocene. Sequence 2 displays the same general characteristics as described by Arthur *et al.* (1989) and Srivastava *et al.* (1989*b*), with a series of high-amplitude reflectors forming pronounced depositional ridges. The new seismic sections allow more detailed observations of the internal structure within Sequence 2 ridge crests to be made.

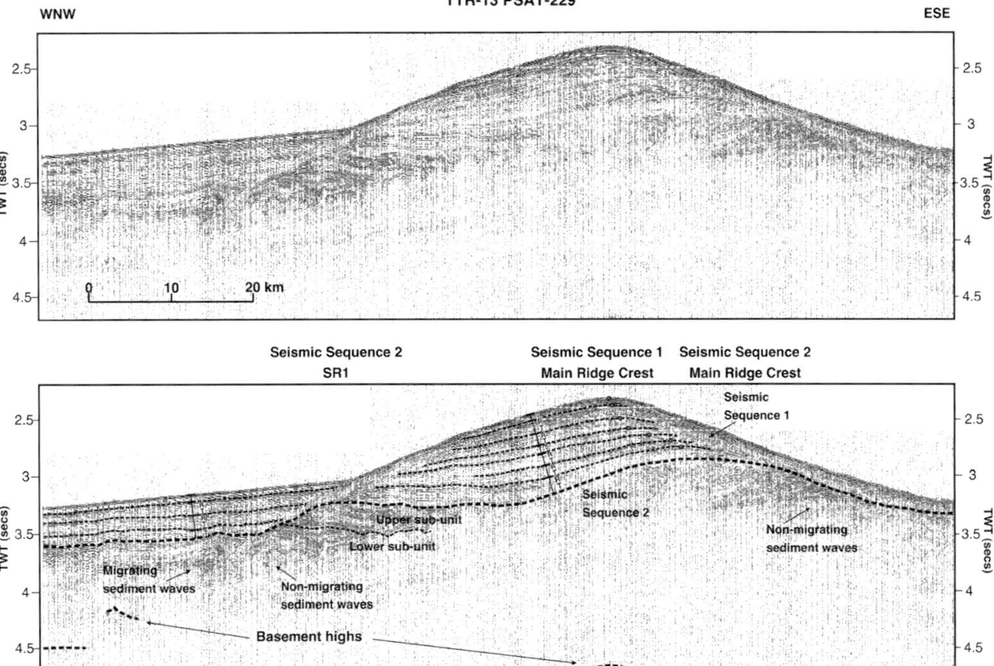

Fig. 7. Line PSAT-229 with uninterpreted section above and interpreted section below. (See Fig. 3 for location.) Within Seismic Sequence 1, T and R indicate alternating transparent and reflective seismic character, and the grey spots mark the position of successive ridge crests to highlight variations in lateral migration.

The main Sequence 2 ridge is composed of north-dipping reflectors that show erosional truncation on the southern, current-facing flank and form a single large drift body (Figs 6 and 7). The sequence here contains numerous sediment waves that appear to be non-migratory (Fig. 6). The degree of lateral migration of the main crest is difficult to determine because of the erosion on the southern flank, but appears to be low. The northern secondary ridge (SR1) has a similar internal structure with approximately SE-dipping reflectors and widespread sediment waves that again appear to be non-migratory. Just downslope of the southeastern flank of SR1, sediment waves are observed, which migrate toward the ridge (Fig. 7). SR1 is composed of two sub-units with the ridge crest stepping upslope, with *c.* 5 km of lateral, approximately westward migration between crests of successive sub-units (Fig. 7). The central secondary ridge crest (SR2) shows a different internal structure again (Fig. 8), being composed of several small, stacked build-ups with a total of 12 km of lateral upslope migration to the NW between successive crests, as previously described by Earley *et al.* (2002) from seismic sections just to the north of PSAT-230. Earley *et al.* (2002) interpreted this stacking pattern as reflecting shallowing of the core of the DWBC, with new build-ups forming upslope of the previous one as the current shallows.

Three very different internal ridge structures are therefore observed, with the main differences between types being the number of internal sub-units and degree of lateral migration between sub-units.

Seismic Sequence 1: Late Pliocene to Pleistocene. This sequence again shows the same general characteristics as described by Srivastava *et al.* (1989*b*), but again the new seismic sections allow the recognition of more detailed features. The sequence is dominated by moderate- to high-amplitude reflectors, parallel to subparallel with the sea floor, forming sedimentary ridges upslope from the Sequence 2 ridges, as described by Srivastava *et al.* (1989*b*). Marked variations in the degree of lateral migration of the drift crests are observed up-sequence, with rapid initial migration and minimal migration toward the top of the sequence (Figs 6–8). This pattern is observed on both the main drift crest and the central secondary ridge.

The new sections reveal cyclicity of reflector amplitude within this upper sequence, with each

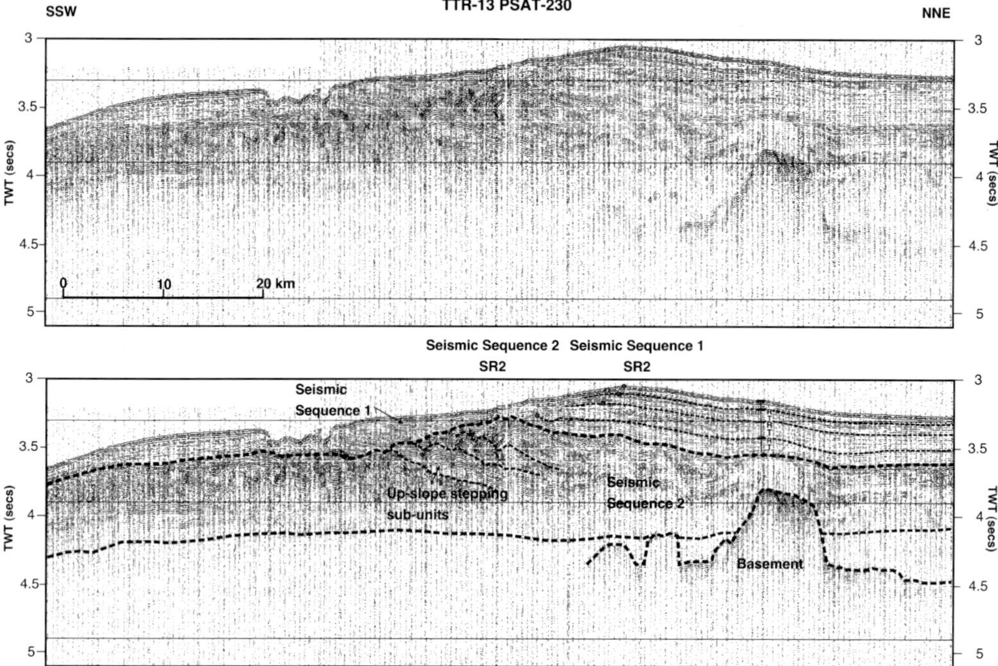

Fig. 8. Line PSAT-230 with uninterpreted section above and interpreted section below. (See Fig. 3 for location.) Within Seismic Sequence 1, T and R indicate alternating transparent and reflective seismic character, and the grey spots mark the position of successive ridge crests to highlight variations in lateral migration.

cycle consisting of a lower section of low- to moderate-amplitude reflectors (marked with a T in Figs 6–8 to denote relatively transparent seismic character) overlain by a high-amplitude section (marked with an R in Figs 6–8, indicating the more reflective seismic character). The number of visible cycles varies, with up to seven cycles being observed within the main Sequence 2 ridge crest (Fig. 7) and a minimum of four cycles observed in the central secondary ridge (Fig. 8). This pattern is similar to that described by Stow *et al.* (2002) from the Faro–Albuferia drift complex in the Gulf of Cadiz, which those researchers interpreted as representing changes in sand content and sedimentation rate linked to variations in bottom-current intensity. As reflectors in this upper sequence of the Eirik Drift are thought to result primarily from changing relative proportions in carbonate and clay (Arthur *et al.* 1989), this pattern of alternating amplitude is likely to reflect long-term changes in surface- and/or bottom-water conditions moderating this balance.

Seismic sequence morphology

Basement and Sequences 1 and 2; basal surfaces. Mapping depth to basement reveals a complex pattern of basement highs (Fig. 9a). On the NW flank of the drift, a series of NW–SE-trending highs occur that are parallel to the trend of magnetic anomalies in this area (Srivastava & Arthur 1989). To the east, the structural pattern becomes less clear. A number of seismic sections of various vintages cross this area (Le Pichon *et al.* 1971; Arthur *et al.* 1989), including the new TTR-13 lines. From these published sections, earlier workers have interpreted the presence of a NE–SW-trending basement high associated with the Farewell Fracture Zone underlying the main crest of the Eirik Drift (Le Pichon *et al.* 1971; Srivastava & Arthur 1989). This high plunges to the SW and is echoed in the plunge of the main drift crest. A number of lines also show basement highs to the north of this structure (e.g. the northern part of PSAT-230 (Fig. 8), the northern end of BGR-1 (Fig. 4) and the northern end of BGR-2 (Fig. 5a)), but the orientation of these structures is not fully resolved. These highs may form a series of approximately NE–SW-trending structures parallel to the Farewell Fracture Zone, although this study suggests that the highs observed on the eastern part of BGR-2 and northern part of PSAT-230 are connected, forming a NW–SE-trending high that is a continuation of the structural pattern

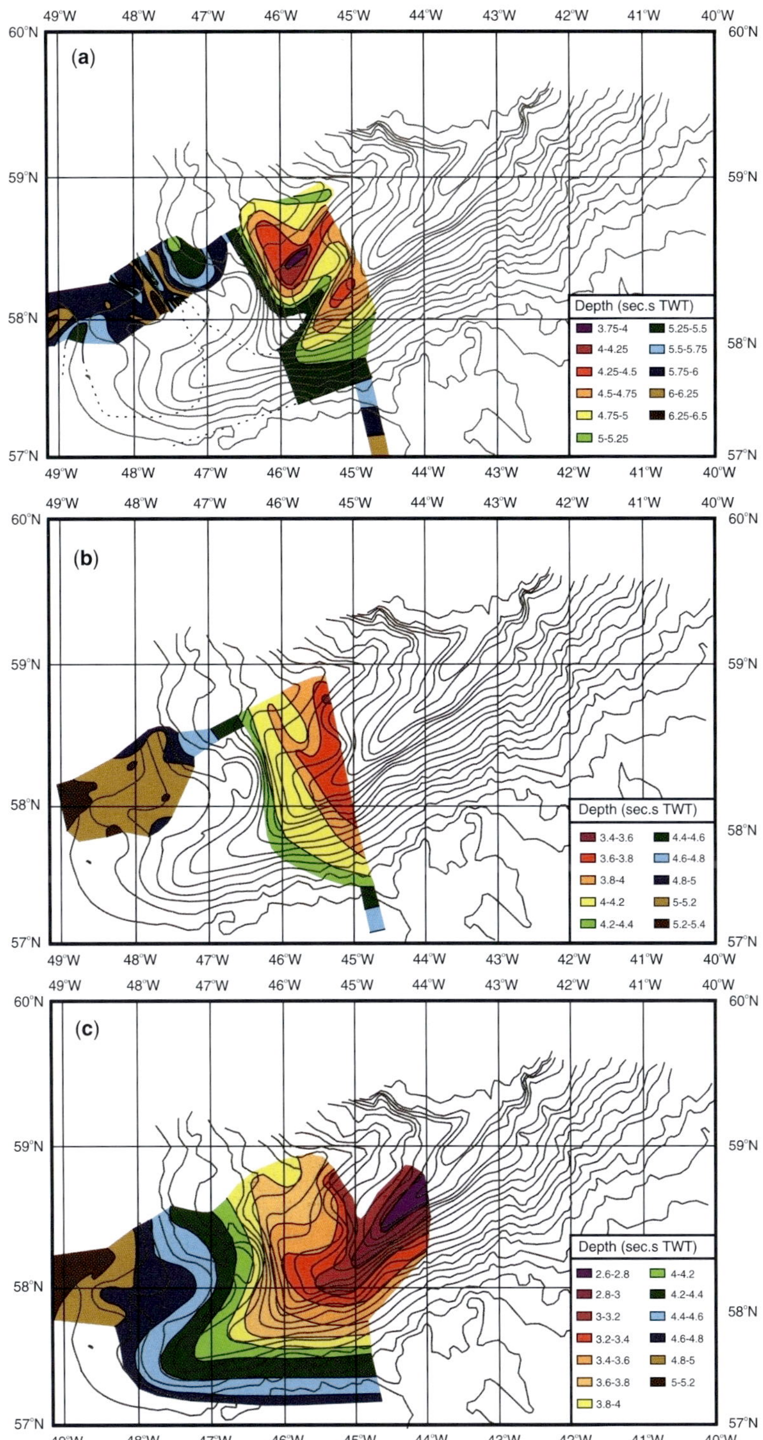

Fig. 9. Depth to basement and seismic sequence basal surfaces (seconds TWT). (**a**) Depth to basement (complied from Srivastava & Arthur 1989, Srivastava *et al.* 1989*b* and new data); (**b**) base Seismic Sequence 2; (**c**) base Seismic Sequence 1.

observed in the west of the study area. Basement topography is very different on the southern flank of the Eirik Drift, with basement sloping steadily to the south.

The basal surface of Seismic Sequence 2 slopes relatively gently to the south and SW (Fig. 9b) but with significant local variations. To the south of the main ridge crest the base of Sequence 2 has a moderately steep and regular slope to the south, echoing that of basement. In the far west of the study area the base of Sequence 2 slopes very gently to the SW. The degree of slope increases to the east toward a relative high overlying the prominent basement structure observed at the northern ends of BGR-2 and PSAT-230. An approximately NW–SE-trending band of relatively steep slope overlies the southern flank of this basement high, with a depression in the base of Sequence 2 overlying the basement depression to the NW of the high. Between the relative high and the very gently sloping area in the west, the degree of slope varies from north to south, with a relatively steep and regular slope in the area of line BGR-2 and a much gentler slope to the south.

Depth to the base of Seismic Sequence 1 (Fig. 9c) shows a very similar pattern to the modern bathymetry, indicating that the structure of the drift was largely formed during the Early to Late Pliocene (Seismic Sequence 2) and has undergone only minor modifications during the Late Pliocene to Pleistocene.

Deposit distribution. The combined thickness of Seismic Sequences 3 and 4 shows trends similar to those of the map of depth to basement and displays thinning over basement highs and thickening into troughs (Fig. 10a). The limited expression of basement topography at the base of Seismic Sequence 2 shows that these units almost entirely fill in the basement topography. Several distinct thickness variations are visible within Seismic Sequence 2 (Fig. 10b), with the most prominent being the regions of increased thickness marking the main and secondary Sequence 2 ridges. A SW-trending area of increased thickness marks the main Sequence 2 ridge crest and is assumed to continue to the toe of the drift. Perpendicular to this, NW-trending zones in increased thickness mark the positions of SR1 and SR2. Between the intersections of these secondary ridges with the main ridge crest is an area of relatively reduced thickness. This relatively thin zone trends to the NW and overlies the basement high and relative high in the base of Seismic Sequence 2 in this area (described above). SR3 is marked by a zone of somewhat less pronounced thickening in the SW of the study area.

The map of Sequence 2 thickness also reveals a zone of thick Pliocene sediments forming a broad SW-trending tongue on the northern flank of the drift. This thick zone lies on trend with a major canyon in the SW Greenland margin and is likely to represent a sequence of Pliocene turbidites with sediments derived from this canyon.

Seismic Sequence 1 (Fig. 10c) shows distinct thinning over the relatively steep south- and SW-facing slopes of all the ridge crests, reflecting increased sediment winnowing or non-deposition resulting from significant, if intermittent, bottom-current activity along these slopes. This thin zone is wide over the southern side of the main ridge crest, reflecting a very thin sequence over the whole southern slope of the main Sequence 2 ridge. Thin zones are narrower and less pronounced overlying the SW-facing slopes of the secondary ridges, particularly SR3.

Sequence 1 is thickest upslope of the main Sequence 2 crest, with the zone of maximum thickness occurring on the main ridge crest to the north of the junction with SR1. Here the sequence reaches over 0.7 s TWT and appears to form one large accumulation upslope of both the main and northern secondary ridge crests (see Fig. 7). In contrast, only moderate thickening is observed upslope of SR2 and SR3. A broad area of thinner sediments is present overlying the zone of probable Pliocene turbidite deposition, suggesting a reduction in turbidite flow at this time.

Discussion

Drift construction and palaeoceanographic history

The history of drift construction and palaeoceanographic changes in the Eirik Drift area is summarized below and in Figure 11. The construction of the Eirik Drift began at 4.5 Ma, as shown by the pronounced sedimentary ridges developed within Seismic Sequence 2, and is thought to result from strong bottom-current activity in conjunction with high sediment input (Arthur *et al.* 1989). The main palaeoceanographic events preceding this phase of drift construction were the onset of the flow of DSOW at 7.5 Ma, following high-latitude cooling and subsidence on the Greenland–Scotland Ridge (Arthur *et al.* 1989; Wright 1998), and increasing bottom-current intensity at 5.6 Ma (Kaminski *et al.* 1989). At 4 Ma a change in the dominant biota reflects a cooling of surface waters (Arthur *et al.* 1989). Ice rafting began in the area at 2.5 Ma (Cremer *et al.* 1989) and was coincident with decrease in bottom-current intensity, upslope migration of drift crests and change in style of

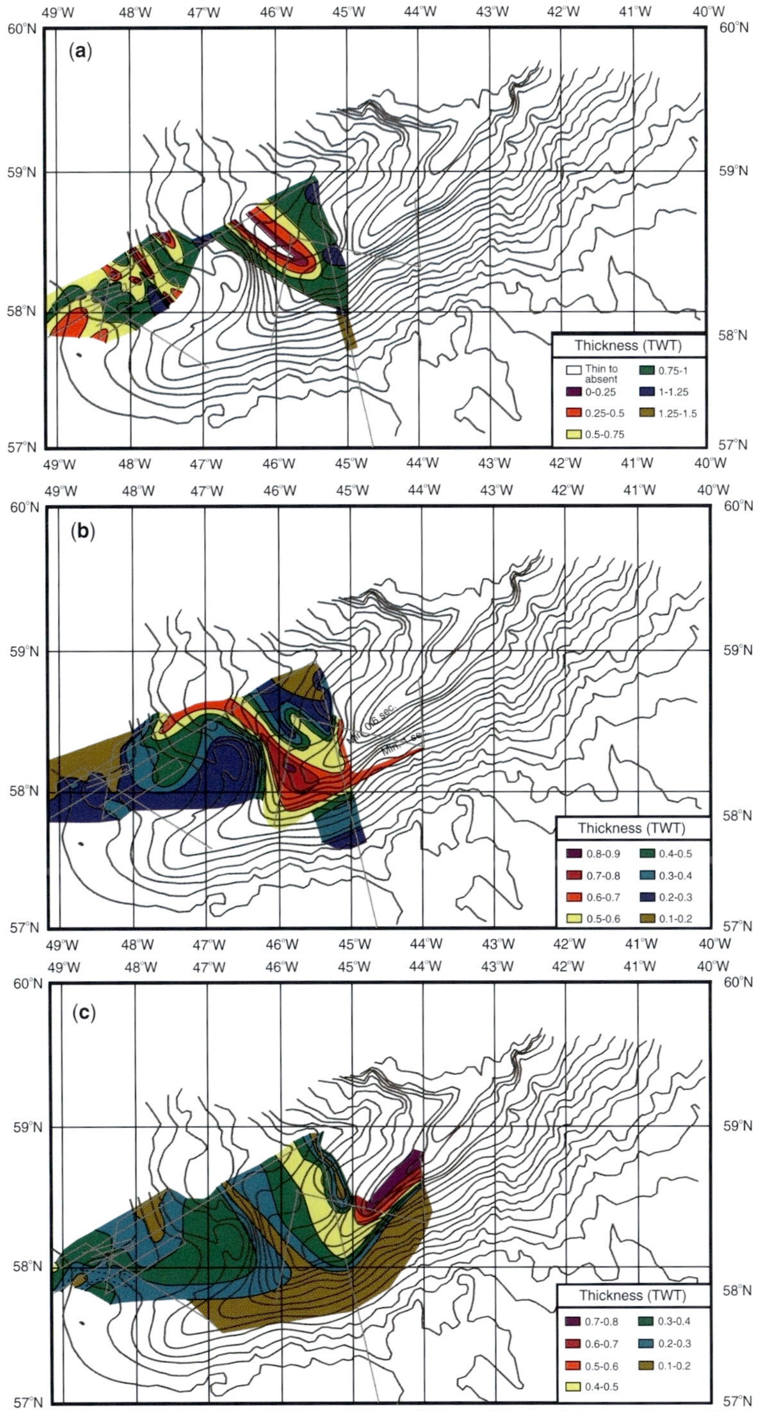

Fig. 10. Seismic sequence thickness distribution (seconds TWT). (**a**) Combined thickness of Seismic Sequences 3 and 4; (**b**) Seismic Sequence 2; (**c**) Seismic Sequence 1.

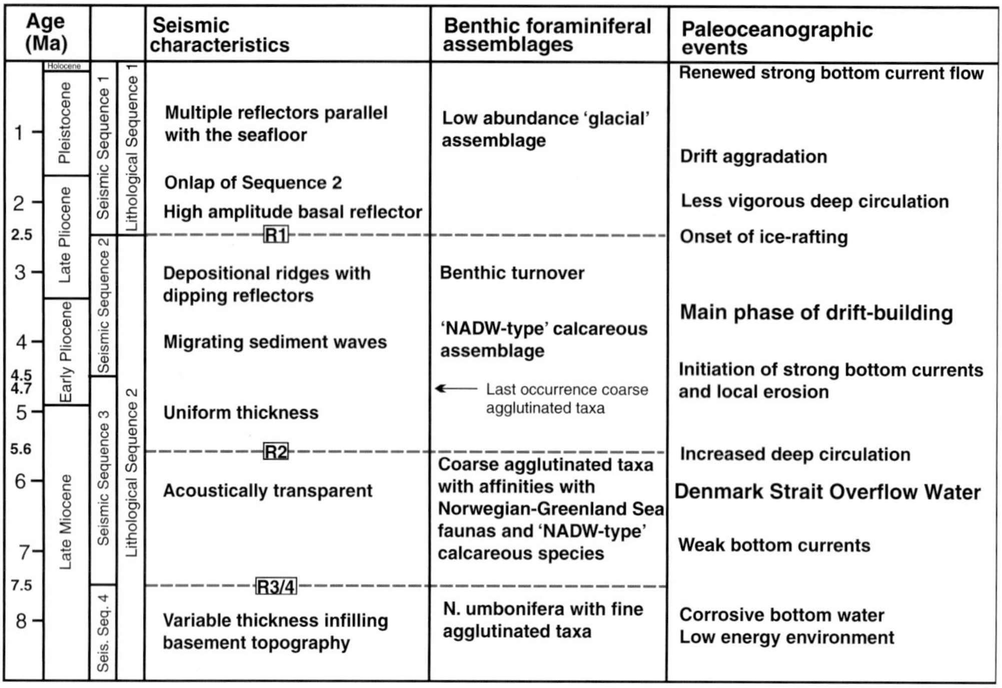

Age (Ma)	Epoch	Seismic sequence	Lithological sequence	Seismic characteristics	Benthic foraminiferal assemblages	Paleoceanographic events
0–2.5 (1, 2)	Holocene, Pleistocene, Late Pliocene	Seismic Sequence 1	Lithological Sequence 1	Multiple reflectors parallel with the seafloor; Onlap of Sequence 2; High amplitude basal reflector	Low abundance 'glacial' assemblage	Renewed strong bottom current flow; Drift aggradation; Less vigorous deep circulation; Onset of ice-rafting
2.5				R1		
2.5–5.6 (3, 4, 4.5, 4.7, 5)	Late Pliocene, Early Pliocene	Seismic Sequence 2; Seismic Sequence 3	Lithological Sequence 2	Depositional ridges with dipping reflectors; Migrating sediment waves; Uniform thickness	Benthic turnover; 'NADW-type' calcareous assemblage; ← Last occurrence coarse agglutinated taxa	Main phase of drift-building; Initiation of strong bottom currents and local erosion
5.6				R2		
5.6–7.5 (6, 7)	Late Miocene	Seismic Sequence 3	Lithological Sequence 2	Acoustically transparent	Coarse agglutinated taxa with affinities with Norwegian-Greenland Sea faunas and 'NADW-type' calcareous species	Increased deep circulation; Denmark Strait Overflow Water; Weak bottom currents
7.5				R3/4		
7.5– (8)	Late Miocene	Seis. Seq. 4		Variable thickness infilling basement topography	N. umbonifera with fine agglutinated taxa	Corrosive bottom water; Low energy environment

Fig. 11. Phases of drift construction and palaeoceanographic events in the Eirik Drift area. Modified from Arthur *et al.* (1989), Cremer *et al.* (1989) and Kaminski *et al.* (1989).

drift sedimentation (Arthur *et al.* 1989). These characteristics represent a major change in THC. Deep-water formation was restricted and intermittent during the Pleistocene, with the main southward-flowing current shifting to intermediate water depths (Glacial North Atlantic Intermediate Water; GNAIW). This shallowing and weakening of contour-current activity led to the observed upslope migration of ridge crests from Sequence 2 to Sequence 1 and the change in depositional style.

Glacial–interglacial cycles are recorded by isotopic, sedimentological and biological variations, with deep-water sites recording high sedimentation rates during interglacial periods when terrigenous and pelagic sediment fluxes were highest (Hall *et al.* 1989) and intermediate-level sites recording low sedimentation rates during interglacials as a result of increased current activity. The modern situation of renewed strong northern-sourced bottom-water flow was re-established during the Holocene (Hillaire-Marcel *et al.* 1994).

Early to Late Pliocene depositional architecture

Comparison of the internal structure of the different Early to Late Pliocene depositional ridges has revealed distinct variations in depositional architecture, with the main differences being the number of internal sub-units and the degree of lateral migration between sub-units. The central NW-trending ridge crest (SR2) contains the greatest number of sub-units and also displays the greatest degree of lateral migration between sub-units. Earley *et al.* (2002) suggested that these sub-units formed as a result of progressive shallowing of the core of the DWBC related to either warming or freshening of the current, an increase in current flux, which may raise the level of the current core in the water column, or an increase in the volume of AABW, which could displace the current upwards. All of these scenarios are plausible and detailed analysis of the Pliocene sedimentary section of the drift would be required to unequivocally determine the cause of current shallowing.

This pattern is not observed in any of the other ridge crests, or indeed at the northern end of SR2 (see line BGR-2, Fig. 5a), raising the question of why such oceanographic changes should be recorded by some ridges and not others. The most probable controlling factor is the initial degree of slope in the area of drift development. The southern end of SR2 formed over one of the most gently sloping areas of base Seismic Sequence 2

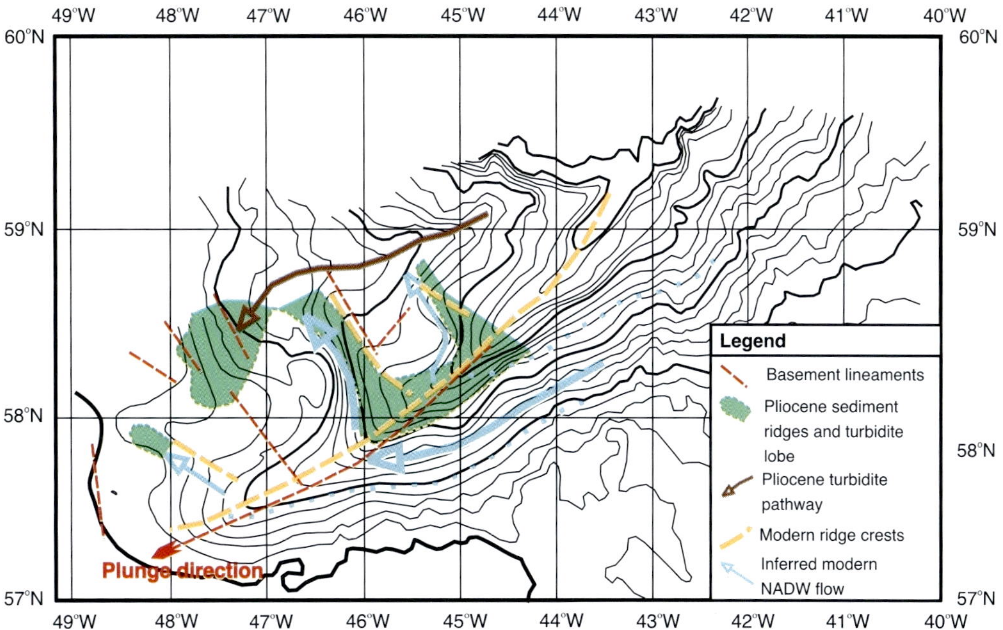

Fig. 12. Summary map showing basement structural trends, Pliocene and modern ridge crests, and the inferred path of Pliocene bottom currents.

topography, which has presumably allowed the DWBC to migrate laterally more freely than in areas of steeper initial slope.

Pleistocene cycles

Analysis of the seismic character of Seismic Sequence 1 has revealed the presence of up to seven cycles of alternating low and high reflector amplitude. It is difficult to assess the exact frequency of these cycles. A maximum of seven cycles are observed within the main ridge crest, with a minimum of four cycles observed within SR1. This difference is presumably due to the expanded nature of the sequence on the main ridge crest with regard to SR1. This raises the possibility that more than seven cycles may exist, but that these could only be resolved by an even further expanded section. As reflectors in this sequence are thought to result from local variations in the proportion of carbonate and clay in the sediments (Arthur *et al.* 1989), and as interglacial periods are times of relatively high carbonate flux and increased winnowing (Hall *et al.* 1989; Hillaire-Marcel *et al.* 1994), it is suggested that these cycles reflect glacial–interglacial variations, with peak interglacial periods being represented by the reflective units in the upper part of each cycle.

Drift morphology; inferred DWBC pathways

This study has demonstrated that the Eirik Ridge is a complex body composed of four ridges that largely reflect Pliocene drift topography, with each ridge seen in the modern bathymetry being approximately underlain by a Pliocene counterpart (Fig. 12). The current system that deposited these Pliocene ridges is assumed to be broadly similar to the modern DWBC off Cape Farewell; that is, flowing south along the SW margin of Greenland before turning north around the distal end of the Eirik Drift and entering the Labrador Sea. The presence of the three NW-trending, coeval Pliocene depositional ridges branching to the NW from the main drift crest suggests that the Pliocene DWBC separated into three strands as it crossed the Eirik Drift, with each strand depositing a separate ridge. The cause of this flow separation appears to have been local variations in sea-bed topography at base Seismic Sequence 2 level, inherited from basement structure. The SW-flowing limb of the DWBC was confined along a relatively steep south-facing slope at base Sequence 2 level, leading to the deposition of the main Sequence 2 ridge crest as one large drift body. The additional SW direction of slope allowed the current to turn to the NW. It is suggested that instabilities within the flow, caused by local variations in the angle of this SW-dipping slope, led to flow separation, with each separated strand depositing one of the secondary ridges. As the modern bathymetry still echoes Pliocene ridge structure, it seems likely that the modern DWBC also separates as it crosses the Eirik Drift.

Conclusion

Review of the existing literature allows the major palaeoceanographic events in the Eirik Drift area to be summarized. The relative timing of these events in the context of global and regional climatic changes has yet to be fully resolved; for example, the relative timing and significance of DWCB flux changes preceding Pliocene warming. More detailed analysis of the Neogene sedimentary section would provide valuable information to help resolve the issue of the relative timings of changes in THC and climatic events; that is, whether THC changes are a cause or consequence of documented climate changes (e.g. Kim & Crowley 2000).

Analysis of the seismic database in the area indicates the following features.

(1) Upslope stacking of multiple Pliocene drift crests reflects shallowing of the DWBC (Earley *et al.* 2002) but is recorded only within part of one ridge, indicating that variations in the degree of slope on which the drift builds form a limit on the degree of lateral migration of sedimentation for a given change in current depth.

(2) The Pleistocene sequence contains approximately seven cycles of reflector amplitude, which appear to be linked to glacial–interglacial variations in carbonate accumulation and deep current strength.

(3) Drift morphology suggests that the DWBC separates into three strands as it turns to the NW around the Eirik Drift and enters the Labrador Sea. It is suggested that this separation was caused by local variations in the degree of slope at base-drift level, causing funnelling of the current.

Continuing sedimentological and isotopic studies aim to unlock the decadal-scale records within the Eirik Drift sequence, so as to determine the sequence of variations in DWBC flux during the Holocene and deglacial period, and examine the relative timing and relationship of these changes in the context of short-term climatic events.

The authors would like to thank the Captain, officers and crew of the R.V. *Professor Logachev* and co-chief scientist M. Ivanov for the acquisition of the new seismic sections during the TTR-13 cruise, which was organized by the UNESCO–IOC Training Through Research Programme. Funding from the UK NERC Rapid Climate Change directed research programme (grant number NER/T/S/2002/00453) is gratefully acknowledged.

References

ALVAREZ, M., PEREZ, F. F., BRYDON, H. & RIOS, A. F. 2004. Physical and biogeochemical transports structure in the North Atlantic subpolar gyre. *Journal of Geophysical Research*, **109**, doi: 10.1029/2003JC002015.

ARTHUR, M., SRIVASTAVA, S. P., KAMINSKI, M., JARRARD, R. & OSLER, J. 1989. Seismic stratigraphy and history of deep circulation and sediment drift development in the Baffin Bay and the Labrador Sea. *In*: SRIVASTAVA, S. P., ARTHUR, M. & CLEMENT, B. (eds) *Proceedings of the Ocean Drilling Program, Scientific Results, 105*. Ocean Drilling Program, College Station, TX, 957–988.

BACON, S. 1998. Decadal variability of the outflow from the Nordic Seas to the deep Atlantic Ocean. *Nature*, **394**, 871–873.

BACON, S., GOULD, W. J. & JIA, Y. 2003. Open-ocean convection in the Irminger Sea. *Geophysical Research Letters*, **30**, 1246.

BROECKER, W. S. 2000. Abrupt climate change: causal constraints provided by the paleoclimate record. *Earth Science Reviews*, **51**, 137–154.

CHOUGH, S. K. & HESSE, R. 1985. Contourites from the Eirik Drift, south of Greenland. *Sedimentary Geology*, **41**, 185–189.

CLARK, P. U., PISIAS, N. G., SOTCKER, T. F. & WEAVER, A. J. 2002. The role of thermohaline circulation in abrupt climate change. *Nature*, **415**, 863–869.

CLARKE, R. A. 1984. *Transport through the Cape Farewell Flemish Cap section.* International Council for the Exploration of the Sea Report, **185**, 120–130.

CREMER, M. 1989. Texture and microstructure of Neogene–Quaternary sediments, ODP sites 645 and 646, Baffin Bay and Labrador Sea. *In*: SRIVASTAVA, S. P., ARTHUR, M. & CLEMENT, B. (eds) *Proceedings of the Ocean Drilling Program, Scientific Results, 105.* Ocean Drilling Program, College Station, TX, 7–20.

CREMER, M., MAILLET, N. & LATOUCHE, C. 1989. Analysis of sedimentary facies and clay mineralogy of the Neogene–Quaternary sediments in ODP site 646, Labrador Sea. *In*: SRIVASTAVA, S. P., ARTHUR, M. & CLEMENT, B. (eds) *Proceedings of the Ocean Drilling Program, Scientific Results, 105*. Ocean Drilling Program, College Station, TX, 71–80.

DICKSON, R. R. & BROWN, J. 1994. The production of North Atlantic Deep Water: sources, rates and pathways. *Journal of Geophysical Research*, **99**, 12319–12341.

EARLEY, R. J., MOUNTAIN, G. S., WRIGHT, J. D. & MANLEY, P. 2002. Bedform evolution on Eirik Drift: hi-res MCS evidence of North Atlantic Deep Water variability along the SW Greenland margin. *EOS Transactions, American Geophysical Union*, **84**(46), Fall Meeting Supplement, Abstract PP151C-0936.

HALL, I. R., BLOEMENDAL, J., KING, J. W., ARTHUR, M. A. & AKSU, A. E. 1989. Middle to Late Quaternary sediment fluxes in the Labrador Sea, ODP leg 105, site 646: a synthesis of rock-magnetic, oxygen-isotopic, carbonate and planktonic foraminiferal data. *In*: SRIVASTAVA, S. P., ARTHUR, M. & CLEMENT, B. (eds) *Proceedings of the Ocean Drilling Program, Scientific Results, 105*. Ocean Drilling Program, College Station, TX, 653–688.

HILLAIRE-MARCEL, C., DE VERNAL, A., BILODEAU, G. & WU, G. 1994. Isotope stratigraphy, sedimentation rates, deep circulation and carbonate events in the Labrador Sea during the last ~200 ka. *Canadian Journal of Earth Sciences*, **31**, 63–89.

KAMINSKI, M. A., GRADSTEIN, F. M., SCOTT, D. B. & MACKINNON, K. D. 1989. Neogene benthic foraminifera biostratigraphy and deep-water history of sites 645, 646 and 647, Baffin Bay and Labrador Sea. *In*: SRIVASTAVA, S. P., ARTHUR, M. & CLEMENT, B. (eds) *Proceedings of the Ocean Drilling Program, Scientific Results. 105*. Ocean Drilling Program, College Station, TX, 731–747.

KENYON, N. H., IVANOV, M. K., AKHMETZHANOV, A. M., KOZLOVA, E. V. & MAZZINI, A. 2004. *Interdisciplinary studies of North Atlantic and Labrador Sea margin architecture and sedimentary processes*. IOC Technical Series, **68**.

KIM, S.-H. & CROWLEY, T. J. 2000. Increasing Pliocene North Atlantic Deep Water: cause or consequence of Pliocene warming? *Paleoceanography*, **15**(4), 451–455.

LE PICHON, X., HYNDMAN, R. D. & PAUTOT, G. 1971. Geophysical study of the opening of the Labrador Sea. *Journal of Geophysical Research*, **76**, 4725–4743.

PICKART, R. S., STRANEO, F. & MOORE, G. W. K. 2003. Is Labrador Sea Water formed in the Irminger Basin? *Deep-Sea Research I*, **50**, 23–52.

RAHMSTORF, S. 2002. Ocean circulation and climate during the past 120,000 years. *Nature*, **419**, 207–214.

ROEST, W. R. & SRIVASTAVA, S. P. 1989. Sea-floor spreading in the Labrador Sea: a new reconstruction. *Geology*, **17**, 1000–1003.

SCHMITZ, W. J., JR 1996. *On the World Ocean Circulation: Volume 1, Some Global Features/North Atlantic Circulation*. Woods Hole Oceanographic Institution, Woods Hole, MA.

SCHMITZ, W. J., JR & MCCARTNEY, M. S. 1993. On the North Atlantic circulation. *Reviews of Geophysics*, **31**, 29–49.

SMITH, W. H. F. & SANDWELL, D. T. 1997. Global sea floor topography from satellite altimetry and ship depth soundings. *Science*, **277**, 1956–1962.

SRIVASTAVA, S. P. & ARTHUR, M. 1989. Tectonic evolution of the Labrador Sea and Baffin Bay: constraints imposed by regional geophysics and drilling results from Leg 105. *In*: SRIVASTAVA, S. P., ARTHUR, M. & CLEMENT, B. (eds) *Proceedings of the Ocean Drilling Program, Scientific Results, 105*. Ocean Drilling Program, College Station, TX, 989–1009.

SRIVASTAVA, S. P. & TAPSCOTT, C. R. 1986. Plate kinematics of the North Atlantic. *In*: VOGT, P. R. & TUCHOLKE, B. E. (eds) *The Geology of North America, Volume M, The Western North Atlantic Region*. Geological Society of America, Boulder, CO, 379–404.

SRIVASTAVA, S. P., ARTHUR, M. & CLEMENT, B. (eds) 1989*a*. *Proceedings of the Ocean Drilling Program, Scientific Results, 105*. Ocean Drilling Program, College Station, TX.

SRIVASTAVA, S. P., LOUDEN, K. E., CHOUGH, S. K., ET AL. 1989*b*. Results of detailed geological and geophysical measurement at ODP Sites 645 in Baffin Bay and 646 and 647 in the Labrador Sea. *In*: SRIVASTAVA, S. P., ARTHUR, M. & CLEMENT, B. (eds) *Proceedings of the Ocean Drilling Program, Scientific Results, 105*. Ocean Drilling Program, College Station, TX, 891–919.

STOW, D. A. V., FAUGÈRES, J.-C., GONTHIER, E. G., ET AL. 2002. Faro–Albuferia drift complex, northern Gulf of Cadiz. *In*: STOW, D. A. V., PUDSEY, C. J., HOWE, J. A., FAUGÈRES, J.-C. & VIANA, A. (eds) *Deep-water Contourite Systems: Modern Drifts and Ancient Series, Seismic and Sedimentary Characteristics*. Geological Society, London, Memoirs, **22**, 137–154.

TUCHOLKE, B. E. & FRY, V. A. 1985. Basement structure and sediment distribution in the Northwest Atlantic Ocean. *AAPG Bulletin*, **69**, 2077–2097.

WRIGHT, J. D. 1998. Role of the Greenland–Scotland Ridge in Neogene climate changes. *In*: CROWLEY, T. J. & BURKE, K. (eds) *Tectonic Boundary Conditions for Climate Reconstructions*. Oxford University Press, Oxford, 192–211.

WRIGHT, J. D. & MILLER, K. G. 1996. Control of North Atlantic Deep Water circulation by the Greenland–Scotland Ridge. *Paleoceanography*, **11**, 157–170.

Ridge and valley systems in the Upper Cretaceous chalk of the Danish Basin: contourites in an epeiric sea

E. V. ESMERODE[1], H. LYKKE-ANDERSEN[2] & F. SURLYK[1]

[1]*Geological Institute, University of Copenhagen, Øster Voldgade 10, DK-1350 Copenhagen K, Denmark (e-mail: estelav@geol.ku.dk)*

[2]*Department of Earth Sciences, University of Aarhus, Finlandsgade 6-8, DK-8200 Aarhus N, Denmark*

Abstract: Extensive low-lying parts of the NW European craton were flooded during the Late Cretaceous transgression, creating a relatively deep epeiric sea with reduced supply of siliciclastic material and insignificant coastal upwelling. The chalk, essentially an oceanic sediment type, was deposited as a pelagic rain of mainly coccolith debris and with local redeposition along structural highs. The study area is located in the eastern part of the Danish Basin, where the bordering Ringkøbing–Fyn High and the inverted Sorgenfrei–Tornquist Zone converge. Multichannel seismic reflection lines show the Chalk Group to be far from the expected flat-lying pelagic succession. A multitude of features of considerable relief, comprising an extensive unconformity, sediment waves, drifts and moats, are recognized. At least two episodes of widespread drift deposition are identified, one in the Santonian–Late Campanian and one in the Maastrichtian, separated by a Top Campanian Unconformity. The structures were formed by strong bottom currents flowing northwestward through the basin parallel to bathymetric contours. A lateral northeastward change, from more depositional to more erosional architecture, indicates a positive current velocity gradient towards the inversion zone, probably as a result of the Coriolis force. The strong similarity between the chalk drifts and modern contourite deposits supports the proposal that the oceanographic conditions linked to continental margins were extended into the Late Cretaceous epeiric sea of NW Europe.

The Late Cretaceous period was characterized by the largest transgression in Phanerozoic Earth history (e.g. Haq *et al.* 1987). The climate had a pronounced greenhouse character and the surrounding low-lying land-masses of NW Europe were flooded, creating a relatively deep and extensive epeiric sea (e.g. Ziegler 1990; Fig. 1). The large extent of the sea, combined with an arid climate in northern Europe, resulted in minimal siliciclastic influx as reflected by the extreme purity of the chalk. This situation was stable for about 30 Ma. Clear waters and high water temperatures, probably with insignificant upwelling, generated nearly optimal conditions for the proliferation of coccolithophorid algae, which are typical of oligotrophic oceanic conditions (Hay 1995; Frakes 1999). The chalk was deposited as a pelagic rain of mainly coccoliths, probably aggregated in pellets, creating a rather monotonous chalk succession with increasingly pure carbonates from Cenomanian to Maastrichtian time. Gravity-driven redeposited chalk is commonly found along structural highs in the Central North Sea (Watts *et al.* 1980; Hardman 1982; Hatton 1986; Kennedy 1987) and in other parts of northern Europe (Gale 1980; Bromley & Ekdale 1987; Evans *et al.* 2003). Slides, slumps, debrites and turbidites are mainly associated with areas affected by halokinetic movements or tectonic inversion (Ziegler 1990). These deposits have received special attention as hydrocarbon reservoirs in the Norwegian North Sea.

The concept of the 'dynamic chalk sea-floor', as an environment subject to the effects of contour-parallel bottom currents, was introduced by Lykke-Andersen & Surlyk (2004) and Surlyk & Lykke-Andersen (in press). During the early depositional stage the unlithified chalk ooze was influenced by even weak currents, which exerted a marked influence on the sea-floor relief. High-resolution 2D seismic data recently acquired in the Kattegat and Øresund area provide valuable information on the development of a pronounced sea-floor topography created by strong contour-parallel currents (Lykke-Andersen & Surlyk 2004; Surlyk & Lykke-Andersen in press).

Modern contourite drifts are in most cases linked to the lower slope and rise of continental margins, where deep-water thermohaline contour currents are characteristic oceanographic elements. Water depth estimation for the chalk sea floor is difficult and inaccurate, as deposition took place well below the photic zone and at the reach of deep storm waves, but the depth values have in general been of the order of several hundreds of metres

From: VIANA, A. R. & REBESCO, M. (eds) *Economic and Palaeoceanographic Significance of Contourite Deposits.* Geological Society, London, Special Publications, **276**, 265–282.
0305-8719/07/$15.00

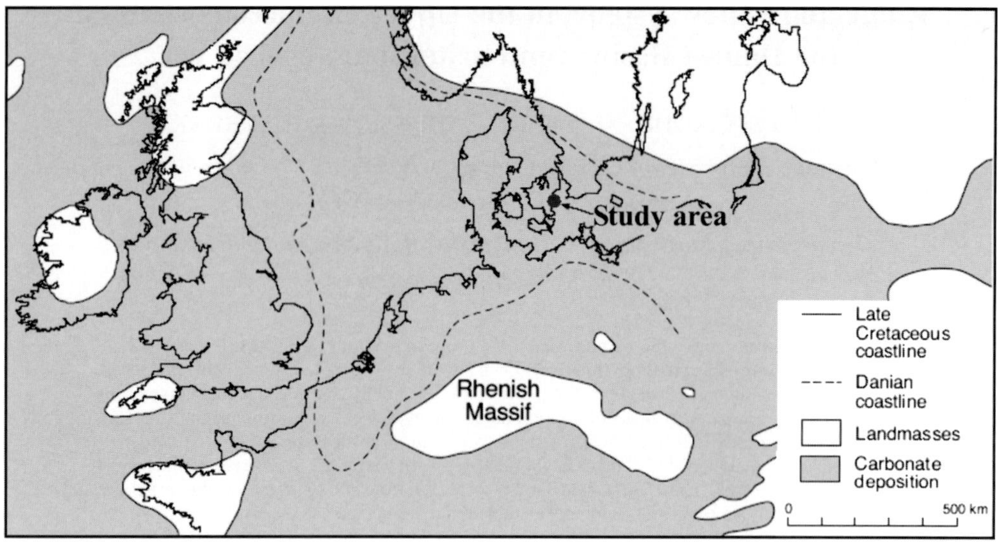

Fig. 1. Palaeogeographical map of NW Europe during the Late Cretaceous showing the large extent of the sea areas compared with the present situation. The continuous line represents the location of the coastline during the Late Cretaceous; the dashed line represents the Danian coastline (modified after Surlyk *et al.* 2003).

ranging up to about 700–800 m (Surlyk & Lykke-Andersen in press). These values are shallower, however, than those of classical continental margin contourite drifts (e.g. Stow & Holbrook 1984; Mezerais *et al.* 1993; Weber *et al.* 1994; Wold 1994; Rebesco *et al.* 1996; Faugères *et al.* 1998; Stoker *et al.* 1998).

The aim of this study is to identify and analyse topographic elements of the Late Cretaceous chalk sea floor in the Øresund area of the Danish Basin, adding to and supplementing the description of the large-scale structures by Lykke-Andersen & Surlyk (2004). In addition, we wish to compare and discuss the similarities between the chalk drifts and Cenozoic–modern contourite drifts.

Geological setting

The Danish Basin was formed by rifting in Late Carboniferous–Early Permian times. Late Cretaceous subsidence was governed by thermal contraction following earlier Mesozoic rift events. Several phases of inversion and uplift of the Sorgenfrei–Tornquist Zone took place in Late Cretaceous–Palaeogene times (Liboriussen *et al.* 1987; Vejbæk & Andersen 2002). The study area is located in the eastern part of the basin, which is bordered by the inverted Sorgenfrei–Tornquist Zone to the NE and the Ringkøbing–Fyn high to the south (Fig. 2). The thickness of the chalk succession increases northwards from less than 1000 m to more than 2000 m in the depocentre along the Sorgenfrei–Tornquist Zone (Liboriussen *et al.* 1987).

Along the northeastern basin margin the chalk passes into skeletal carbonate sands, greensand and local siliciclastic clay, silt and sand. Small bryozoan mounds developed in relatively shallow areas of the basin, represented by the Stevns Klint outcrop, as a response to a relative sea-level drop during the late Maastrichtian (Surlyk 1997). At the Maastrichtian–Danian boundary biogenic carbonate production was stopped and the Fish Clay (K–T boundary clay) was deposited (Christensen *et al.* 1973; Alvarez *et al.* 1984). A sea-level fall in the earliest Danian is in Stevns Klint expressed by an erosional hardground, which truncates the crests of the uppermost Maastrichtian bryozoan mounds and the intervening lowermost Danian carbonates (Rosenkrantz 1938; Surlyk 1997). The succeeding Danian sedimentation comprises bryozoan wackestone and rudstone passing basinwards into chalk, but water depths in the eastern Danish Basin were still of the order of several hundreds of metres (Surlyk & Håkansson 1999). The end of the Danian is characterized by an abrupt facies change from carbonates to terrigenous siliciclastic deposits.

Deep high-resolution 2D seismic data recently collected in the Øresund area reveal the presence of a WNW–ESE-trending ridge-and-valley system, which is parallel to the Sorgenfrei–Tornquist Zone and the axis of the Danish Basin (Lykke-Andersen & Surlyk 2004). This system has kilometre-scale

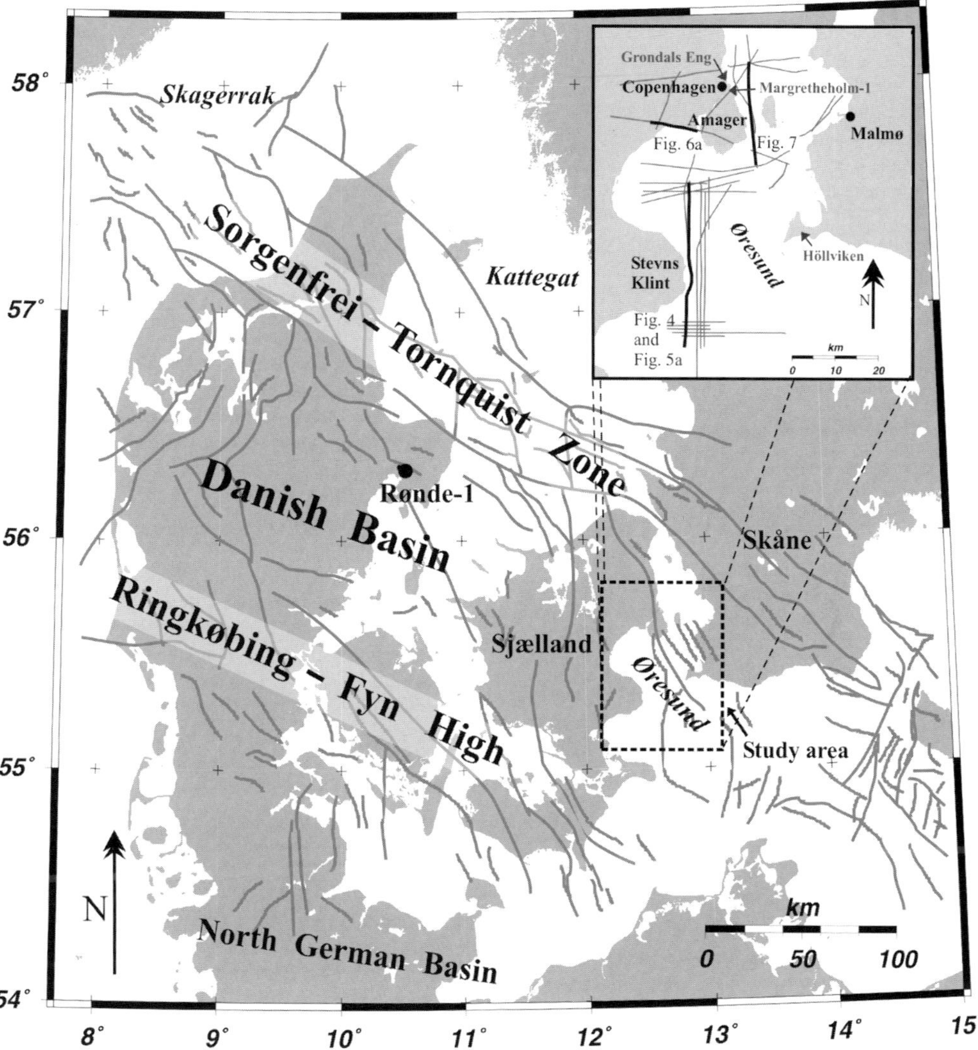

Fig. 2. Map of the study area showing the position of the main structural elements, seismic grid and boreholes used in this study. The seismic survey comprises 29 lines, four of which are onshore lines. The locations of the seismic lines illustrated in Figures 4, 5a, 6a and 7 are shown in the survey grid map.

wavelengths and amplitudes up to about 150 m. A surface expression of the ridges and valleys is exposed in Stevns Klint as gentle highs and intervening lows outlined by the elevation of the K–T boundary. This has generally been regarded as the result of post-Danian Laramide folding, following Rosenkrantz (1938). However, the seismic data indicate that the chalk succession is not folded, as seen by the flat and undisturbed, only slightly northward-tilted nature of the Base Chalk reflector. The ridge-and-valley system is thus of depositional nature and not of tectonic origin. It is interpreted as formed by bottom currents, flowing from ESE to WNW parallel to the contours of the Sorgenfrei–Tornquist Zone (Lykke-Andersen & Surlyk 2004; Surlyk & Lykke-Andersen in press).

Database and methods

Seismic dataset

The seismic grid comprises onshore and offshore profiles acquired between 1999 and 2000 during the campaigns DA-99, DA-00, FL-00 and HGS-00 in the Øresund region between Sjælland to the west and southern Skåne to the east (Fig. 2). The survey consists of 29 high-resolution 2D seismic sections covering a total length of *c.* 520 km. Four onshore seismic lines were collected on the island of Amager and in southern Copenhagen.

The Base Chalk reflector marks the acoustic impedance inversion at the sharp contact between Lower Cretaceous claystones and indurated Upper Cretaceous chalk. Reflection quality of the monotonous chalk succession is remarkably good compared with the underlying deposits. However, the shallowest intervals are significantly affected by noise from both the water column and the sea bed, which in most cases hinders interpretation of the top 100 m. Vertical seismic resolution ranges from 10 m in the shallowest parts to 30 m in the deepest, at the Base Chalk reflector.

Borehole data

Only a few boreholes are available in the study area (Fig. 2). The information from the Grøndals Eng boring, drilled in Copenhagen in 1894–1907 (Bonnesen *et al.* 1913; Ravn 1913), and the Höllviken boreholes, drilled in SW Skåne in 1941–1944 (Brotzen 1945), has been used with a certain degree of confidence when tying the seismic unit boundaries to lithostratigraphic boundaries. The well Margretheholm-1 was drilled in 2002–2003 on Amager, south of Copenhagen, aiming at a deeper sandstone geothermal water reservoir, and downhole measurements of the chalk were performed in only the Turonian–upper Campanian part. This provided valuable information on the physical properties of the lowermost 900 m of the Chalk Group in the area. Rønde-1, drilled in eastern Jylland (Bang 1971; Stenestad 1971), offered good lithostratigraphical and biostratigraphical data despite its distance from the study area. The signatures of gamma-ray, sonic and resistivity logs of Margretheholm-1 and Rønde-1 show a strong similarity and allow a good correlation. This indicates that there may be only small lithological changes along the axis of the Danish Basin, and justifies the use of Rønde-1 as a lithostratigraphical reference for the study area.

Data analysis

The seismic stratigraphic subdivision of the chalk is based on the presence of a significant number of reflector terminations (onlaps, downlaps, toplaps and erosional truncation) supported by seismic facies analysis. The seismic stratigraphic sequence boundaries were tied to lithostratigraphic and chronostratigraphic information obtained from the Margretheholm-1 logs and the Grøndals Eng, Höllviken and Rønde-1 boreholes (Lykke-Andersen & Surlyk 2004). Key stratigraphic surfaces were mapped and used as reference levels for an approximate dating of major topographic sea-floor features. Two-way travel time (TWT) to depth conversion was made using an equation derived from the chalk velocity expression for the Danish Basin, $V = 2421 + 1.07z$ (Japsen 1998). Lateral correlation was undertaken to pinpoint the time intervals when the formation of a strong sea-floor relief was more widespread.

Stratigraphy

The top part of the Lower Cretaceous succession is formed by Albian reddish brown marine marlstone, marking the onset of marine conditions and the beginning of the major transgression that dominated during the Late Cretaceous (Fig. 3). This episode is expressed in the sedimentary record as a sharp transition from clay-dominated deposits to almost pure biogenic calcium carbonate. The total thickness of the Chalk Group drilled at Rønde-1 is 1858 m. Cenomanian and Turonian limestones are separated by a 3 m thick layer of calcareous dark grey shale. The overlying thick Coniacian succession comprises hard limestone rich in flint and marlstone in the lowest part of the succession, whereas the higher parts are virtually devoid of flint and clay. The Santonian–lower Maastrichtian succession consists of *c.* 1300 m of interbedded limestone and marl. The upper Maastrichtian is composed of indurated to soft chalk, flint being present in only the uppermost part of the succession (Stenestad 1971). The thick lower Danian succession consists of bryozoan limestone with flint and is overlain by chalk resembling the upper Maastrichtian chalk (Bang 1971). The Danian top of the Chalk Group is represented by a sharp lithological facies change to Selandian siliciclastic deposits.

Results

Seismic stratigraphy

The Chalk Group is far from regularly and horizontally bedded as would be expected for a pelagic white coccolithic chalk. The strong and continuous reflectivity in most of the intervals of the chalk reveals a complex, but well-layered, internal architecture (Fig. 4). The lateral extent of the reflectors in general is limited and reflector terminations such as downlaps and truncations are abundant, pointing to a rather dynamic depositional setting and hampering regional correlation of individual morphological elements.

The seismic unit subdivision for the Chalk Group adopted in the present work is nearly the same as

Age Ma	Stratigraphy			Seismic units Lykke-Andersen and Surlyk (2004)	Höllviken Brotzen (1945)	Rønde-1 Stenestad (1971) Bang (1971)
61.7	Palaeocene	Selandian				dark grey calcareous clay
65.5	Palaeocene	Danian			bryozoan limestone	white chalk bryozoan limestone
	Upper Cretaceous	Maastrichtian	upper	Unit 6 Unit 5	white chalk with interbedded clay and flint at the top	white chalk with flint
70.6	Upper Cretaceous	Maastrichtian	lower	Unit 4 Unit 3	light grey argillaceous limestone	light grey limestone
83.5	Upper Cretaceous	Campanian		Unit 2	white and grey limestone with intercalations of marl	light grey limestone
85.8	Upper Cretaceous	Santonian		Unit 1	limestone with interbedded clay	marlstone with interbedded limestone
89.3	Upper Cretaceous	Coniacian		Unit 1	grey limestone with interbedded clay	white to light grey limestone
93.5	Upper Cretaceous	Turonian		Unit 1	white limestone with flint	greenish grey limestone
99.6	Upper Cretaceous	Cenomanian			conglomerate and phosphoritic sandstone	dark grey shale light grey limestone
	Lower Cretac.	Albian			sandy shales	brown marine marlstone

Fig. 3. Stratigraphy of the Upper Cretaceous–Lower Palaeocene and lithological description of wells Höllviken and Rønde-1.

that defined by Lykke-Andersen & Surlyk (2004). The main difference is the subdivision of Lykke-Andersen & Surlyk's Unit 1 into two subunits of differing seismic facies.

The Base Chalk reflector is strong and flat with negative acoustic impedance contrast, tilted northwards and dipping *c.* 0.5° over at least 50 km (light blue in Fig. 4). The low-amplitude Turonian–lower Santonian part of the succession appears as a sheet, draping the Base Chalk. The upper Santonian interval in contrast thickens markedly northwards from around 100 m to almost 200 m. The overlying Campanian interval shows stronger reflectivity in the lower part of the succession gradually decreasing upwards. It is bounded at its top by a marked unconformity in the northern part of the study area, where a large part of the Campanian succession has been removed by erosion. The lower and upper Maastrichtian units are characterized by higher reflector amplitudes and a northward thickening. This thickening is probably a result of synsedimentary northward tilting (Lykke-Andersen & Surlyk 2004). The lower Maastrichtian interval comprises a main

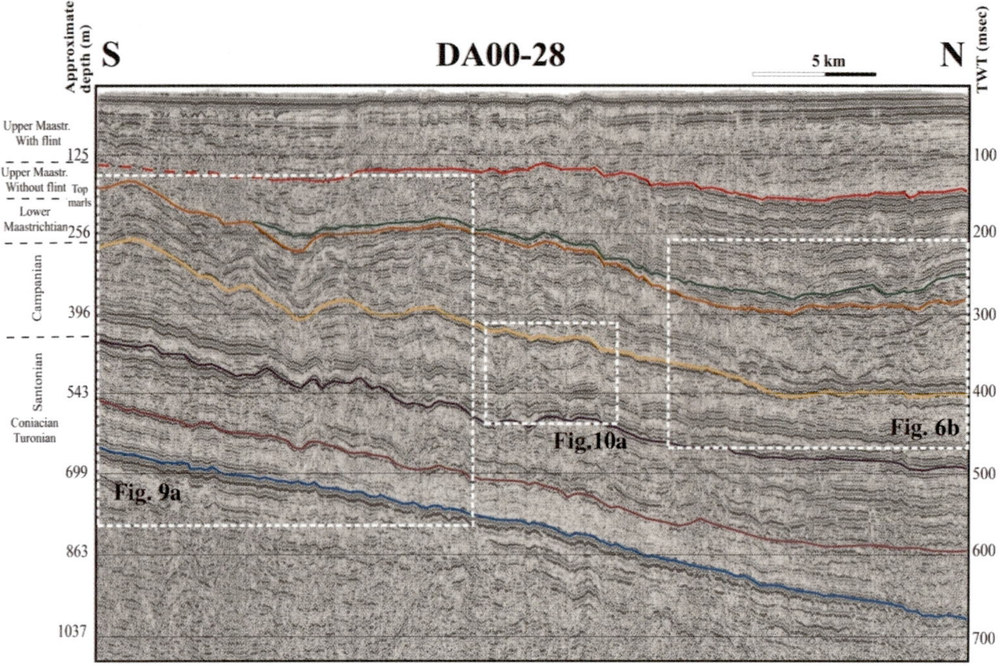

Fig. 4. The north–south-oriented seismic profile DA00-28 located offshore Stevns Klint (position shown in Fig. 2) shows the seismic stratigraphic subdivision of the Chalk Group in the study area. The Base Chalk reflector (light blue) is rather flat and dips northwards about 0.5°. It is overlain by a Turonian–lower Santonian succession of nearly constant thickness and weak reflectivity. The upper Santonian unit thickens markedly towards the north, in clear contrast to the overlying Campanian unit. The lower Maastrichtian succession is subdivided into a lower thick unit of variable thickness with a pronounced wavy reflection pattern, and an upper thin unit that seems to pinch-out southwards. The boundary between the lower and upper Maastrichtian units represents the top of marly chalk, overlain by upper Maastrichtian white chalk with remarkably good reflector quality. The dashed squares mark the location of the details of the seismic profile shown in Figures 6b, 9a and 10a.

unit, thickest in the centre of the profile and showing signs of local slope-failure deposits, overlain by a very thin wedge-shaped unit. The lower–upper Maastrichtian boundary coincides with the top of the marly chalk. The upper Maastrichtian interval can be subdivided into two thick units showing markedly different acoustic responses especially in the northern part. The strong and continuous reflectivity pattern of the lower unit in the north becomes more irregular and diffuse towards the south. Correlation to the boreholes suggests that the unit consists of chalk without flint. The upper unit shows poor reflectivity throughout the area and represents chalk with flint. The weak acoustic response is not simply due to shallow noise, because if this were the case the vertical change of seismic facies observed in the northern part would be gradual rather than the sharp contact observed between the two units. Therefore changes of the physical properties account for the variation of the acoustic response.

Main morphological elements

The sea-floor relief of the chalk in the Øresund area is dominated by the presence of a major ESE–WNW-trending ridge-and-valley system formed by bottom currents (Lykke-Andersen & Surlyk 2004). Here we focus on the large number of minor drifts, channels and scours of the system, as seen by complex structures involving erosional truncation and combined lateral migration and aggradation. These structures are laterally extensive and their dimensions are typically a few kilometres in wavelength and up to 100 m in amplitude. The large interline separation of the seismic survey grid hinders a more precise determination of their shapes and trends. The structures of depositional origin are especially concentrated in the middle Santonian to Campanian and in the upper Maastrichtian intervals. At least one major unconformity of late Campanian–early Maastrichtian age is associated with the major ridge-and-valley system, separating two intervals of dominant drift sedimentation.

The majority of the mapped structures are recognized on the north–south-oriented seismic lines, whereas those mapped on west–east- and SW–NE-oriented lines show a much smoother relief. Thus, the structures seem to have a preferential WNW–ESE direction, proving them to be elongate rather than mounded. This is in agreement with observations from the Kattegat Sea, where the Maastrichtian topographic elements are elongate and parallel to the Sorgenfrei–Tornquist Zone (Surlyk & Lykke-Andersen in press).

Top Campanian Unconformity. An important unconformity appears in the seismic profiles and can be readily traced across the entire study area. Correlations with the records of the Grøndals Eng and Rønde-1 boreholes suggest a late Campanian–early Maastrichtian age of the prominent reflector. The Top Campanian Unconformity (TCU) is illustrated by the yellow reflector in Figure 5a. In the southern part of the study area erosion was least pervasive and bedding is more conformable, whereas the northern part is characterized by erosional truncations. In the northern area, closer to the inversion zone, the unit beneath the TCU is characterized by erosional scours or channels and minor truncations (Fig. 6a). This unit is the only one within the Chalk Group that thins northwards, partly as a result of at least one erosive event affecting the northern area. The most dramatic expression of this event is the presence of a 60 m deep scour or channel downcutting into parallel, well-layered deposits (Fig. 6b). The scour has a well-layered fill, which in turn is overlain by two levee-like drifts with *c.* 25 m crest-to-trough relief.

The gamma-ray log from the Margretheholm-1 well does not indicate a facies change at the Campanian–Maastrichtian boundary, nor do the sedimentary logs from the Rønde-1 borehole. The presence of a hardground is not recorded by the density and sonic records either, which display rather constant values. However, small troughs in the self-potential and resistivity curves may indicate the presence of slightly coarser material at and above the unconformity surface, which may represent reworked and winnowed deposits.

The TCU divides two units where sedimentation predominated over erosion and where drift formation was widespread. Deposition influenced by relatively weak bottom currents during the Campanian was succeeded by winnowing and erosion that led to the formation of the TCU. The large erosional area indicates the presence of strong or more focused bottom currents during the late Campanian–early Maastrichtian, coinciding with the Late Cretaceous sea-level highstand (Haq *et al.* 1987). The bottom-current system possibly transgressed further into the epeiric sea and may have been constricted and amplified by embayments and irregularities of the coast. Deflection to the right of the northwesterly flowing bottom current by the Coriolis force possibly piled the water against the southwesterly dipping slope of the Sorgenfrei–Tornquist Zone, creating a lateral positive gradient of the current velocity and its erosive potential. The regional-scale erosive event at the late Campanian–early Maastrichtian transition removed large amounts of chalk ooze from the study area. This material was probably deposited down-current and along its SW margin, where the damping of the current velocity allowed settling of suspended and bedload-transported material. Intense current erosion affecting extensive areas of the sea floor is a feature typical of contour-current deposits and contourite successions are characterized by the presence of extensive unconformity surfaces as a result of multiple episodes of reactivation of the bottom currents (Faugères & Stow 1993; Faugères *et al.* 1998). The succession overlying the TCU was deposited under slower bottom currents, or possibly the main axis of the current was shifted to an area outside our survey grid. Weaker current conditions, with less erosive capacity, resulted in infilling of topographic lows and reactivation of drift formation.

A modern analogue to the TCU is found in the present sea-bed morphology of the Barra Fan, in the NW UK margin (Fig. 5b). In both the Chalk Group and the Barra Fan a zone of low deposition is seen to pass laterally into a zone dominated by winnowing and erosion, from south to north and from east to west, respectively. Erosional processes affecting the northern part of the upper Campanian–lower Maastrichtian chalk were possibly related to higher current velocities, similar to the stronger arm of the Deep Northern Boundary Current, which affects the western part of the Barra Fan.

Complex mounded drift. A prominent topographic relief in the upper Maastrichtian succession is observed in the seismic profiles offshore Copenhagen. It comprises a WNW–ESE-trending mounded sedimentary prism at least 15 km long and up to 160 m thick (Fig. 7). The 2D seismic lines in this area are too widely spaced to allow a more precise determination of the 3D shape of the drift, part of which is located north of the survey grid. The unit is characterized by a good reflector quality, both internally and externally, and the unit-bounding reflectors are laterally coherent. The lower Maastrichtian succession beneath the drift is scoured, as seen by a series of truncated reflectors forming a moat prior to deposition of

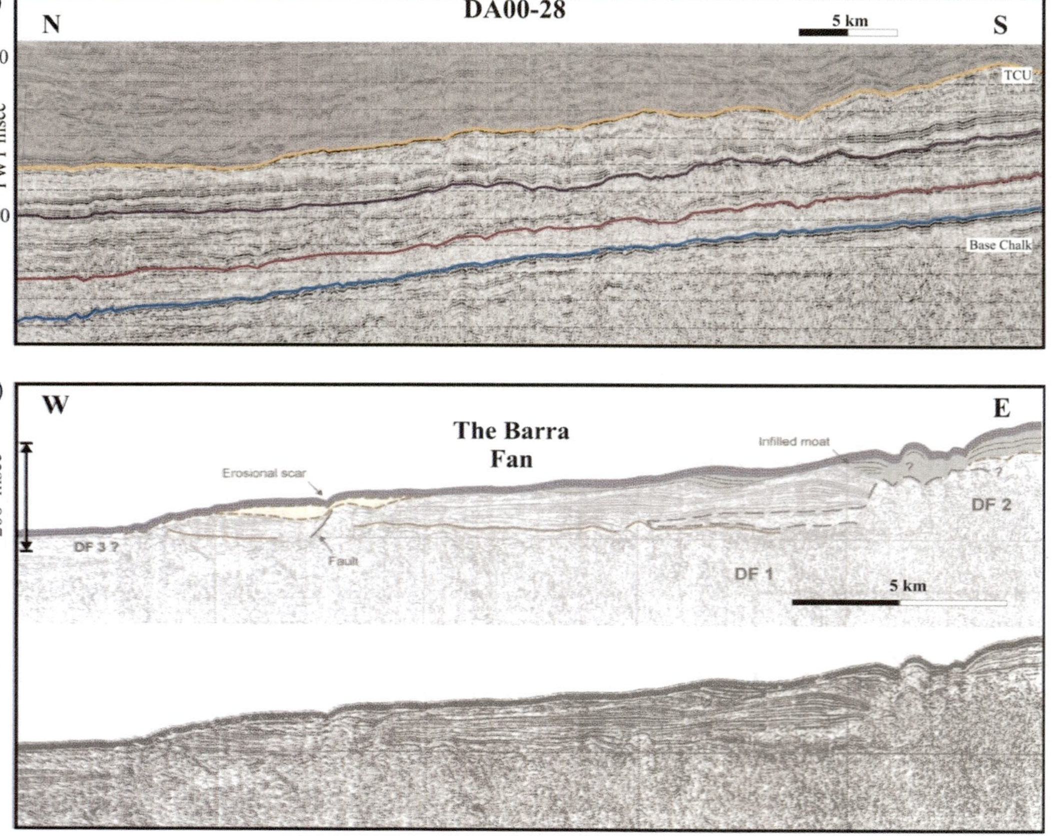

Fig. 5. (**a**) The seismic line DA00-28 (location shown in Fig. 2) illustrates the different nature of the Top Campanian Unconformity (TCU), represented by the yellow reflector, across the study area. In the southern part of the study area the TCU is conformable and signs of erosion are rare, whereas in the northern part the TCU is characterized by erosional scours or channels and minor truncations. The north–south orientation (opposite to the south–north orientation of Fig. 4) allows direct comparison with the section in Figure 5b. (**b**) Seismic profile and interpretation from the Barra Fan in the eastern Rockall Trough (NW of the UK margin) at depths of about 2000 m. The fan is characterized by a zone of drift deposition to the east, passing westwards into a zone of winnowing and/or erosion. Its morphology results from two bottom-current pathways related to the Deep Northern Boundary Current. Erosional scours appear in the distal part of the drift, under the main axis of one of the strong bottom currents (modified after Knutz *et al.* 2002).

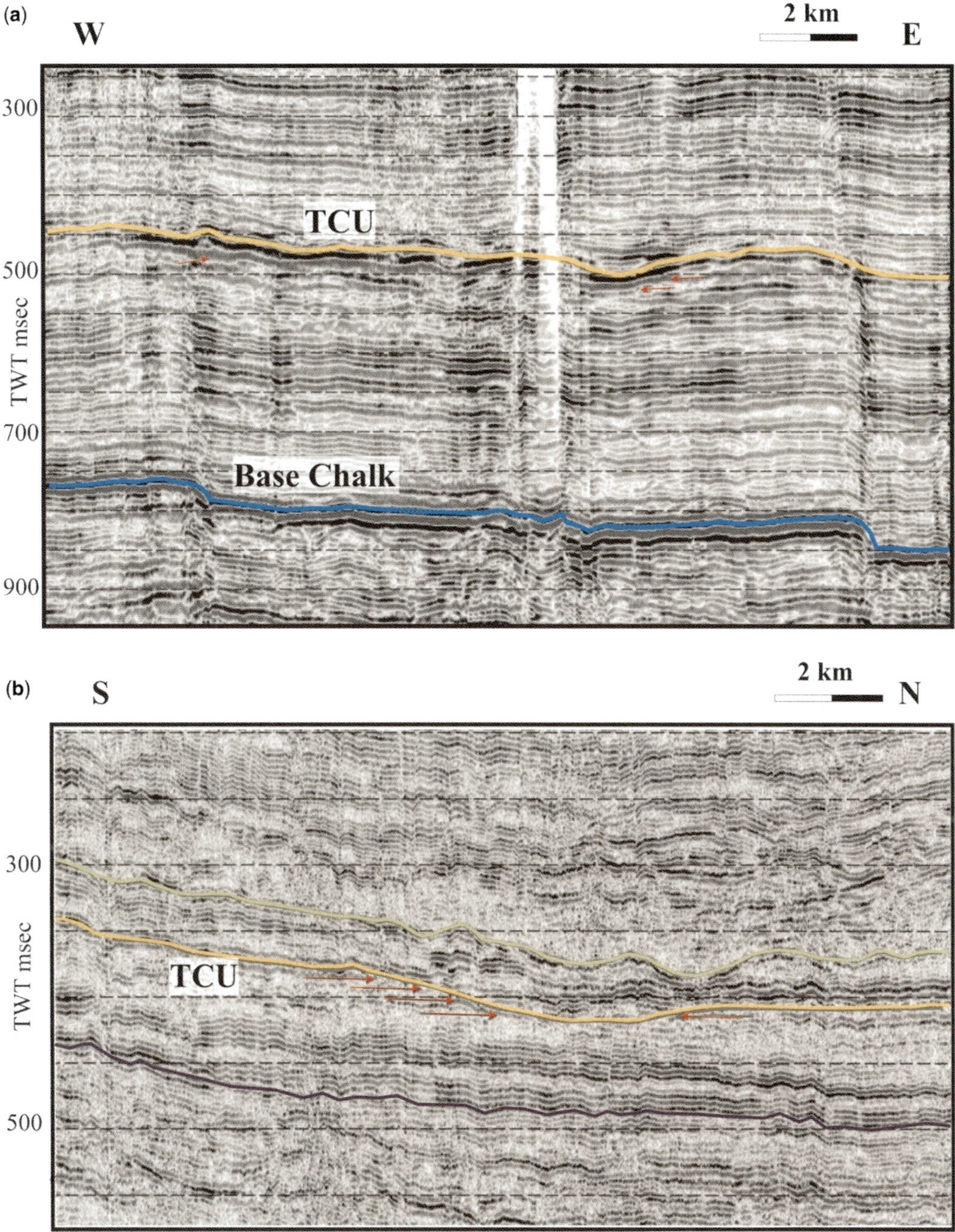

Fig. 6. (**a**) Detail of the TCU (location shown in Fig. 2), represented by the yellow reflector, showing erosional incision and consequent truncation of the reflectors of the lower unit. (**b**) Detail of the TCU (location shown in Fig. 4), illustrating a scour or channel downcutting through about 60 m of parallel, well-layered deposits.

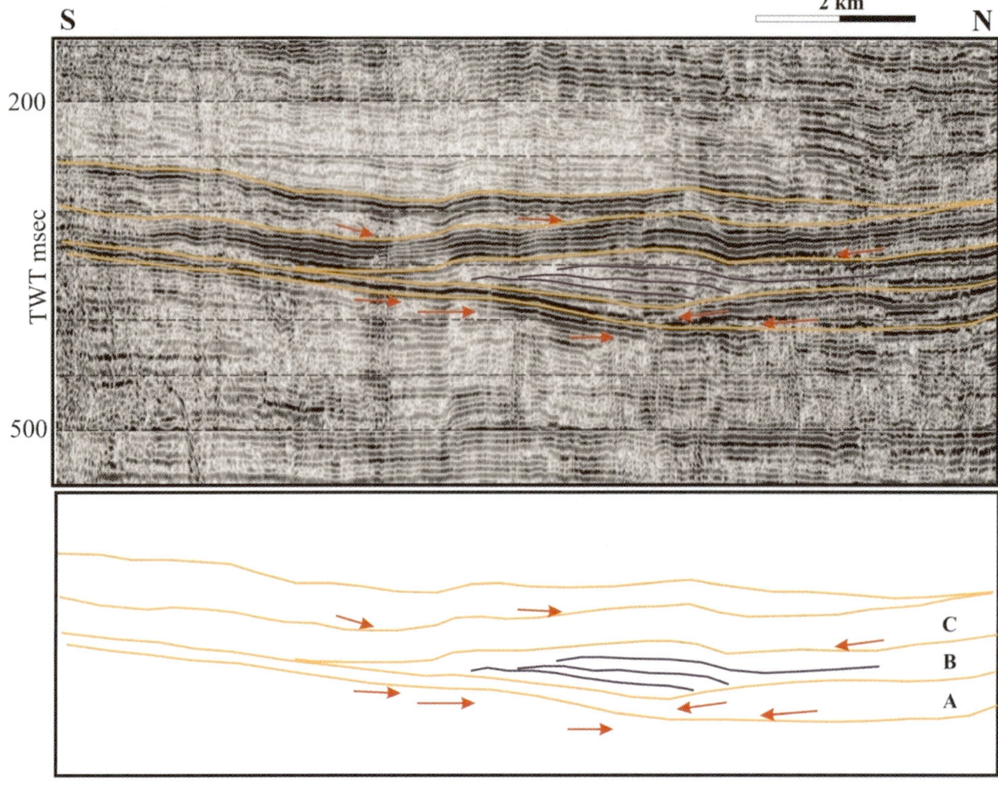

Fig. 7. Mounded drift in the upper Maastrichtian interval. The drift is at least 15 km long and up to 160 m thick, and comprises a system of complex associations of reflector terminations. It is interpreted as formed by three units with differing seismic facies and shifting laterally in opposite directions. (See Figs 2 and 8 for location.)

the drift. To ease the interpretation we subdivide the drift into seismic units A–C based on their migration patterns. Unit A is wedge-shaped, downlaps the moat towards the south and thins in the same direction. This unit may represent an early stage of the drift dominated by a SSW migration trend. Unit B consists of a system of strong reflectors, showing low-angle lateral migration in the opposite direction to those of Unit A. Its formation involves a rapid southward displacement of the whole system. The lateral shift of downlapping reflectors results from a combination of lateral migration and aggradation towards the NNE. The thickness of Unit B decreases strongly towards the south until it pinches-out. Unit C is characterized by a lack of reflector terminations, and the unit drapes the entire structure and shows a rather constant thickness across the northernmost part of the study area. The drift is overlain by a northward-migrating wedge, which downlaps onto the top surface and pinches-out further north than Units A and B. A general migration trend of the moat towards the SSW is interpreted from the adjacent seismic lines.

To the south the drift is flanked by a region characterized by very low sedimentation rates, as seen by the accentuated thinning of the lower upper Maastrichtian succession. The lack of reflector truncations indicates that this is the result of non-deposition rather than erosion. The area where Units A and B pinch-out marks the northern edge of this region, also observed in the isochron map for the lower upper Maastrichtian unit (Fig. 8).

The moat that underlies the mounded drift was formed by an erosional episode at the early–late Maastrichtian transition. This event seems to have mainly affected the northernmost part of the area. The high amplitude and continuous nature of the reflectors that characterize the seismic units of the drift point towards a bottom-current influence and rule out the possibility of downslope failure origin. The formation of such a complex internal architecture is probably linked to highly variable current velocities during the early late Maastrichtian. Thus, the opposing migration directions of Units A and B represent a period when the current axis shifted laterally across the Øresund area. The

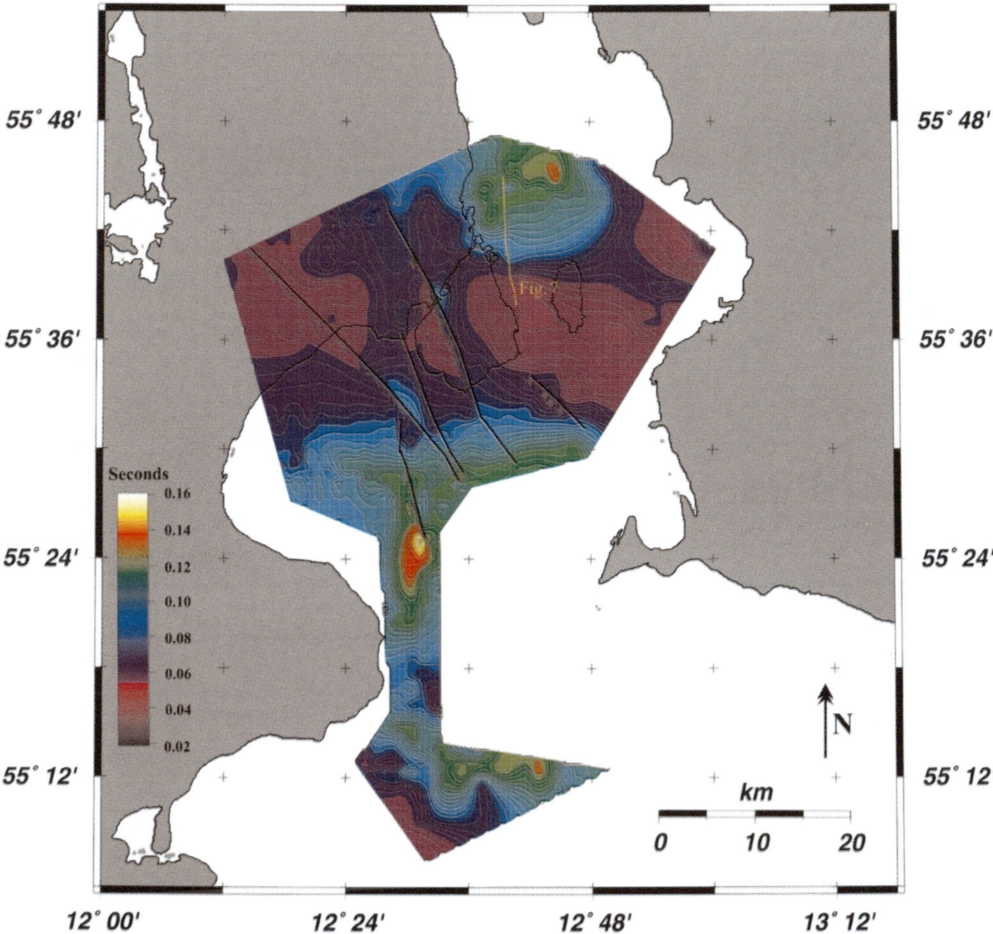

Fig. 8. Isochron map of the lower upper Maastrichtian succession in the study area. The mounded drift is located in the green area in the upper part of the figure. Location of the profile in Figure 7 is shown in the map by the orange line. South of the drift a belt with a thin lower upper Maastrichtian succession is observed across the area, with a WNW–ESE orientation.

deposition of Unit C, as a drape covering the seafloor topography, possibly indicates an episodic decrease in current velocity. The presence of a condensed unit south of the drift observed in the isochron map is possibly linked to the formation of the drift and results from the interplay between current winnowing and localized sedimentation (Fig. 8).

The southwestward migration of the moat, away from the Sorgenfrei–Tornquist Zone, marks the onset of a palaeoceanographic change possibly resulting from a drop in relative sea level. This may have been of eustatic nature or caused by uplift of the Sorgenfrei–Tornquist Zone displacing the current system towards the axis of the Danish Basin.

Sediment wave complex. In the southern part of the study area, offshore Stevns Klint, the upper Santonian–lower Maastrichtian succession is characterized by a large and persistent valley system associated with a complex of upslope-migrating sediment waves (Fig. 9a). In the lower part of the succession the valley appears as erosional, incised into the lower Santonian deposits (marked by an arrow in Fig. 9a). Upwards it evolves into a depositional valley forming part of the larger ridge-and-valley system described by Lykke-Andersen & Surlyk (2004). The main system and the crests of the sediment waves trend roughly parallel to the Sorgenfrei–Tornquist Zone, as indicated by the offshore seismic data and by onshore evidence from the Stevns peninsula.

(a)

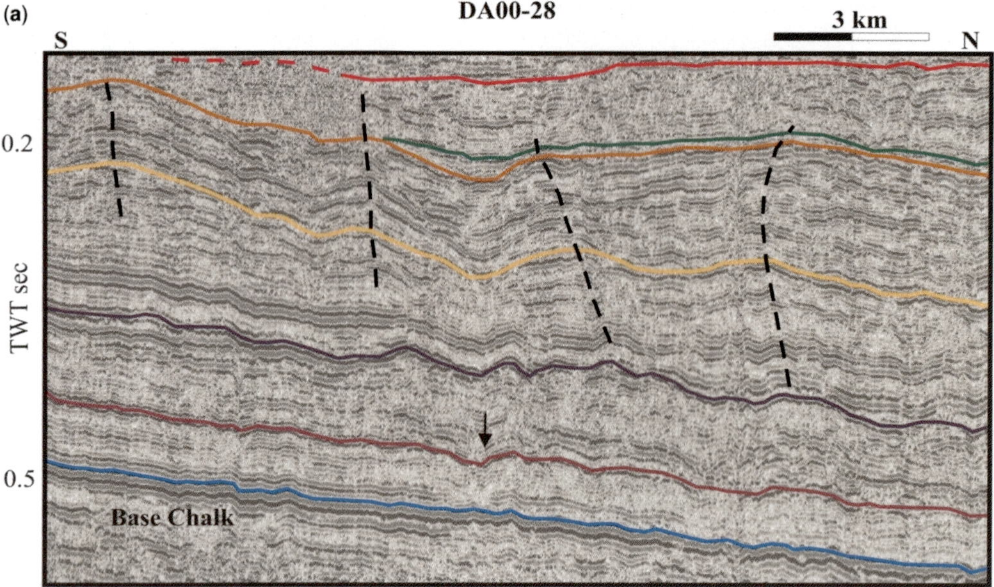

(b)

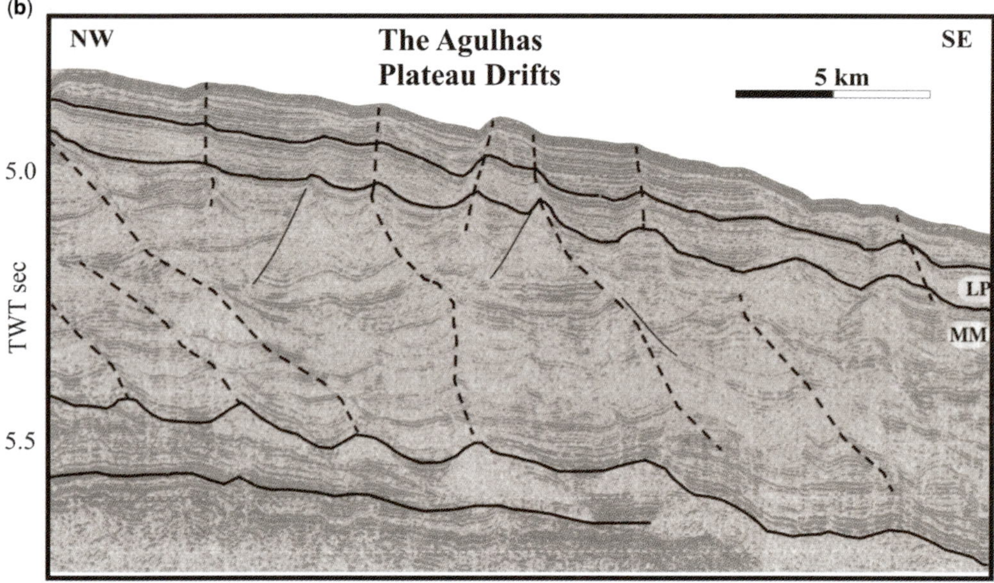

Fig. 9. (**a**) Detail of sediment wave complex in the southern part of the study area (see Fig. 4 for location). The wave complex seems to start in the late Santonian as an incised valley, evolving into a system of sediment waves and intervening valleys in the early Maastrichtian and reaching wavelengths of 2 km and amplitudes of up to 30 m. A general upslope migration trend can be observed. The location of the erosional valley is marked by the arrow. (**b**) Example of a sediment wave system from the Agulhas Plateau located at depths of about 2000–3000 m. The Agulhas Plateau drifts consist of dunes, drifts, channels and erosional unconformities formed as a result of the circulation of the Agulhas Current and the Antarctic Bottom Water (modified after Uenzelmann-Neben 2002).

The sediment waves are characterized by asymmetrically climbing forms, about 2 km in wavelength and up to 30 m in amplitude. The vertical resolution of the seismic data is of sufficient quality to show that downslope resedimentation did not play an important role in the formation of the wave forms. This is seen by the presence of well-layered structures with good internal reflectivity, showing combined lateral migration and aggradation growth patterns, which in turn suggest a current origin. This is supported by the close resemblance to the wave fields associated with well-known contourite drifts such as the drifts of the Rockall Trough (Howe *et al.* 2002), the Barra Fan (Knutz *et al.* 2002; Stow *et al.* 2002*b*), the Agulhas Plateau (Uenzelmann-Neben 2002), the Weddell Sea (Pudsey 2002), and the Bahama Outer Ridge (Flood 1994) (see also Faugères *et al.* 1999).

The internal architecture of each of the large-scale bed-forms shows a systematic onlap onto the flank of the adjacent upslope wave as the waves migrate southwestwards (Fig. 9a). Some of the waves build up from topographic lows of the Top Santonian sea floor, and display clear examples of topographic inversion from concave-up structures in the lower part of the succession to convex-up structures in the upper part.

A slight reinforcement of the wave pattern of sedimentation seems to occur above the TCU, as seen by higher-relief waves and more numerous reflector onlaps. In the upper Maastrichtian the wave-type sedimentation is poorly defined and apparently vanishes in the shallowest intervals, possibly because of shallow noise masking.

The formation of the sediment wave complex was related to the presence of the erosional valley. In the lower Santonian succession the material had spilled over the valley walls and formed levee-like deposits, which triggered the wavy sedimentation pattern as a feedback loop. The sediment waves probably built up as a result of the flow turbulence created by the changing sea-bed morphology owing to their own sedimentation.

The sediment wave sedimentation pattern indicates long-term preferential sediment deposition (Flood 1994). Sediment transport direction can generally be inferred from the migration direction. If the main sediment source was the contour-parallel current, the wave system would have prograded in a direction parallel to the flow. However, in the present case, migration was oblique to the bathymetric contours. Updip climbing indicates high rates of sediment input (Locker & Laine 1992). The mapped sediment wave field may therefore reflect the combination of downslope sediment supply from surrounding topographic highs and drift formation by alongslope flow. In some cases development of wave fields associated with a drift has been ascribed to the effect of internal waves of a weakly structured water column (Flood 1994). This was probably the case during the late Cenomanian major transgression, which broke down the shelf-edge front and dismantled the water column stratification (Hay 1995; Gale *et al.* 2000). The collapse of the wave-like type of sedimentation pattern in the late Maastrichtian may reflect the onset of less stable oceanographic conditions and possibly changes in the sediment source area, or result from shallow noise masking.

A Cenozoic sediment wave system that is still active is found in the contourite drifts of the Agulhas Plateau, south of South Africa (Uenzelmann-Neben 2002), and resembles the geometries of the chalk sediment waves both internally and externally (Fig. 9b). The Agulhas drifts are, however, found at depths of 2000–3000 m, whereas the chalk sea floor was some hundreds of metres deep, possibly up to 800 m in localized areas. Despite the difference of depth, the formational process seems to be very similar, although probably not of the same magnitude, as nannofossil ooze is more easily reworked than argillaceous materials.

Elongate moat–drift system. The seismic profiles offshore Stevns Klint show the presence of a drift situated on the southern side of a moat in the Campanian marly chalk (Fig. 10a). The absence of reflector truncation in the sediment units beneath the moat–drift complex suggests that the moat is of non-depositional rather than erosional origin. The wavelength of the drift is more than 4 km and it is about 100 m thick. The internal reflectors of the drift are weaker than those of the units above and below, but show good continuity. The seismic resolution does not allow us to discern whether the northern flank of the drift shows progressive southward pinch-out of the reflectors or if it represents an angular unconformity. Crest-to-moat height may have reached a maximum value of 70 m and correlation to the closest seismic lines seems to indicate an elongate west–east-trending shape. The moat–drift system is overlain by an infill succession, which onlaps the northern flank of the drift and shows a similar internal acoustic response. Reflectors that at present appear as horizontal in the infill succession are likely to represent an original very low angle inclined bedding when post-depositional tilting is taken into consideration. Inclined bedding has been observed in other fine-grained contourites and in some cases represents truncated clinoform successions (Duan *et al.* 1993). The uppermost part of the drift and the prograding inclined reflectors of the moat fill succession have been truncated during the erosive episode that gave rise to the extensive TCU.

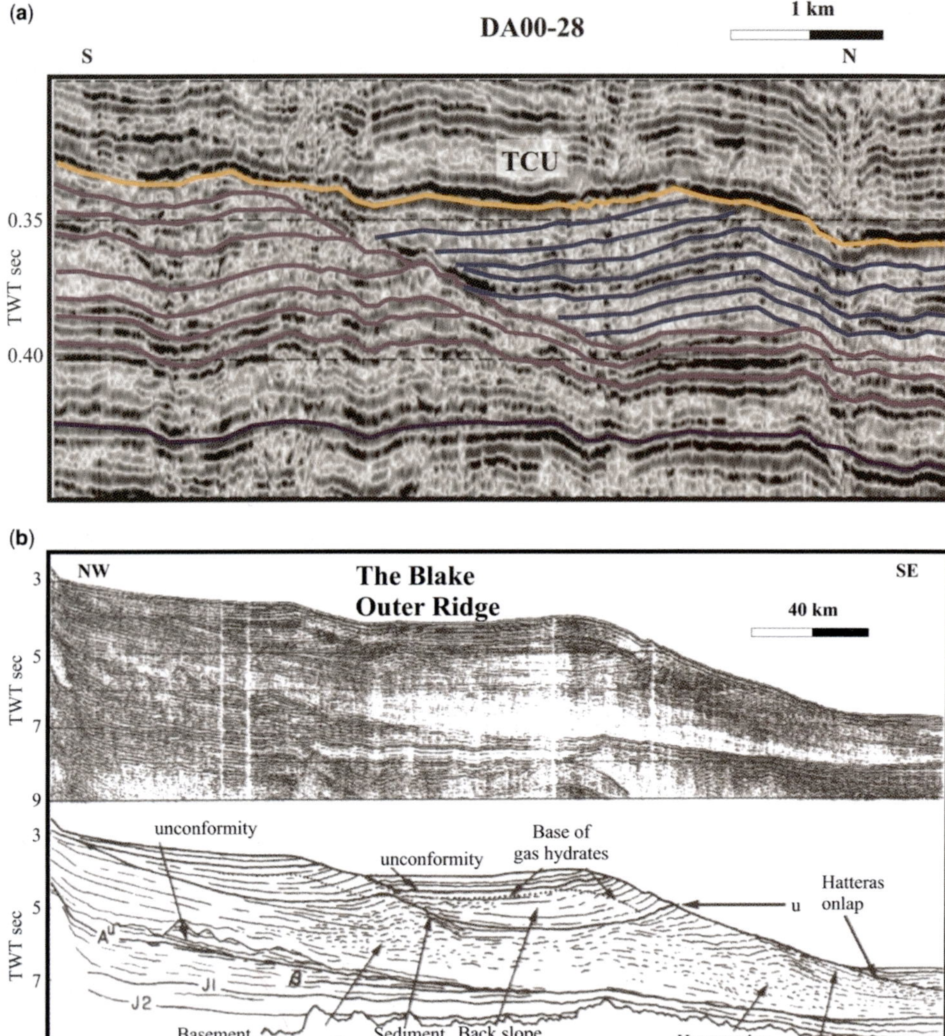

Fig. 10. (**a**) Detail of elongate moat–drift system within Campanian marly chalk offshore Stevns Klint (see Fig. 4 for location). The drift has a wavelength of more than 4 km and is up to 100 m thick. The moat infill succession is well layered and onlaps the northern flank of the drift. (**b**) Example of an elongate mounded drift from the Blake Outer Ridge on the southeastern continental margin of North America. The drift is located at 2000–5000 m on the Hatteras Abyssal Plain and its deposition is affected by changes in intensity and/or position of the Western Boundary Undercurrent (Markl & Bryan 1983; Haskell & Johnson 1993; Faugères *et al.* 1999). The NW–SE seismic profiles across the Blake Outer Ridge show that the eastern part of the drift is onlapped by sediments of the Hatteras Abyssal Plain (modified after Faugères *et al.* 1999) u, unconformity.

The sigmoidal progradational–aggradational pattern of the lithologically homogeneous Campanian moat–drift system indicates that its development is probably related to bottom-current activity rather than a slump origin. The bottom currents flowing through this area were, however, not strong enough to erode, but only caused localized deposition. Thus, the growth of the drift was directly related to sediment starvation and winnowing in the adjacent moat. The succeeding fill of the moat by a southwards prograding–aggrading succession may have been the result of the attenuation of the current velocity or the lateral shift of its main axis. The upper part of the system was later truncated as current velocity increased during the major late Campanian–early Maastrichtian erosive event. The presence of large-scale unconformities associated with sedimentary drifts is a typical feature observed in contour-current deposits and is probably the result of a drastic change towards enhanced current intensity (Stow *et al.* 2002*a*; Martin-Chivelet *et al.* 2003).

Unlike the chalk structures described in the previous sections, the general migration direction of the moat–drift complex towards the south seems to indicate the effect of an easterly flowing current. This could be explained by the presence of a minor eddy detached from the main flow and flowing anticlockwise. However, the limited amount of data does not allow any firm conclusion.

A possible modern analogue of the depositional pattern observed in the chalk moat–drift system is seen in the Blake Outer Ridge of the southeastern continental margin of North America (Fig. 10b). In this area, non-steady contour-current conditions led to the formation of a series of angular unconformities and related terraces west of the drift crest, as the flank of the drift was onlapped by abyssal sediments (Markl & Bryan 1983). The infill succession of the chalk moat–drift system could also represent simple onlap or the onlap by a southward-prograding drift. In any of these cases, the formation of the entire systems seems to be strongly influenced by bottom-current activity.

Deposition of the chalk drifts

The chalk succession in the Øresund area has traditionally been considered as gently folded during Palaeocene inversion tectonics (Rosenkrantz 1938). This hypothesis was based on the undulating shape of the earliest Danian erosion surface exposed in Stevns Klint. It has recently been shown that the chalk in this area is not folded and that the relief of the erosion surface mirrors the actual sea-floor relief (Lykke-Andersen & Surlyk 2004). In the north–south seismic sections the Base Chalk surface appears as a strong and flat, slightly northward-dipping reflector, which extends unbroken for more than 50 km and is disrupted only by the system of major faults related to the Carlsberg Fault in the northern part of the study area (Lykke-Andersen & Surlyk 2004; Nielsen & Thybo 2004).

The mapped mounded features described here could represent biogenic mounds similar to the Danian bryozoan mounds that crop out in Stevns Klint. However, the structures are one order of magnitude larger, have a ridge-like shape and are formed by almost pure chalk without framework. Therefore a purely biogenic origin seems unlikely, and comparison with modern contourite systems shows a great similarity.

The broad range of sea-floor features and morphologies observed may be explained by the combined action of long- and short-term fluctuations of the current system. Long-term variations in flow direction and intensity may be due to the tectonic opening and closure of seaways. Short-term variations are related to relative sea-level fluctuations and associated lateral shifts of the cores of the contour-parallel currents and current velocity.

The hydrodynamic processes affecting the sea bed partly result from the interaction between the circulation of geostrophic currents and the shallow circulation controlled by atmospheric conditions. During pervasive low atmospheric pressures or under eddies of powerful currents, large amounts of kinetic energy can be propagated to the bottom and lead to the formation of benthic storms (Hollister & McCave 1984). The sediment resuspension that takes place during these storms triggers the formation of dense nepheloid layers and the generation of bedforms (Faugères & Stow 1993; Hollister 1993; Stow *et al.* 2002*a*). Minute clay-sized coccolith fragments show no cohesion, and even mild agitation may resuspend the thixotropic nannofossil ooze and form nepheloid layers at the reach of contour currents (e.g. Bramwell *et al.* 1999). A competent current can transport the suspended sediment in the water column for very long distances until conditions are more tranquil and deposition predominates over erosion or bedload transport. Current velocities between 8 and 20 cm s^{-1} have been suggested for conditions of non-deposition and erosion of Maastrichtian nannofossil ooze (see Surlyk & Lykke-Andersen in press). High sedimentation rates and drift formation occur principally along the edges of the flow axis, as a result of the damping of the shear stress within the benthic boundary layer (e.g. Knutz & Cartwright 2003).

The northwestward-flowing current, parallel to the contours of the Danish Basin, generated a deep moat and a multitude of subsidiary drifts along the slope of the inverted Sorgenfrei–Tornquist Zone in the Kattegat Sea (Surlyk & Lykke-Andersen in press). The study area is located SE of this area and is not directly connected to the prominent SW-dipping slope of the inversion zone. However, the mounded and elongate drifts and the wavy sea-floor relief that characterize the Santonian–upper Campanian succession indicate that along-slope currents strongly sculpted the sea bottom. The dominance of sediment wave-type deposition in this period of sea-level rise is the result of the combination of continuous sedimentation and a stable current system of low erosive potential. The current reinforcement during the late Campanian–early Maastrichtian, which generated the TCU, not only winnowed or prevented deposition, but also eroded the previous deposited chalk. This main erosive event, probably synchronous over the entire area, was more pervasive in the northern part than in the south. Thus, the sharp erosional scours observed in the north are nearly absent in the south, where the succession is rather conformably layered. The late Santonian–late Maastrichtian sediment wave system continued throughout this period and does not seem to have been affected by the erosive event. The horizontal velocity gradient in the Øresund area caused by the dextral deflection of the Coriolis force is interpreted as responsible for the lateral variation of current-produced bedforms. Lower current velocities in the Maastrichtian allowed for reactivation of drift sedimentation in the area, as seen by the widespread distribution of drifts and their larger relief. However, the formation of the moat beneath the upper Maastrichtian complex mounded drift is a sign of significant erosion in the north of the area at the early–late Maastrichtian boundary. Formation of the mounded drift and the absence of sediment waves in the upper Maastrichtian chalk may have been the result of less stable oceanographic conditions and a general southwestward shift of the current axis. This instability could be caused by further inversion and uplift of the Sorgenfrei–Tornquist Zone (Liboriussen *et al.* 1987; Vejbæk & Andersen 2002).

Contourite drifts are commonly defined as sedimentary drifts formed by deep-sea bottom currents driven by the thermohaline circulation (e.g. Stow *et al.* 2002*a*). A general consensus limits the minimum water depth for the formation of contourite drifts to 300 m, above which superficial currents exert a major influence (Stow *et al.* 2002*a*). This has led to attempts to provide a nomenclature system for drifts formed at shallower depths by bottom currents different from thermohaline currents (Faugères & Stow 1993; Stow *et al.* 2002*a*). The classical contourite depositional model has been mainly constructed for Pleistocene–Holocene oceanographic conditions, in which geostrophic currents are restricted to oceanic basins. In the Late Cretaceous geostrophic currents dominated much of the relatively deep epeiric sea of NW Europe, and therefore the oceanographic framework of the chalk drifts resembles that of modern contourite drifts. The classification of the chalk drifts as contourite drifts is also supported by comparison of some of the topographic elements of the chalk sea floor with published studies of well-known Pleistocene–modern drifts. The shape of a drift, its location in the basin and its composition are the main characteristics used as modifiers before the name contourite drift, and the drifts observed in this study are essentially epeiric elongate–mounded chalk contourite drifts.

Conclusions

(1) During the Late Cretaceous a geostrophic current system connected to the global circulation trends flowed towards the NW parallel to the contours of the Sorgenfrei–Tornquist Zone and the axis of the Danish Basin.

(2) The chalk succession in the Øresund region is characterized by the presence of numerous structures that represent real topographic elements of the Late Cretaceous sea floor and were formed by along-slope currents.

(3) The migration trends of the mapped structures and the widespread drift sedimentation during the Santonian–late Campanian indicate a time interval of decreased current velocity, dominated by localized sedimentation, possibly linked to a stage of general sea-level rise.

(4) Large-scale erosion in the late Campanian–early Maastrichtian resulted in the formation of the Top Campanian Unconformity. The erosive event was more pervasive in the northeastern part of the study area (i.e. closer to the SW-dipping slope of the Sorgenfrei–Tornquist Zone). This supports the idea of a main contour current flowing northwestward through the area with a positive lateral velocity gradient towards the slope, generated by the Coriolis force.

(5) Destabilization of the oceanographic framework in the late Maastrichtian is inferred from the collapse of the sediment wave-like sedimentation and formation of drifts displaying complex internal geometries.

(6) The Late Cretaceous current system is analogous to thermohaline current systems flowing parallel to the contours of continental margins and responsible of modern contourite deposition. The slope of the Sorgenfrei–Tornquist Zone acted as an analogue to a continental margin during the Late Cretaceous sea-level highstand.

(7) The physical oceanographic framework associated with continental margins and the formation of classical contourite drifts expanded into the European Late Cretaceous epeiric sea.

The present study was funded by the Danish Natural Science Research Council. Offshore seismic data were collected onboard the research vessels *Dana* and *Flyvefisk*. Onshore and offshore seismic data and well-log data for Margretheholm-1 were kindly placed at our disposal by DONG E&P A/S. The seismic data were processed by Nørmark, E. We thank Uenzelmann-Neben, G. and Roveri, M. for critical reading of the manuscript.

References

ALVAREZ, W., KAUFFMAN, E. G., SURLYK, F., ALVAREZ, L. W., ASARO, F. & MICHEL, H. V. 1984. Impact theory of mass extinctions and the invertebrate fossil record. *Science*, **223**, 1135–1141.

BANG, I. 1971. Palæocæn i Rønde nr. 1. *In*: *Dybeboringen Rønde nr.1 på Djursland.* Danmarks Geologiske Undersøgelse, III Række, **39**, 50–52.

BONNESEN, E. P., BØGGILD, O. B. & RAVN, J. P. J. 1913. *Calsbergfondets dybdeboring i Grøndals Eng ved København 1894–1907 og dens videnskabelige resultater.* Museum de Minéralogie et de Géologie de l'Université de Copenhague, Communications Géologiques.

BRAMWELL, N. P., CAILLET, G., MECIANI, L., JUDGE, N., GREEN, M. & ADAM, P. 1999. Chalk exploration, search for a subtle trap. *In*: FLEET, A. J. & BOLDY, S. A. R. (eds) *Petroleum Geology of Northwest Europe; Proceedings of the 5th Conference.* Geological Society, London, 911–938.

BROMLEY, R. G. & EKDALE, A. A. 1987. Mass transport in European Cretaceous chalk; fabric criteria for its recognition. *Sedimentology*, **34**, 1079–1092.

BROTZEN, F. 1945. *De geologiska resultaten från borrningarna vid Höllviken. Preliminär rapport. Del I: Kritan.* Sveriges Geologiska Undersökning, Series C, **465**.

CHRISTENSEN, L., FREGERSLEV, S., SIMONSEN, A. & THIEDE, J. 1973. Sedimentology and depositional environment of lower Danian fish clay from Stevns Klint, Denmark. *Bulletin of the Geological Society of Denmark*, **22**, 193–212.

DUAN, T., GAO, Z., ZENG, Y. & STOW, D. A. V. 1993. A fossil carbonate contourite drift on the Lower Ordovician palaeocontinental margin of the middle Yangtze Terrane, Jiuxi, northern Hunan, southern China. *Sedimentary Geology*, **82**, 271–284.

EVANS, D. J., HOPSON, P. M., KIRBY, G. A. & BRISTOW, C. R. 2003. The development and seismic expression of synsedimentary features within the Chalk of southern England. *Journal of the Geological Society, London*, **160**, 797–813.

FAUGÈRES, J. C. & STOW, D. A. V. 1993. Bottom-current-controlled sedimentation—a synthesis of the contourite problem. *Sedimentary Geology*, **82**, 287–297.

FAUGÈRES, J. C., IMBERT, P., MEZERAIS, M. L. & CREMER, M. 1998. Seismic patterns of a muddy contourite fan (Vema Channel, South Brazilian Basin) and a sandy distal turbidite deep-sea fan (Cap Ferret system, Bay of Biscay): a comparison. *Sedimentary Geology*, **115**, 81–110.

FAUGÈRES, J. C., STOW, D. A. V., IMBERT, P. & VIANA, A. 1999. Seismic features diagnostic of contourite drifts. *Marine Geology*, **162**, 1–38.

FLOOD, R. D. 1994. Abyssal bedforms as indicators of changing bottom current flow: examples from the U.S. East Coast continental rise. *Paleoceanography*, **9**, 1049–1060.

FRAKES, L. A. 1999. Estimating the global thermal state from Cretaceous sea surface and continental temperature data. *In*: BARRERA, E. & JOHNSON, C. C. (eds) *Evolution of the Cretaceous ocean–climate system.* Geological Society of America, Special Papers, **332**, 49–57.

GALE, A. S. 1980. Penecontemporaneous folding, sedimentation and erosion in Campanian chalk near Portsmouth, England. *Sedimentology*, **27**, 137–151.

GALE, A. S., SMITH, A. B., MONKS, N. E. A., YOUNG, J. A., HOWARD, A., WRAY, D. S. & HUGGETT, J. M. 2000. Marine biodiversity through the Late Cenomanian–Early Turonian: palaeoceanographic controls and sequence stratigraphic biases. *Journal of the Geological Society, London*, **157**, 745–757.

HAQ, B. U., HARDENBOL, J. & VAIL, P. R. 1987. Chronology of fluctuating sea levels since the Triassic. *Science*, **235**, 1156–1167.

HARDMAN, R. F. P. 1982. Chalk reservoirs of the North Sea. *Bulletin of the Geological Society of Denmark*, **30**, 119–137.

HASKELL, B. J. & JOHNSON, T. C. 1993. Surface sediment response to deepwater circulation on the Blake Outer Ridge, western North Atlantic; paleoceanographic implications. *Sedimentary Geology*, **82**, 133–144.

HATTON, I. R. 1986. Geometry of allochthonous Chalk Group members, Central Trough, North Sea. *Marine and Petroleum Geology*, **3**, 79–98.

HAY, W. W. 1995. Cretaceous paleoceanography. *Geologica Carpathica*, **46**, 257–266.

HOLLISTER, C. D. 1993. The concept of deep-sea contourites. *Sedimentary Geology*, **82**, 5–11.

HOLLISTER, C. D. & MCCAVE, I. N. 1984. Sedimentation under deep-sea storms. *Nature*, **309**, 220–225.

HOWE, J. A., STOKER, M. S., STOW, D. A. V. & AKHURST, M. C. 2002. Sediment drifts and contourite sedimentation in the northeastern Rockall Trough and Faroe–Shetland Channel, North Atlantic Ocean. *In*: STOW, D. A. V., PUDSEY, C. J., HOWE, J. A., FAUGÈRES, J. C. & VIANA, A. (eds) *Deep-water Contourite Systems; Modern Drifts and Ancient Series, Seismic and Sedimentary Characteristics.* Geological Society, London, Memoirs, **22**, 65–72.

JAPSEN, P. 1998. Regional velocity–depth anomalies, North Sea Chalk: a record of overpressure and Neogene uplift and erosion. *AAPG Bulletin*, **82**, 2031–2074.

KENNEDY, W. J. 1987. Sedimentology of Late Cretaceous–Paleocene chalk reservoirs, North Sea Central Graben. *In*: BROOKS, J. & GLENNIE, K. W. (eds) *Petroleum Geology of North West Europe.* Graham & Trotman, London, 469–481.

KNUTZ, P. C. & CARTWRIGHT, J. A. 2003. Seismic stratigraphy of the West Shetland Drift: implications for late Neogene paleocirculation in the Faeroe–Shetland gateway. *Paleoceanography*, **18**, 1–17.

KNUTZ, P. C., JONES, E. J. W., AUSTIN, W. E. N. & VAN WEERING, T. C. E. 2002. Glacimarine slope sedimentation, contourite drifts and bottom current pathways on the Barra Fan, UK North Atlantic margin. *Marine Geology*, **188**, 129–146.

LIBORIUSSEN, J., ASHTON, P. & TYGESEN, T. 1987. The tectonic evolution of the Fennoscandian border zone in Denmark. *Tectonophysics*, **137**, 21–29.

LOCKER, S. D. & LAINE, E. P. 1992. Paleogene Neogene depositional history of the middle United States Atlantic continental rise—mixed turbidite and contourite depositional systems. *Marine Geology*, **103**, 137–164.

LYKKE-ANDERSEN, H. & SURLYK, F. 2004. The Cretaceous–Palaeogene boundary at Stevns Klint, Denmark: inversion tectonics or sea-floor topography? *Journal of the Geological Society, London*, **161**, 343–352.

MARKL, R. G. & BRYAN, G. M. 1983. Stratigraphic evolution of Blake Outer Ridge. *AAPG Bulletin*, **67**, 666–683.

MARTIN-CHIVELET, J., FREGENAL-MARTINEZ, M. A. & CHACON, B. 2003. Mid-depth calcareous contourites in the latest Cretaceous of Caravaca (Subbetic zone, SE Spain). Origin and palaeohydrological significance. *Sedimentary Geology*, **163**, 131–146.

MEZERAIS, M. L., FAUGÈRES, J. C., FIGUEIREDO, A. G. & MASSE, L. 1993. Contour current accumulation off the Vema Channel mouth, southern Brazil Basin—pattern of a contourite fan. *Sedimentary Geology*, **82**, 173–187.

NIELSEN, L. & THYBO, H. 2004. Location of the Carlsberg fault zone from seismic controlled-source fan recordings. *Geophysical Research Letters*, **31**, L07621, doi:10.1029/2004GL019603.

PUDSEY, C. J. 2002. The Weddell Sea: contourites and hemipelagites at the northern margin of the Weddell Gyre. *In*: STOW, D. A. V., PUDSEY, C. J., HOWE, J. A., FAUGÈRES, J. C. & VIANA, A. (eds) *Deep-Water Contourite Systems; Modern Drifts and Ancient Series, Seismic and Sedimentary Characteristics*. Geological Society, London, Memoirs, **22**, 289–303.

RAVN, J. P. J. 1913. Calsbergfondets dybeboring i Grøndals Eng ved København 1894–1907 og dens videnskabelig resultater. *In*: BONNESEN, E. P., BØGGILD, O. B. & RAVN, J. P. J. (eds) *Museum de Minéralogie et de Géologie de l'Université de Copenhague, Communications Géologiques*, **3**, 91–105.

REBESCO, M., LARTER, R. D., CAMERLENGHI, A. & BARKER, P. F. 1996. Giant sediment drifts on the continental rise west of the Antarctic Peninsula. *Geo-Marine Letters*, **16**, 65–75.

ROSENKRANTZ, A. 1938. Bemærkninger om det østsjællandske Daniens stratigrafi og tektonik. *Meddelelser fra Dansk Geologisk Forening*, **9**, 199–212.

STENESTAD, E. 1971. Øvre Kridt i Rønde nr. 1. *In*: *Dybeboringen Rønde nr. 1 på Djursland*. Danmarks Geologiske Undersøgelse, III Række, **39**, 53–60.

STOKER, M. S., AKHURST, M. C., HOWE, J. A. & STOW, D. A. V. 1998. Sediment drifts and contourites on the continental margin off northwest Britain. *Sedimentary Geology*, **115**, 33–51.

STOW, D. A. V. & HOLBROOK, J. A. 1984. North Atlantic contourites; an overview. *In*: STOW, D. A. V. & PIPER, D. J. W. (eds) *Fine-Grained Sediments, Deep-Water Processes and Facies*. Geological Society, London, Special Publications, **15**, 245–256.

STOW, D. A. V., FAUGÈRES, J. C., HOWE, J. A., PUDSEY, C. J. & VIANA, A. R. 2002*a*. Bottom currents, contourites and deep-sea sediment drifts; current state-of-the-art. *In*: STOW, D. A. V., PUDSEY, C. J., HOWE, J. A., FAUGÈRES, J. C. & VIANA, A. (eds) *Deep-Water Contourite Systems; Modern Drifts and Ancient Series, Seismic and Sedimentary Characteristics*. Geological Society, London, Memoirs, **22**, 7–20.

STOW, D. A. V., ARMISHAW, J. E. & HOLMES, R. 2002*b*. Holocene contourite sand sheet on the Barra Fan slope, NW Hebridean margin. *In*: STOW, D. A. V., PUDSEY, C. J., HOWE, J. A., FAUGÈRES, J. C. & VIANA, A. (eds) *Deep-Water Contourite Systems; Modern Drifts and Ancient Series, Seismic and Sedimentary Characteristics*. Geological Society, London, Memoirs, **22**, 99–119.

SURLYK, F. 1997. A cool-water carbonate ramp with bryozoan mounds; Late Cretaceous–Danian of the Danish Basin. *In*: JAMES, N. P. & CLARKE, J. A. D. (eds) *Cool-Water Carbonates*. SEPM, Special Publications, **56**, 293–307.

SURLYK, F. & HÅKANSSON, E. 1999. Maastrichtian and Danian strata in the southeastern part of the Danish Basin. *In*: PEDERSEN, G. K. & CLEMMENSEN, L. B. (eds) *19th Regional European Meeting of Sedimentology. Field Trip Guidebook*. Geological Museum of the University of Copenhagen, Contributions to Geology, **829**, 29–58.

SURLYK, F. & LYKKE-ANDERSEN, H. in press. Contourite drifts, moats and channels in the Late Cretaceous chalk of the Danish Basin. Sedimentology.

SURLYK, F., DONS, T., CLAUSEN, C. K. & HIGHAM, J. 2003. Upper Cretaceous. *In*: EVANS, D., GRAHAM, C., ARMOUR, A. & BATHURST, P. (eds) *The Millenium Atlas: Petroleum Geology of the Central and Northern North Sea*. Geological Society, London, 213–233.

UENZELMANN-NEBEN, G. 2002. Contourites on the Agulhas Plateau, SW Indian Ocean: indications for the evolution of currents since Palaeogene times. *In*: STOW, D. A. V., PUDSEY, C. J., HOWE, J. A., FAUGÈRES, J. C. & VIANA, A. (eds) *Deep-water Contourite Systems; Modern Drifts and Ancient Series, Seismic and Sedimentary Characteristics*. Geological Society, London, Memoirs, **22**, 271–288.

VEJBÆK, O. V. & ANDERSEN, C. 2002. Post mid-Cretaceous inversion tectonics in the Danish Central Graben—regionally synchronous tectonic events? *Bulletin of the Geological Society of Denmark*, **49**, 129–144.

WATTS, N. L., LAPRÉ, J. F., VAN SCHIJNDEL GOESTER, F. S. & FORD, A. 1980. Upper Cretaceous and lower Tertiary chalks of the Albuskjell area, North Sea; deposition in a slope and a base-of-slope environment. *Geology*, **8**, 217–221.

WEBER, M. E., BONANI, G. & FUTTERER, K. D. 1994. Sedimentation processes within channel–ridge systems, southeastern Weddell Sea, Antarctica. *Paleoceanography*, **9**, 1027–1048.

WOLD, C. N. 1994. Cenozoic sediment accumulation on drifts in the northern North Atlantic. *Paleoceanography*, **9**, 917–941.

ZIEGLER, P. A. 1990. *Geological Atlas of Western and Central Europe*. Shell Internationale Petroleum, The Hague, 131–142.

Are there Middle Jurassic contourites in the Tarnovo depression (Southern Moesian platform margin)?

GEORGE GEORGIEV & NIKOLA BOTOUCHAROV

Sofia University, Department of Geology, 15 Tzar Osvoboditel Bd., 1504 Sofia, Bulgaria
(e-mail: gigeor@gea.uni-sofia.bg)

Abstract: The first boreholes in the Tarnovo depression, located in the central Southern Moesian platform margin zone (Bulgaria), drilled at the end of the 1970s, display intervals of unusually thin, irregular and lens-like interbedding of shales and siltstones in the Middle Jurassic succession. Their specific lithological, log and seismic features, which resemble or distinguish them from contourites, have been studied and described in detail. The depositional setting in the Tarnovo depression during the Late Aalenian–Early Bajocian was favourable for the accumulation of deeper-water sediments, influenced by bottom currents. These were deposited along the hanging walls of major growth faults, which have a decisive geodynamic importance for Early–Middle Jurassic basin evolution throughout the Southern Moesian platform margin zone. Some lithological characteristics of the studied sediments correspond to diagnostic criteria for contourites. However, some other features, such as tractive indications, thin fine-grained laminae and wavy facies alternation, are also indicative of fine-grained turbidites and/or reworked bottom-current deposits.

Active petroleum exploration took place at the end of the 1970s and in the 1980s in the Tarnovo depression, located in the central Southern Moesian platform margin zone (Fig. 1a–c). The main target was the thick (more than 500 m) Lower–Middle Jurassic succession, drilled unexpectedly by the first two wells (Tchapaevo-1 and Resen-1), located in the northern depression zone. The presence of unusually thin, irregular and lens-like interbeds of Middle Jurassic shale and siltstones was observed during the field description of large well-core intervals, and they were described as 'contourites' by Nachev *et al.* (1981). Subsequent intensive seismic acquisition (1979–1981) and drilling in this area resulted in more than 15 new wells during the next 10 years (Fig. 1c) and many new data regarding the presence and extent of these deposits.

The main aim of this paper is to present the results of a more comprehensive study of these specific Middle Jurassic sediments, buried at a depth of 2500–3800 m. All available well, log and seismic data were used for this purpose. We have tried to recognize and describe the lithological, log and seismic features of identified bottom-current deposits, which can be used as criteria to determine whether or not they are contourites.

The most debatable use of the term 'contourite' refers to marine deposits formed by currents that are not of thermohaline origin. Although the initial definition of 'contourite' was applied widely to various kinds of ocean-floor deposits (Rebesco *et al.* 2007), this term should now be used only for relatively deeper-water bottom-current sediments (e.g. greater than 500 m), deposited or significantly reworked by stable geostrophic currents (Faugères & Stow 1993). Given a lack of precisely defined palaeobathymetry in the study area, we will use the more general term 'bottom-current deposits' instead of 'fossil contourites'. The definition includes contourites *sensu stricto* and deposits of all bottom-current types, although there are several examples that have been described as fossil contourites deposited under shallow-water conditions (Stow *et al.* 1998).

Geological setting

Regional tectonic framework

The territory of Bulgaria is located on the East European continental margin and covers parts of the northern periphery of the Alpine orogen and its foreland, the Moesian platform (Fig. 1a). This mainly Mesozoic platform occupies the southern part of Romania and the northern part of Bulgaria. The Moesian platform, as a promontory of the European platform, is separated on its northeastern side from the Scythian platform by the North Dobrogea orogenic belt. The geological boundaries of the platform are well defined by the leading edge of the surrounding Alpine orogen. It is bordered to the north and west by the Southern Carpathians, which docked onto the northern platform margin during the Middle Miocene with a southerly

From: Viana, A. R. & Rebesco, M. (eds) *Economic and Palaeoceanographic Significance of Contourite Deposits.* Geological Society, London, Special Publications, **276**, 283–298.

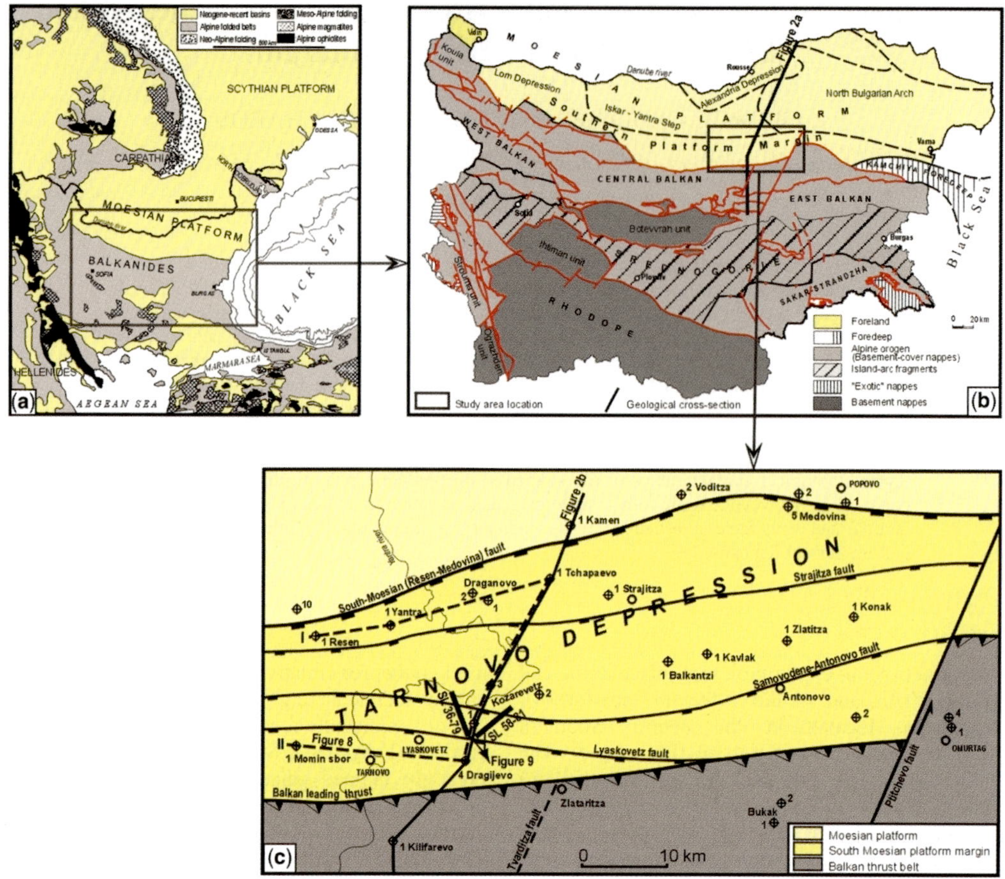

Fig. 1. (**a**) General setting of the Moesian platform in relation to the western Black Sea and surrounding Alpine fold belts; (**b**) tectonic sketch map of Bulgaria with location of the study area (modified from Tzankov *et al.* 1996; Georgiev & Dabovski 1997); (**c**) tectonic sketch of the study area with database (deep wells and interpreted seismic lines).

vergence; and to the south by the Balkan thrust belt (Fig. 1), which is a system of stacked north-verging thrust sheets, formed by multiphase compression, which culminated during the Middle Cretaceous and Middle Eocene. Towards the Black Sea, only a broad transitional zone can be outlined where the relatively undeformed platform succession has been downfaulted to the Western Black Sea basin (Tari *et al.* 1997).

The Moesian platform is composed of up to 4–6 km thick, relatively undeformed, dominantly shallow-marine Mesozoic sediments that rest on a gently folded Palaeozoic and older basement. Major unconformities, occurring at the base of the Triassic, Jurassic, Callovian, Upper Cretaceous and Middle Eocene sequences, are related to major compressional events within the Alpine thrust belt (Georgiev & Dabovski 1997).

Southern Moesian platform margin (SMPM)

The Mesozoic–Cenozoic evolution of the SMPM was governed by geodynamic processes in the northern Peri-Tethyan shelf system. The SMPM was repeatedly affected by notrth–south-directed intra-continental extensions and failed rifting cycles, during the Late Permian–Early Triassic, the Late Triassic, the Early Jurassic and the Late Cretaceous. These were interrupted and followed by compressional events, which caused strong platform margin shortening that was ultimately overprinted by the Alpine orogeny (Georgiev & Dabovski 2000; Georgiev *et al.* 2001*a*). Mid-Cretaceous and Mid-Eocene compression resulted in the formation of the Balkan thrust belt, over the southern edge of the Moesian platform (Figs 1a, b and 2a).

The present-day SMPM is an west–east-trending 20–40 km wide subsided zone, lying between the Balkan thrust front and the southern edge of the Moesian platform, marked by the South Moesian fault (Figs 1b, c and 2a). The margin is characterized by a complex south-deepening monoclinal structure (Fig. 2a). West of the transverse Tvarditza–Ptitchevo strike-slip fault zone (Fig. 1b and c), the sedimentary record presents at least three main stages in the platform margin separation: Early–Middle Jurassic, Tithonian–Valanginian and Late Cretaceous–Cenozoic. During these stages the subsidence of this marginal platform zone accelerated, leading to the deposition of various thicker and deeper-water facies sediments (Fig. 2a).

The Tarnovo depression is defined as a small tectonic unit in the central part of SMPM developed mainly during the Early–Middle Jurassic stage (Georgiev & Dabovski 1997).

Early–Middle Jurassic basin evolution and lithostratigraphy

The initial and major stage in the differentiation of the SMPM west of the transverse Tvarditza–Ptitchevo strike-slip fault zone was in the Early–Middle Jurassic (Figs 1c, 2 and 3). In this zone the Lower–Middle Jurassic sequence is more than 200 m thick, and up to 1500 m in the basin depocentre (Fig. 3), whereas in other parts of the Moesian platform it is strongly reduced in extent and thickness (<150 m).

After a regional depositional break at the end of the Triassic (Middle Norian), caused by the onset of the Early Cimmerian orogeny, Early Jurassic sedimentation resumed in the southern Moesian platform domain as the result of a more regional rifting event (Stampfli *et al.* 2001). Two Early–Middle Jurassic sedimentary basins developed in the East Srednogorie–Balkan rift zone (Georgiev *et al.* 2001*a*) and in its northern shoulder; that is, in the central zone of the SMPM (Fig. 3). These two basins are located respectively east and west of the transverse Tvarditza–Ptitchevo strike-slip fault zone. In both basins sedimentation lasted until the Late Bathonian, when the Mid-Cimmerian orogeny culminated (Georgiev *et al.* 2001*a*).

The Early–Middle Jurassic Moesian platform marginal basin was formed on deeply and irregularly eroded pre-Jurassic relief (Tari *et al.* 1997, fig. 10a–e). Widespread folding of the pre-Jurassic sequence underneath the Moesian platform is traditionally attributed to the Early Cimmerian orogeny at the end of the Triassic (Georgiev & Atanasov 1993). Numerous folds are interpreted as fault-bend folds involving various Palaeozoic décollement levels (Tari *et al.* 1997) and this north-vergent foreland thrust–fold belt is characterized by such folds. In a wider palaeotectonic scenario, this thrust–fold belt represents the frontal part of the Mediterranean Cimmerides propagating into the European foreland.

The lithostratigraphy of the Lower–Middle Jurassic sequence in the Tarnovo depression is shown in Figure 4 (Georgiev 1983; Sapunov & Tchoumatchenco 1987). Within the sequence, the Ozirovo and Etropole Formations occupy a major position in terms of stratigraphic duration and thickness (Georgiev 1983).

Jurassic sedimentation began in the Sinemurian with thin very shallow-water clastic deposits (<50 m), the Batchichtene and Kostina Formations (Fig. 4). Pliensbachian–Early Aalenian sedimentation, recorded by the Ozirovo Formation (up to 300 m), consists of two carbonate units, the Dolni Lukovit and Suhindol Members, separated by a clay–silty unit, the Bukorovo Member. The Ozirovo Formation reflects mainly shallow marine conditions. However, the observed increase of clay content in the carbonates testifies to some deepening of the depositional environment in the axial basin zone.

The Etropole Formation is also heterogeneous and comprises three units (Fig. 4). The lower Stefanetz and the upper Shipkovo Members are dominated by shale, whereas the middle Lopyan Member is mostly sandy, especially in its upper part. Basin evolution reached its deepest-water pelagic stage in Late Aalenian–Early Bajocian times, when the sedimentary record is dominated by shale, turbiditic and bottom-current deposits (Etropole Formation, Stefanetz Member), with thicknesses up to 120 m. The shale-dominated Stefanetz Member changes slowly upwards through irregular interbedding of shale and siltstones, and becomes dominated by coarse sandstones in the upper part of the Lopyan Member. This upward slow and gradual facies transition records a shallowing of the depositional environment. In some localities the basin shallowing even led to the short appearance of dry land and erosion, as indicated by the succession drilled in the well Momin Sbor 1 (Fig. 1c).

The Early Bathonian sedimentary record is dominated by shales (Shipkovo Member) and marls (Bov Formation), which marked some deepening in the depositional environment. During the late Bathonian the facies environment became very shallow water and the basin expanded over all of NE Bulgaria (Fig. 1b). This resulted in the deposition of sandy limestones (Polaten Formatiom) grading northeastward into thin silty shales (Esenitza Formation).

The Early–Middle Jurassic basin structure and depositional setting in the Tarnovo depression is

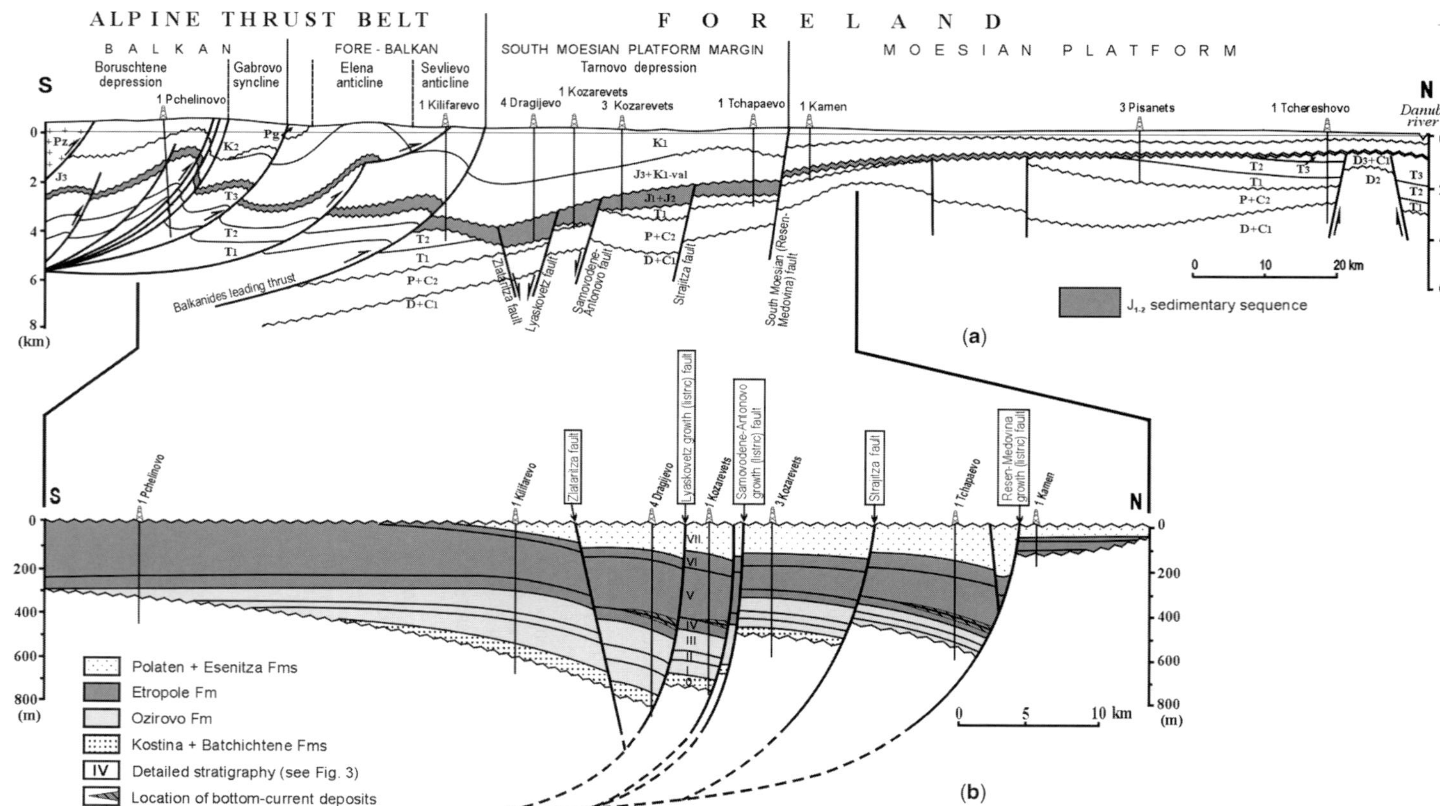

Fig. 2. **(a)** Geological cross-section across the south Moesian platform domain and adjacent Balkan thrust belt (location is shown in Fig. 1b); **(b)** Early–Middle Jurassic basin structure with location of sedimentary drifts.

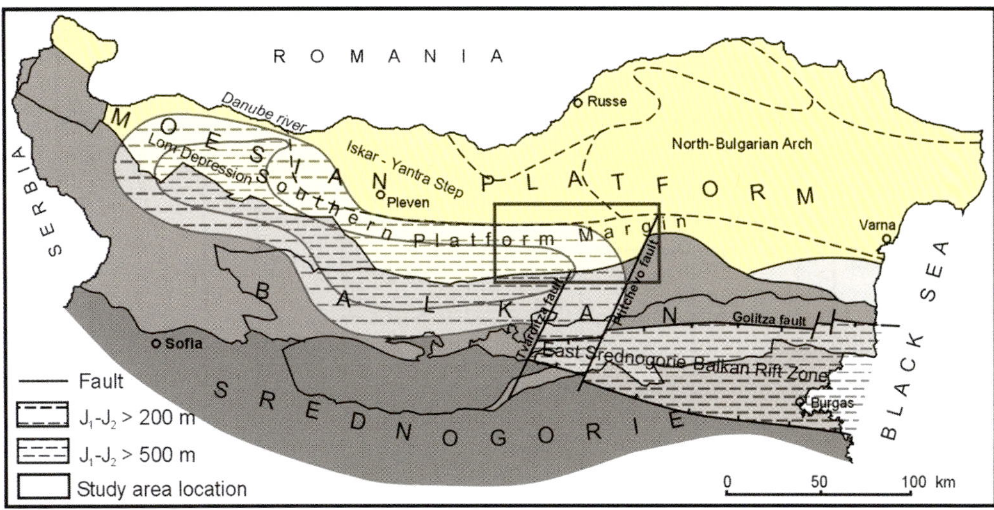

Fig. 3. Thickness of Lower–Middle Jurassic units (>200 m) and sedimentary basins in the southern part of the Moesian platform.

shown in Figure 2b. A typical platform marginal extensional basin was initiated, developed and then ceased during the Early–Middle Jurassic (Fig. 3). The west–east elongated basin mainly covers the western and central segments of the SMPM. To the west the basin overlies the entire Lom depression and to the east the basin prolongation is flanked by the Ptitchevo dextral fault (Figs 1b, c and 3). The basin is about 300 km in length and 80 km in width (Fig. 3).

Early Jurassic rifting strongly affected the periphery of the southern Moesian platform and induced fault disruption (Fig. 2b), with a system of growth faults and tilted blocks formed at this time. The South Moesian growth fault acted as the main north-bounding border fault. The incipient marginal basin has complex half-graben geometry with southward shallowing. The structure of its northern subsided zone is complicated by a system of thetic and antithetic faults (Fig. 2b). Some faults, such as the South Moesian, Samovodene–Antonovo and Lyaskovets faults, manifest typical growth development during the Early–Middle Jurassic and consequently are of decisive influence on the depositional environment. The main basin depocentres are located along the southern hanging walls of the main growth faults, where the thickness and variety of facies within the Lower–Middle Jurassic sequence are greatest. The basin bathymetry is controlled by the subsidence rate (Fig. 5), which is greater along the main growth faults, as well as global sea-level changes. In general, during the Early–Middle Jurassic, global sea level rose (Haq *et al.* 1987). Local maxima in sea-level rise, occurring in late Toarcian and late Aalenian–early Bathonian times (Fig. 4), are recorded in the sedimentary succession by deeper water shaly-dominated deposits.

Contourite diagnostic features and study methods

The German physical oceanographer Georg Wüst was the first, in 1930, to recognize abyssal sediments influenced by bottom currents. Later, many workers (Bouma 1972, 1973; Hollister & Heezen 1972; Hesse 1975; Stow & Lovell 1979; Faugères & Stow 1993; Nelson *et al.* 1993; Stanley 1993; Stow *et al.* 1998; Faugères *et al.* 1999; among many others) have tried to distinguish contourites from other deep-water sediments, mainly by their lithological and seismic features. The problem of defining diagnostic criteria of contourites *sensu lato* (Table 1) is not completely resolved. Unequivocal recognition of their geological record still lacks conclusive agreement despite the great research effort during the last decade (Viana 2007). In many cases, deep-water sediments are misinterpreted as contourites instead of fine-grained turbidites or reworked turbidites. Often the primary sedimentary structures are not preserved and contourites may mimic sediments deposited by other processes. Consequently, bottom currents may rework to a greater or lesser extent other types of sediments if there is an interaction between different mechanisms of deposition.

Some of the lithological features used to define contourites are not unquestionable and they can be

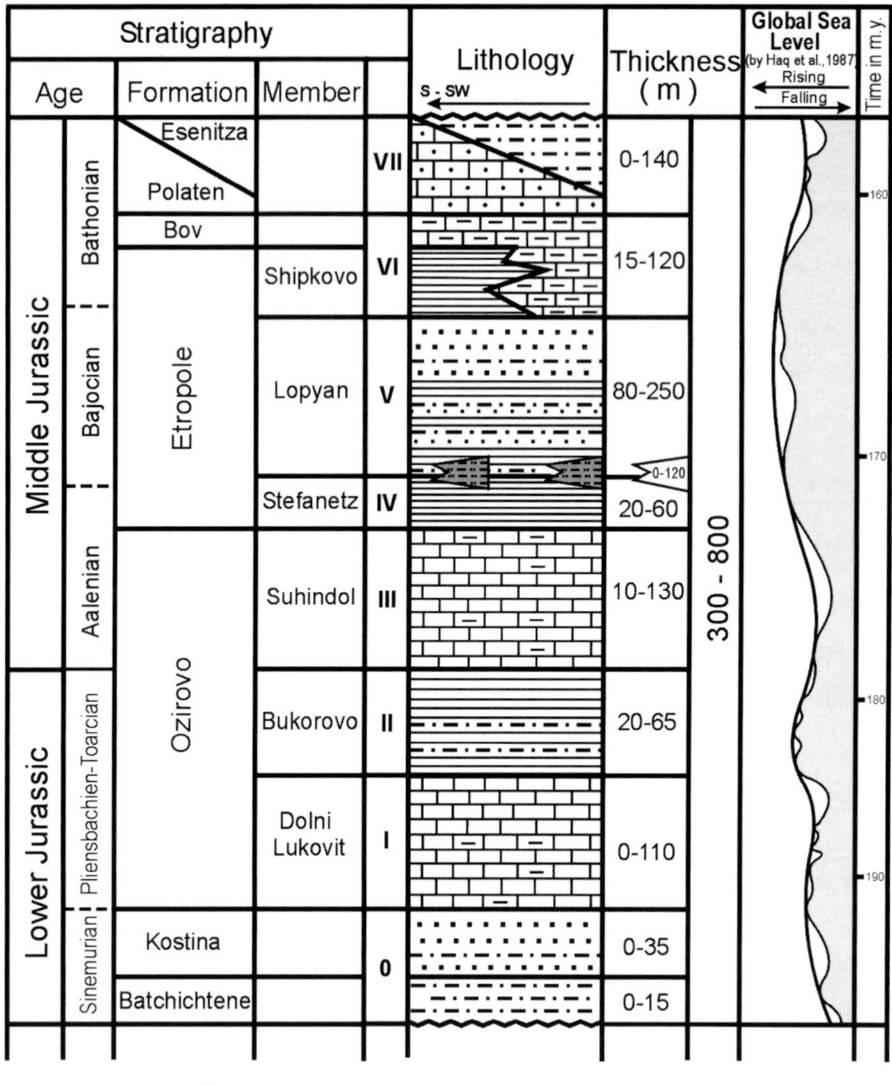

Fig. 4. Lithostratigraphy of the Lower–Middle Jurassic units in the Tarnovo depression with global sea-level changes.

observed in other deep-water deposits. For instance, sand–silt ripple-laminated facies have been identified in bottom-current reworked turbidite series and barely recognized in the present oceans. Many thin-bedded sandstones with regular to lenticular bedding, sharp tops and bases, internal lamination, cross-lamination and fading ripples are often interpreted as fossil contourites (Stow *et al.* 1998). However, they can be identified as reworked sequences and fine-grained turbidites as well, particularly if fading ripples are climbing (Piper & Stow 1991). Stanley (1988) described similar facies as progressively winnowed and reworked turbidites.

Seismic criteria for identifying contourites (Table 1) are highly variable and sometimes difficult to define. The variety of contourites is found mainly in their geometry and seismic facies. Generally, contourites are considered as elongated and/or mounded deposits with an alongslope disposition, which is one of the most important diagnostic features. A frequent phenomenon is the interbedding of the contourites with other types of deep-water facies, which complicates their undisputed recognition. Other kinds of

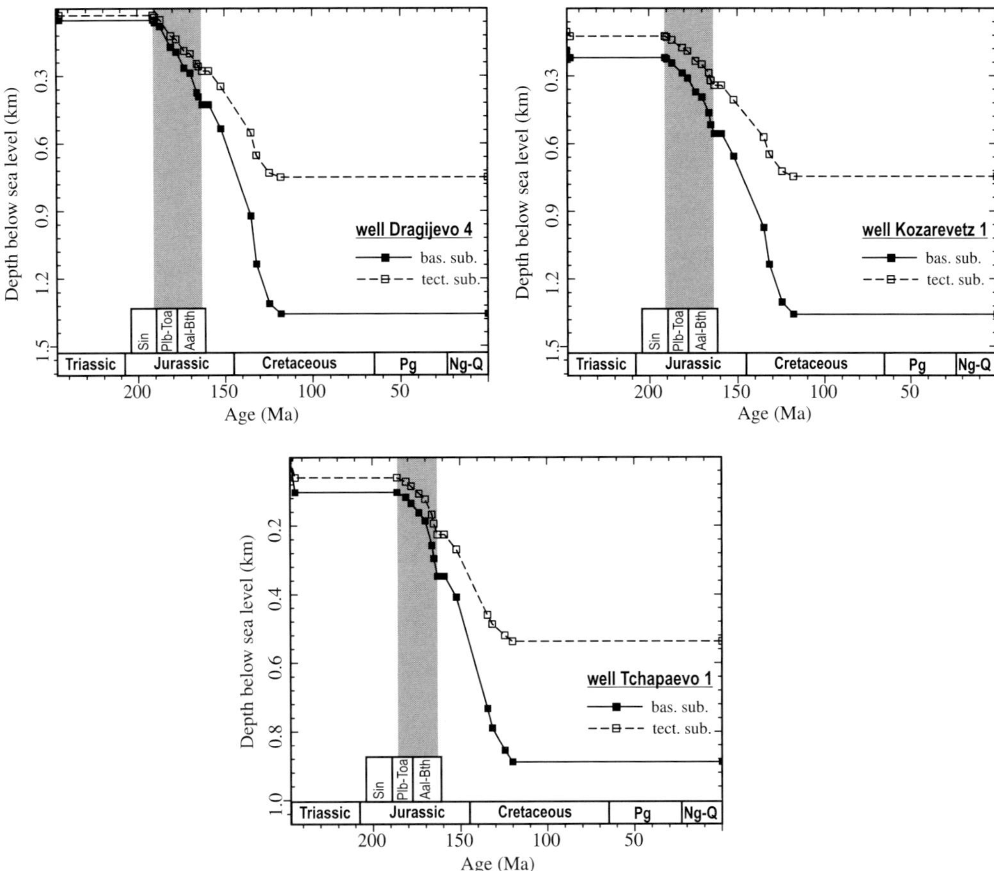

Fig. 5. Tectonic subsidence (tect. sub.) and basement subsidence (bas. sub.) modelling according to some drilled sections in the Tarnovo depression (location is shown in Figs 1c and 2). Shaded areas mark the Early–Middle Jurassic evolution. The geological time scale is from Harland *et al.* (1990).

depositional systems (turbidites, reworked deep-water sediments, etc.) can often form sedimentary bodies with seismic features and facies similar to those of typical contourites.

At the scale of drift geometry, depositional units and seismic facies, many of the contourite criteria are very similar to those for turbidites (Faugères *et al.* 1999). Nevertheless, some features, such as major discontinuities that can be traced across the whole drift, lenticular and convex-upward depositional geometry, and typical progradation–aggradation styles of these units are specific to contourites.

The main target of the present study is the intervals with lens-like thin alternation of black shale and gray siltstones with tractive structures, established locally in the lower part of the Etropole Formation (upper part of the Stefanetz Member and lower part of the Lopyan Member) within the Late Aalenian–Early Bajocian succession (Figs 2b and 4). They have been studied by detailed lithological observation and description of well-core intervals, integrative log-core analysis using the methods of Muromtsev (1984) and correlations, seismic recognition of sediment drifts, and comparison with the already defined criteria for recognition of contourites (Table 1).

It is possible to trace intervals with intercalation of shale, siltstones and sandstones, identified in current-influenced deeper-water deposits, using the spontaneous polarization (SP) log-facies methods of Muromtsev (1984), considering also gamma-ray and caliper logs. The main log-facies features for their identification are the shape of SP anomalies and log-curve components: top line, bottom line and sideline (Muromtsev 1984). The SP anomaly shape (top and bottom lines) reflects the type of lithological transition from underlying to overlying sediments. Tilted, somewhat wavy top lines mark a

Table 1. *Main characteristic features of contourites*

Feature	Contourites
Occurrence	In deep-water settings and along slope where bottom currents are or have been active
Bedding	Normally <5–10 cm
Layer contacts of sand beds	Upper and lower contacts normally sharp
Graded bedding	Normal and reverse grading occasionally
Massive bedding	Absent
Foreset bedding	Common, often with parallel lamination
Sedimentary structures, vertical order and lamination	Homogeneous or bioturbated throughout, with few primary structures remaining; may show ripple-laminated reverse grading near top, with sharp or erosive contacts; lack of a vertical sequence of internal structures; coarse lag concentrations; parallel, parallel–cross, irregular horizontal lamination; ripple laminated and reverse facies sequence; lenticular starved ripples, subparallel lenses and thin discontinuous laminae
Texture and sorting	Mostly silty mud to sand sized; low values of skewness; textural variations indicating alongslope transport; usually well to very well sorting; poorly to well-sorted trend in muddy contourites
Fabric	Grain orientation well developed; indication of alongslope transport at time of final deposition; alignment not well preserved in fossil contourites
Composition	Typical mixed composition
Grain size	Relatively fine grained (very fine sand to silt), showing evidence of reworking, transport and deposition of sediments
Sequence	Typically arranged in decimetre-scale cycles of grain size and/or compositional variation with muddy–sandy contourites; partial sequences also
Matrix	Usually 0–5%
Inclusions	Inclusions other than organic never observed
Microfossils, plant and skeletal remains	Rare and usually worn or broken
Seismic	Top surface of drifts may show a regular wavy reflector pattern; marked erosional discontinuity; depositional units are generally lenticular in shape with a convex-up geometry of seismic reflectors; upward and oblique stacking of units and lateral migration indicating current direction; generally transparent or structureless appearance of seismic facies; moderate- to high-amplitude subparallel reflectors

The summary is after Hollister & Heezen (1972), Hesse (1975), Stow (1979), Stow & Lovell (1979), Faugères & Stow (1993), Nelson *et al.* (1993), Stanley (1993), Stow *et al.* (1998) and Faugères *et al.* (1999).

relatively gradual but uneven change of lithology that can be explained by hydrodynamic fluctuations in the basin. The sidelines reflect the facies features.

Results

Well-core description

All Middle Jurassic core intervals with lens-like or thin alternating layers of black shale and grey siltstones drilled in the study area (Fig. 1c) have been extensively field-surveyed and described in detail.

In some intervals of the Upper Aalenian–Lower Bajocian succession there is a sedimentary facies possessing characteristic features that resemble those of contourites (Fig. 6; Table 1). The observed layer alternation of silts, shale and thin-bedded fine-grained sandstones without a definite order of sedimentary structures also suggests the bottom-current nature of the deposits. The thickness of alternating layers is usually less than 5 cm (very often less than 1 cm) and the ratio between shale and clastic layers is variable (Fig. 6). The average size of the clasts is very fine to fine grained and they are relatively well sorted. An additional indication for contourites is the presence of tractive structures, including horizontal lamination and climbing ripples (also observed in turbidites; Stanley 1988), with internal discontinuous thin shale seams.

The above-mentioned contourite features are present in the core samples from the well Tchapaevo 1 (Figs 6 and 7), in which the core-drilled range is the largest. They are observed in a large

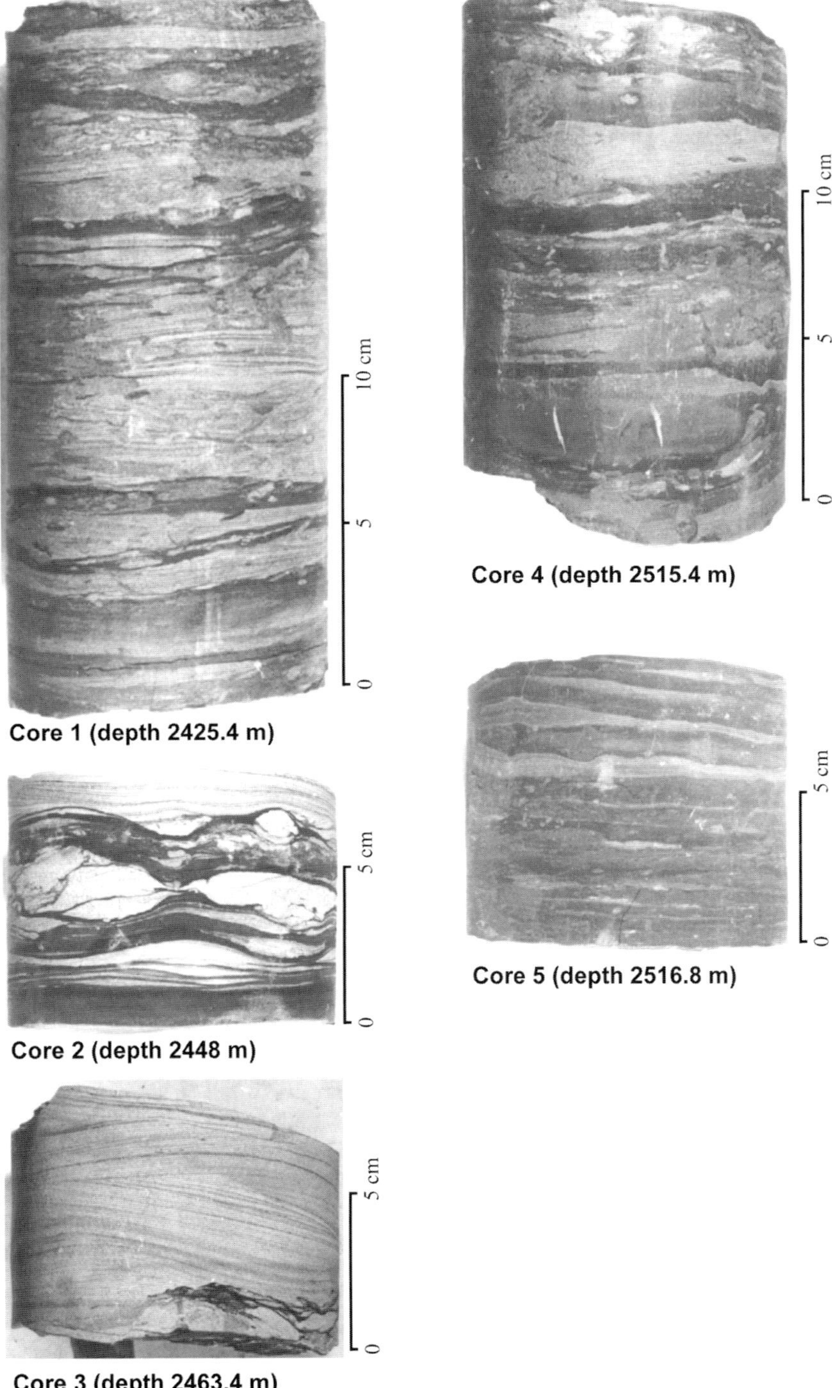

Fig. 6. Some contourite features in the Middle Jurassic core samples from the Tchapaevo 1 well (location is shown in Figs 1c and 7).

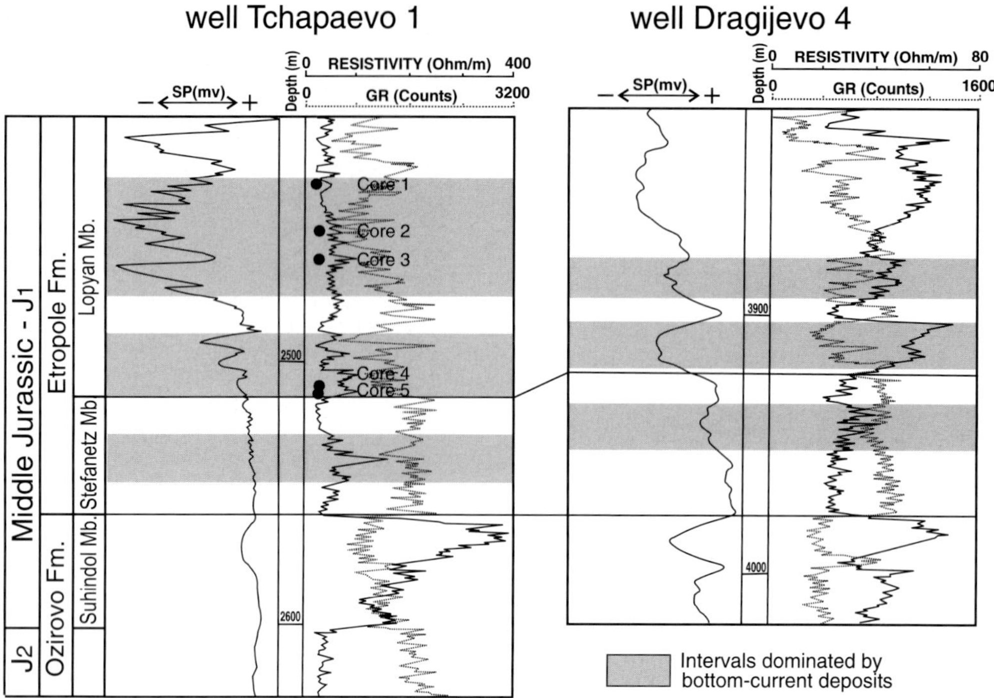

Fig. 7. Log characterization of bottom-current intervals (location of wells is shown in Fig. 1c).

depth interval in many of the drilled wells (Figs 7 and 8), with maximum thickness up to 120 m. Thin and wavy alternation between shale, siltstones and some fine sandstone layers is also present. The observed reverse-graded laminae and indications of traction are typical of bottom-current disintegration, transportation and redeposition of fine-grained deposits.

Some of the described lithological features could also be recognized in typical bottom-current reworked deposits. The presence of thin, fragile mud laminae within tractive structures denotes periods of settling during dominant bottom-current processes. The shale material is probably a feature of slow, deep-water sedimentation by bottom currents along the southern hanging walls of the

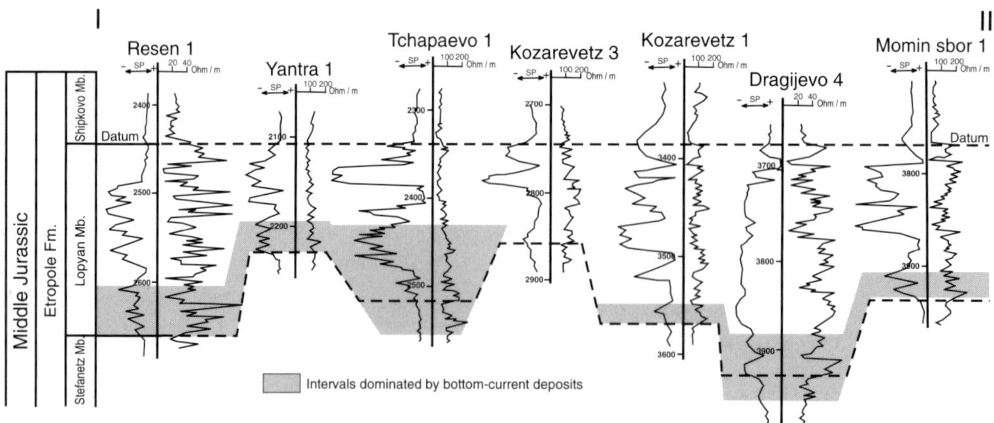

Fig. 8. Log correlation of the bottom-current intervals along line I–II (location is shown in Fig. 1c).

main growth faults, active during the Early–Middle Jurassic. It also might be indicative of decreasing current velocity.

The other possible interpretation is that the laminae consist of rapidly deposited, very fine-grained floccules from turbidity currents. Although the sedimentary facies described for the well Tchapaevo 1 are not exclusive to contourites, two features can be emphasized: precise well-core observations show reverse-graded beds and there are also some indications for bi-directional cross-lamination, which is less typical for turbidites.

Log analysis and interpretation

Detailed log analysis, interpretation and correlation of the drilled Middle Jurassic sections (Figs 7 and 8) have been carried out using the SP log-facies diagnostic criteria of Muromtsev (1984) along with resistivity log, gamma-ray, caliper log and core data. The SP curves of the wells Tchapaevo 1, Dragijevo 4 (Fig. 7) and others along the line I–II (Figs 1 and 8) show a log response diagnostic for the presence of discontinuous sands and silts in shale, observed in the cores (Fig. 6). Log curves identify specific sediment interactions, described by Muromtsev (1984) as a slope facies influenced by strong currents. They are typically characterized by a weak SP negative anomaly with the form of an isosceles triangle.

Special attention has been paid to the log analysis of the wells Tchapaevo 1 and Dragijevo 4 (Fig. 7), because they penetrated the largest depth interval with bottom-current deposits and their drilled core intervals are sufficient for integrative interpretation. The SP log response from the upper part of the Stefanets Member and the lower part of the Lopyan Member in the Etropole Formation resembles the slope facies model of Muromtsev (1984). The tilted, somewhat wavy top lines of SP negative anomalies characterize a relatively gradual change of lithology through thin, irregular and lens-like layer alternation. This can be explained by an inconsistency of palaeohydrodynamics in the depositional settings. In both sections, the anomaly sidelines are either missing or show jagged to wavy shapes. They indicate an irregular alternation of fine-grained layers, predominantly shale and silts with some sandy layers, and relatively rapid changes in depositional setting, caused by fluctuating bottom currents. The observed anomaly bottom lines are analogous to the top lines and indicate the same kind of transition to the underlying shale sequence.

The general electrometric features of the Stefanets Member upper interval and the Lopyan Member lower interval correspond to relatively fine-grained intertwined facies probably affected by bottom currents, according to the slope facies model of Murmontsev (1984). The Stefanetz Member comprises mainly shale. The appearance in its upper part of thin discontinuous silt layers and their increasing occurrence upwards mark the transition to the Lopyan Member and indicate changes in depositional setting, involving more active bottom-current hydrodynamics and the deposition of more clastic material. As shown by the recorded log response this process continues (or is initiated in some parts of the basin) in the lowermost intervals of the Lopyan Member (Fig. 8), where the bottom-current intervals can be observed. Subsequently, the basin starts to shallow and the depositional setting changes from relatively deep water and shaly to shallow water and coarse-grained clastic deposits in the upper part of the Lopyan Member (Figs 4, 7 and 8).

Along the log correlation line I–II some changes in the stratigraphic position and thickness of drilled bottom-current intervals can be traced (Fig. 8). Their absence in the well Kozarevetz 3 section (Fig. 2b) corresponds to the higher intra-basinal position of this location. The Well Kilifarevo 1, not included in the correlation table (Fig. 8), has the same characteristics (Figs 1c and 10).

Seismic characterization

Numerous seismic criteria for the recognition of contourite drifts (Table 1) have recently been defined (Faugères & Stow 1993; Faugères *et al.* 1999). There has been a wide-ranging discussion about these criteria and they have been clarified and modified, but are still not definitively accepted. The main reason for this is the use of mainly shallow seismic data from modern oceans, without enough examples for fossil deep buried contourites, as in our case.

We demonstrate two transverse seismic sections, SL 36-79 and SL 58-81 (Figs 1c and 9), from the last seismic acquisition (1979–1981) in the study area, with the aim of defining the seismic characteristics of drilled contourite (bottom-current) intervals and to clarify their basinal extent (Fig. 10). The seismic lines traverse the deepest (most subsided) basinal zones, drilled by the wells Kozarevetz 1 and 3 and Dragijevo 4 (Figs 1 and 2).

The vague large-scale features (i.e. drift scale) are indicative of variable current regime conditions. The hydrodynamic conditions have probably been stable for only short time periods. The presence of relatively extended, less discernible bottom-current drift is indicated on seismic section SL 36-79. A small-scale sediment mound has been distinguished on seismic section SL 58-81, characterized by an irregular shape and discontinuity of the

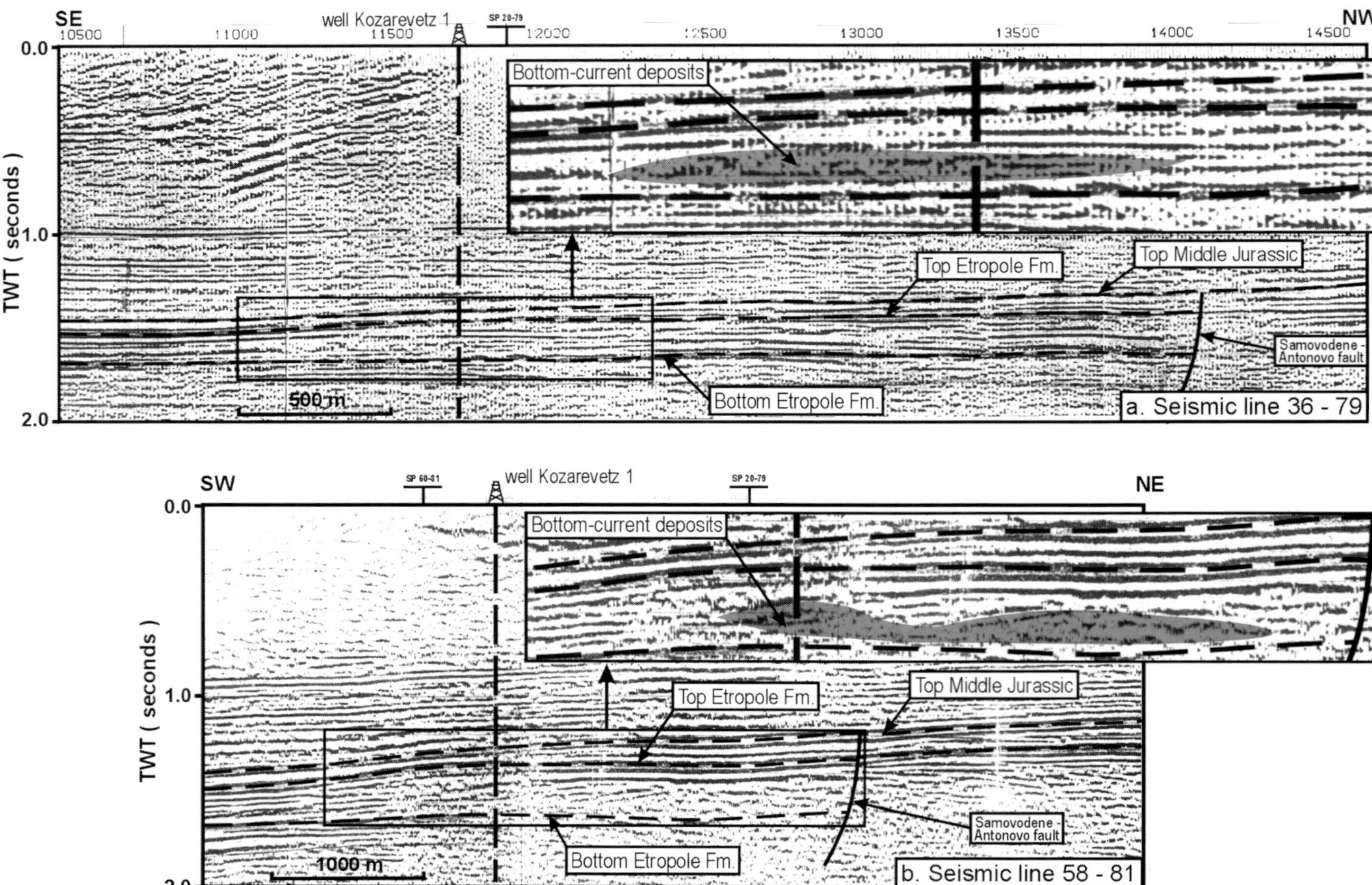

Fig. 9. Seismic features and interpretation of sediment drifts (location of seismic lines is shown in Fig. 1c).

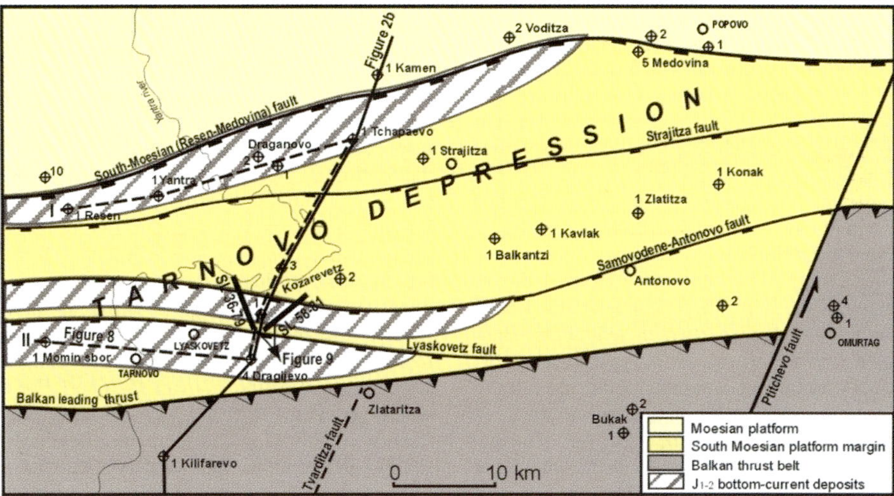

Fig. 10. Extent of the Middle Jurassic sediment drifts in the lower part of the Etropole Formation (see details in Figs 2b, 3, 7, 8 and 9).

reflections at the base, as well as within the deposits. The sediment drifts from both seismic profiles might be interpreted as a possible channel–levee system or mixed drift system with predominant down-current migration and a trend parallel to the slope. The drifts are overlain by very slightly undulating parallel reflections. The mounded geometry is due to the interaction between hydrodynamic activity and structural control.

For the recognition of contourite drifts from medium-scale criteria (i.e. depositional seismic units) we make the following observations. On seismic section SL 36-79, a slightly convex-up seismic unit can be outlined, and a typical downlapping pattern, low-amplitude and relatively transparent seismic record are also observed. On seismic section SL 58-81, the seismic features are similar; however, the deposited mound has a more or less discernible sigmoidal progradational reflector pattern. Additionally, there is an apparent downlap to the base of the sediment drift.

Unquestionable small-scale criteria (i.e. seismic facies) for the recognition of contourite deposits *sensu lato* on the interpreted seismic sections cannot be defined. This is mainly due to the wide variety of seismic facies in the study area and the lack of modern seismic acquisition and processing; the available seismic data are from 1979–1981.

Discussion

The location and extent of the studied Middle Jurassic bottom-current drifts in the Tarnovo depression are shown in Figs 2b, 4, 8 and 10, and include all the results from our integrative well-core–log–seismic study, interpreted in the light of our basin structure and geodynamic model for the Early–Middle Jurassic basin (Fig. 2). Although the seismic data are not so informative and some of the geological and palaeogeographical parameters are not entirely clarified, a depositional model for bottom-current sediments can be constructed.

Three elongate bottom-current zones can be traced along the hanging walls of the Resen–Medovina, Samovodene–Antonovo and Lyaskovetz growth faults (Fig. 10), which had a decisive geodynamic influence on the basin evolution (Fig. 2). The lithological, log and seismic features of established bottom-current deposits in the lower intervals of the Etropole Formation (Figs 4, 6 and 7) resemble to some extent contourites *sensu lato* (Table 1). Also, they correspond to fine-grained turbidites and bottom-current reworked turbidites as described in the literature (Piper & Stow 1991; Stanley 1993).

The described Late Aalenian–Early Bajocian bottom-current drifts are related to the deepest-water stage in the Early–Middle Jurassic basin evolution, which is recorded by the shale sequence of the Stefanetz Member and the lower part of the Lopyan Member. The deepening of the Early–Middle Jurassic platform marginal basin (Fig. 3) is related to the continuing synrift stage, which caused the development of growth faults. Along their southern hanging walls, the basinal environment during this stage was the deepest, and also had fluctuating palaeohydrodynamics in depositional settings. The assumed palaeocurrents along the hanging walls of the main Resen–Medovina, Samovodene–Antonovo and Lyaskovetz growth faults, directed westward to the basinal depocentre (Figs 3 and 10), as well as the complex basin bottom morphology (Fig 2b), determined variable depositional settings. The Middle Jurassic sediment drifts discussed here, according to their tectonic position in the basin (Figs 2b and 10), resemble to some extent the confined drifts described by Faugères *et al.* (1999).

Our structural and evolutionary model (Fig. 2b) also suggests the deposition of fine-grained turbidites or reworked deeper-water sediments. Basin dynamics was the major factor for the formation of bottom-current deposits and could have led to the interaction of gravitational displacements and currents along with other slope processes. The syn-depositional growth faults and block subsidence, together with sea-level changes, could have initiated turbidite flows on the northeastern basin slopes (Figs 3, 10 and 11). During the short time of downslope slide deposition and thereafter, the active bottom currents, directed westward along the hanging walls of the Resen–Medovina, Samovodene–Antonovo and Lyaskovetz growth faults, led to reworking of these sediments. Such interactions formed the various litho-facies relationships, masking to some extent the sedimentary record of bottom-current activity. During the Late Aalenian–Early Bajocian deepest basin stage, at a time of relatively high sea level, the bottom-current activity led to the reworking of the deposited deeper-water sediments. More active bottom-current dynamics winnowed part of the deposited Stefanetz Member sediments and redeposited and mixed them with the newly formed, relatively fine coarse-grained sediments in the lower interval of the Lopyan Member. The along-slope sediment drifts described here could be formed in this way, built up by silt–clayey and shaley components and small amount of sands. These thin bottom-current layers are located in the uppermost levels of the Stefanetz Member and the lowermost levels of the Lopyan Member, forming an overall sequence with a thickness of 30–120 m (Figs 4, 7 and 8).

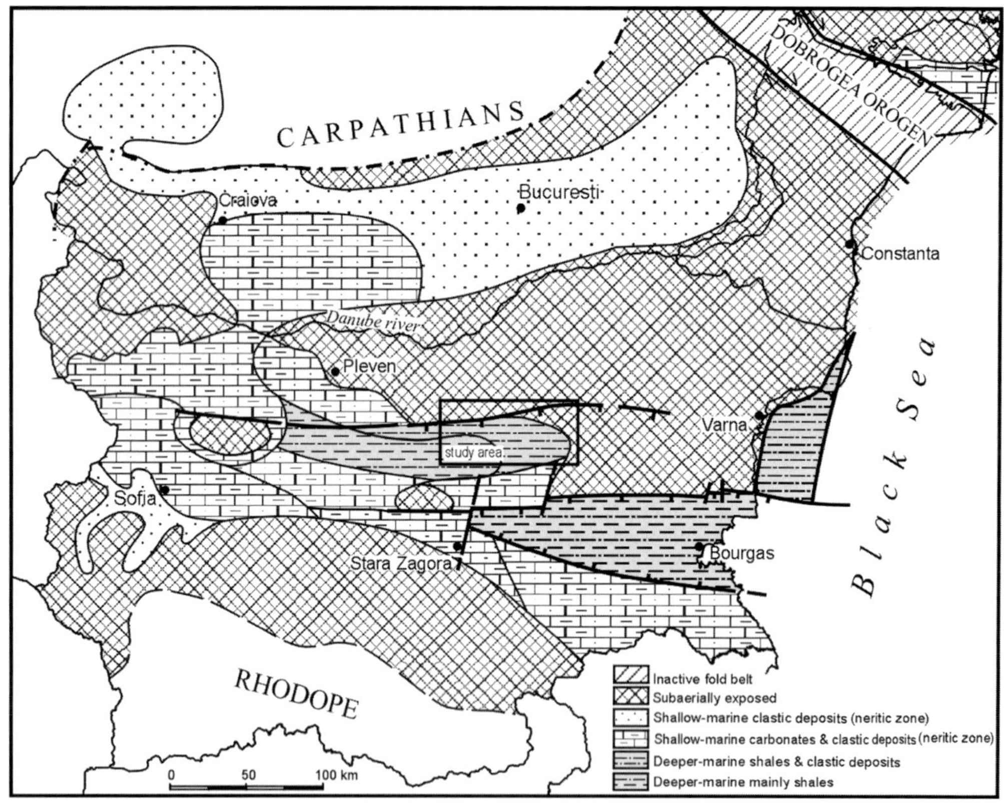

Fig. 11. Late Aalenian–Early Bajocian palaeogeographical sketch for the Moesian platform (modified from Georgiev *et al.* 2001*b*).

Conclusions

(1) A typical Early–Middle Jurassic platform marginal basin developed in the west–central zone of the SMPM, westward of the transverse Ptitchevo strike-slip fault (Figs 1 and 3). The Tarnovo depression is located in the eastern part of this basin. The deepest-water stage in basin evolution occurred during the Late Aalenian–Early Bajocian, when shale, turbiditic and bottom-current sediments up to 120 m in thickness were deposited.

(2) The analysis of Middle Jurassic cores from the wells drilled in the Tarnovo depression shows the presence of thin beds of silt, shale and sand facies in the lower intervals of the Etropole Formation (upper part of the Stefanetz Member and lower part of the Lopyan Member). These sediments are often closely intercalated and show complex structural features.

(3) The geological and depositional settings in the Tarnovo depression during the Late Aalenian–Early Bajocian were favourable for the accumulation of deeper-water sediments, influenced by bottom currents. These were deposited along the hanging walls of the main Resen–Medovina, Samovodene–Antonovo and Lyaskovetz growth faults, which had a decisive geodynamic influence on Early–Middle Jurassic basin evolution.

(4) Some lithological characteristics of the studied deeper-water Middle Jurassic sediments correspond to recognition criteria for contourites (Table 1). They could have been deposited by bottom palaeocurrents along the hanging walls of the main growth faults, directed westward to the basinal depocent (Figs 3 and 10). The sediment drifts, according to their tectonic position in the basin (Figs 2, 10 and 11), resemble to some extent the confined drifts described by Faugères *et al.* (1999).

(5) On the other hand, the tractive indications (including horizontal lamination and climbing, fading ripples) and thin, fragile, very fine-grained laminae within sedimentary structures, as well as the presence of wavy facies alternation, are also indicative of fine-grained turbidites and/or reworked bottom-current deposits. The syndepositional growth faults and block subsidence, together with sea-level changes, could have initiated south–SW-directed turbidite flows in the northeastern basin slopes (Figs 3, 10 and 11). Subsequently, active, west-directed bottom currents along the hanging walls of the main growth faults could have reworked them.

(6) Detailed study of the Middle Jurassic deep-water bottom-current sediments in the Tarnovo depression should continue with more precise lithological facies analyses, and especially if more seismic and drilling data become available in the future.

The authors thank Stampfli, G., Magalhães, P. and Mutti, E. for constructive reviews of an earlier version of this paper. We are grateful to Rebesco, M., Viana, A. and Piper, D. for helpful ideas and advice. Many thanks so especially to Stephenson, R. who greatly helped us with English language editing.

References

BOUMA, A. H. 1972. Fossil contourites in Lower Niesenflysch, Switzerland. *Journal of Sedimentary Petrology*, **42**, 917–921.

BOUMA, A. H. 1973. Contourites in Niesenflysch, Switzerland. *Eclogae Geologicae Helvetiae*, **66**(2), 315–323.

FAUGÈRES, J. C. & STOW, D. A. V. 1993. Bottom-current-controlled sedimentation: a synthesis of the contourite problem. *Sedimentary Geology*, **82**, 287–297.

FAUGÈRES, J. C., STOW, D. A. V., IMBERT, P. & VIANA, A. 1999. Seismic features of contourite drifts. *Marine Geology*, **162**, 1–38.

GEORGIEV, G. V. 1983. Geological preconditions for oil and gas perspectives of Lower–Middle Jurassic sediments from the southern part of Moesian Platform in Northeast Bulgaria. *Petroleum and Coal Geology*, **18**, 20–32.

GEORGIEV, G. & ATANASOV, A. 1993. The importance of the Triassic–Jurassic unconformity to the hydrocarbon potential of Bulgaria. *First Break*, **11**, 489–497.

GEORGIEV, G. & DABOVSKI, C. 1997. Alpine structure and petroleum geology of Bulgaria. *Geology and Mineral Resources*, **8–9**, 3–7.

GEORGIEV, G. & DABOVSKI, C. 2000. Rifting and thrusting in Southern Moesian Platform Margin—implications for petroleum geology. In: *EAGE 62nd Conference & Technical Exhibition, SECC, Glasgow, Scotland, Extended Abstracts, Volume 2*, 18.

GEORGIEV, G., DABOVSKI, C. & STANISHEVA-VASSILEVA, G. 2001*a*. East Srednogorie–Balkan Rift Zone. *In*: ZIEGLER, P. A., CAVAZZA, W., ROBERTSON, A. H. F. & CRASQUIN-SOLEAU, S. (eds), *Peri-Tethys Memoir 6: PeriTethyan Rift/Wrench Basins and Passive Margins*. Mémories du Muséum National d'Histoire Naturelle, **186**, 259–293.

GEORGIEV, G., DABOVSKI, C. & SEGHEDI, A. 2001*b*. Moesian Platform—paleogeographic maps. *In*: STAMPFLI, G., BOREL, G., CAVAZZA, W., MOSSAR, J. & ZIEGLER, P. (eds) *The Paleotectonic Atlas of the PeriTethyan Domain*. IGCP Project 369 (CD-ROM). European Geophysical Society.

HAQ, B., HARDENBOL, J. & VAIL, P. R. 1987. The new chronostratigraphic basis of Cenozoic and Mesozoic sea level cycles. *In*: ROSS, C. A. & HAMAN, D. (eds) *Timing and Depositional History of Eustatic Sequences: Constraints on Seismic Stratigraphy*. Cushman Foundation for Foraminiferal Research, Special Publications, **24**, 7–13.

HARLAND, W. B., ARMSTRONG, R. L., COX, A. V., CRAIG, L. E., SMITH, A. G. & SMITH, D. G. 1990. *A Geological Time Scale*. Cambridge University Press, Cambridge.

HESSE, R. 1975. Turbiditic and non-turbiditic mudstone of Cretaceous flysch sections of the East Alps and other basins. *Sedimentology*, **22**, 387–416.

HOLLISTER, C. D. & HEEZEN, B. C. 1972. Geologic effects of ocean bottom currents. *In*: GORDON, A. L. (ed.) *Studies in Physical Oceanography, 2*. Gordon and Breach, New York, 37–66.

MUROMTSEV, V. S. 1984. *Electrometric Geology of the Sand Bodies. Lithological Oil and Gas Traps*. Nedra, Leningrad.

NACHEV, I., GEORGIEV, G., ZHELEV, S., CHALAKOV, N., ZHELEVA, C. & VAVILOVA, M. 1981. The Jurassic system in part of Northeast Bulgaria. *In*: MANDEV, P. & NACHEV, I. (eds) *Geology and Petroleum Potential of NE Bulgaria*. Technika, Sofia, 25–35.

NELSON, C. H., BARAZA, J. & MALDONADO, A. 1993. Mediterranean undercurrent sandy contourites, Gulf of Cadiz, Spain. *Sedimentary Geology*, **82**, 103–131.

PIPER, D. J. W. & STOW, D. A. V. 1991. Fine-grained turbidites. *In*: EINSELE, G., SEILACHER, A. & RUCKEN, A. (eds) *Sequence and Event Stratigraphy*. Springer, Berlin, 360–375.

REBESCO, M., CAMERLENGHI, A., VOLPI, V., ET AL. 2007. Interaction of processes and importance of contourites: insights from the detailed morphology of sediment Drift 7, Antarctica. *In*: VIANA, A. R. & REBESCO, M. (eds) *Economic and Palaeoceanographic Significance of Contourite Deposits*. Geological Society, London, Special Publications, **276**, 95–110.

SAPUNOV, I. & TCHOUMATCHENCO, P. 1987. Geological development of Northeast Bulgaria during the Jurassic. *Paleontology, Stratigraphy and Lithology*, **24**, 3–56.

STAMPFLI, G. M., MOSAR, J., FAVRE, P., PILLEVUIT, A. & VANNAY, J. C. 2001. Permo-Mesozoic evolution of the Western Tethys realm: the Neo-Tethys–East Mediterranean basin connection. *In*: ZIEGLER, P. A., CAVAZZA, W., ROBERTSON, A. H. F. & CRASQUIN-SOLEAU, S. (eds) *Peri-Tethys Memoir 6: PeriTethyan Rift/Wrench Basins and Passive Margins*. Mémoires du Muséum National d'Histoire Naturelle, **186**, 51–108.

STANLEY, D. J. 1988. Turbidites reworked by bottom currents: Upper Cretaceous examples from St. Croix, US Virgin Islands. *Smithsonian Contributions to Marine Science*, **22**, 79.

STANLEY, D. J. 1993. Model for turbidite-to-contourite continuum and multiple processes transport in deep marine settings: examples in the rock record. *Sedimentary Geology*, **82**, 241–255.

STOW, D. A. V. 1979. Distinguishing between fine-grained turbidites and contourites on the Nova Scotian deep water margin. *Sedimentology*, **26**, 371–387.

STOW, D. A. V. & LOVELL, J. P. B. 1979. Contourites: their recognition in modern and ancient sediments. *Earth-Science Reviews*, **14**, 251–291.

STOW, D. A. V., FAUGÈRES, J. C., VIANA, A. & GONTHIER, E. 1998. Fossil contourites: a critical review. *Sedimentary Geology*, **115**, 3–31.

TARI, G., DICEA, O., FAULKERSON, J., GEORGIEV, G., POPOV, S., STEFANESCU, M. & WEIR, G. 1997. Cimmerian and Alpine stratigraphy and structural evolution of the Moesian Platform (Romania/Bulgaria). *In*: ROBINSON, A. G. (ed.) *Regional and Petroleum Geology of the Black Sea and Surrounding Region*. American Association of Petroleum Geologists, Memoirs, **68**, 63–90.

TZANKOV, T., ZAGORCHEV, I., DABOVSKI, C., BOYANOV, I., HAYDOUTOV, I. & YANEV, S. 1996. *Tectonic maps of Bulgaria at a scale of 1:50000*. National Fund for Earth Sciences, Project 26/91, Sofia.

VIANA, A. R., ALMEIDA, W. JR, NUNES, M. C. V. & BULHÕES, E. M. 2007. The economic importance of contourites. *In*: VIANA, A. R. & REBESCO, M. (eds) *Economic and Palaeoceanographic Significance of Contourite Deposits*. Geological Society, London, Special Publications, **276**, 1–24.

Pelagic carbonate ooze reworked by bottom currents during Devonian approach of the continents Gondwana and Laurussia

HEIKO HÜNEKE

Institute of Geography and Geology, University of Greifswald, D-17487 Greifswald, Germany (e-mail: hueneke@uni-greifswald.de)

Abstract: Givetian and lower Frasnian carbonates of pelagic carbonate-platform and distal slope-apron settings in the Harz Mountains of Germany (Herzyn Limestone Formation), the eastern Moroccan Central Massif (Ziar–Mrirt Nappe), and the Carnic Alps in Austria–Italy (Valentin and Pal Limestone Formation) show strong evidence for bottom-current activity during deposition. Calcarenites, laminated calcisiltites, and mottled calcisiltites and calcilutites can be distinguished, which are similar to recent calcareous bioclastic contourites. They combine faint structures caused by current action with pervasive bioturbation. Calcarenites are mostly represented by styliolinid grainstones to packstones with rarely preserved parallel lamination and ripple cross-lamination. Laminated calcisiltites are particularly rich in non-carbonate components with a higher density than calcite such as conodonts and phosphatic intraclasts. Relics of coarsening-upward to fining-upward micro-sequences a few centimetres thick are preserved in the Moroccan record. Erosional surfaces, hardgrounds and condensed phosphates are more typical of the Harz Mountains and the Carnic Alps. The bottom-current influenced facies build up strongly condensed and reduced sequences that occur at the same stratigraphic interval in different areas of central Europe and NW Africa. Variations in rate of accumulation, magnitude of erosion and microfacies, which are found across the three regions, are compatible with a contourite interpretation. The widespread current-induced reworking of calcareous sediments and phosphate formation during the Givetian and early Frasnian as well as the associated erosion marked by pronounced hiatuses all signal a major palaeocirculation event. Thermohaline currents were intensified by the acceleration of flows constricted in narrow oceanic passages between the approaching continental plates Laurussia and Gondwana. Areas affected were the southeastern Rhenish Sea shelf, which occupied the distal passive margin of Laurussia, the disintegrated northern continental margin of Gondwana, whose sedimentary record is now preserved in the Moroccan Meseta, and deep marginal plateaux of the Noric Terrane in the western part of the Prototethys. Thus, the occurrence of fossil calcareous contourites confirms a very advanced convergence between Gondwana and Laurussia and the minor terranes between during Middle and Late Devonian times.

In Devonian times, pelagic lime muds were deposited on a large scale on continental rises and terraces of the rifted northern continental margin of Gondwana, the southern margin of Laurussia and minor continental plates sandwiched between them, such as the Armorican and Noric terranes. These successions are well exposed today in the Variscan massifs of Europe and North Africa, and allow detailed examination of the sedimentary response to changes in oceanic circulation and basin-forming processes in ocean basins between continental plates. The overall palaeogeography of the oceanic passages between Gondwana and Laurussia changed during the Devonian as a result of northward drift of the two mega-plates (e.g. Ziegler 1989; Scotese & McKerrow 1990; Golonka 2002; Blakey 2003) and continuing convergence along several subduction zones (Franke 2000; Matte 2001; Winchester *et al.* 2002).

The upper Givetian and lower Frasnian limestones under consideration, which were deposited offshore, far from terrigenous detrital influence, are usually characterized by more condensed or even reduced successions in comparison with older and younger pelagic successions, and in many cases by hiatuses and pronounced facies changes. The sedimentary environmental conditions of these deposits are reconstructed here, based on published and new data from the Moroccan Central Massif, the Harz Mountains in Germany and the Carnic Alps in Austria (Schönlaub 1980, 1985, 1992; Göddertz 1982; Lottmann 1990; Lazreq 1992; Joachimski *et al.* 1994; Hüneke 1995, 1997, 1998; Walliser *et al.* 1995*b*, 1999; Becker & House 1999*a*; Hüneke & Reich 2000; Buchholz *et al.* 2001; Pohler & Schönlaub 2001).

This study presents lower to upper Devonian stratigraphic successions, focusing on microfacies and biostratigraphic data for Givetian and Frasnian strata from three different palaeogeographical settings of the closing seaway between Gondwana and Laurussia. The main aim is to evaluate the

From: Viana, A. R. & Rebesco, M. (eds) *Economic and Palaeoceanographic Significance of Contourite Deposits*. Geological Society, London, Special Publications, **276**, 299–328.
0305-8719/07/$15.00

sedimentary characteristics, biostratigraphic framework and palaeoceanographic criteria indicative of distinct bottom-current influence during the Givetian and Frasnian, and to discuss the information about palaeocirculation preserved in the oceanic drifts and corresponding hiatuses that are closely linked to past palaeoceanographic changes.

Study areas and methods

Macro- and microfacies features of bottom-current reworked hemipelagic and pelagic limestones have been studied in Devonian basin and pelagic platform settings. The Givetian to Frasnian calcareous sediments discussed in this paper are mainly from the nappe of Ziar–Mrirt at the eastern margin of the Moroccan Central Massif (sections Mrirt M, MG and MT), from the Harzgerode Zone of the Harz Mountains in Germany (sections Antoinettenweg, Neue Mühle and Eselsstieg), and from the Rauchkofel Nappe in the Carnic Alps of Austria (sections Wolayer Gletscher, Valentin Törl) (Fig. 1). The three-stage approach of Stow *et al.* (1998, 2002) to identification of fossil contourites is followed here.

The successions investigated were logged on a centimetre scale. The microfacies analysis is based on continuous bed-by-bed sampling and complete documentation by thin sections. The term 'component' is used to describe both matrix and grains, whereas the term 'particle' refers to the coarser fraction of bimodal limestones such as wackestones and packstones. Bedding-parallel oriented thin sections were used for measurements of fossil orientation and palaeocurrent analysis. Biostratigraphic correlation is based on conodonts (Ziegler & Klapper 1982; Bultynck 1987; Klapper *et al.* 1987; Sandberg *et al.* 1989; Ziegler & Sandberg 1990; Walliser *et al.* 1995*a*).

The Devonian time scale relies heavily on data of Tucker *et al.* (1998) but also considers data provided by Sandberg & Ziegler (1996), which are based on estimated durations of conodont zones. A revised Devonian time scale based on new radiometric age data and global correlations was recently proposed and is followed herein (see Weddige 2003).

Geological setting

The sediments of the Ziar–Mrirt nappe in Morocco (Fig. 1) were deposited somewhere along the disintegrated northern continental margin of Gondwana, in southern parts of the Prototethys (Fig. 2). The primary plate-tectonic relation to the West African craton is difficult to reconstruct because of dextral strike-slip faulting during and subsequent to the Variscan collision (Matte 2001; Piqué 2001). The depositional setting evolved through the period under consideration from a hemipelagic basin to a pelagic carbonate platform. Sedimentation was strongly influenced by faulting in Famennian times, caused by uplift and deformation of the Midelt Zone (Walliser *et al.* 1999) and/or onset of rifting of the Azrou–Khenifra Basin in the eastern Central Massif (Hüneke 2001). In this way, a pelagic carbonate platform–half-graben system was created.

The Herzyn Limestone Formation of the Harz Mountains (Fig. 1) was deposited on a marginal plateau in the distal Rhenish shelf sea of Laurussia bounded to the south by the small Rhenohercynian Ocean (Lütke 1990; Hüneke 1998) (Fig. 2). Interpretation of the plate-tectonic and geodynamic evolution of this Rhenohercynian basin is controversial (see Schwab 1979; Wachendorf 1986; Franke 1995, 2000; Wachendorf *et al.* 1995; Oncken *et al.* 2000). Rifting during the Emsian and disintegration of the passive margin caused drowning of the external carbonate shelf and gave way to condensed Eifelian–Famennian accumulation on pelagic carbonate platforms. A failed rift protected the marginal plateau from siliciclastic influence from the Caledonian orogen to the north (Oncken *et al.* 1999). Water depth, based on facies, microfossil assemblages and preservation, was in the range of 200–1000 m. The nearest land lay to the north, now in an area NW of the Rhenish Slate Mountains some 300–400 km distant, but which may have been at least twice as far away during the Givetian to Frasnian (50% orogenic shortening on average).

Both the Valentin Limestone and the overlaying Pal Limestone Formation of the Rauchkofel Nappe in the Carnic Alps (Fig. 1) were deposited in a basin plain to distal slope-apron setting (Pohler & Schönlaub 2001) on a continental terrace of the Noric Terrane (the South Alpine–Dinarid Terrane) in the southern part of the Prototethys (Frisch & Neubauer 1989; Ziegler 1989; Läufer *et al.* 2001) (Fig. 2). An associated shallow-water carbonate platform is very well exposed in the Kellerwand and Cellon nappes (Bandel 1969, 1972; Schönlaub 1971, 1985, 1992; Kreutzer 1990, 1992). The Noric unit is suggested to have been a promontory of Gondwana at least during Early–Middle Devonian times and to have been separated from Africa at a later stage to form a terrane (Frisch & Neubauer 1989).

Results at outcrop

The calcareous facies associations of the Givetian and/or Frasnian are briefly described and documented by key sections (Tables 1–3). Some of the

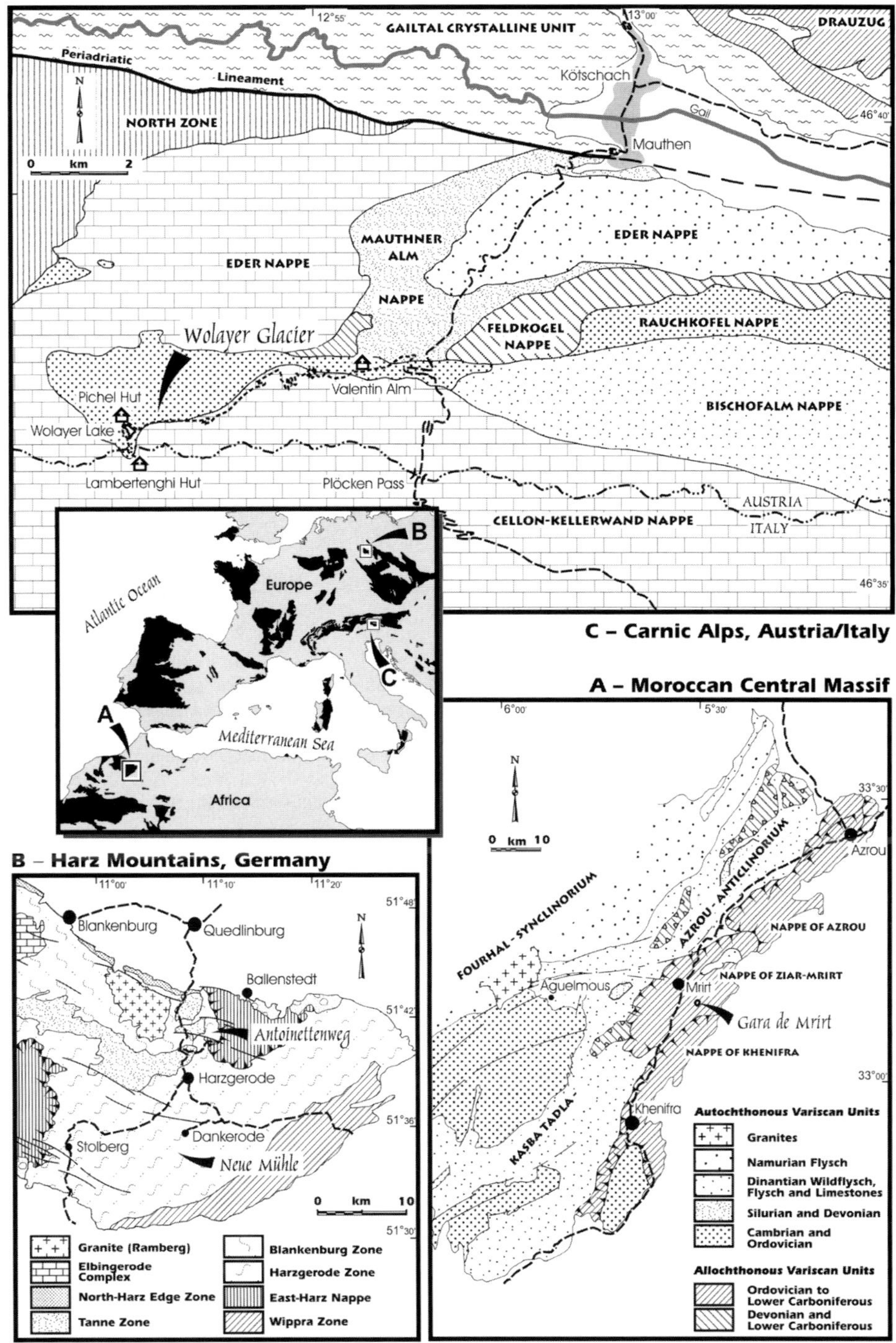

Fig. 1. Field site maps with geological background information (simplified from Schönlaub 1985; Faik 1988; Bouabdelli 1989; Hüneke 1998). The inset regional map shows the location of the three main study areas.

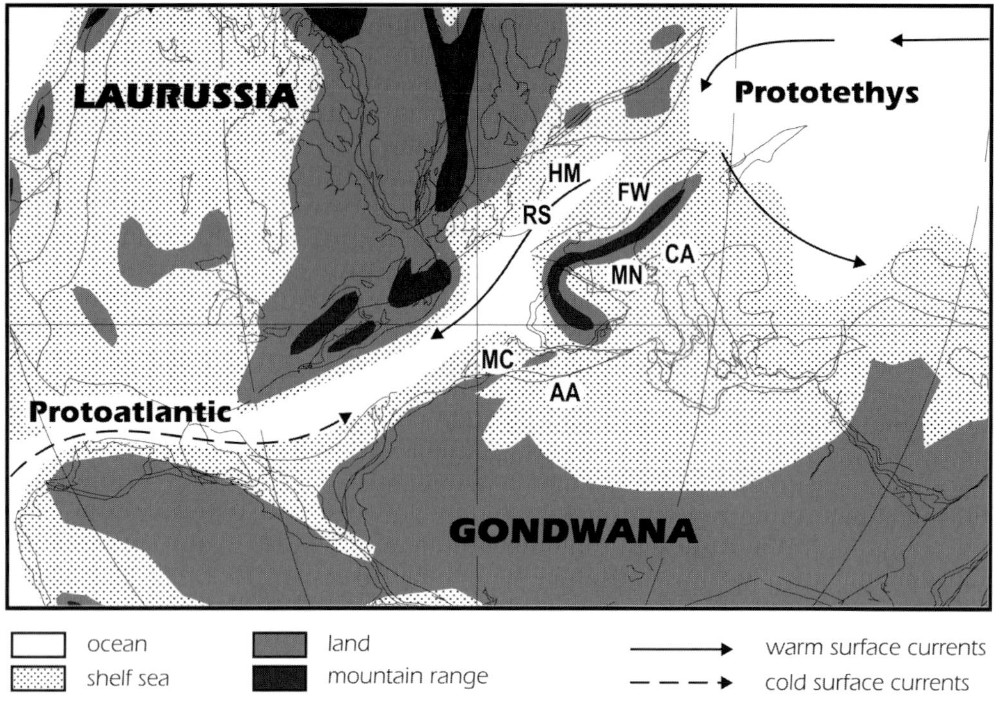

Fig. 2. Palaeogeographical reconstruction of the area between Laurussia and Gondwana during the Givetian–Frasnian (370 Ma), showing expected pattern of oceanic surface circulation as well as distribution of deep sea, shelf sea, land and mountain ranges. Map provided by W. Kiessling (see Copper 2002), based on the reconstruction of Golonka (2002). The areas investigated and discussed were part of Gondwana (MC, Moroccan Central Massif; AA, Anti-Atlas), the Armorican terrane assemblage (MN, Montagne Noire; FW, Frankenwald), the Noric terrane (CA, Carnic Alps), and Laurussia (RS, Rhenish Slate Mountains; HM, Harz Mountains).

features that are interpreted as most diagnostic of fossil contourites are highlighted below. Microsequences are best preserved and discussed in the Moroccan successions.

Moroccan Central Massif

Sections M, MG and MT of the 13 localities at Gara de Mrirt studied in detail by Lazreq (1992), Becker & House (1999*a*) and Walliser *et al.* (1999) display the best evidence for fossil contourites (Hüneke 2001). Only the early Frasnian calcareous record (early *hassi*–early *rhenana* Zones) of these successions is discussed here (see Fig. 3), although parts the Eifelian and Givetian record of predominantly siliciclastic composition may also show bottom-current influence. However, in the latter case sediment facies characteristics are equivocal and biostratigraphic age control is insufficient.

Facies and structures. The lower Frasnian limestones show indistinct lamination and a distinct mixed composition. The calcareous sediments are characterized by early diagenetic lithification and preserve numerous erosional surfaces (corrasional surfaces), which are associated with pristine and condensed phosphatic sediments. Calcarenites, laminated calcisiltites and mottled calcisiltites and calcilutites are distinguished, similar to recent contourites (Table 1, Fig. 4).

Microfacies and composition. Calcarenites are represented by styliolinid grainstones to packstones with rarely preserved horizontal and cross-lamination. In many layers, the conical tests of styliolinids show a preferred parallel orientation (Fig. 4: 0.0–0.5 cm, 1.5–2.0 cm). Cross-laminated units exhibit styliolinid tests that are oriented parallel to foreset (Fig. 4: 4.5–7.5 cm; Fig. 5). Calcarenites occur in irregular laminae and lenses (<2 cm) or continuous layers (5–25 cm). Erosional contacts appear to be more common at the top of the layers and are commonly mineralized and altered to phosphoritic hardgrounds (Fig. 4: 0.5 cm, 20.5 cm). In some cases, it is difficult to differentiate

Table 1. *Main characteristics of contourite facies in the Ziar–Mrirt Nappe, Moroccan Central Massif (early Frasnian record)*

	Mottled calcisiltites and calcilutites	Laminated calcisiltites	Calcarenites
Structure	Dominantly homogeneous; bioturbational mottling common; irregular silt laminae rare	Irregular thin lamination; mottled or distinct burrows; erosive surfaces very common (mostly at base); normal grading common, reverse grading rare; sharp–gradational contacts; phosphatic hardgrounds at the base common	Irregular lamination, cross-laminated or homogeneous; indistinct mottling; erosive upper contacts common; gradational lower contacts; phosphatic hardgrounds at the top common
Texture	Dominantly silty mud; bioclastic wackestones; poorly sorted	Muddy silt; fine to medium sand at the base of some laminae; poorly to well sorted; weakly bimodal or flat grain-size distributions; bioclasts heavily fragmented	Dominantly medium to coarse sand; styliolinid grainstones and packstones; groundmass is commonly orthosparite (cement) or calcilutite and calcisiltites (muddy matrix)
Fabric	Unclear; locally circular grain orientation as a result of bioturbation	Unclear	Locally grain orientation (long axis of styliolinid cones) parallel and perpendicular to the bottom current
Composition	Biogenic material from pelagic sources	Dominantly biogenic material from pelagic sources; phosphate intraclasts common; conodonts common; rare terrigenous material	Biogenic material from pelagic sources, mainly styliolinids
Sequence	Typically arranged in centimetre-scale micro-sequences of grain-size, structure and compositional variation similar to the standard mud–silt–sand contourite sequence of Stow *et al.* (2002); partial sequences are common		

between contacts that are of primary depositional nature and those caused by bioturbation.

Laminated calcisiltites are particularly rich in non-carbonate components with a higher density than calcite (conodonts, phosphatic intraclasts) and quartz grains (Fig. 4: 33.5–34.5 cm). Carbonate components are intensively fragmented. An irregular and wavy lamination is the most typical feature of the calcisiltites (Fig. 4: 31.5–33.5 cm). Grading is commonly present (Fig. 4: 21.0–21.5 cm, 38.0–39.5 cm). Although alternate normal and reverse grading is evident on a lamina scale, the former is much more common. There is a well-developed distribution grading mostly above erosional surfaces and hardgrounds. Laminated calcisiltites form a distinctive facies that comprises either lenses and infillings of shallow scours or, more commonly, thin continuous sediment layers <3 cm thick (Fig. 4: 10.5–11.0 cm, 40.5–42.0 cm). At certain horizons, laminated units clearly pass laterally into indistinct burrow-mottled calcisiltites. Isolated pockets and lenticular tube-like structures both filled with calcisiltites occur above and below the laminated units.

Mottled calcisiltites and calcilutites show a bimodal grain-size distribution similar to pelagic limestones. Loosely packed bioclastic and styliolinid wackestones are most common, but loosely packed lithoclastic wackestones, densely packed styliolinid wackestones and homogeneous calcilutites occur as well. The calcisiltites as well as the calcilutites include a large number of biomorpha and bioclasts: styliolinids, planktonic tentaculites, thin-shelled trilobites and brachiopods, cephalopods, ostracodes, tiny gastropods, mollusc shells, disarticulated crinoid ossicles, and rare bryozoan branches, solitary rugose corals, sponge spicules and conodonts. The dominant feature of this facies is a thorough bioturbational mottling (Fig. 4: 13.5–14.5 cm).

Sequences. The arrangement of the three main facies and the transitions vary considerably, but two characteristic types of vertical micro-sequences are recognized: coarsening-upward micro-sequences and fining-upward micro-sequences (Hüneke 2001). These micro-sequences are mostly between 2 and 10 cm thick and are in many cases

Table 2. *Main characteristics of contourite facies in the Herzyn Limestones, Harz Mountains (late Givetian and early Frasnian record)*

	Mottled calcisiltites and calcilutites	Laminated calcisiltites	Calcarenites
Structure	Dominantly homogeneous; bioturbational mottling common; irregular silt laminae rare; isolated pockets filled with calcisiltites rare; mostly gradational contacts	Thinly laminated; distinct burrows; erosional surfaces common at the base (=corrasional surfaces); normal size-grading and mineralogical grading common, sharp–gradational contacts; phosphorites and phosphatic hardgrounds at the base	Vague parallel laminated or homogeneous; indistinct mottling; erosive upper contacts common; erosive lower contacts rare; autochthonous phosphorites at the top common; isolated pockets and lenticular tube-like structures both filled with calcisiltites
Texture	Dominantly silty mud; bioclastic wackestones; poorly sorted	Muddy silt; fine to medium sand at the base of some laminae (conodonts, phosphate lithoclasts, quartz, feldspar); poorly to moderately sorted; bioclasts heavily fragmented	Dominantly medium to coarse sand; styliolinid grainstones and packstones; groundmass is commonly orthosparite (cement) or calcilutite and calcisiltites (muddy matrix)
Fabric	Unclear	Unclear	Locally grain orientation (long axis of styliolinid cones) parallel to bedding
Composition	Dominantly biogenic material from pelagic sources, benthic biogenic material rare	Dominantly biogenic material from pelagic sources; phosphate intraclasts common; conodonts common; some laminae with terrigenous material (quartz silt, feldspar grains and clay aggregates)	Biogenic material from pelagic sources, mainly styliolinids, phosphate intraclasts common
Sequence	Common partial fining-upward micro-sequences similar to the standard mud–silt–sand contourite sequence of Stow *et al.* (2002)		Unclear

separated by an omission surface. Hardgrounds, erosional surfaces and microbially encrusted surfaces represent periods of shorter or longer sedimentation cessation.

The complete coarsening-upward micro-sequence begins at the base with mottled calcilutites and calcisiltites, gradually passes upwards into calcarenites and is ideally topped by a phosphoritic hardground. It displays a tendency towards an upward increase in grain size and percentage of bioclasts, and a decrease in the intensity of burrow mottling. The complete fining-upward micro-sequence shows a gradual upward change from laminated calcisiltites into mottled calcisiltites and calcilutites. It is accompanied by a tendency towards an upward increase in bioturbational structures. Although the complete fining-upward micro-sequence is not exactly the reverse of the coarsening-upward micro-sequence, the centimetre-scale cycles of grain-size, structure and compositional variation are similar to the mud–silt–sand contourite sequence of Stow *et al.* (2002).

Harz Mountains

The upper Givetian to lower Frasnian succession (*disparilis–jamieae* Zones) of the Herzyn Limestone Formation is considered to be of bottom-current origin (Hüneke 1997, 2001; Hüneke & Reich 2000). The Antoinettenweg section and the Neue Mühle section are presented here as well-documented localities with detailed stratigraphic age control (Fig. 6).

Facies and structures. The main facies are laminated calcisiltites and burrow-mottled calcisiltites

Table 3. *Main characteristics of contourite facies in the Rauchkofel Nappe, Carnic Alps (late Givetian and early Frasnian record)*

	Mottled calcisiltites and calcilutites	Laminated calcisiltites	Styliolinid calcarenites	Peloidal calcarenites
Structure	Dominantly diffuse bioturbational mottling common; gradational contacts	Lenses and infillings of shallow scours or, more commonly, thin continuous sediment layers; sharp and gradational contacts; bioturbation and distinct burrows; erosive surfaces common at the base; normal size-grading common	Mostly homogeneous; indistinct mottling; gradational contacts common; more rarely erosive upper and lower contacts; isolated pockets filled with calcisiltites	Parallel lamination common; normal grading, locally burrow-mottling and diffuse bioturbation; base and top contacts commonly erosive
Texture	Dominantly silty mud; bioclastic wackestones; poorly sorted	Muddy silt; fine to medium sand at the base of some laminae (conodonts, fish remains, and phosphate lithoclasts); poorly sorted; bioclasts heavily fragmented	Dominantly medium to coarse sand; styliolinid grainstones and packstones; groundmass is commonly orthosparite (cement) or calcilutite and calcisiltites (muddy matrix)	Dominantly fine to medium sand; sorting moderate; peloidal grainstones; groundmass very variable
Fabric	Unclear	Unclear	Locally unimodal grain orientation (long axis of styliolinid cones) parallel to bedding)	Isometric grains do not indicate preferred orientation
Composition	Dominantly biogenic material from pelagic sources, locally fragmented; plus material from benthic and resedimented sources	Dominantly biogenic material from pelagic sources; phosphate intraclasts and conodonts common; some laminae with material from benthic and resedimented sources	Biogenic material from pelagic sources, mainly styliolinids, phosphate intraclasts common; locally material from benthic and resedimented sources	Peloids, cortoids, lithoclasts and biogenic material from benthic and resedimented sources
Sequence	Partial coarsening-upward and fining-upward micro-sequences similar to the standard mud–silt–sand contourite sequence of Stow *et al.* (2002)			Does not occur within standard cyclic micro-sequence

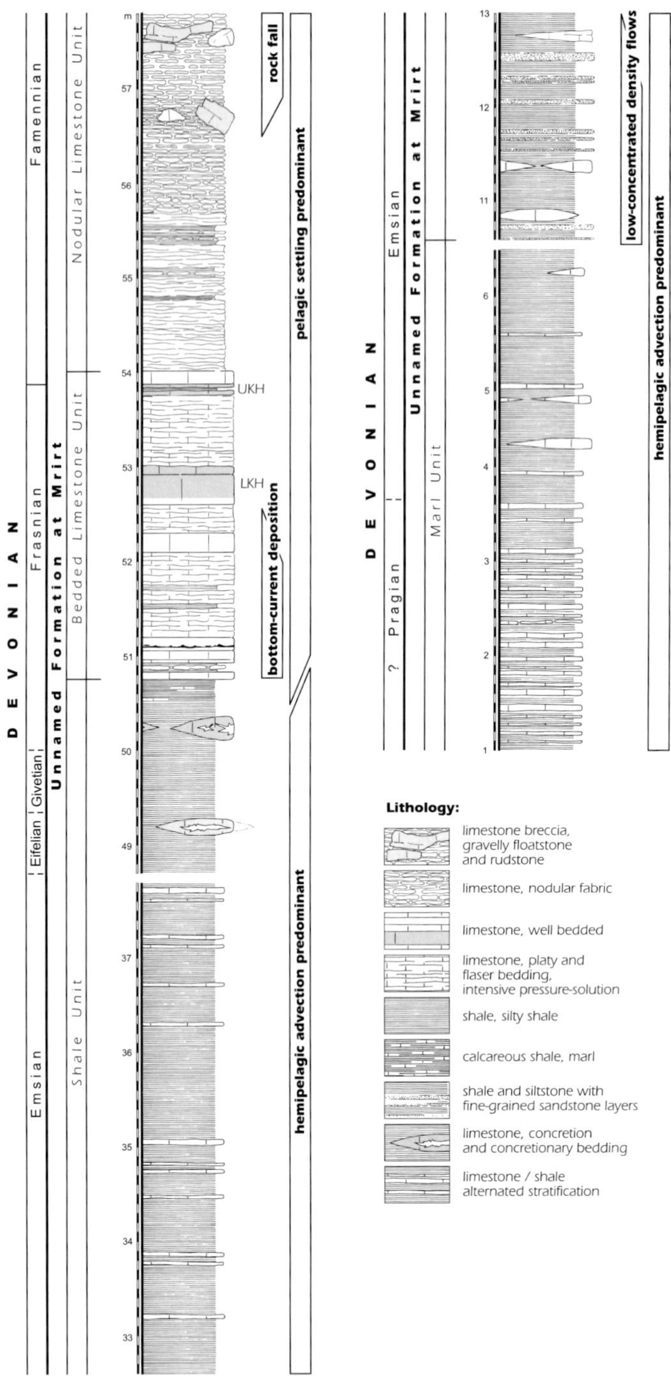

Fig. 3. Lithostratigraphic section of the formation at Mrirt, Ziar–Mrirt Nappe of the Moroccan Central Massif, (type section 'Mrirt M'). Stratigraphic age, lithological units, thickness and principal depositional processes are shown.

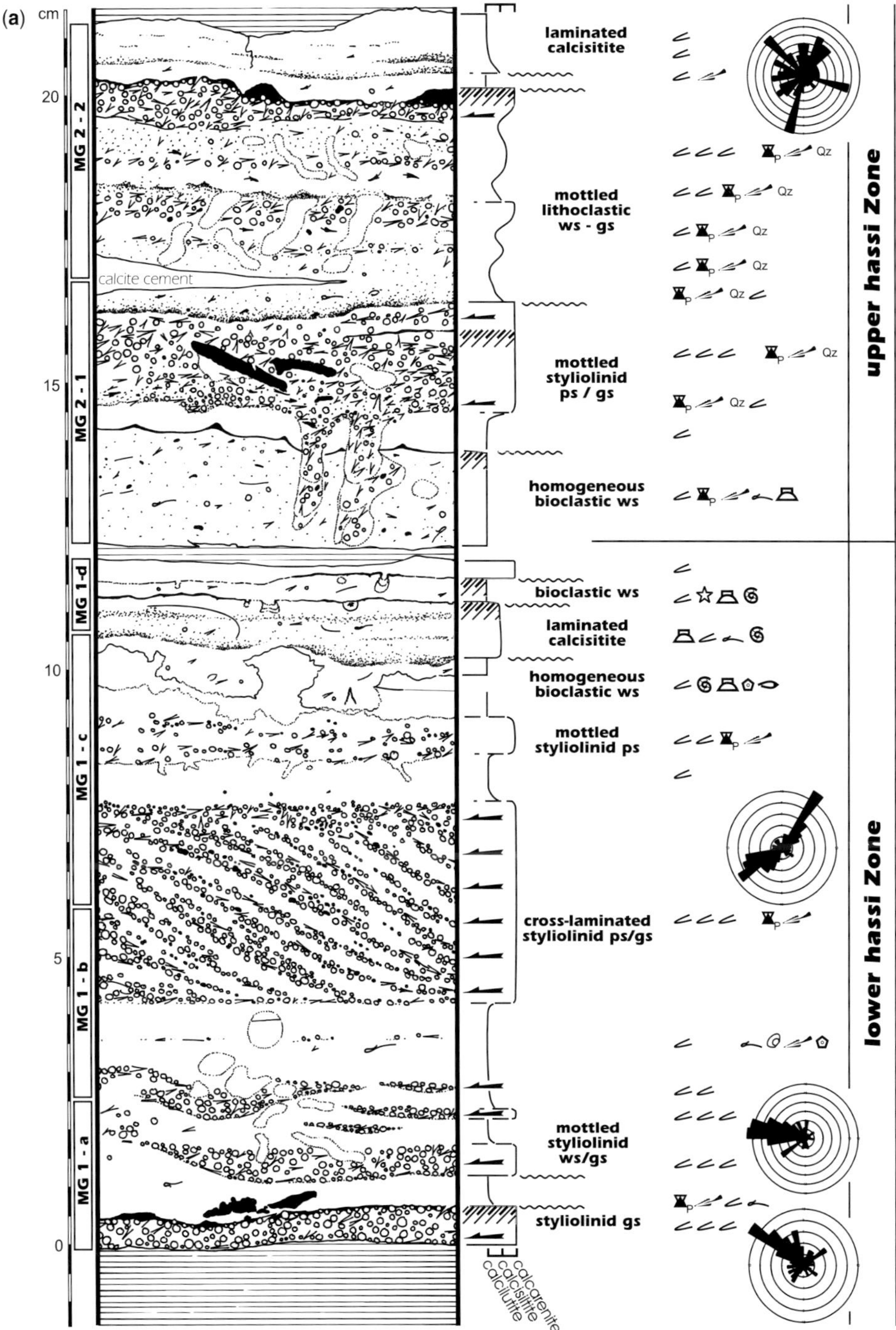

Fig. 4. Microfacies features and palaeocurrent data for lower Frasnian limestones from sections MG and M at the southwestern slope of the Gara de Mrirt (Moroccan Central Massif), from Hüneke (2001). (Note that skeletal grains are shown magnified.) Reference to the Devonian conodont zonation (Ziegler & Sandberg 1990; Ziegler 1996) is indicated at the right-hand side. Line drawing from thin sections.

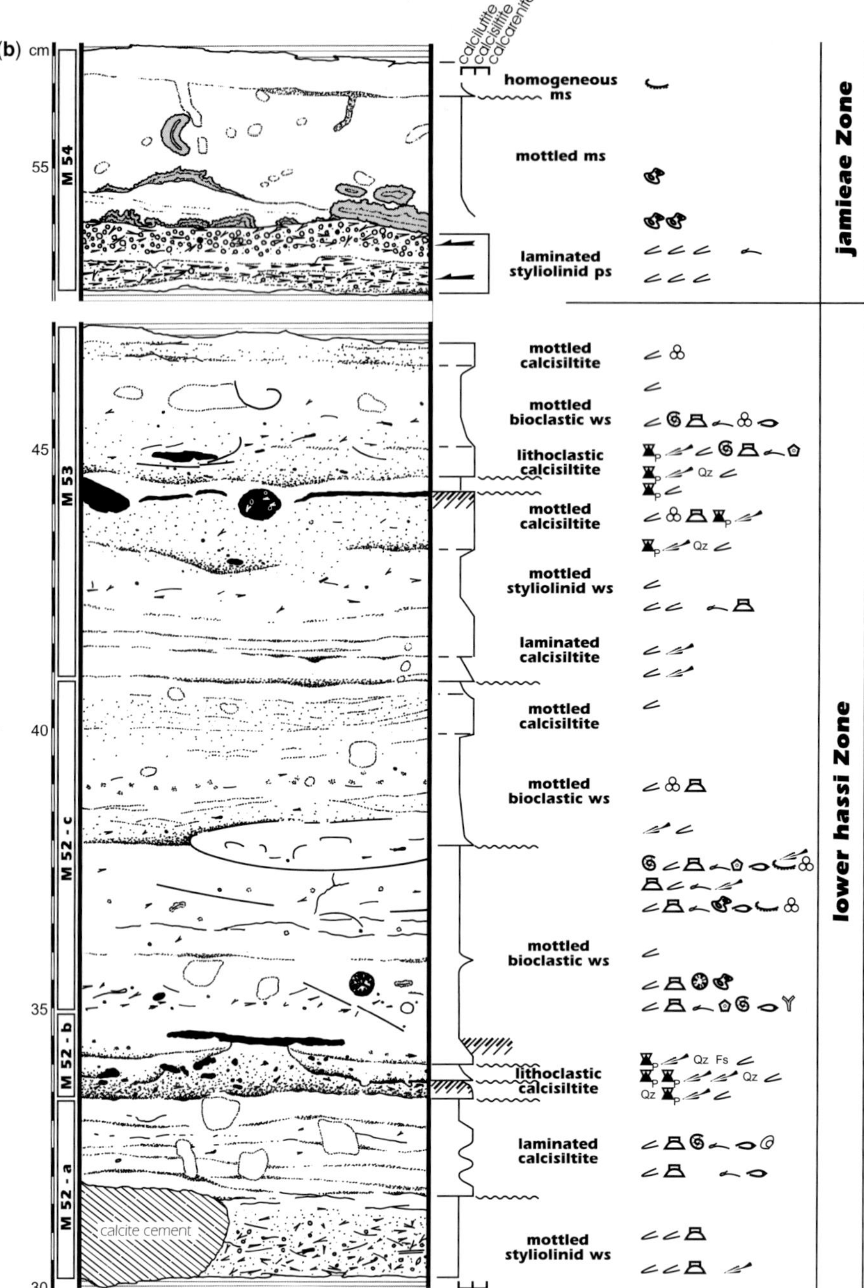

Fig. 4. (*Continued*)

and calcilutites, the latter showing some irregular, discontinuous lamination. Irregularly laminated and highly bioturbated calcarenites are less common (Table 2).

Microfacies and composition. Calcarenites are represented by styliolinid grainstones to packstones, which mainly occur immediately below and occasionally above the oldest phosphoritic

(c) **depositional textures**

mudstone (homogeneous)
wackestone (lamination)
wackestone, sparse (homogeneous)
wackestone, packed (homogeneous)
grain- to packstone (lamination)
grain- to packstone (cross lamination)
grain- to packstone (homogeneous)
calcisiltite (homogeneous)
calcisiltite (lamination)
calcisiltite (normal grading)
calcisiltite (inverse grading)

discontinuity surfaces

stromatoloid encrustation
omission surface
erosional surface
hardground (phosphatic)
condensed phosphates

bioturbation

biogenic traces
biodeformational structures (mottling)
homogenized

lithoclasts and oncoids

oncoids
limestone intraclasts
phosphate intraclasts

particle types

corals, solitary rugose
crinoid ossicles
echinoderms
bryozoans
brachiopods, thick-shelled
brachiopods, thin-shelled
trilobites, thin-shelled
molluscs
styliolinids
foraminifers
calcispheres
cephalopods
ostracodes
ostracodes, entomozoid
conodonts
oncoids, micritic
cortoids
peloids
intraclasts (C-carbonatic)
intraclasts (P-phosphatic)
Qz quartz

graphic symbols

erosional contact (corrasion)
sharp (irregular) contact
phosphatization
current / wave oriented styliolinid cones (visual examination)
gradational contact
recrystallization

abbreviations

ms - mudstone
ws - wackestone
ps - packstone
gs - grainstone
fs - floatstone
rs - rudstone

current / wave orientation of styliolinid cones; petals point to the taper end of styliolinid cones (measurements in thin sections parallel to bedding)

The azimuth data are measured as angle between the present North and the perpendicular projection of the long shell axis to the plane of stratification.

Fig. 4. (*Continued*)

hardground or stratigraphic hiatus (Fig. 7: 0.0–4.5 cm). They form irregular laminae and locally distinctive beds more than 20 cm thick. The styliolinid shells are always predominant and well preserved. A vague parallel lamination is occasionally recognizable in some of the sampled beds, but a preferred parallel orientation of the conical styliolinid tests is rarely evident. Bioturbation is probably intensive, although in many of the calcarenites it is visible only as an indistinct mottling. It is not clear whether the absence of lamination within homogeneous calcarenites is in every case due to bioturbation or if it is primary.

Laminated calcisiltites cover the uneven relief of phosphorites, phosphoritic hardgrounds or erosional surfaces (corrasional surfaces) and include lithoclasts consisting of phosphorite slabs up to some centimetres in length and styliolinid

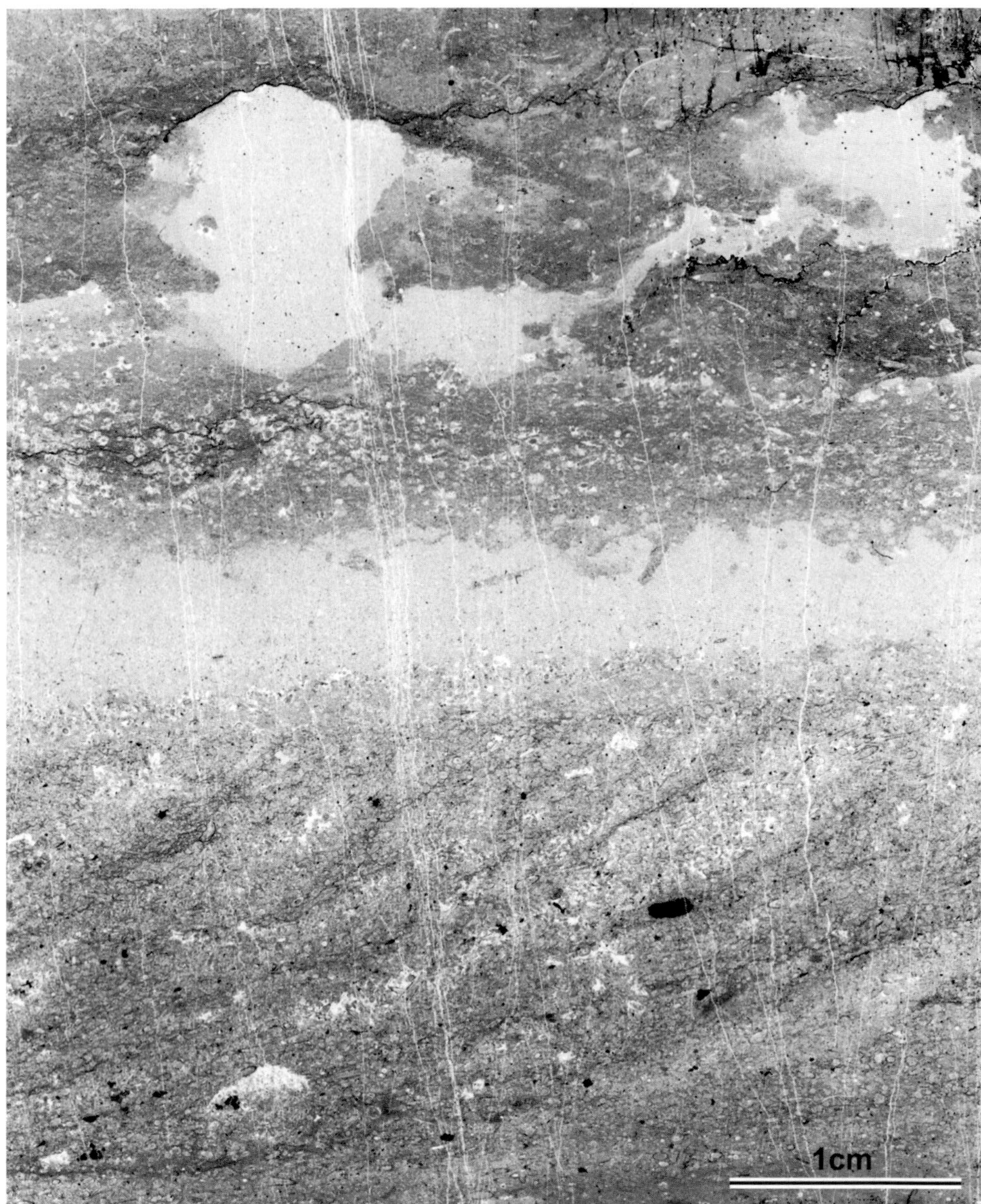

Fig. 5. Succession of cross-laminated styliolinid packstones–grainstones, mottled styliolinid wackestones, bioclastic wackestones and calcisiltites (from base to top) at Gara de Mrirt (Moroccan Central Massif). Note well-preserved cross-lamination in lower part and palimpsest biogenic traces in upper part of the figure. Sample MG 1 (see Fig. 4: 4.0–12.0 cm), thin-section perpendicular to bedding (lower *hassi* Zone). Scale bar represents 1 cm.

grainstones (Fig. 7: 5.5–15 cm; Fig. 8). Conodonts, Muellerisphaerida and phosphorite intraclasts may predominate at the base and within lower parts of the normally graded laminae (see Hüneke & Reich 2000). These grains have a high density compared with calcite components and display mineralogical grading. In other cases, quartz silt, feldspar grains and clay aggregates prevail (up to 40%). In contrast, the upper parts of graded laminae commonly exhibit a well-developed

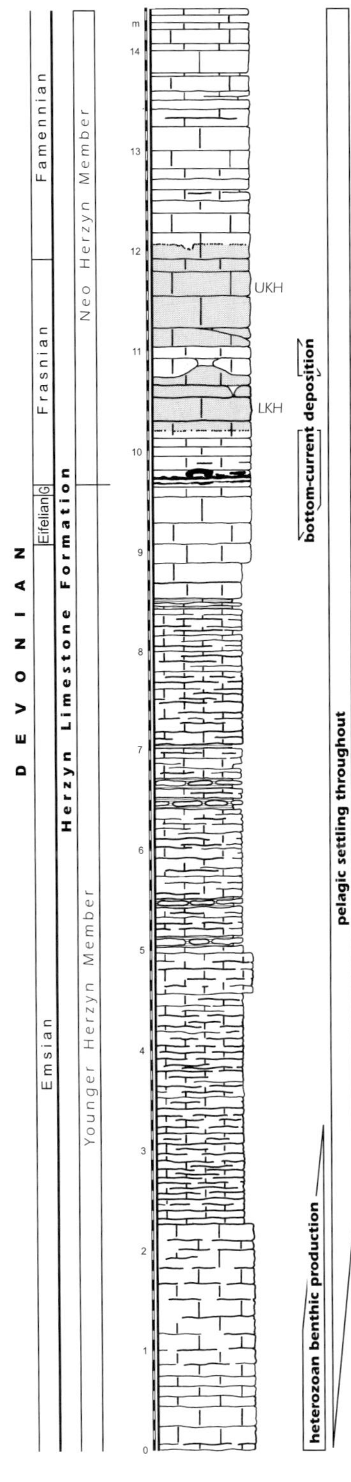

Fig. 6. Lithostratigraphic section of the Herzyn Limestone Formation, Harzgerode Zone of the Harz Mountains (type section 'Antoinettenweg'). Stratigraphic age, lithological units, thickness and principal depositional processes are shown. (See Fig. 3 for a lithological legend.)

distribution grading. The bulk of calcisiltitic particles are derived from mechanically abraded and crushed tests of pelagic organisms, such as styliolinids, molluscs, thin-shelled trilobites, brachiopods and ostracodes. A discontinuous lamination marked by pinching and swelling or the horizontal alignment of flattened lenses of calcisiltite is the most typical feature of the calcisiltites and is caused by an alternation of grain size, composition, or both. Laminated beds may pass laterally into mottled calcisiltites with only few primary features preserved. Isolated pockets and lenticular tube-like structures both filled with calcisiltite occur above and below the laminated calcisiltites.

Mottled calcisiltites and calcilutites mainly occur in the upper part of the reduced sequence and are most commonly in gradational contact with the laminated calcisiltites (Fig. 7: 15.0–20.5 cm). Bioclastic and styliolinid wackestones, which include a typical pelagic fauna, are the most common deposits. The dominant feature of this facies is a thorough burrow mottling. Where lamination is preserved, it is mostly very fine or less well defined, disorganized and wavy. It may be caused by slight changes of grain size and proportion, type and size of bioclastic grains. Fragmentation of bioclasts is most intensive within thoroughly mottled areas.

Sequences. There are partial fining-upward micro-sequences, which display a gradual upward change from laminated calcisiltites to mottled calcisiltites and calcilutites. Contacts between and within divisions vary from gradational to sharp or erosional, and the complete micro-sequence is typically 2–30 cm thick. At the base, graded calcisiltite laminae generally fill the relief above an erosional surface, phosphoritic hardground or phosphorite. Content grading occurs at the base, whereas distribution grading and graded lamination are typical in the middle part. Conodonts and phosphorite intraclasts are particularly common within the basal laminae immediately above the omission surface (constituting up to 40%). Partial coarsening-upward micro-sequences are not well documented.

Carnic Alps

The Rauchkofel Nappe in the Carnic Alps of Austria preserves a continuous Devonian succession of pelagic and periplatform carbonates (Fig. 9) (see Schönlaub *et al.* 2004*a*). The upper Givetian to lower Frasnian calcareous succession (*semialternans*–lower *rhenana* Zones) of the Middle Devonian Valentin Limestone and the overlying Late Devonian Pal Limestone Formations is interpreted as a bottom-current deposit (Schönlaub *et al.* 2004*b*). The Wolayer Gletscher and Valentin Törl sections are biostratigraphically well dated

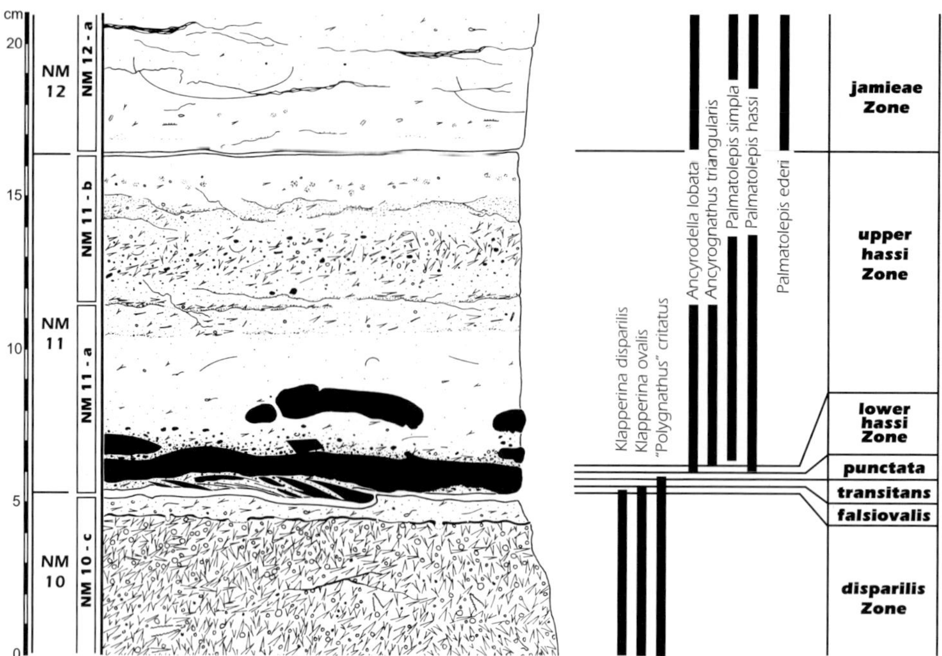

Fig. 7. Microfacies features and stratigraphic subdivision of upper Givetian to lower Frasnian limestones from section NW of Neue Mühle (Harz Mountains) from Hüneke & Reich (2000; reproduced by permission of Schweizerbart). (Note that skeletal grains are shown magnified.) Reference to the Devonian conodont zonation (Ziegler & Klapper 1982; Ziegler & Sandberg 1990; Weddige & Ziegler 1996; Ziegler 1996) is indicated at the right-hand side. Line drawing from thin sections. A microfacies legend is given in Figure 4.

(Göddertz *in* Schönlaub 1980, 1985; Göddertz 1982; Joachimski *et al.* 1994) and present the best evidence of fossil contourites in this area (Fig. 10).

Facies and structures. The facies comprise burrow-mottled calcisiltites and calcilutites, laminated calcisiltites and calcarenites (Table 3, Fig. 11). Calcirudite lag deposits are also identified.

Microfacies and composition. Calcarenites are represented by two facies types. Styliolinid grainstones and packstones comprise pelagic biogenic material (Fig. 11: 7.0 cm, 13.0–14.5 cm), whereas peloidal grainstones are composed of benthic and resedimented material (Fig. 11: 7.0–11.5 cm, 26.5–34.0 cm). Top and basal contacts of the calcarenites are commonly erosional. They occur in irregular laminae and lenses (<2 cm) or continuous layers (2–25 cm). The styliolinid grainstones and packstones consist almost exclusively of well-preserved tests of nektoplanktonic styliolinids, and show vertical gradational contacts to densely packed styliolinid wackestones. The conical tests of styliolinids rarely show preferred unimodal comet-shaped current orientations but internal lamination has not been observed in the styliolinid grainstones. The peloidal grainstones also include crinoid ossicles, cortoids, calcispheres, lithoclasts of bioclastic wackestones, styliolinids and some coral fragments. Lamination and normally graded layers, locally disturbed by burrow-mottling and more rarely diffuse bioturbation, are distinctive features of the peloidal grainstones.

Clearly laminated calcisiltites are rare within the bottom-current deposited facies although calcisiltites form a distinctive facies that comprises either lenses and infillings of shallow scours or, more commonly, thin continuous sediment layers <3 cm thick. Normally graded calcisiltites above a hiatus, comprising the *disparilis* to *punctata* Zones, contain a high proportion of fine- to medium-grained sand, which includes conodonts, fish remains and phosphate lithoclasts in addition to calcite bioclastic fragments (Fig. 11: 17.0–18.5 cm). The grain size and shape of phosphorite intraclasts are highly variable (fine silt to pebbles); the feature indicates that they are locally derived. Contacts are either sharp or gradational, with the basal contacts most commonly abrupt. Isolated pockets and lenticular tube-like structures both filled with calcisiltites occur above

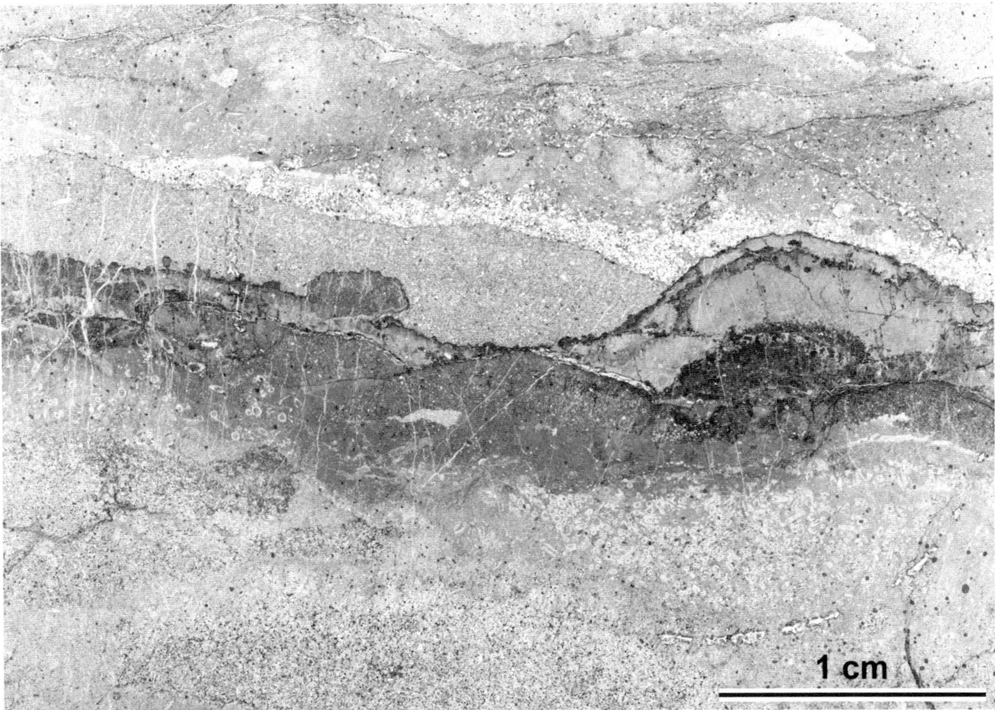

Fig. 8. Condensed phosphate embedded in a discontinuous carbonate succession at Antoinettenweg (Harz Mountains). Mottled styliolinid wackestones and calcisiltites (*falsiovalis* Zone) gradually merge into a condensed phosphorite, which shows a multitude of *in situ* phosphatized laminae and a complex micro-stratigraphy (*transitans–hassi* Zones). Discontinuous calcisiltite patches are preserved on top of the phosphorite and above a sharp erosional surface (*jamieae* Zone). Note the isolated pockets filled with calcisiltites in the lower part of the figure and the indistinct biogenic mottling in the upper part. Sample A 33, thin section perpendicular to bedding. Scale bar represents 1 cm.

and below the laminated units. Sediment may have been displaced vertically by burrowing over several centimetres.

The mottled calcisiltite and calcilutite facies exhibits an irregular, subparallel stratification on a millimetre to centimetre scale, and flat to irregular boundaries, although a diffuse bioturbation is common throughout (Fig. 11: 14.5–16.5 cm, 22.0–26.5 cm). Contacts between calcisiltite and calcilutite layers are gradational. In general, the grain-size distribution is bimodal. Recognizable fossils include styliolinids, planktonic tentaculites, thin-shelled trilobites and brachiopods, cephalopods, ostracodes, many of which are fragmented and probably reworked, as well as some disarticulated crinoid ossicles, bryozoan branches, solitary rugose corals, sponge spicules and conodonts. Besides the typical pelagic biogenic grains, styliolinid wackestones locally include peloids, cortoids and parathuramminid foraminifers, again revealing resedimented material from calciturbidites of shallow-water origin. Larger shells are commonly embedded convex-up. With a continuously decreasing proportion of calcisiltite, the facies become similar to the older and younger pelagic and periplatform limestones.

Sequences. Partial coarsening-upward micro-sequences of mottled calcilutites and calcisiltites passing upwards into styliolinid calcarenites and fining-upward micro-sequences of laminated calcisiltites gradually giving way to mottled calcisiltites and calcilutites are preserved, similar to the mud–silt–sand contourite sequence of Stow *et al.* (2002). These sequences are mostly 2–20 cm thick and are in many cases separated by an omission surface. Calcarenites composed of peloidal grainstones are not part of the cyclic sequence.

Regional evidence for bottom-current activity

This section highlights the differences between the Givetian–Frasnian facies interpreted as contourites

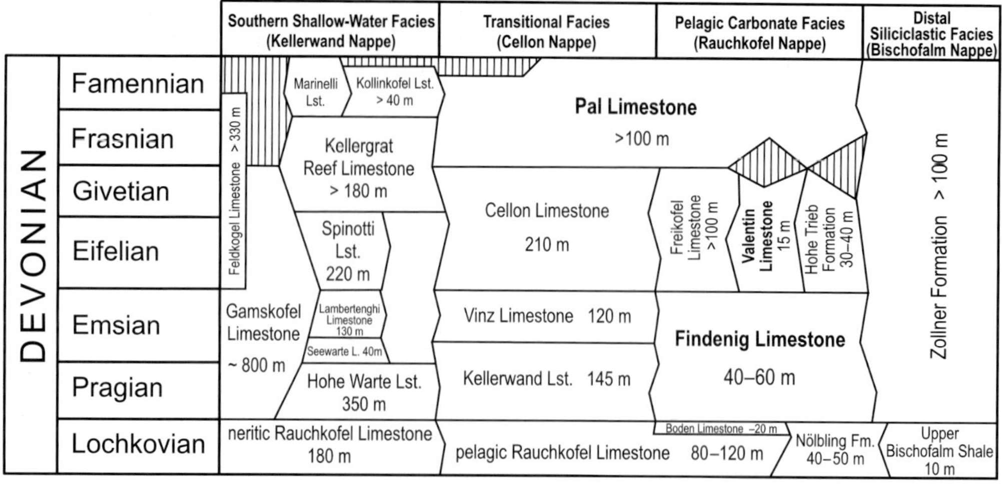

Fig. 9. Lithostratigraphic scheme for the Devonian of the Carnic Alps (Schönlaub *et al.* 2004*a*). The units are arranged on a proximal (left) to distal (right) transect. Vertical hatching indicates major hiatuses.

and older and younger parts of the Devonian successions. It focuses on the presence of hiatuses, phosphorites and other features that support a bottom-current activity. Regional trends and accumulation rates are examined on the basis of the whole Emsian to Famennian record studied. Conodont age determinations prove essential for achieving a correct perspective of the time dimensions of sediment accumulation, degree of condensation, non-deposition and erosion.

Moroccan Central Massif

Pragian to Givetian strata of the Ziar–Mrirt nappe mainly consist of hemipelagic shales and siltstones (<80 m) with rare intercalations of calcareous shale and limestone (Fig. 3). Thin beds of turbidity-flow deposits are restricted to the lower Emsian. Platy limestones with thin shale intercalations of Frasnian and earliest Famennian age represent a condensed sedimentary record (<10 m) that is typical of many Devonian pelagic platforms as defined by Santantonio (1994). They are overlain by nodular limestones and marly shales of Famennian age, which include thick limestone conglomerates and breccias (<20 m) in the southeastern part of the Mrirt unit, suggesting active basin-bounding faults probably related to an extensional event (Hüneke 2001). (For an alternative interpretation see Walliser *et al.* (1999).)

Scour-and-fill structures along the stratigraphic contact between the Givetian shales and siltstones and the lower Frasnian laminated limestones indicate local erosion of siliciclastic muds prior to deposition of calcareous bottom-current deposits. However, a hiatus has not been biostratigraphically proved. The contourite facies gradually pass upwards into typical bioturbated pelagic limestones.

Sedimentation rates and hiatuses. The accumulation rate curve (Fig. 12) is tentative for the Lower to Middle Devonian part of the succession, as biostratigraphic control is very limited within the siliciclastic record. Nevertheless, rare biostratigraphic data point to net accumulation of more than 2.0 m Ma^{-1} on average during the Emsian, which decreased to <0.5 m Ma^{-1} during the Eifelian, Givetian and earliest Frasnian. Erosive omissions cannot be ruled out. The Frasnian and lower Famennian limestones (*hassi–crepida* Zones), which are biostratigraphically dated bed by bed, exhibit very low accumulation rates varying between 0.5 and 2.0 m Ma^{-1}. The normal and resedimented pelagic facies of the upper Famennian units give accumulation rates at least 20 times higher (>40 m Ma^{-1}).

Associated phosphates and facies variation. The associated phosphorites and phosphoritic hardgrounds provide additional evidence of temporarily raised hydraulic energy. Pristine phosphates are common within the bottom-current deposited parts of the Moroccan successions. These autochthonous formations lack signs of any reworking and occur as discrete phosphatized laminae and as isolated phosphatized biogens. They have undergone one cycle of phosphogenesis. Discrete phosphatized burrow systems belong to this type. More rarely, there are condensed phosphates consisting of only two or

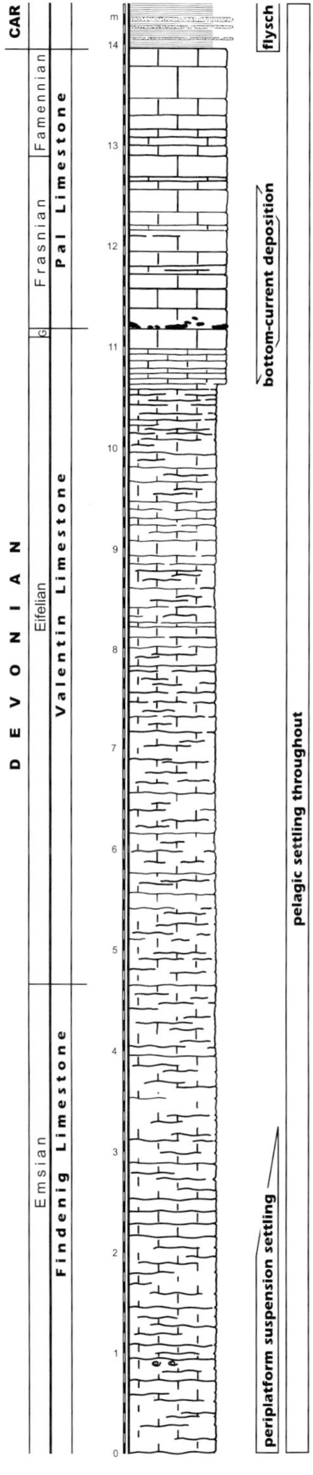

Fig. 10. Lithostratigraphic section of the Valentin Limestone and Pal Limestone formations, Rauchkofel Nappe of the Carnic Alps (type section 'Wolayer Glacier'). Stratigraphic age, lithological units, thickness and principal depositional processes are shown. (See Fig. 3 for a lithological legend.)

three accreted phosphate laminae (Fig. 4: 00.5 cm, 20.5 cm). These are more or less autochthonous and have experienced one or more phases of sediment reworking, such as winnowing, erosion or lateral transport (Föllmi 1996). Commonly, they mark the boundary between two conodont zones.

Harz Mountains

The sedimentary record of mostly pelagic Middle and Upper Devonian limestones in the eastern Harz Mountains is preserved as olistoliths up to 100 m or more in diameter and as coherent units within the Harzgerode and Blankenburg tectonosedimentary units (Lutzens 1991; Günther & Hein 1999). The primary depositional area of the Herzyn Limestone Formation is interpreted as representing a pelagic carbonate platform (Harzgerode Schwelle) at distal parts of the passive continental margin of Laurussia (Lütke 1990; Hüneke 1998). Sediments considered of bottom-current origin form a strongly condensed carbonate succession of late Givetian to early Frasnian age (*semialternans–jamieae* Zones). Condensed phosphatic sediments are related to a number of hiatuses (e.g. *hermanni-cristatus–disparilis* Zones and *transitans*–lower *hassi* Zones) indicating a regional disconformity (Hüneke 1997, 1998; Hüneke & Reich 2000). Underlying Eifelian and lower Givetian pelagic deposits consist of well-bioturbated bioclastic wackestones rich in styliolinid shells, showing a typical flaser fabric (Fig. 6). Overlying upper Frasnian and Famennian pelagic deposits are mostly bioturbated bioclastic wackestones with minor intercalations of current-redeposited calcisiltites. Cephalopod limestone is a term commonly used for this well-bedded facies in Germany.

Sedimentation rates and hiatuses. The accumulation rate curve, which clearly illustrates the depositional conditions prevailing in the Harz Mountains, shows several characteristic steps (Fig. 13). The accumulation rate during the early Emsian was about 10 m Ma^{-1} and above. During the late Emsian condensation is clearly documented and the net sediment accumulation rate decreased rapidly to 0.1 m Ma^{-1} and below until the Givetian (*semialternans* Zone). During the late Givetian and early Frasnian (*hermanni-cristatus–jamieae* Zones), carbonate accumulation was repeatedly interrupted by erosion and precipitation of condensed phosphates. A single intraclastic limestone bed is preserved from the lower Frasnian *falsiovalis* Zone and the bottom-current reworked facies essentially occurs above the hiatus. Finally, rather low accumulation rates were typical during the late Frasnian and Famennian, with values between 1 and 0.1 m Ma^{-1}. Clearly,

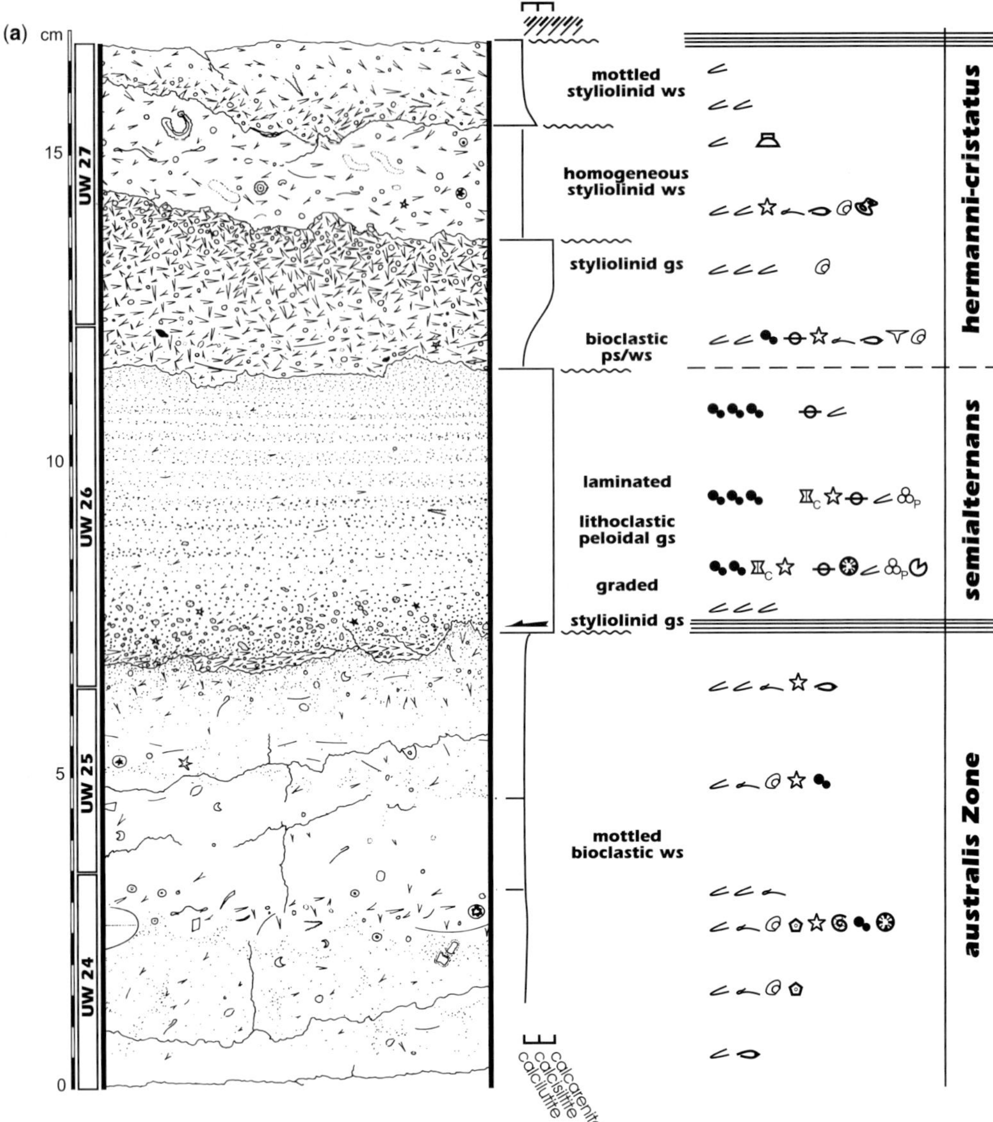

Fig. 11. Microfacies features and stratigraphic subdivision of upper Eifelian to lower Frasnian limestones from the Wolayer Glacier section of the Rauchkofel Nappe in the Carnic Alps of Austria. (Note that skeletal grains are shown magnified.) Reference to the Devonian conodont zonation (Bultynck 1987; Ziegler & Sandberg 1990; Walliser *et al.* 1995*a*; Weddige & Ziegler 1996; Ziegler 1996) is indicated at the right-hand side. Line drawing from thin sections. A microfacies legend is given in Figure 4.

the lower rates apply where the thickness is least and bottom-current intensity was presumably stronger and hence non-deposition or erosion most likely.

Hiatuses are evident where closely spaced biostratigraphic sampling has been carried out and comprise different stratigraphic ranges at the various localities. At Neue Mühle the carbonate accumulation was disrupted during the *falsiovalis* to early *hassi* Zone time by the formation of phosphates (Fig. 7), whereas at Antoinettenweg the sedimentation of carbonates had stopped already during the *semialternans* Zone and onset of phosphatization continued until the late *hassi* Zone

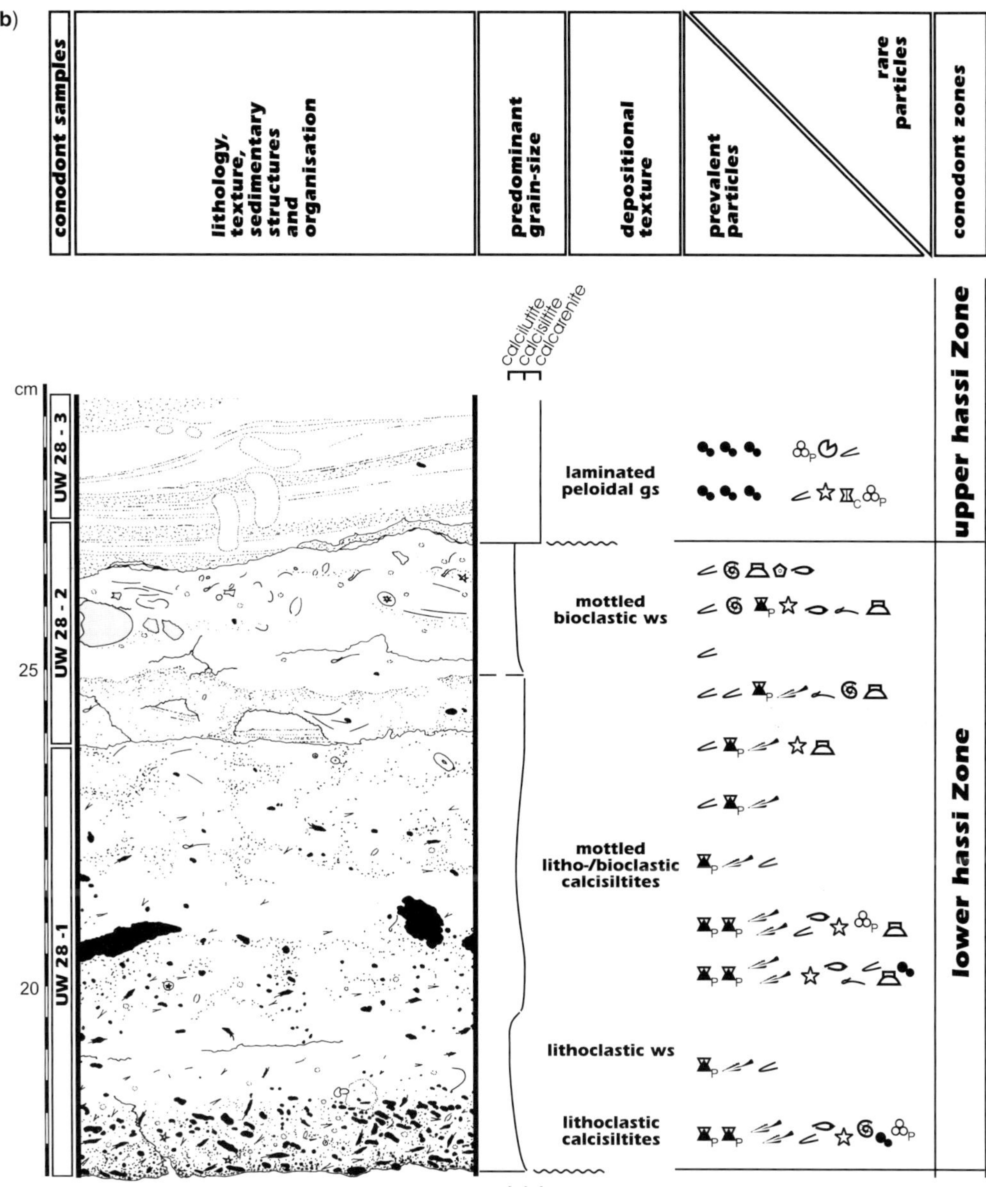

Fig. 11. (*Continued*)

time. Only from *falsiovalis* Zone time is a thin carbonate deposit preserved (Fig. 8). Hiatuses are most intense and widespread in the upper Givetian and lower Frasnian sequences, representing exactly the time period during which contourite accumulation occurred. They did not form during any of the rest of the late Emsian to Famennian pelagic sedimentation.

In a detailed outcrop study, Buchholz *et al.* (2001) recognized an erosional disconformity and several hiatuses within a succession of hemipelagic and pelagic limestones in the Langestal of the Clausthaler Kulmfaltenzone as well. This area is clearly part of the autochthon of the Harz Mountains (Walliser & Alberti 1983; Wachendorf 1986; Franke 1995). The reduced succession was

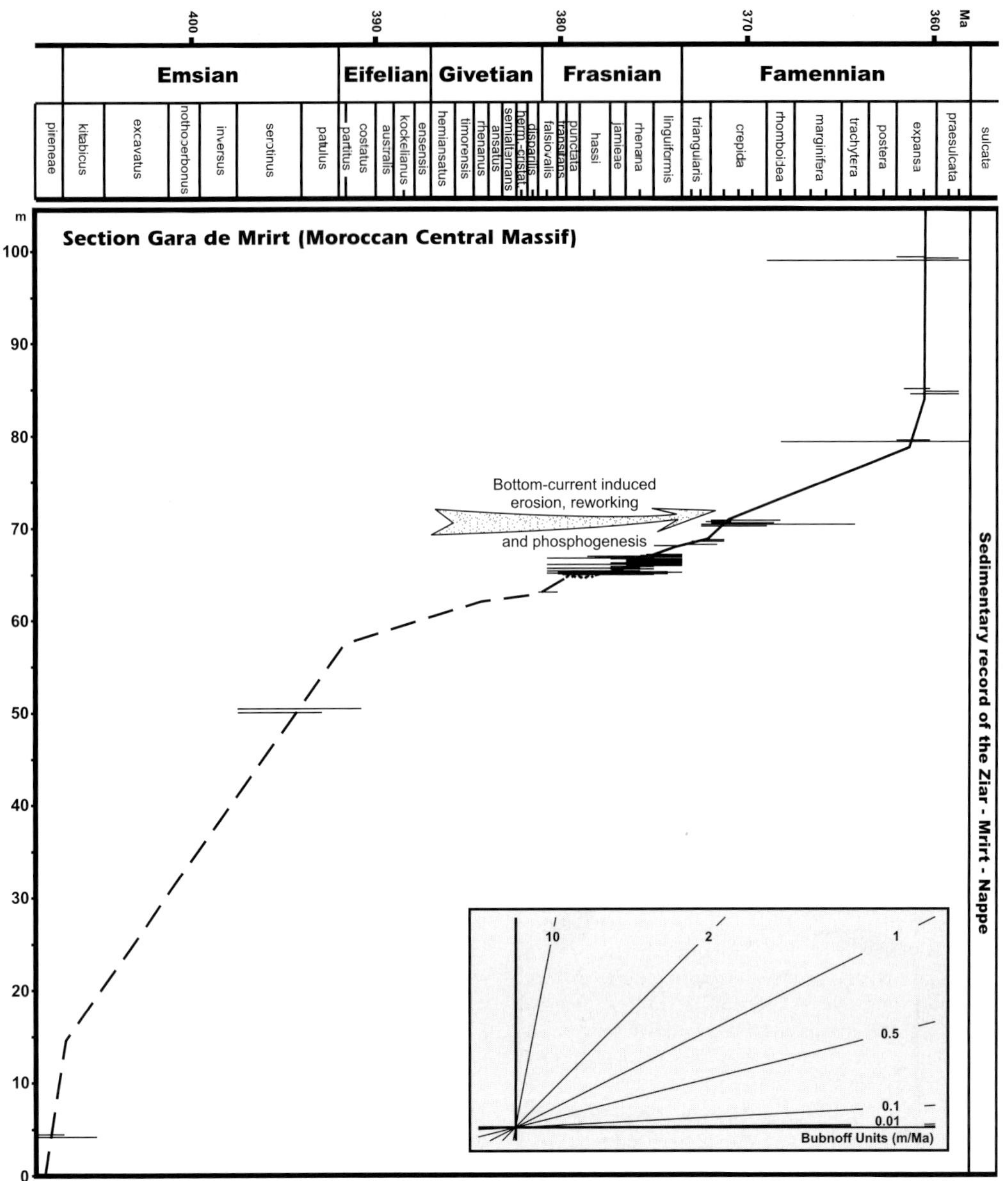

Fig. 12. Periods of condensation during the accumulation of hemipelagic to pelagic sediments of the nappe of Ziar–Mrirt in the Moroccan Central Massif exemplified at the sections Mrirt I, III, M and MG (see Hüneke 2001). The cumulative sediment thickness is plotted against the Devonian time scale (see Weddige 2003). The accumulation rate curve includes biostratigraphic age control (horizontal bars) and hiatuses (wavy horizontal lines). Conodont data from Walliser *et al.* (1999) are supplemented with the author's own data.

deposited on a pelagic carbonate platform (Westharz-Schwelle). Givetian styliolinid wackestones and packstones include a hiatus covering the *semialternans* to upper *hermanni-cristatus* Zones in the southern part of the Langestal, in which the range of the gap diminishes towards the north of the Langestal area. The erosional relief is covered by a thin shale seam and a 28 cm thick styliolinid packstone bed, which gradually passes into styliolinid wackestones at its upper part (component grading). A second hiatus, which spans the *falsiovalis* to upper *hassi* Zones, is

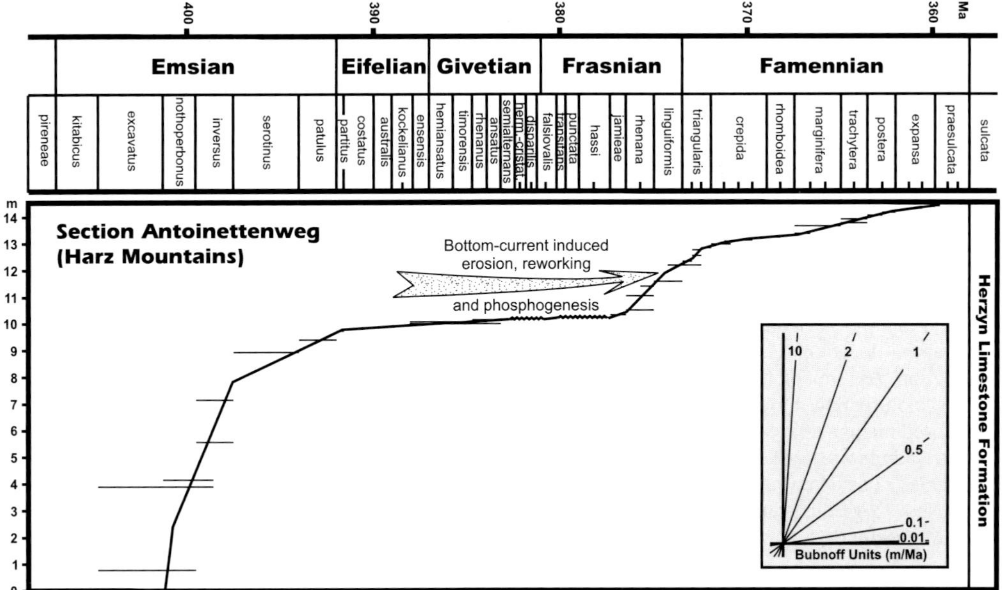

Fig. 13. Periods of condensation during the accumulation of the Herzyn Limestones in the Harz Mountains exemplified at the Antoinettenweg section (Hüneke 1995). The cumulative sediment thickness is plotted against the Devonian time scale (see Weddige 2003). The accumulation rate curve includes biostratigraphic age control (horizontal bars) and hiatuses (wavy horizontal lines).

documented immediately below Frasnian cephalopod limestones.

Associated phosphates and facies variation. Autochthonous phosphates were most extensive and widespread during the Givetian and early Frasnian. Within the condensed phosphates, several phases of phosphogenesis are indicated by several generations of accreted phosphate (Fig. 8). Internal boundaries separating the phosphate phases are discernible by relics of borings, micro-stromatolitic overgrowths and inclusion of different detrital particles. Complex internal micro-stratigraphic relationships with low-angle truncations and scours exhibit multiple phases of intense sediment reworking. The amalgamated condensed phosphate beds locally comprise up to six conodont zones, which represent *c.* 3.5 Ma (see Sandberg & Ziegler 1996; Weddige 2003).

Such local differences are part of the nature of condensed phosphates, which are formed under non-steady depositional conditions. One genetic cycle comprises the colonization of the sea bottom by benthic microbes under eutrophic conditions, their sudden burial by a migrating bed or a suspension load (styliolinid grainstones or calcisiltites), the early diagenetic phosphatization within the sediment, and the exposure of the phosphatized laminae by hydrodynamically induced erosion (Föllmi 1989, 1990; Kidwell 1991). These processes are controlled by fluctuations in velocity of the transporting bottom current or shifts of the current axis as well as palaeobathymetric position of the oxygen minimum layer. Consequently, local conditions are crucial for the preservation of individual sediment layers and the microstratigraphy of the succession.

Carnic Alps

The Middle to Upper Devonian sedimentary record of the Rauchkofel nappe in the Carnic Alps consists of an up to 230 m thick succession of pelagic and more rarely periplatform limestone deposits (Fig. 10). Distal turbiditic limestone intercalations occur at only a few levels. They were derived from peritidal carbonate platform and related slope-apron settings of the Kellerwand and Cellon nappes (Fig. 9) (Bandel 1969, 1972; Schönlaub 1971, 1985, 1992; Kreutzer 1990, 1992). Detailed conodont stratigraphic investigations have revealed one or more hiatuses separating the Middle Devonian Valentin Limestone from the overlying Upper Devonian Pal Limestone (Göddertz *in* Schönlaub 1980, 1985; Göddertz 1982; Joachimski *et al.* 1994). A bed that includes granule phosphate intraclasts and fish remains represents the boundary between the two lithostratigraphic units. Similar to the situation in the Harz Mountains, the reddish Middle Devonian limestones show the characteristic diagenetic fabric of solution stringers or flasers, whereas the predominantly greyish Upper Devonian limestones preserve distinct bedding. Generally, well-bioturbated bioclastic wackestones with numerous remains of styliolinids and cephalopods prevail. However, the upper Givetian to lower Frasnian deposits (*semialternans*–lower *rhenana* Zones) bounding the hiatuses are interpreted as an extremely reduced succession of bottom-current redeposited pelagic limestones, periplatform limestones and turbidites on the basis of their distal slope-apron location, current indicators, particle association, features of traction flow coupled with intense bioturbation, biogenic allochthonous phosphatic sediments, mixed conodont faunas (stratigraphic admixtures) and relic micro-sequences. The facies considered as being of bottom-current origin show a gradual transition into the overlying pelagic limestones.

Sedimentation rates and hiatuses. The net accumulation rate within the Rauchkofel Nappe of the Carnic Alps is calculated on the basis of detailed conodont stratigraphic data of Schönlaub (1980, 1985) and Göddertz (1982) at the Wolayer Glacier section, and the author's own studies (Fig. 14). Again, the curve shows a steep incline for the Emsian part of the succession (2–10 m Ma^{-1}). Condensation started during the early Eifelian and the net sediment accumulation rapidly decreased to <0.5 m Ma^{-1} (*australis* Zone). During the late Eifelian to early Frasnian (*kockelianus–punctata* Zone time), condensation and sediment reworking prevented a sustained accumulation. Only thin limestone layers of the contourite facies are preserved in the upper Givetian (*semialternans* and *hermanni-cristatus* Zones). A continuous record without biostratigraphically recognizable gaps continues from the upper Frasnian (lower *hassi* Zone) onwards, starting with allochthonous phosphatic sediments and accumulation rates of <0.5 m Ma^{-1}. The remaining part of the condensed Frasnian to Famennian succession is characterized by values between 0.2 and 10 m Ma^{-1}.

A disadvantage of the successions studied in the Carnic Alps is the bedding-parallel orientation of stylolite seams, which usually follow discontinuity surfaces. Erosional surfaces and hardgrounds are preserved only in very rare cases. Nevertheless, hiatuses representing longer periods are recognizable with the help of conodont biostratigraphy. The succession of the Rauchkofel Nappe west of the Wolayer Lake exhibits breaks in accumulation during *kockelianus* to *ansatus* Zone time and during *disparilis* to *punctata* Zone time (Fig. 11: 6.5–7.0 cm, 16.5–17.0 cm). Over a lateral distance of some hundred metres, the biostratigraphic gaps comprise a slightly different range (Göddertz 1982). Hiatuses are most extensive and widespread in the Givetian and lower Frasnian. Thus, they occur at the base and within the lowermost part of the contourite unit. There is no biostratigraphic indication for hiatuses in any of the older and younger parts of the Devonian pelagic succession.

The Eifelian flaser limestones are the diagenetic product of intensively bioturbated pelagic carbonate muds. Nevertheless, indistinct bedding, based on more subtle changes in composition, is still preserved. Ferruginous coatings around bioclasts at some stratigraphic levels point to longer periods of sea-floor exposure. Abrasion and bottom-current induced reworking started not before the late Eifelian and continued at least until the middle Frasnian. From the Givetian period contourites are preserved only in relicts, whereas during the early Frasnian the velocity and erosive capacity of the current was obviously diminished and gave way to a sustained accumulation of thin but continuous bottom-current deposits. At some stratigraphic levels these sediments include a high proportion of particles that suggest a shallow-water source area (peloids, cortoids, crinoids and fragments of rugose corals); they are, however, interpreted as bottom-current deposits. In the case of the Carnic Alps, the redeposited calcareous material was partly derived from bottom-current reworked calciturbidites or periplatform carbonates that were supplied from a shallow-water carbonate platform. Kreutzer (1990, 1992) documented facies types with such particle associations from the successions of the Kellerwand and Cellon Nappes (cortoid grainstones, ostracode and parathuramminid packstones). Styliolinid grainstone laminae included at the base of peloidal grainstone layers (Fig. 11: 6.5–7.0 cm), erosional down-cutting of bioclastic wackestones to packstones with pelagic composition (Fig. 11: 11.0–12.0 cm) and subsequent inversely graded transitions into styliolinid grainstones (Fig. 11: 12.0–14.0 cm) are features in favour of deposition from a bottom current in the studied successions of the Rauchkofel Nappe. Finally, the late Frasnian and Famennian record of generally mud-supported limestones is interpreted as intensively bioturbated pelagic deposits, which do not show clear indications of current-induced reworking and include a typical pelagic fauna without any shallow-water derived particles.

Associated phosphates and facies variation. Allochthonous phosphates are particularly common within calcisiltites of the lower *hassi* Zone above the hiatus comprising the *disparilis* to *punctata* Zones. Faintly rounded granule phosphate lithoclasts are densely packed at the base of the

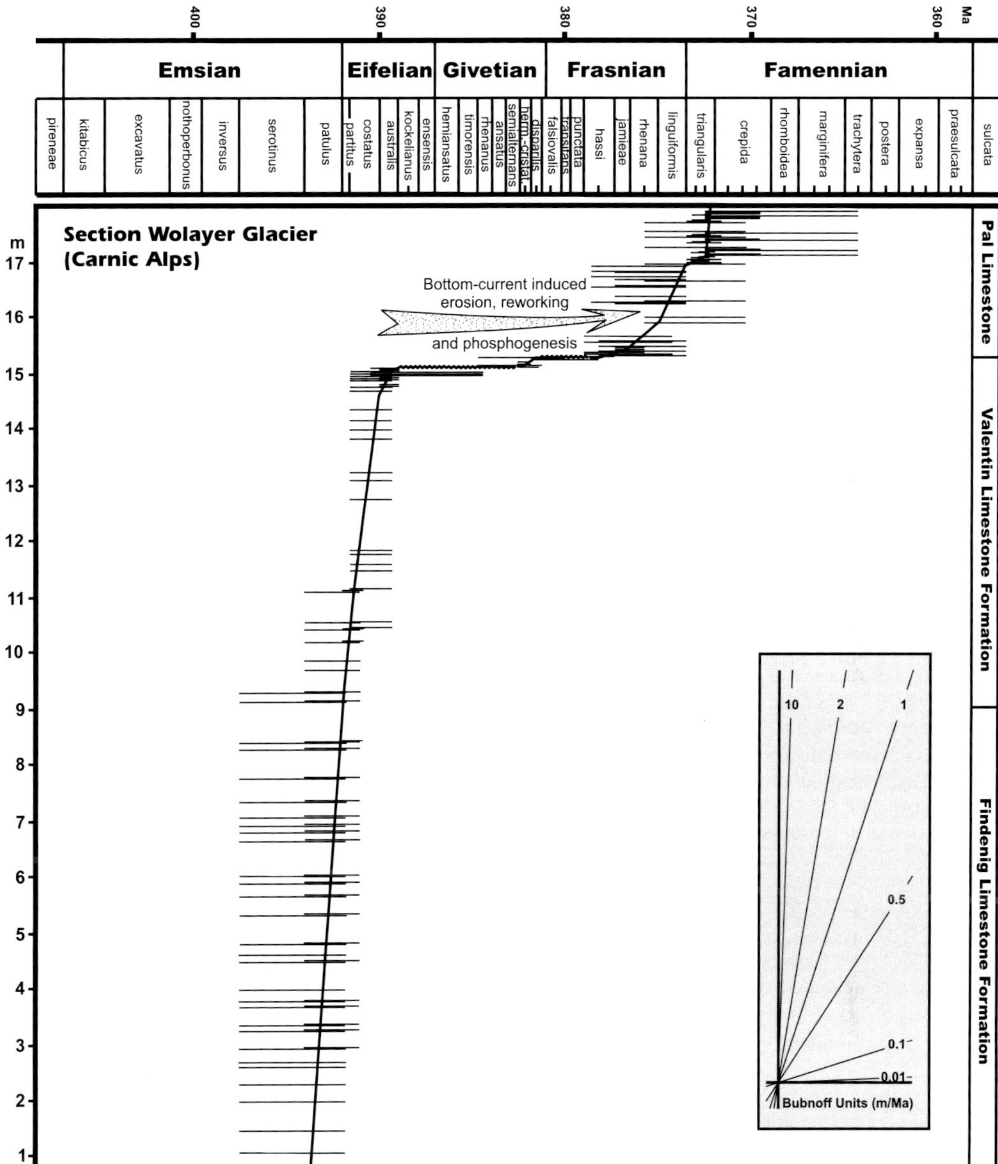

Fig. 14. Periods of condensation during the accumulation of the Findenig, Valentin and Pal Limestone formation of the Rauchkofel Nappe in the Carnic Alps exemplified at the Wolayer Glacier section (representative section). The accumulative sediment thickness is plotted against the Devonian time scale (see Weddige 2003). The accumulation rate curve includes biostratigraphic age control (horizontal bars) and hiatuses (wavy horizontal lines). Conodont data from Schönlaub (1980, 1985) and Göddertz (1982) are supplemented with the author's own data. Reprinted from Hüneke (2000), with permission of Elsevier.

calcisiltite layer, but larger phosphate pebbles do not occur until some centimetres above the hiatus (Fig. 11: 17.0–22.0 cm). The mixed conodont fauna of this phosphorite bed documents an intensive and long-lasting condensation during the *transitans* to early *hassi* Zone time; conodonts from the preceding *disparilis* and *falsiovalis* Zones are not preserved. However, the sedimentary characteristics of the bed point to a short-term depositional event during the early *hassi* Zone time, which is indicated by the youngest elements of the conodont fauna included (stratigraphic admixture). The phosphorites are clearly allochthonous intraclasts that display a rather short transport distance and document phases of phosphatization over a longer period during the preceding *disparilis* to *punctata* Zones. They result from reworking of originally condensed phosphates and subsequent transport by sediment-carrying bottom currents.

Palaeoceanographic setting

During the Devonian, the investigated depositional areas were part of the oceanic realm between Gondwana and Laurussia, spilling warm waters from the NE low latitudes and cool waters from the SW high latitudes (Fig. 15). The sandwiched Armorican terrane assemblage (Franke 2000; Tait *et al.* 2000; Matte 2001) separated a Protoatlantic in the north and west from a Prototethys in the south and east (Ziegler 1989; Winchester *et al.* 2002). During the Givetian and Frasnian (*c.* 388–375 Ma), Gondwana and Laurussia drifted northward and rotated clockwise (Scotese & Barrett 1990; Torsvik *et al.* 1996) covering a palaeolatitude from about 35°S on the cool northern margin of Gondwana to near the equator at the warm southern margins of Laurussia. The somewhat faster rotation rate of Laurussia led to the first contact between Laurussia and the Central European promontory of Gondwana during the Late Devonian (Ziegler 1989; Golonka 2002). This initial contact marks the onset of the Variscan orogeny.

Devonian reef models suggest strong thermal stratification during global warm episodes, especially during the Eifelian and Givetian (Copper 2002). The nature of bottom circulation within the oceanic passage between Gondwana and Laurussia during the Devonian is not well known. However, the period witnessed constriction of the main surface flow from the Prototethys to the Protoatlantic (Fig. 15). It seems most likely that continued constriction of the oceanic seaways led to an intensification of the westward-directed oceanic current, which was therefore capable of influencing sedimentation at greater depths. Alternatively, there may have been a deeper counterflow to the east. According to the principle of continuity, a broad and thin surface current that has to pass a narrow oceanic strait may accelerate or swell to greater depth (Brown *et al.* 1991). In both circumstances, the sediment accumulation in pelagic depositional environments would have been strongly affected, especially on the top of topographic heights.

The suturing of Gondwana and Laurussia during the late Devonian diverted ocean currents with global consequences (Heckel & Witzke 1979; Oczlon 1990; Copper 2002; Goddéris & Joachimski

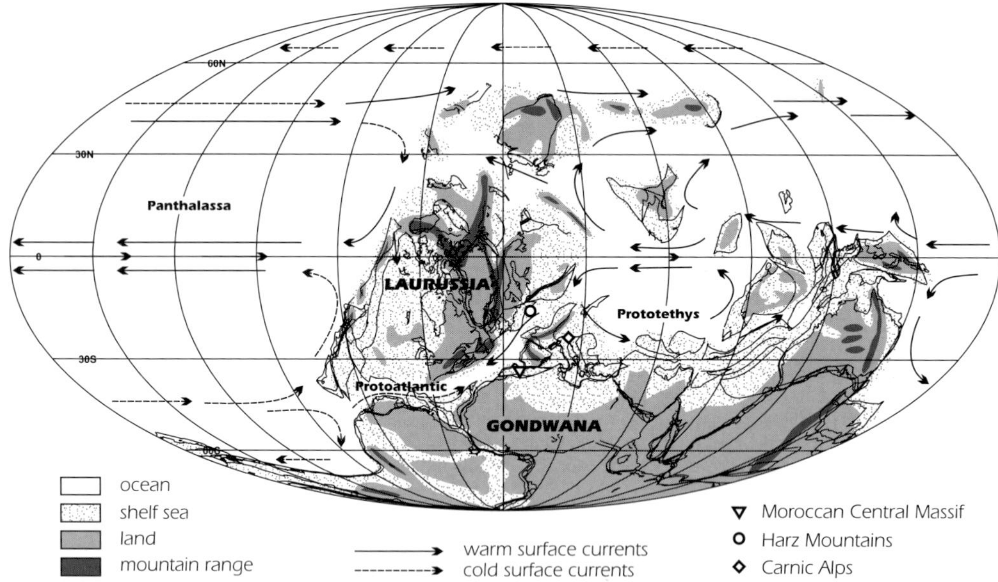

Fig. 15. Palaeogeographical reconstruction of the Earth during the Givetian–Frasnian transition (370 Ma), showing expected pattern of oceanic surface circulation as well as the distribution of deep sea, shelf sea, land and mountain ranges. Map provided by W. Kiessling (see Copper 2002), based on reconstruction of Golonka (2002). Locations of the areas investigated are indicated. Reprinted from Hüneke (2000), with permission of Elsevier.

2004; Hüneke 2006). Whereas warm surface currents were apparently restricted to continental platforms, there was weak or no horizontal exchange between bathyal water reservoirs of the Prototethys and Panthalassa oceanic basins (Fig. 15). The thermohaline circulation of the Panthalassa ocean with upwelling can be compared with the modern oceans, whereas the Prototethys ocean is interpreted as having been stratified with dysoxia below 100 m depth through the formation of warm saline deep waters in extended epicontinental seas (Goddéris & Joachimski 2004). Heckel & Witzke (1979) and Oczlon (1990) postulated a south Equatorial Surface Current along the southern margin of Laurussia being deflected anti-clockwise to the North Gondwana Current at the contact between Gondwana and Laurussia. Warm oceanic surface water of the Prototethys south of the equator was increasingly held captive. Conodont apatite $\delta^{18}O$ values show a significant decrease for the early Frasnian, indicating a major rise in seawater temperature from about 23 °C to 31 °C in low latitudes, only temporarily interrupted by two late Frasnian cooling events (Joachimski *et al.* 2004).

Discussion

Bottom currents affect to a greater or lesser extent ambient sedimentation by other processes (pelagic, hemipelagic and turbiditic) so that a mix of characteristics is the common result (Stow *et al.* 2002). Consequently, it is important to discuss the most relevant additional criteria for current-induced sedimentation.

Phosphates

The current strength, varying through time, is reflected in the stratification type of phosphate beds (Föllmi *et al.* 1991; Glenn *et al.* 1994). Once phosphatic laminae have formed on the sea bed, they may either be buried by calcareous sediments, transferred below the zone of main phosphogenesis and remain pristine, or experience reworking and become concentrated into phosphorite accumulations, depending on the intensity of hydraulic energy. Biological sediment-reworking processes, sediment properties and the sediment-binding capacity of microbial colonies are also important for the type of stratification (Föllmi 1996).

The preservation potential of calcareous bottom-current deposits was highest at the Gara de Mrirt in Morocco, where pristine and more rarely condensed phosphates with a single phase of reworking are a constituent part of a stratigraphically complete succession. Pristine phosphates originate in sedimentary regimes characterized by minimal sediment reworking, whereas condensed phosphates form in areas in which hydraulic energy levels were intermittently or persistently high and where sediment accumulation is balanced by winnowing or erosion. The amalgamated condensed phosphatic beds of the Harz Mountains are indicative of the poor preservation potential of calcareous bottom-current deposits in these sections (e.g. Fig. 8). Multi-event winnowed condensed beds preferentially occur in more distal shelf settings where sediment accumulation rates are generally low and energy conditions oscillate in a more steady way (Föllmi *et al.* 1991). In such a situation, geostrophic currents are powerful enough to induce sediment bypass over long periods. The allochthonous phosphates of the Carnic Alps indicate preferentially accumulative conditions in a marginal location of the main current system from the middle Frasnian onward, which were suitable for preservation of phosphates as well as calcareous bottom-current deposits.

The Protoatlantic was still open during the Givetian, spilling cool waters to the NE, across the continental microplates of Armorica and the Noric terranes (Fig. 15). Because of the effect of topographic constriction, bottom currents were here accelerated, reaching velocities that were sufficient to erode, transport and redistribute fine-grained sediment. The association with pristine and condensed phosphorites is probably an indication of an upwelling of cool nutrient-rich water at the submarine topographic barrier of the continental microplates. In such conditions, the bottom-shaping currents were probably intermediate or bathyal cool currents from the SW. On the other hand, phosphatic sediments that occur in current-dominated environments are not necessarily related to upwelling (Föllmi *et al.* 1991).

Hiatuses

The sediments deposited during late Givetian to early Frasnian time that were studied here are usually characterized by distinct condensation or even reduced successions with stratigraphic gaps in comparison with older and younger pelagic and hemipelagic deposits (e.g. Figs 13 and 14). This is also the case for many offshore marine depositional areas far from detrital terrigenous influence between Gondwana and Laurussia, such as, for example: many successions in the Tafilalt of the Anti-Atlas (Becker & House 1999*b*, *c*, 2001; Bultynck & Walliser 1999; Walliser 1999); in the Montagne Noire the La Serre successions (Feist 1985; Lottmann 1990); in the Rhenish Slate Mountains, for example, the succession at Martenberg (Ziegler 1966; House & Ziegler 1977); in the Harz Mountains the autochthonous successions of the Upper Harz (Lottmann 1990; Buchholz *et al.* 2001) as well as successions of the resedimented

Herzyn Limestones in the Lower Harz (Hüneke 1995, 1997; Hüneke & Reich 2000); successions in the Rauchkofel Nappe of the Carnic Alps (Schönlaub 1980, 1985; Göddertz 1982; Joachimski *et al.* 1994); and in the Frankenwald, for example, the succession at Schübelberg with a hiatus of a much greater time span (Hüsken 1993). This compilation is incomplete, but shows the importance of sediment bypassing or erosive processes in subtidal and bathyal depositional environments during late Givetian to early Frasnian times. Although the individual hiatuses in each case have been interpreted in different ways, an omission of the accumulation caused by emersion is not a plausible explanation in any of these successions. The inferred stratigraphic position of the hiatuses within the successions studied in Morocco, Germany and Austria is similar to erosional discontinuities in deep-sea sediment drifts of modern oceans (Stow *et al.* 2002). They are typically widespread both at the base and within the drift, extending across the accumulation as a whole.

In modern oceans, there is a general lack of detrital terrigenous siliciclastic and carbonate material available for accumulation in depositional areas that are far from land, at great water depths, and in areas of low biological productivity. Therefore, condensed successions do not necessarily indicate sediment reworking by bottom currents. However, unconformities and hiatuses of reduced successions are generally the result of sediment removal by gravity flows or current-induced erosion, or arise in areas with permanent sediment bypass. In many cores from the North Atlantic, there are hiatuses of mid–late Eocene and early–mid Miocene age (Stow 1982; Keller & Baron 1983). At places where sediment removal by gravity flows or slumps and slides can ruled out, these hiatuses are interpreted as the result of initiation or intensification of deep oceanic circulation associated with the Antarctic Bottom Water and the Norwegian Sea Overflow Water, respectively. The sediments adjacent to such hiatuses commonly represent accumulations under the influence of bottom currents (Stow 1982).

Mixed conodont faunas

Mixed conodont faunas are indirect indications of reworking or redeposition of pelagic sediments. Stratigraphic admixtures of older allochthonous conodont elements in calcareous bottom-current deposits are a common phenomenon in some of the successions investigated (e.g. Schönlaub 1980, 1985; Göddertz 1982; Joachimski *et al.* 1994). Amalgamated conodont faunas are rare and characterize condensed phosphate beds. Conodonts and sedimentary phosphate clasts are particularly prone to be concentrated by winnowing processes to form (parautochthonous) lag deposits, because of the high density of francolite (2.9–3.2 g cm^{-3}) relative to other sediment particles such as calcareous skeletal grains, feldspar and quartz (1.5–2.7 g cm^{-3}).

Conclusions

There is considerable evidence that demonstrates bottom-current influence on sedimentation during the Givetian and early Frasnian accumulation of otherwise hemipelagic and pelagic carbonate oozes in the nappe of Ziar–Mrirt of the Moroccan Central Massif, of the Herzyn Limestone Formation in the Harz Mountains of Germany, and of the Valentin Limestone and Pal Limestone Formations in the Carnic Alps of Austria.

(1) Calcarenites, laminated calcisiltites, and mottled calcisiltites and calcilutites are distinguished that are similar to recent and fossil contourites (see Stow *et al.* 1986, 1996, 1998). Calcarenites are mainly represented by styliolinid grainstones to packstones with rarely preserved horizontal and cross-lamination. Laminated calcisiltites are particularly rich in non-carbonate components with a higher density than calcite (conodonts, phosphatic intraclasts). Grain fragmentation and reworking of older bioclasts are evident. Both calcarenites and laminated calcisiltites show generally faint structures indicative of current-induced deposition, together with bioturbation throughout. Burrow mottled calcisiltites and calcilutites document predominant pelagic settling. Parts of coarsening-upward and fining-upward micro-sequences are best preserved in the Moroccan record. These micro-sequences are mostly between 2 and 10 cm thick and in many cases include an omission surface, which separates the top of the coarsening-upward micro-sequence from the base of the fining-upward micro-sequence. The vertical facies variation can be interpreted in terms of fluctuations in velocity of the transporting bottom currents or shifts of the current axes (Hüneke 2001).

(2) Overall, very low net accumulation rates prevailed during bottom-current induced deposition, sediment bypass or erosion. Therefore, strongly condensed successions with distinctive reduced intervals and biostratigraphic hiatuses (in the record of the Harz Mountains and the Carnic Alps), which occur mainly at the base of the bottom-current deposits, characterize the Givetian and early Frasnian record, in contrast to the more expanded and complete sequences of older and younger ages. In addition, erosional surfaces and hardgrounds are widespread. There are marked variations in thickness, accumulation

rates, erosive amounts and microfacies of the principal fossil contourite units documented across the three regions studied. Stratification types of the associated phosphatic sediments provide further evidence of temporarily raised, long-lasting hydraulic energy. These features are best compatible with moat areas of contourite drifts that experienced higher-velocity bottom currents and more marked erosion (Faugères *et al.* 1993; Stow *et al.* 2002).

(3) Bottom-current induced deposition contemporaneously occurred in different settings of the narrow oceanic passageways (Protoatlantic–Prototethys) between the approaching continents Gondwana and Laurussia. Areas affected were the SE Rhenish Sea shelf, which occupied the distal passive margin of Laurussia, the disintegrated northern continental margin of Gondwana, whose sedimentary record is now preserved in the Moroccan Meseta, and deep marginal plateaux of the Noric Terrane (South Alpine–Dinarid terrane) in the western part of the Prototethys. Whereas the accumulation in the Harz Mountains and the nappe of Ziar–Mrirt occurred on pelagic carbonate platforms, the redeposition in the Carnic Alps occurred in proximity to a carbonate slope-apron system. The simultaneity of the occurrence and the likelihood of bottom currents according to the palaeogeographical reconstruction both independently support the case for fossil contourites.

Further work is required on these examples as well as on comparable Devonian successions elsewhere in the Protoatlantic–Prototethys region to reconstruct the shape and 3D geometry of the entire sedimentary bodies and to find reliable indications of palaeocurrent directions.

I thank H. P. Schönlaub and S. Pohler for introducing me to the geology of the Carnic Alps. Many thanks also go to G. K. B. Alberti and O. H. Walliser for supporting preparation of fieldwork in Morocco. A. El Hassani provided excellent scientific and logistic support in Morocco. The German Research Foundation (DFG) financed the project and made equipment available for fieldwork in Morocco (Hu 804/1-1). Additional financial support was contributed by the Geological Society of Germany (DGG). I acknowledge the thorough reviews by F. Surlyk and J. J. Veevers and the editorial comments by A. R. Viana, which greatly improved the paper. J. Fanning improved the English of the manuscript.

References

BANDEL, K. 1969. Feinstratigraphische und biofazielle Untersuchungen unterdevonischer Kalke am Fuß der Seewarte (Wolayer See, Zentrale Karnische Alpen). *Jahrbuch der Geologische Bundesanstalt, Wien*, **112**, 197–234.

BANDEL, K. 1972. Palökologie und Paläogeographie im Devon und Unterkarbon der zentralen Karnische Alpen. *Palaeontographica*, **A141**, 1–117.

BECKER, R. T. & HOUSE, M. R. 1999*a*. Sedimentary and faunal succession of the allochthonous Upper Devonian at Gara d'Mrirt (eastern Moroccan Meseta). *In*: EL HASSANI, A. & TAHIRI, A. (eds) *The North-Western Moroccan Meseta*. Excursion Guidebook Part II, Joint Meeting IGCP 421/SDS, Errachidia, Rabat, Morocco, **3**. Institut Scientifique, Rabat, 133–139.

BECKER, R. T. & HOUSE, M. R. 1999*b*. Late Givetian and Frasnian ammonoid succession at Bou Tchrafine (Anti Atlas, southern Morocco). *In*: EL HASSANI, A. & TAHIRI, A. (eds) *Tafilalt and Maider (Eastern Anti Atlas)*. Excursion Guidebook Part I, Joint Meeting IGCP 421/SDS, Errachidia, Rabat, Morocco. Institut Scientifique, Rabat, 28–37.

BECKER, R. T. & HOUSE, M. R. 1999*c*. Devonian ammonoid succession at Jebel Amelane (western Tafilalt, southern Morocco). *In*: EL HASSANI, A. & TAHIRI, A. (eds) *Tafilalt and Maider (Eastern Anti Atlas)*. Excursion Guidebook Part I, Joint Meeting IGCP 421/SDS, Errachidia, Rabat, Morocco. Institut Scientifique, Rabat, 55–64.

BECKER, R. T. & HOUSE, M. R. 2001. New data on Devonian hypoxic and eustatic events in southern Morocco. *In*: JANSEN, U., KÖNIGSHOF, P., PLODOWSKI, G. & SCHINDLER, E. (eds) *Mid-Palaeozoic Bio- and Geodynamics—the North Gondwana/Laurussia Interaction*. Joint Meeting IGCP 421/SDS, Frankfurt/Main, Forschungsinstitut Senckenberg, Frankfurt am Main, 12–13.

BLAKEY, R. C. 2003. Carboniferous–Permian global palaeogeography of the assembly of Pangaea. *Abstracts, International Congress on Carboniferous and Permian Stratigraphy*. University of Utrecht, 57–58.

BOUABDELLI, M. 1989. *Tectonique et sédimentation dans un bassin orogénique: le sillon viséen d'Azrou-Khenifra (Est du Massif hercynien central du Maroc)*. Thèse Doctorat ès Sciences, Université Strasbourg.

BROWN, J., COLLING, A., PARK, D., PHILLIPS, J., ROTHERY, D. & WRIGHT, J. 1991. *Ocean Circulation*. Pergamon, Oxford.

BUCHHOLZ, P., LUPPOLD, F. W. & STOPPEL, D. 2001. Conodontenstratigraphie und Sedimentologie der Cephalopodenkalke im Langestal (Eifelium–Frasnium, Oberharz). *Braunschweiger Geowissenschaftliche Arbeiten*, **24**, 109–131.

BULTYNCK, P. 1987. Pelagic and neritic conodont successions from the Givetian of pre-Sahara Morocco and the Ardennes. *Bulletin de l'Institut Royal des Sciences Naturelles de Belgique, Sciences de la Terre*, **57**, 149–181.

BULTYNCK, P. & WALLISER, O. H. 1999. Emsian to Middle Frasnian sections in the northern Tafilalt. *In*: EL HASSANI, A. & TAHIRI, A. (eds) *Tafilalt and Maider (Eastern Anti Atlas)*. Excursion Guidebook Part I, Joint Meeting IGCP 421/SDS, Errachidia, Rabat, Morocco. Institut Scientifique, Rabat, **6**, 1–19.

COPPER, P. 2002. Silurian and Devonian reefs: 80 million years of global greenhouse between two ice ages. *In*: KIESSLING, W., FLÜGEL, E. & GOLONKA, J. (eds) *Phanerozoic Reef Patterns*. Society for Sedimentary Geology, Special Publications, **72**, 181–239.

FAIK, F. 1988. *Le Paléozoique de la région de M'rirt (est du Maroc Central): évolution stratigraphique et structurale*. Thèse du Doctorat de 3ème cycle, Université Toulouse.

FAUGÈRES, J. C., MÉZERAIS, M. L. & STOW, D. A. V. 1993. Contourite drift types and their distribution in the North and South Atlantic Ocean basins. *Sedimentary Geology*, **82**, 189–205.

FEIST, R. 1985. Devonian stratigraphy of the southeastern Montagne Noire (France). *Courier Forschungs-Institut Senckenberg*, **75**, 331–352.

FÖLLMI, K. B. 1989. *Evolution of the mid-Cretaceous triad: platform carbonates, phosphatic sediments, and pelagic carbonates along the northern Tethys margin*. Lecture Notes in Earth Sciences, **23**, Springer, Berlin.

FÖLLMI, K. B. 1990. Condensation and phosphogenesis: example of the Helvetic mid-Cretaceous (northern Tethyan margin). *In*: NOTHOLT, A. J. G. & JARVIS, I. (eds) *Phosphorite Research and Development*. Geological Society, London, Special Publications, **52**, 237–252.

FÖLLMI, K. B. 1996. The phosphorus cycle, phosphogenesis and marine phosphate-rich deposits. *Earth-Science Reviews*, **40**, 55–124.

FÖLLMI, K. B., GARRISON, R. E. & GRIMM, K. A. 1991. Stratification in phosphatic sediments: illustrations from the Neogene of California. *In*: EINSELE, G., RICKEN, W. & SEILACHER, A. (eds) *Cycles and Events in Stratigraphy*. Springer, Berlin, 492–507.

FRANKE, W. 1995. Rhenohercynian foldbelt: autochthon and nonmetamorphic nappe units, stratigraphy. *In*: DALLMEYER, R. D., FRANKE, W. & WEBER, K. (eds) *Pre-Permian Geology of Central and Eastern Europe*. Springer, Berlin, 33–49.

FRANKE, W. 2000. The mid-European segment of the Variscides: tectono-stratigraphic units, terrane boundaries and plate tectonic evolution. *In*: FRANKE, W., HAAK, V., ONCKEN, O. & TANNER, D. (eds) *Orogenic Processes: Quantification and Modelling in the Variscan Belt*. Geological Society, London, Special Publications, **179**, 35–62.

FRISCH, W. & NEUBAUER, F. 1989. Pre-Alpine terranes and tectonic zoning in the eastern Alps. *In*: DALLMEYER, R. D. (ed.) *Terranes in the Circum-Atlantic Paleozoic Orogens*. Geological Society of America, Special Papers, **230**, 91–100.

GLENN, C. R., FÖLLMI, K. B., RIGGS, S. R., *ET AL*. 1994. Phosphorus and phosphorites: sedimentology and environment of formation. *Eclogae Geologiae Helveticae*, **87**, 747–788.

GODDÉRIS, Y. & JOACHIMSKI, M. M. 2004. Global change in the Late Devonian: modelling the Frasnian–Famennian short-term carbon isotope excursions. *Palaeogeography, Palaeoclimatology, Palaeoecology*, **202**, 309–329.

GÖDDERTZ, B. 1982. *Zur Geologie und Conodontenstratigraphie der Rauchkofelböden und des Rauchkofel in den Zentralen Karnischen Alpen*. Diploma thesis, University of Bonn.

GOLONKA, J. 2002. Plate-tectonic maps of the Phanerozoic. *In*: KIESSLING, W., FLÜGEL, E. & GOLONKA, J. (eds) *Phanerozoic Reef Patterns*. Society for Sedimentary Geology, Special Publications, **72**, 21–75.

GÜNTHER, K. & HEIN, S. 1999. Die Olisthostrome des Mittelharzes nordwestlich von Bad Lauterberg, eine Folge akkretionärer Prozesse. *Neues Jahrbuch für Geologie und Paläontologie, Abhandlungen*, **211**, 355–410.

HECKEL, P. H. & WITZKE, B. J. 1979. Devonian world palaeogeography determined from distribution of carbonates and related lithic palaeoclimatic indicators. *In*: HOUSE, M. R., SCRUTTON, C. T. & BASSET, M. G. (eds) *The Devonian System*. Special Papers in Palaeontology, **23**, 99–123.

HOUSE, M. & ZIEGLER, W. 1977. The goniatite and conodont sequences in the early Upper Devonian at Adorf, Germany. *Geologica et Palaeontologica*, **11**, 69–108.

HÜNEKE, H. 1995. Early Devonian (Emsian) to Late Devonian (Famennian) stratigraphy and conodonts of the Antoinettenweg section in the Lower Harz Mountains (Germany). *Courier Forschungs-Institut Senckenberg*, **188**, 99–131.

HÜNEKE, H. 1997. Die Herzynkalke im Randgebiet der Selke-Einheit (Harz)—schrittweiser Übergang von neritischer Akkumulation zu pelagischer Kondensation während des Devons. *Neues Jahrbuch für Geologie und Paläontologie, Abhandlungen*, **205**, 209–264.

HÜNEKE, H. 1998. Die Herzynkalk-Entwicklung im Randgebiet der Selke-Einheit (Harz)—Abbild der Subsidenz im östlichen Rhenischen Trog während des Devons. *Zeitschrift der Deutschen Geologischen Gesellschaft*, **149**, 381–430.

HÜNEKE, H. 2000. Erosion and deposition from bottom currents. *Palaeogeography, Palaeoclimatology, Palaeoecology*, **234**, 146–167.

HÜNEKE, H. 2001. *Gravitative und strömungsinduzierte Resedimente devonischer Karbonatabfolgen im Marokkanischen Zentralmassiv (Rabat–Tiflet-Zone, Decke von Ziar–Mrirt)*. Habilitation thesis, University of Greifswald.

HÜNEKE, H. 2006. Erosion and deposition from bottom currents during the Givetian and Frasnian: response to intensified oceanic circulation between Gondwana and Laurussia. *Palaeogeography, Palaeoclimatology, Palaeoecology*, **234**, 146–167.

HÜNEKE, H. & REICH, M. 2000. Muellerisphaerida (incertae sedis) from condensed carbonates (Givetian/Frasnian) of the Harz Mountains. *Neues Jahrbuch für Geologie und Paläontologie, Abhandlungen*, **218**, 201–241.

HÜSKEN, T. C. 1993. *Zur Faziesausbildung und Stratigraphie (Conodonten) der obersilurisch–unterdevonischen Karbonate der Bayerischen und*

Thüringischen Faziesreihe des Frankenwaldes. Doctoral thesis, University of Hamburg.

JOACHIMSKI, M. M., BUGGISCH, W. & ANDERS, T. 1994. Mikrofazies, Conodontenstratigraphie und Isotopengeochemie des Frasne/Famenne-Grenzprofils Wolayer Gletscher (Karnische Alpen). *Abhandlungen der Geologischen Bundesanstalt von Österreich*, **50**, 183–195.

JOACHIMSKI, M. M., VAN GELDERN, R., BREISIG, S., BUGGISCH, W. & DAY, J. 2004. Oxygen isotope evolution of biogenic calcite and apatite during the Middle and Late Devonian. *International Journal of Earth Sciences*, **93**, 543–553.

KELLER, G. & BARRON, J. A. 1983. Paleoceanographic implications of Miocene deep-sea hiatuses. *Geological Society of America Bulletin*, **94**, 590–610.

KIDWELL, S. M. 1991. Condensed deposits in siliciclastic sequences. *In*: EINSELE, G., RICKEN, W. & SEILACHER, A. (eds) *Cycles and Events in Stratigraphy.* Springer, Berlin, 682–695.

KLAPPER, G., FEIST, R. & HOUSE, M. R. 1987. Decision on the boundary stratotype for the Middle/Upper Devonian series boundary. *Episodes*, **10**, 97–101.

KREUTZER, L. H. 1990. Mikrofazies, Stratigraphie und Paläogeographie des Zentralkarnischen Hauptkammes zwischen Seewarte und Cellon. *Jahrbuch der Geologische Bundesanstalt, Wien*, **133**, 275–343.

KREUTZER, L. H. 1992. Palinspastische Entzerrung und Neugliederung des Devon in den neuen Untersuchungen. *Jahrbuch der Geologische Bundesanstalt, Wien*, **135**, 261–272.

LÄUFER, A., HUBICH, D. & LOESCHKE, J. 2001. Variscan geodynamic evolution of the Carnic Alps (Austria/Italy). *International Journal of Earth Sciences*, **90**, 855–870.

LAZREQ, N. 1992. The Upper Devonian of M'rirt (Morocco). *Courier Forschungs-Institut Senckenberg*, **154**, 107–123.

LOTTMANN, J. 1990. Die pumilio-Events (Mittel-Devon). *Göttinger Arbeiten zur Geologie und Paläontologie*, **44**, 1–98.

LÜTKE, F. 1990. The Rhenian shelf sea–ocean–volcanic arc domain with special reference to the structural development of the Harz Mountains. *In: Abstracts, International Conference on Terranes in the Circum-Atlantic Paleozoic Orogens*, Göttingen, Germany, 163–166.

LUTZENS, H. 1991. Flysch, Olisthostrome und Gleitdecken im Unter- und Mittelharz. *Zeitschrift für Geologische Wissenschaften*, **19**, 617–623.

MATTE, P. 2001. The Variscan collage and orogeny (480–290 Ma) and the tectonic definition of the Armorica microplate: a review. *Terra Nova*, **13**, 122–128.

OCZLON, M. 1990. Ocean currents and unconformities: the north Gondwana Middle Devonian. *Geology*, **18**, 509–512.

ONCKEN, O., VON WINTERFELD, C. & DITTMAR, U. 1999. Accretion of rifted passive margin: the Late Palaeozoic Rhenohercynian fold and thrust belt (Middle European Variscides). *Tectonics*, **18**, 75–91.

ONCKEN, O., PLESCH, A., WEBER, J., RICKEN, W. & SCHRADER, S. 2000. Passive margin detachment during arc–continent collision (Central European Variscides). *In*: FRANKE, W., HAAK, V., ONCKEN, O. & TANNER, D. (eds) *Orogenic Processes: Quantification and Modelling in the Variscan Belt.* Geological Society, London, Special Publications, **179**, 199–216.

PIQUÉ, A. 2001. *Geology of Northwest Africa.* Borntraeger, Berlin.

POHLER, S. & SCHÖNLAUB, H. P. 2001. Shelf to basin transect of Lower Devonian carbonates of the Carnic Alps, Austria. *In*: JANSEN, U., KÖNIGSHOF, P., PLODOWSKI, G. & SCHINDLER, E. (eds) *Mid-Palaeozoic Bio- and Geodynamics—the North Gondwana/Laurussia Interaction.* Joint Meeting IGCP 421/SDS, Frankfurt/Main, Forschungsinstitut Senckenberg, Frankfurt am Main, 78–79.

SANDBERG, C. A. & ZIEGLER, W. 1996. Devonian conodont biochronology in geologic time calibration. *Senckenbergiana Lethaea*, **76**, 259–265.

SANDBERG, C. A., ZIEGLER, W. & BULTYNCK, P. 1989. New standard conodont zones and early Ancyrodella phylogeny across Middle–Upper Devonian boundary. *Courier Forschungs-Institut Senckenberg*, **110**, 195–230.

SANTANTONIO, M. 1994. Pelagic carbonate platforms in the geologic record: their classification, and sedimentary and paleotectonic evolution. *Bulletin*, **78**, 122–141.

SCHÖNLAUB, H.-P. 1971. Die fazielle Entwicklung im Altpaläozoikum der Karnischen Alpen. *Zeitschrift der Deutschen Geologischen Gesellschaft*, **122**, 97–111.

SCHÖNLAUB, H.-P. 1980. Carnic Alps. Field Trip A. *In*: SCHÖNLAUB, H.-P. (ed) *Second European Conodont Symposium ECOS II.* Abhandlungen der Geologischen Bundesanstalt von Österreich, **35**, 5–57.

SCHÖNLAUB, H.-P. 1985. Das Paläozoikum der Karnischen Alpen. *In*: SCHÖNLAUB, H.-P. (ed.) *Arbeitstagung der Geologischen Bundesanstalt 1985*. Geologische Bundesanstalt, Wien, 34–52.

SCHÖNLAUB, H.-P. 1992. Stratigraphy, biogeography and paleoclimatology of the Alpine Palaeozoic and its implications for plate movements. *Jahrbuchder Geologische Bundesanstalt, Wien*, **135**, 381–418.

SCHÖNLAUB, H.-P., HISTON, K. & POHLER, S. 2004*a*. The Paleozoic of the Carnic Alps. *In*: SCHÖNLAUB, H.-P. (ed.) *Field Trip Carnic Alps 2004*. Geologische Bundesanstalt, Wien, 2–32.

SCHÖNLAUB, H.-P., JOACHIMSKI, M. M. & HÜNEKE, H. 2004*b*. Stop 7—Wolayer 'Glacier' section. *In*: SCHÖNLAUB, H.-P. (ed.) *Field Trip Carnic Alps 2004*. Geologische Bundesanstalt, Wien, 59–68.

SCHWAB, M. 1979. Zum Deckenbau der Variszíden (Harz–Rheniden–Südwestural). *Zeitschrift für Geologische Wissenschaften*, **7**, 1131–1155.

SCOTESE, C. R. & BARRETT, S. F. 1990. Gondwana's movement over the South Pole during the Palaeozoic; evidence from lithological indicators of climate. *In*: MCKERROW, W. S. & SCOTESE, C. R. (eds) *Palaeozoic Palaeogeography and*

Biogeography. Geological Society, London, Memoirs, **12**, 75–85.

SCOTESE, C. R. & MCKERROW, W. S. 1990. Revised world maps and introduction. *In*: MCKERROW, W. S. & SCOTESE, C. R. (eds) *Palaeozoic Palaeogeography and Biogeography*. Geological Society, London, Memoirs, **12**, 1–21.

STOW, D. A. V. 1982. Bottom currents and contourites in the North Atlantic. *Bulletin de l'Institut de Géologie du Bassin d'Aquitaine*, **31**, 151–166.

STOW, D. A. V., FAUGÈRES, J. C., & GONTHIER, E. 1986. Facies distribution and drift growth during the late Quaternary, Faro Drift, Gulf of Cadiz. *Marine Geology*, **72**, 71–100.

STOW, D. A. V., READING, H. G. & COLLINSON, J. D. 1996. Deep seas. *In*: READING, H. G. (ed.) *Sedimentary Environments: Processes, Facies and Stratigraphy*. Blackwell, Oxford, 395–453.

STOW, D. A. V., FAUGÈRES, J. C., VIANA, A. R. & GONTHIER, E. 1998. Fossil contourites: a critical review. *Sedimentary Geology*, **115**, 3–31.

STOW, D. A. V., FAUGÈRES, J. C., HOWE, J. A., PUDSEY, C. J. & VIANA, A. R. 2002. Bottom currents, contourites and deep-sea sediment drifts: current state-of-the-art. *In*: STOW, D. A. V., PUDSEY, C. J., HOWE, J. A., FAUGÈRES, J. C. & VIANA, A. R. (eds) *Deep-water Contourite Systems: Modern Drifts and Ancient Series, Seismic and Sedimentary Characteristics*. Geological Society, London, Memoirs, **22**, 7–20.

TAIT, J. A., SCHÄTZ, M., BACHTADSE, V. & SCOFFEL, H. C. 2000. Palaeomagnetism and Palaeozoic palaeogeography of Gondwana and European terranes. *In*: FRANKE, W., HAAK, V., ONCKEN, O. & TANNER, D. (eds) *Orogenic Processes: Quantification and Modelling in the Variscan Belt*. Geological Society, London, Special Publications, **179**, 21–34.

TORSVIK, T. H., SMETHURST, M. A., MEERT, J. G., *ET AL.* 1996. Continental break-up and collision in the Neoproterozoic and Palaeozoic: a tale of Baltica and Laurentia. *Earth-Science Reviews*, **40**, 229–258.

TUCKER, R. D., BRADLEY, D. C., VERSTRAETEN, C. A., HARRIS, A. G., EBERT, J. R. & MCCUTCHEON, S. R. 1998. New U–Pb zircon ages and the duration and division of Devonian time. *Earth and Planetary Science Letters*, **158**, 175–186.

WACHENDORF, H. 1986. Der Harz—variszischer Bau und geodynamische Entwicklung. *Geologisches Jahrbuch, Hannover*, **A91**, 1–71.

WACHENDORF, H., BUCHHOLZ, P. & ZELLMER, H. 1995. Fakten zum Harz-Paläozoikum und ihre geodynamische Interpretation. *Nova Acta Leopoldina, NF 71*, **291**, 119–150.

WALLISER, O. H. 1999. The Jebel Mech Irdane section. *In*: EL HASSANI, A. & TAHIRI, A. (eds) *Tafilalt and Maider (Eastern Anti Atlas)*. Excursion Guidebook Part I, Joint Meeting IGCP 421/SDS, Errachidia, Rabat, Morocco. Institut Scientifique, Rabat, 65–71.

WALLISER, O. H. & ALBERTI, H. 1983. Flysch, olistostromes and nappes in the Harz Mountains. *In*: MARTIN, H. & EDER, F. W. (eds) *Intracontinental Fold Belts*. Springer, Berlin, 145–170.

WALLISER, O. H., BULTYNCK, P., WEDDIGE, K., BECKER, R. T. & HOUSE, M. R. 1995*a*. Definition of the Eifelian–Givetian stage boundary. *Episodes*, **18**, 107–115.

WALLISER, O. H., EL HASSANI, A. & TAHIRI, A. 1995*b*. Sur le Dévonien de la Meseta marocaine occidentale: comparaisons avec le Dévonien allemand et événements globaux. *Courier Forschungs-Institut Senckenberg*, **188**, 21–30.

WALLISER, O. H., EL HASSANI, A. & TAHIRI, A. 1999. Mrirt: a key area for the Variscan Meseta of Morocco. *In*: EL HASSANI, A. & TAHIRI, A. (eds) *The North-Western Moroccan Meseta*. Excursion Guidebook Part II, Joint Meeting IGCP 421/SDS, Errachidia, Rabat, Morocco. Institut Scientifique, Rabat, 110–131.

WEDDIGE, K. 2003. M. y. calibration by Weddige *ET AL.* 2002. *Senckenbergiana Lethaea*, **83**, 223, column 0 001dm03, 229, column 0 001ds03.

WEDDIGE, K. & ZIEGLER, W. 1996. Conodonten-Zonen, globale; aktueller Stand; Mitteldevon. *Senckenbergiana Lethaea*, **76**, 278, column B 030dm96.

WINCHESTER, J. A., PHARAOH, T. C. & VERNIERS, J. 2002. Palaeozoic amalgamation of Central Europe: an introduction and synthesis of new results from recent geological and geophysical investigations. *In*: WINCHESTER, J. A., PHARAOH, T. C. & VERNIERS, J. (eds) *Palaeozoic Amalgamation of Central Europe*. Geological Society, London, Special Publications, **201**, 1–18.

ZIEGLER, P. A. 1989. *Evolution of Laurussia: a Study in Late Palaeozoic Plate Tectonics*. Kluwer, Dordrecht.

ZIEGLER, W. 1966. Eine Verfeinerung der Conodontengliederung an der Grenze Mittel-/Oberdevon. *Fortschritte in der Geologie des Rheinlandes und von Westfalen*, **9**, 647–676.

ZIEGLER, W. 1996. Conodonten-Zonen, globale; aktueller Stand; Oberdevon. *Senckenbergiana Lethaea*, **76**, 282, column B 030ds96.

ZIEGLER, W. & KLAPPER, G. 1982. The *disparilis* conodont zone, the proposed level for the Middle–Upper Devonian boundary. *Courier Forschungs-Institut Senckenberg*, **55**, 463–491.

ZIEGLER, W. & SANDBERG, C. A. 1990. The Late Devonian standard conodont zonation. *Courier Forschungs-Institut Senckenberg*, **121**, 1–115.

Hydrodynamic modelling of bottom currents and sediment transport in the Canyon São Tomé (Brazil)

JOSÉ A. M. LIMA[1], OSMAR O. MOLLER, Jr[2], ADRIANO R. VIANA[3] & RAFAEL PIOVESAN[2]

[1]*Petróleo Brasileiro S.A., Petrobras, Research Center, CENPES, Rio de Janeiro, RJ, 21.941-598, Brazil (e-mail: jlima@petrobras.com.br)*

[2]*Departamento de Física, Fundação Universidade do Rio Grande—FURG, Av. Italia Km 8, Mail Box 474, Rio Grande, RS, 96201-900, Brazil*

[3]*Petróleo Brasileiro S.A., Petrobras, E&P-Exploration, 65, Chile Av., Rio de Janeiro, RJ, 20031-912, Brazil*

Abstract: This study presents the hydrodynamic modelling of ocean currents along the Canyon São Tomé, Campos Basin, Brazil, and their impact on sediment transport. The objective is to develop a tool to simulate the interaction between bottom currents and the submarine physiography, and to depict the relative importance of any individual current forcing mechanism as a sediment-reworking agent. This paper presents the evolution of along-channel currents over a tidal cycle and the simulation of a turbidity current. The resultant sediment transport under the combination of turbidity and oceanic currents is also simulated. This model is a step forward towards the understanding of the geometric and textural modifications imposed by bottom currents upon gravity-driven deposits, which is of great importance for the oil industry.

The main channels of active sediment transport from continental shelves to the deep ocean are continental slope canyons. Currents flowing along the floors of submarine canyons were investigated especially by Francis Shepard during the 1970s (Shepard & Marshall 1973, 1978; Shepard *et al.* 1974, 1979). More recent studies and numerical simulations include Petruncio *et al.* (1998, 2002). The Campos Basin, situated on the SE Brazilian shelf, has many canyons distributed over its entire region from latitude 21°S to 23°S. Deep-water deposits contain variable proportions of sand- and clay-based sediments, as a function of the available sources at the canyon head and the dynamic characteristics of the currents through the canyon. It is very important to understand both the hydrodynamics through the canyon that causes bottom currents and their contribution to sediment transport. Interaction between oceanic currents and turbidity current can also be simulated to better understand the relative impact of each process in developing deep-water coarse-grained accumulations, which are of fundamental importance as hydrocarbon reservoirs. Many researchers, including Stow *et al.* (1998), Viana (1998), Viana & Faugères (1998), Viana *et al.* (1998) and Kidd (1999), have studied the interaction of bottom currents and sediment deposits.

The hydrodynamics of the SE Brazilian shelf and continental slope is very complex. At the low-frequency end of the spectrum, it is under the influence of distinct western boundary currents that alternate in flow direction at different depth levels. At the high-frequency end, it is affected by tidal currents that can drive along-channel currents in the canyons. Some workers, such as Evans *et al.* (1983), Campos *et al.* (1995), Lima (1997) and Castro & Miranda (1998), have provided an overview of the oceanic circulation in the region. It is also important to evaluate the possible effects of density-driven currents, or turbidity currents, on the along-canyon sediment transport.

This paper presents the application of a 3D hydrodynamic model to Canyon São Tomé to evaluate bottom currents and their effect on sediment transport. The following sections describe the numerical model implementation, results associated with tidal-driven currents, the simulation of turbidity currents and associated sediment transport.

The study region and hydrodynamic model

The study site is situated on the SE Brazilian shelf, which has a fairly complex bathymetry. Figure 1 presents a bathymetric map of the region with a box showing the location of the Campos Basin. The green colour indicates the coastline, and it can be seen that the general orientation of the coastline in the region changes from NE–SW to

From: VIANA, A. R. & REBESCO, M. (eds) *Economic and Palaeoceanographic Significance of Contourite Deposits.* Geological Society, London, Special Publications, **276**, 329–342.

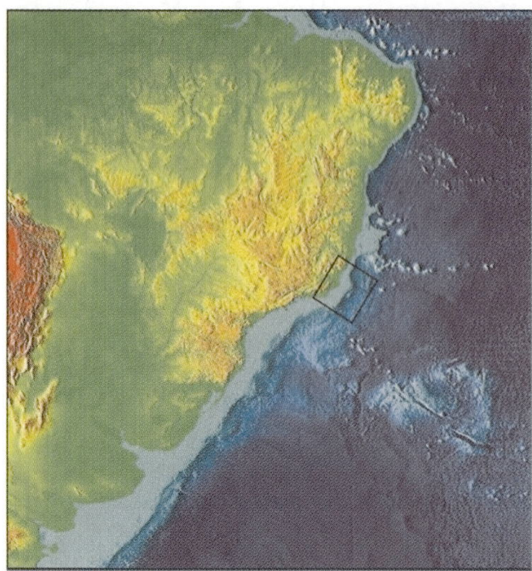

Fig. 1. Bathymetric map of the east and SE South America Atlantic margin. Continuous outline indicates the Campos Basin region. Topography and bathymetry data from NOAA.

east–west. The Campos Basin is situated on the NE–SW continental margin and its continental slope has many canyons, as shown in Figure 2, where Canyon São Tomé is indicated by the black outline. Figure 3 shows the detailed bathymetry of Canyon São Tomé. It can be seen that many sediment deposits at the canyon's mouth have been reworked by the action of tidal and low-frequency bottom currents. A 20 km scale bar indicates the dimensions of the canyon.

The Princeton Ocean Model (POM) was used to study the hydrodynamics of the Canyon São Tomé. It is a fully 3D, primitive equation model described by Blumberg & Mellor (1987). It has an in-built turbulence closure model, as proposed by Mellor & Yamada (1974), that is able to simulate the complex flows of turbulent boundary layers such as the oceanic mixed layer and bottom boundary layer. This is particularly important to study bottom currents, which are our main concern in this study.

The POM grid used here has 109×109 horizontal grid points with 16 vertical sigma levels. A telescopic grid was designed in such a way that the canyon bathymetry was detailed in the centre with a horizontal grid spacing of 400 m, and the grid spacing gradually increased to 20 km at the model boundaries. Thus, undesired boundary effects in

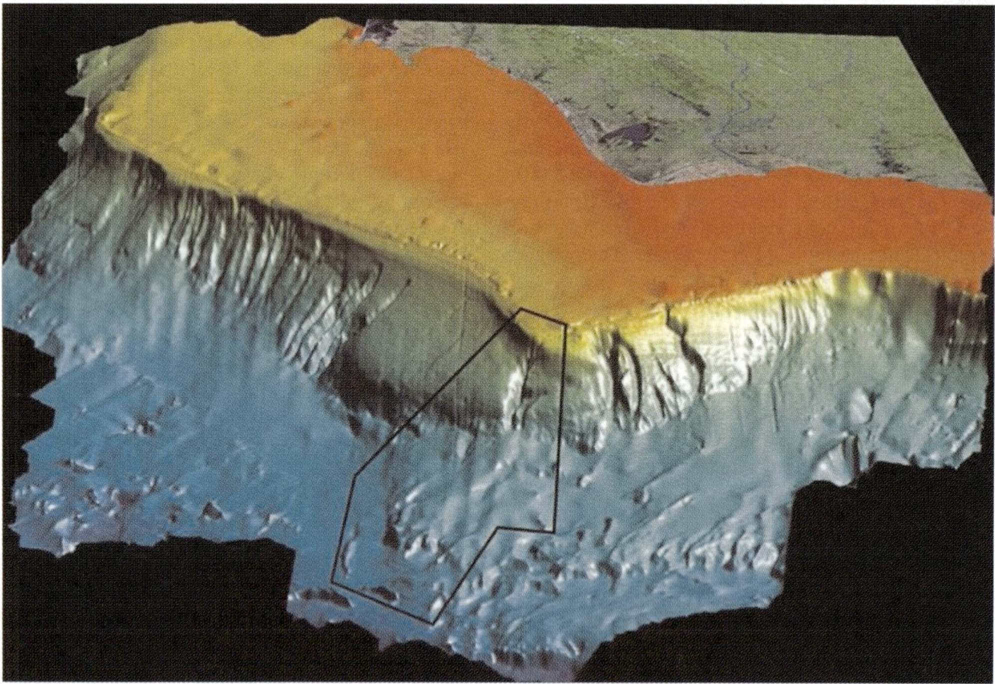

Fig. 2. A 3D view of the SE Brazilian margin from Vitoria in the north to Cabo Frio in the south. Many canyons can be observed in the Campos Basin. Canyon São Tomé is shown by the continuous outline. Bathymetric database from Petrobras. Landsat TM satellite image from NASA databank.

Fig. 3. Canyon São Tomé, Campos Basin, Brazil. The arrows indicate complex sediment deposits distributed at the mouth of the canyon. These deposits are supposed to have been partially reworked by the action of tidal and low-frequency currents. Sea-floor physiography from Petrobras 3D seismic data and sidescan sonar.

the centre of model were avoided. Figure 4 shows the telescopic grid and Figure 5 shows a close-up view of the canyon bathymetry.

Experiments with tidal-driven currents

Tidal-driven currents are important hydrodynamic mechanisms to drive sediment transport in oceanic canyons. The POM was implemented with open boundary conditions to simulate tidal forcing along the Canyon São Tomé. The semi-diurnal tidal components are predominant in the region, and the M_2 and S_2 components were selected to be implemented at the model boundary points with amplitudes and phases provided by the 'Finite Element Solution (FES95)' of Le Provost *et al.* (1994).

Another problem when implementing oceanic models with open boundaries is the choice of an appropriate set of conditions to be applied at the boundaries. Many different conditions were tested for the Canyon São Tomé model, and the best results were achieved with partially clamped conditions for the elevation, radiative condition with constant phase speed for external velocities, radiative conditions with an Orlanski scheme for internal velocities, and relaxing conditions for temperature and salinity. These boundary conditions have been described by Palma & Matano (1998, 2000).

The model runs to simulate tidal motion covered periods of 45 days to avoid the initial oscillations caused by model initialization and inertial currents. The local inertial period is 31 h. Thus, the results for the first 10 days were discarded and the model results were analysed from the tenth day onward.

Figure 6 presents the results for tidal elevations at three grid points: the first is on the continental shelf, the second is the middle of the canyon and the third is at the canyon's mouth. The semi-diurnal tidal behaviour is very clear for the 19 days that the model simulated.

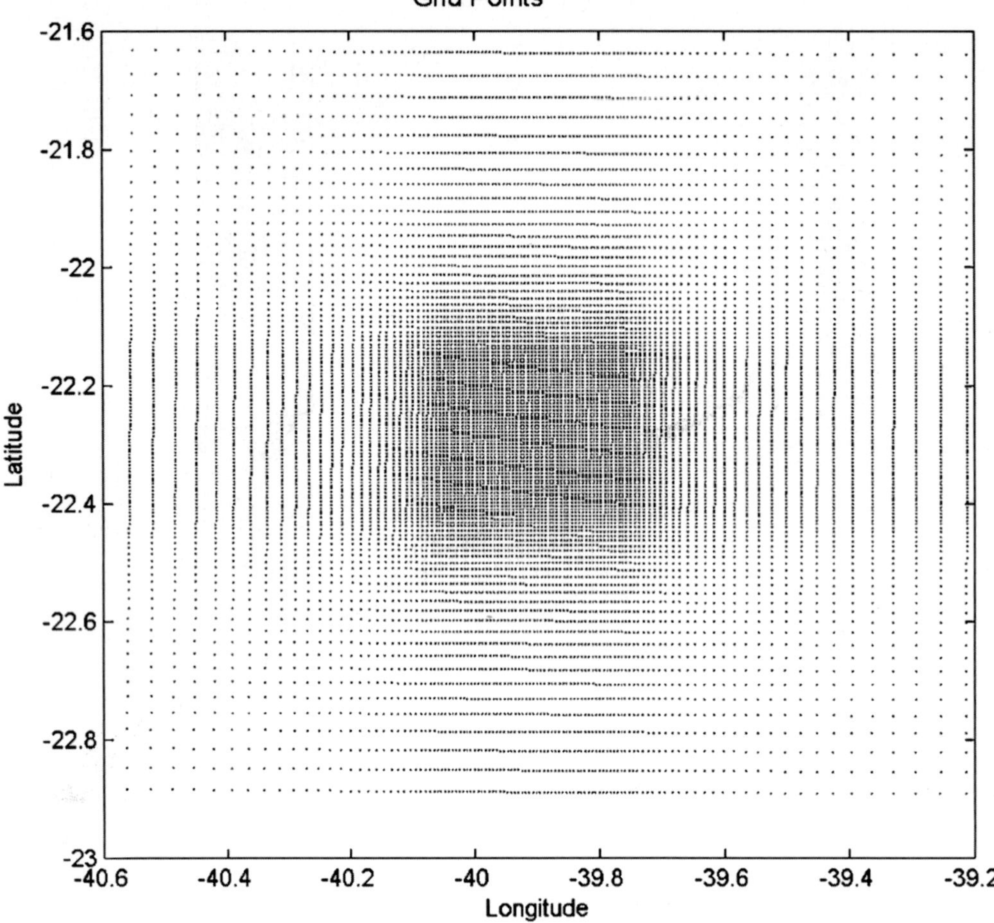

Fig. 4. The telescopic grid used to implement the hydrodynamic model, with 109 × 109 grid points.

Figure 7 presents the distribution of mean kinetic energy in the grid for a complete tidal cycle calculated with bottom currents (5 m above the bottom). An intensification of kinetic energy along the head and the main channel can be observed. This is an indication of active zones of sediment transport. Figures 8–10 present the contours of along-channel U-velocity components.

The sediment transport model

A sediment transport model was implemented based on the formulation for suspended and bed-load transport proposed by Van Rijn (1993). The model was carefully adapted to estimate the transport based on the bottom currents provided by the hydrodynamic model described above. The component of sediment transport as a result of gravity wave motion, which is present in the original work by Van Rijn, was not considered.

For each grid point, the total sediment transport is the sum of the transport of suspended sediments q_s plus the bed-load transport q_b:

$$\boldsymbol{q} = \boldsymbol{q}_s + \boldsymbol{q}_b.$$

The suspended sediment transport is calculated by the integration of the bottom boundary layer velocity v_R multiplied by the sediment concentration c at vertical increments dz. Within the bottom boundary layer, an adaptable number of layers is used to calculate the suspended transport. The suspended load is given by

$$q_s = \int_a^h v_R c \, dz.$$

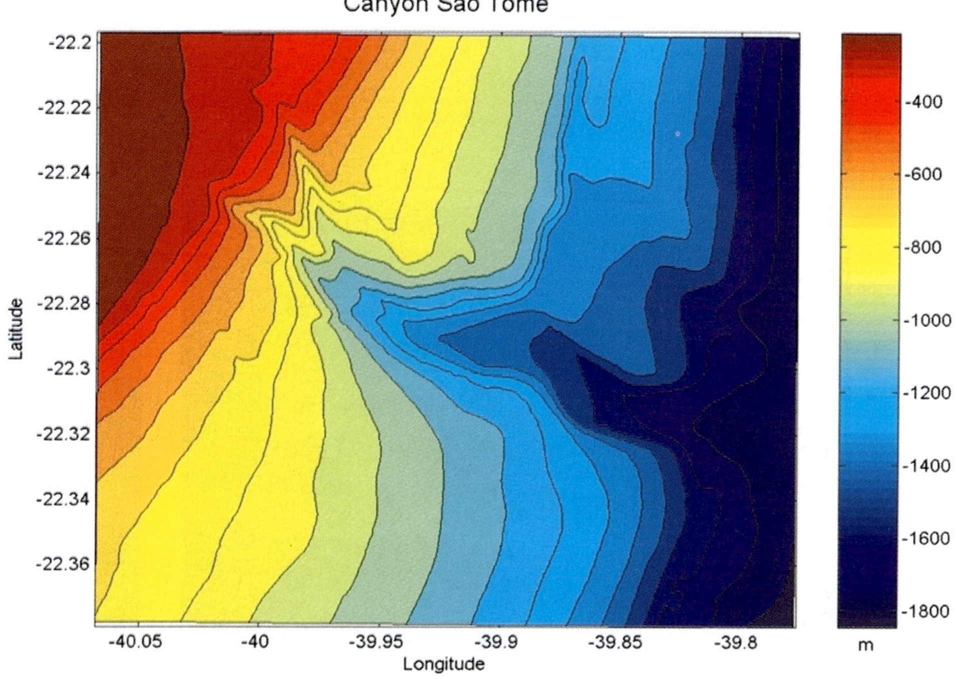

Fig. 5. Detailed view of the Canyon São Tomé bathymetry as projected at the centre of the model grid.

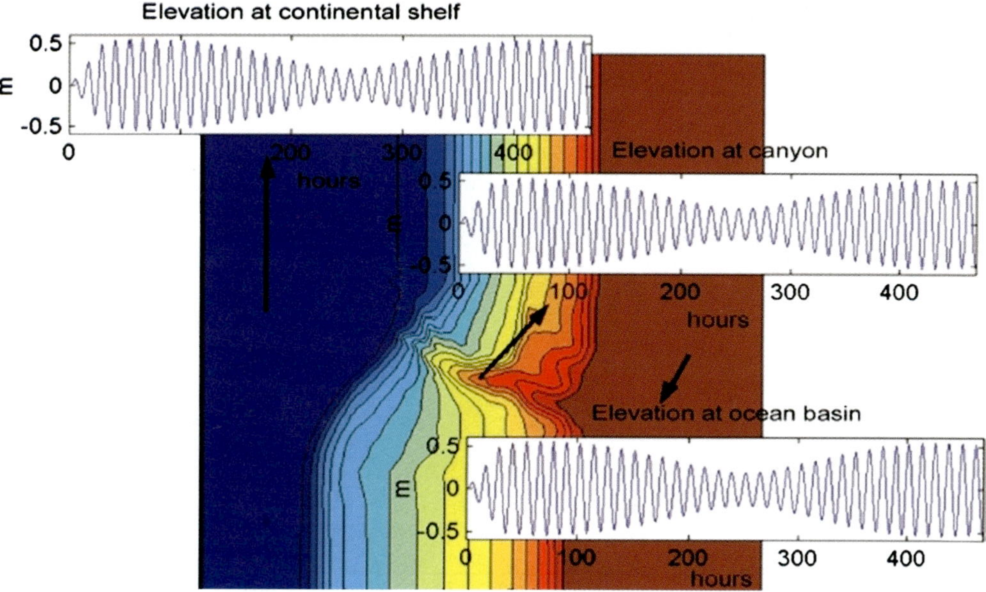

Fig. 6. Results of tidal elevations at different grid points of the model.

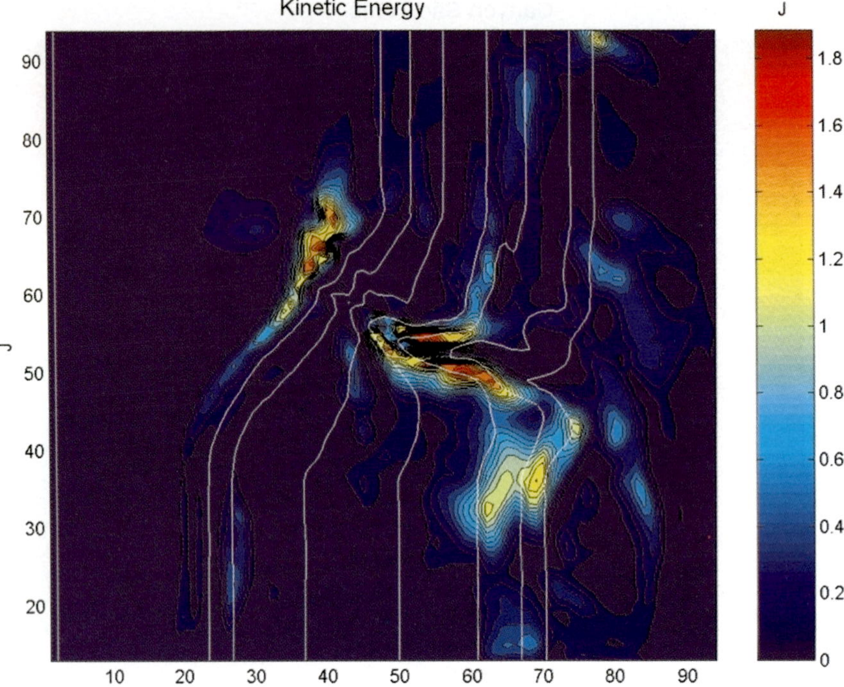

Fig. 7. Mean kinetic energy calculated with bottom currents over an entire tidal cycle. Red indicates higher values of kinetic energy.

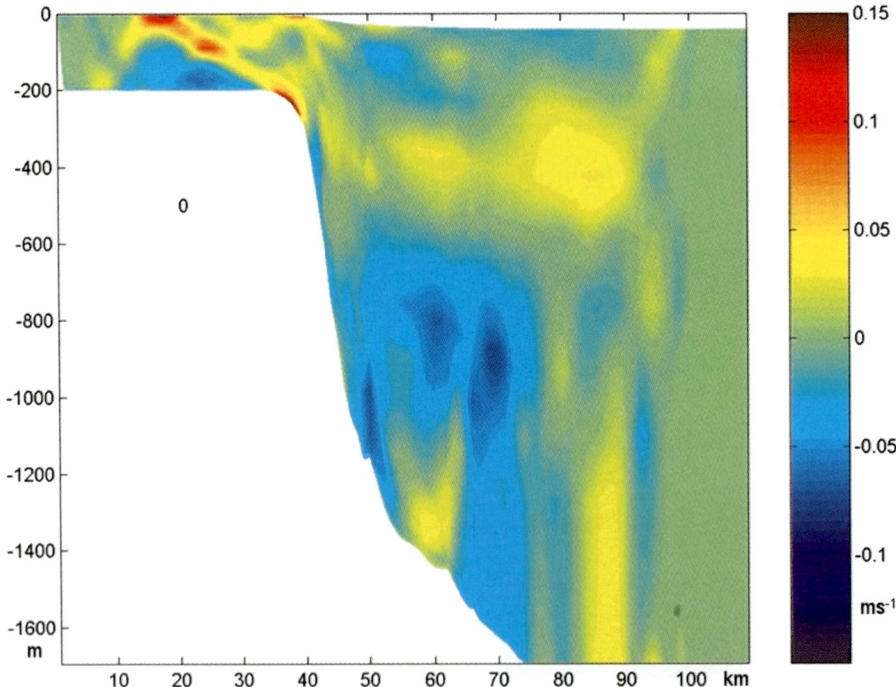

Fig. 8. The along-channel U-velocity component at the beginning of a semi-diurnal tidal cycle (0 h).

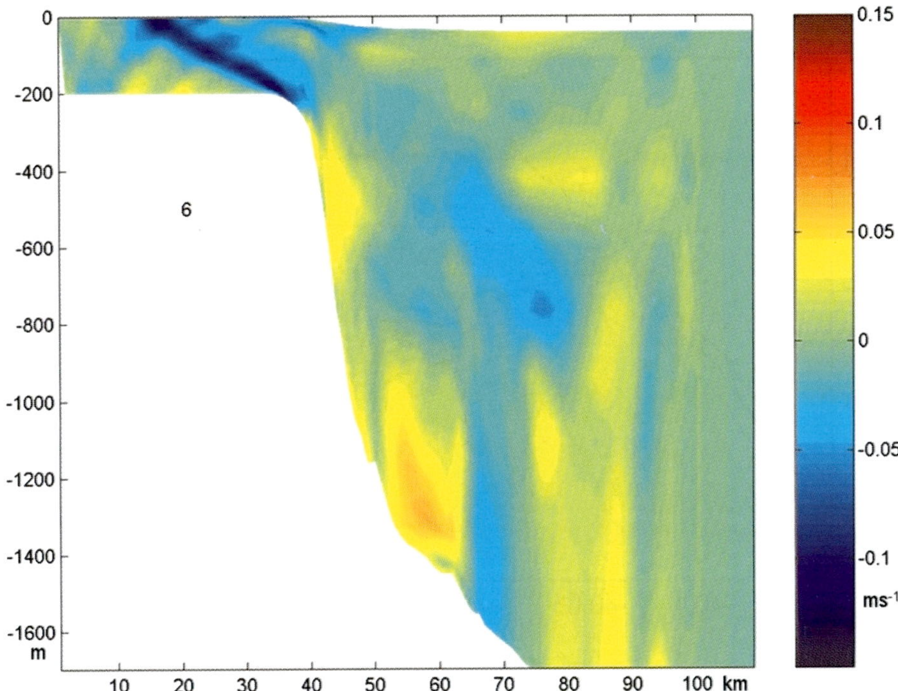

Fig. 9. The along-channel U-velocity component at the half-interval of a semi-diurnal tidal cycle (6 h).

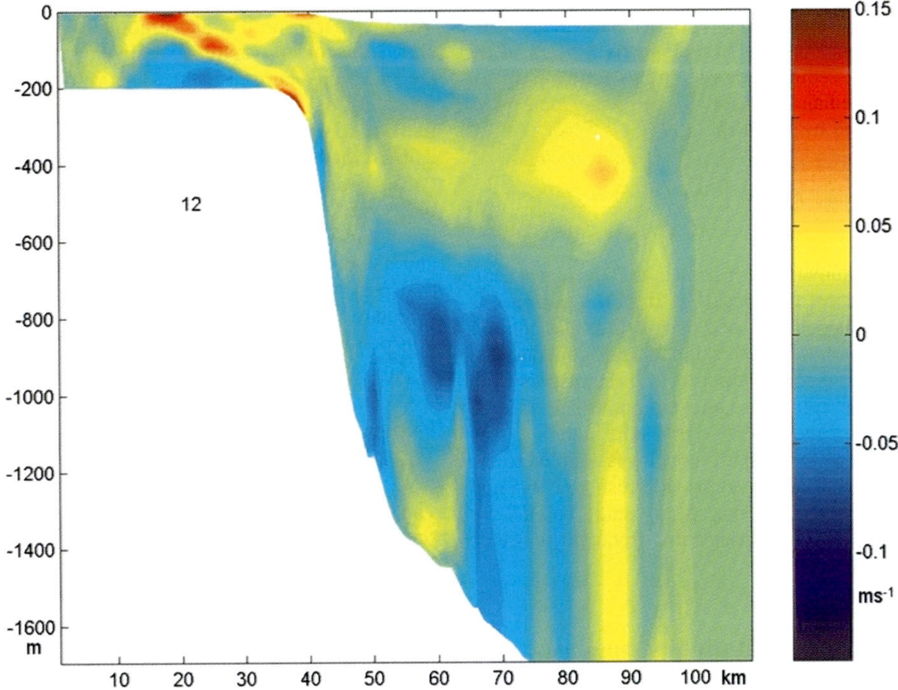

Fig. 10. The along-channel U-velocity component at the end of a semi-diurnal tidal cycle (12 h).

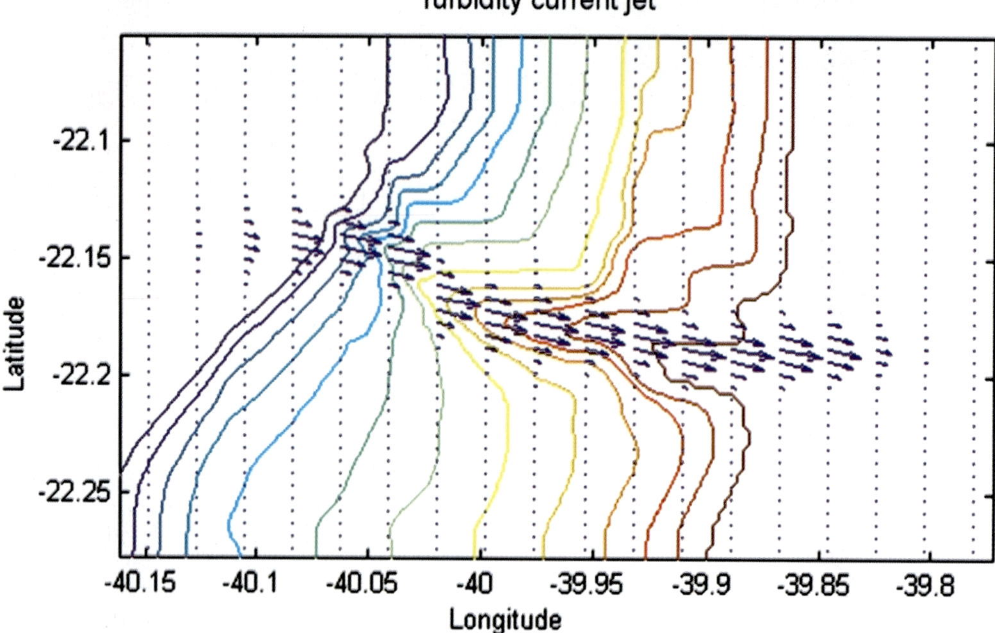

Fig. 11. Simulation of an intense flow along the main axis of the Canyon São Tomé. Flows like these may be associated with turbidity currents.

The bed-load transport is calculated using a formulation that takes into account many parameters, such as the instantaneous bottom shear stress $\tau_{b,cr}$, sediment density ρ_s, particle parameter D_*, the median diameter d_{50} and water density ρ, as described by Van Rijn (1993):

$$q_b = 0.25\gamma\rho_s d_{50} D_*^{-0.3}\left[\frac{\tau'_{b,cw}}{\rho}\right]^{0.5}\left[\frac{\tau'_{b,cw}-\tau_{b,cr}}{\tau_{b,cr}}\right]^{1.5}.$$

For the numerical experiment that evaluates sediment transport as a result of turbidity currents, a transport model was used that permits the computation of both the local sediment transport rate and the advective contribution from the transport from an adjacent cell. The Quickest numerical scheme was used for the advective transport model.

The sediment transport model was coupled to the hydrodynamic model to provide results associated with the bottom currents. Many study cases were simulated for tidal currents, low-frequency currents and turbidity currents.

Experiments with turbidity currents

For the particular case of turbidity currents, an along-channel jet with enough strength to cause a significant amount of suspended and bed-load sediment transport was simulated. Figure 11 shows the current vectors associated with an idealized turbidity current event. The velocity field is affected from the canyon head at the 200 m bathymetric contour to the 1500 m contour at the canyon mouth.

The sediment transport associated with this turbidity current event was calculated using the sediment transport model described above. Figures 12 and 13 show the results of the integrated sediment transport for 18 h, and shows the sand deposits originally situated at the head of the canyon being transported downwards to the canyon mouth.

Experiments with sediment transport as a result of western boundary currents

The SE Brazilian continental slope is affected by several western boundary currents. The Brazil Current jet transports hot and saline water towards the south on the surface. An intermediate countercurrent transports Antarctic Intermediate Water (AAIW) to the north. Below it, another western boundary current transports North Atlantic Deep Water (NADW) towards the south. Thus, the sediment deposits are subjected to several distinct

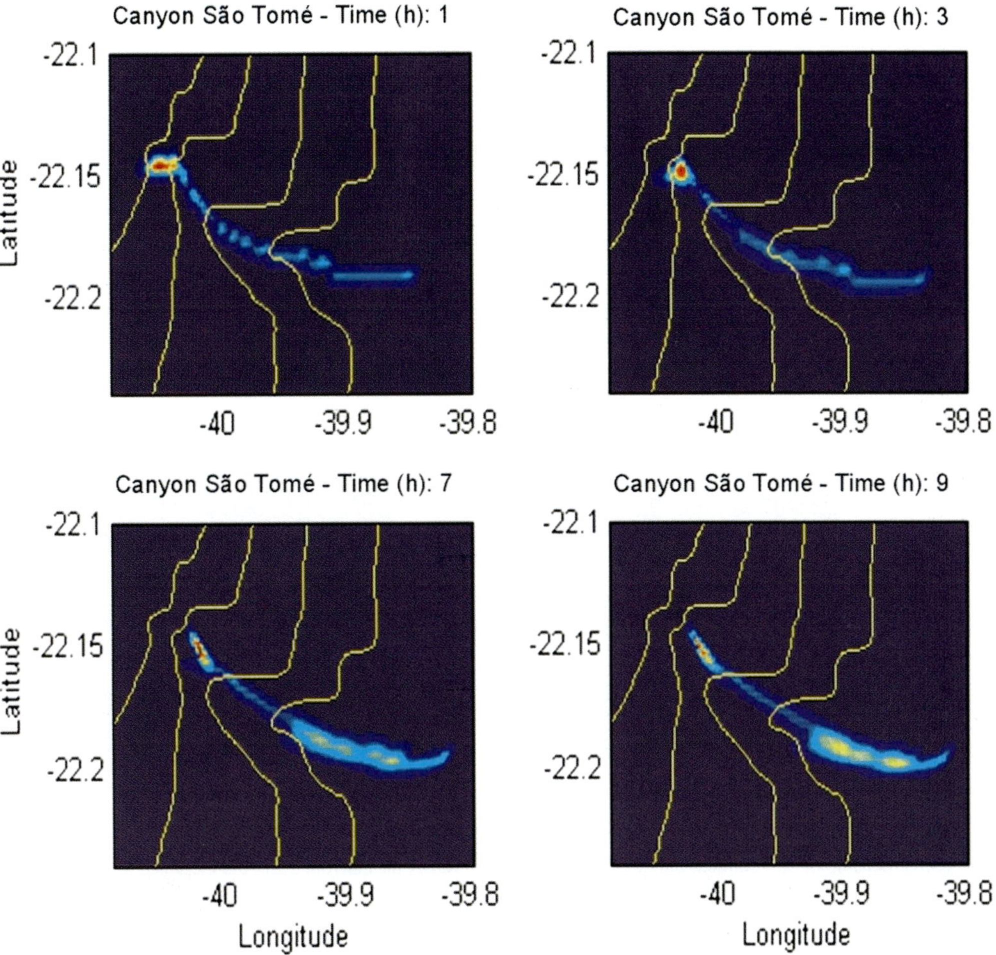

Fig. 12. Sediment transport in Canyon São Tomé as caused by intense along-channel flows. The bright colours indicate the sediment plume.

currents that are responsible for an active sediment transport. The POM was used to investigate the action of western boundary current jets on the sediment deposits, particularly on sand deposits carried down the canyons by the action of turbidity currents, as simulated above.

The open boundary conditions of the regional model were carefully selected to establish an oceanic flow that would reproduce the western boundary currents. The model used no gradient condition for the surface elevation, Flather radiative conditions for the external velocities, Orlanski radiative conditions for the internal velocities, and relaxing conditions for temperature and salinity. These boundary conditions have been described by Palma & Matano (1998, 2000).

The model was initialized using temperature and salinity fields measured by conductivity–depth–temperature (CDT) profiler transects in the Campos Basin. The measurements extended to water depths greater than 3000 m. Thus, it was possible to investigate the structure of all three jets listed above. Figures 14 and 15 show temperature and salinity cross-sections measured in a west–east direction at 22°S, in the northern Campos Basin. The data presented have been interpolated in the sigma-coordinate grid of the hydrodynamic model. The model was initialized in diagnostic mode to hold the thermohaline structure fixed, and the velocity field was gradually built up in geostrophic balance with the prescribed density structure. Figure 16 shows a cross-section of the

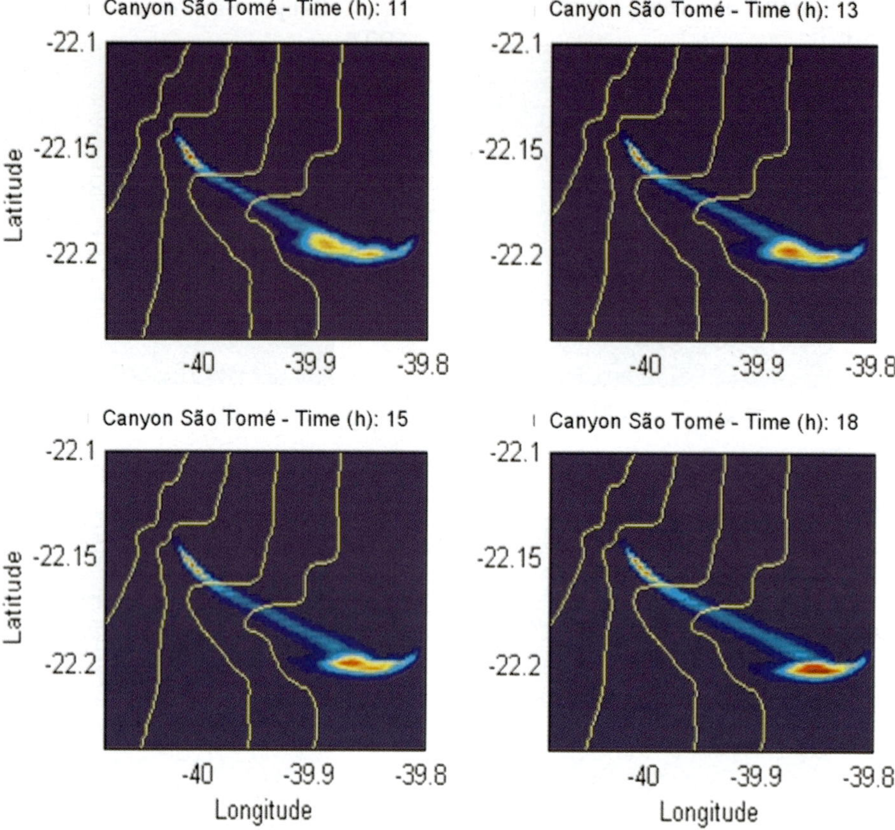

Fig. 13. Simulation of the sediment plume as it reaches the mouth of the canyon, using the modelled hydrodynamic currents.

velocity field, with the Brazil Current (blue), AAIW jet (red) and NADW jet (green) clearly established.

The residual sediment transport on the lower continental slope was estimated using the same sediment transport model applied to the turbidity current experiment. Figure 17 presents the hotspots with the highest rates of sediment transport in water depths from 1500 to 2000 m in the Campos Basin. It can be seen that the hotspots occur close to inflection points of the isobaths at 22.2°S and 23.4°S. The NADW jet is constrained in these regions by the bottom topography and accelerates, imposing a higher shear stress on the sediments, which will be responsible for a higher bed-load transport rate.

Figure 17 also explains the southern residual sediment transport acting on the sand deposits that arrive at the mouth of Canyon São Tomé, as observed in Figure 3. The net sediment transport has a southerly direction in this portion of the continental slope (from 1600 to 2000 m) because of the action of the NADW jet, which flows towards the south.

Conclusions

This paper presents the development of a coupled hydrodynamic–sediment transport model that can be used as a diagnostic tool by exploratory geologists and geotechnical engineers. The model is able to simulate complex bottom-flow conditions associated with canyons with complex bathymetry, such as the Canyon São Tomé, situated in the Campos Basin, Brazil, and deep western boundary currents.

The hydrodynamic model was based on the Princeton Ocean Model, as described by Blumberg & Mellor (1987), and the sediment transport was based on the formulation of Van Rijn (1993). It is important to point out that the sediment is considered as a passive tracer in the model presented in this paper. In future developments, the sediment plume can be also used as an active component of the turbulent boundary condition that drives the hydrodynamic model, because it affects the density of the surrounding water.

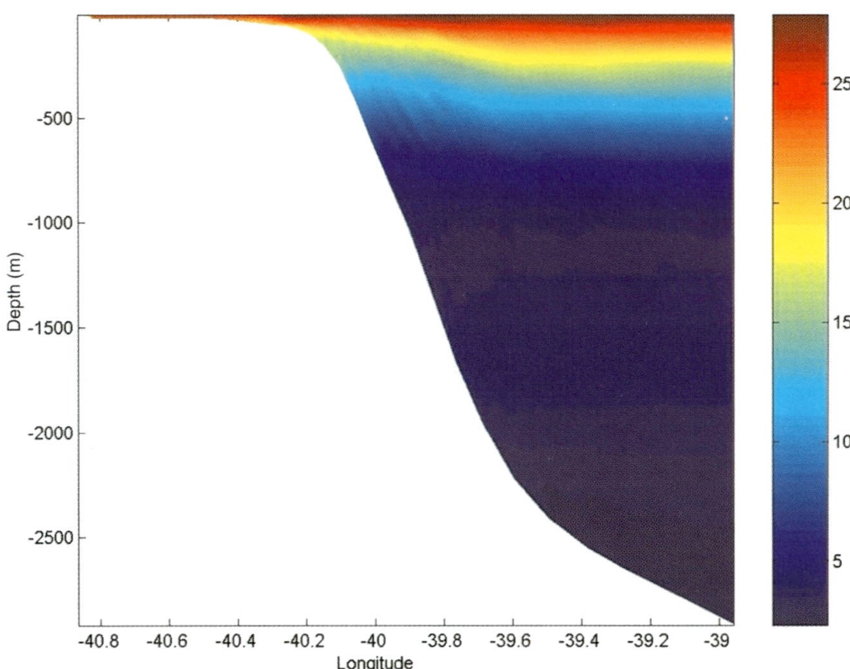

Fig. 14. Measured temperature cross-section used to initialize the hydrodynamic model.

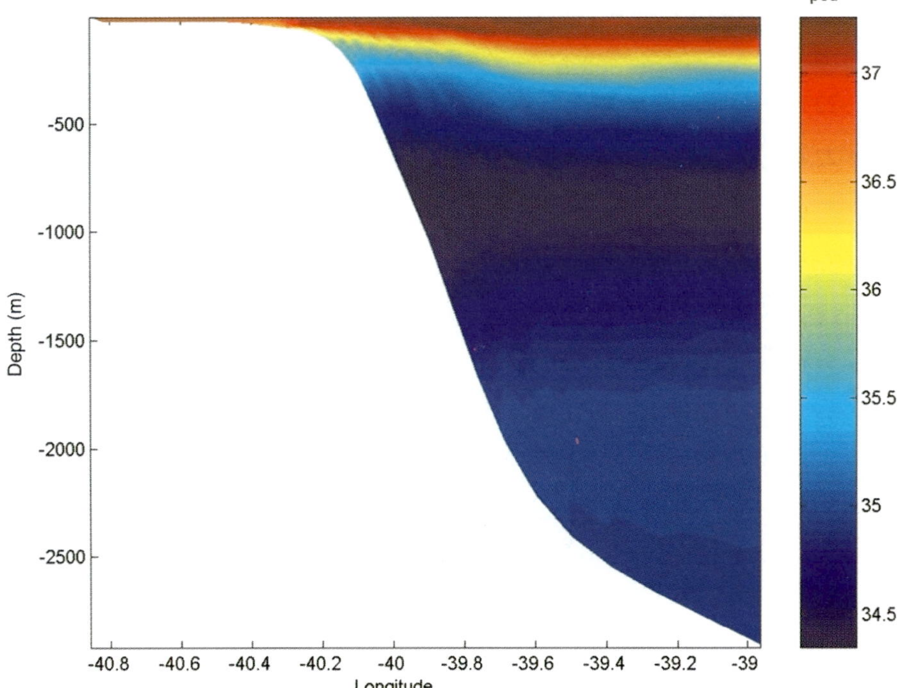

Fig. 15. Measured salinity section used to initialize the hydrodynamic model.

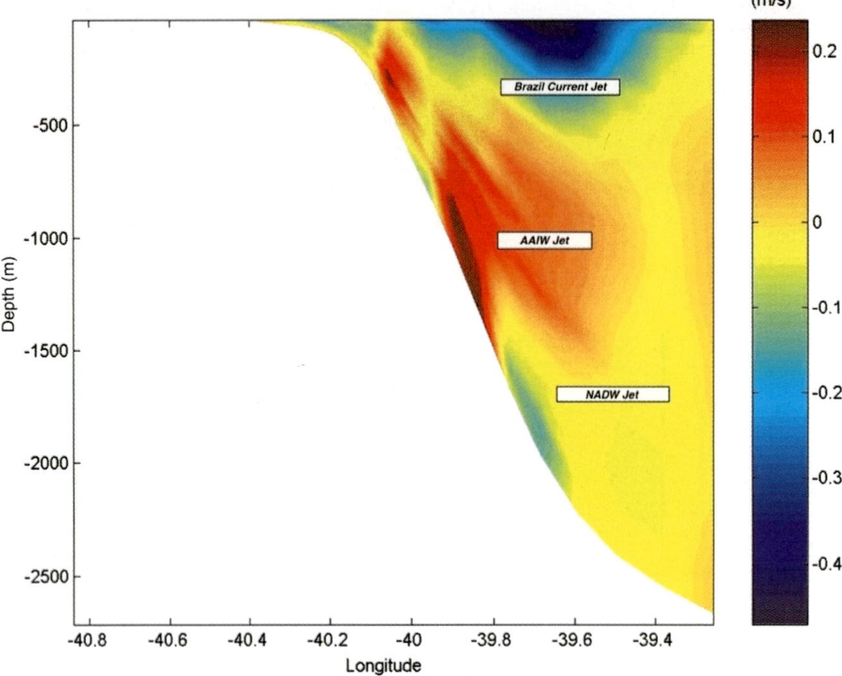

Fig. 16. Cross-section of modelled along-shelf velocity field in geostrophic balance with the density field for the prescribed initial temperature and salinity conditions.

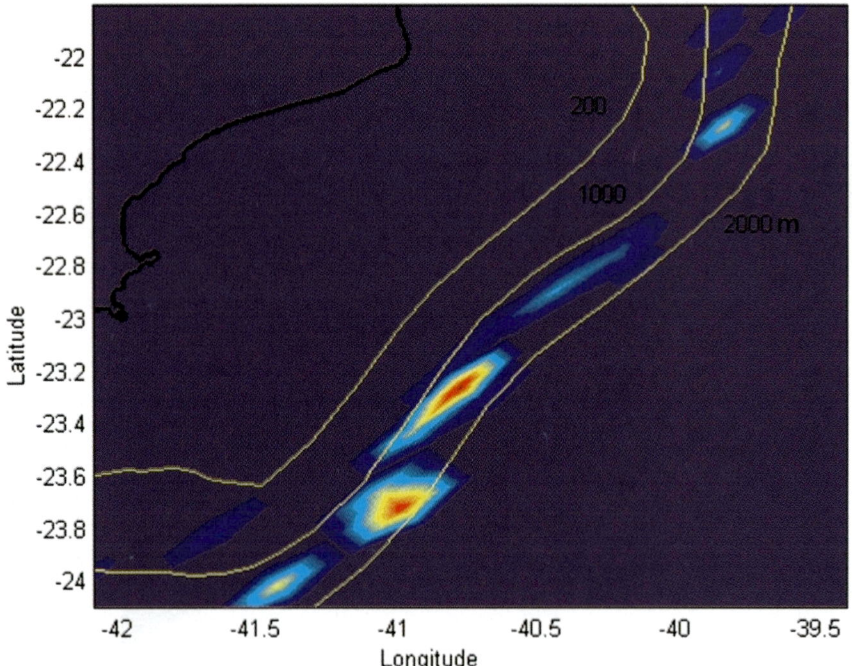

Fig. 17. Hotspots with highest residual sediment transport rates at water depths from 1500 to 2000 m in the Campos Basin.

The results show that the hydrodynamic model was able to simulate the tidal currents in Canyon São Tomé, including indication of hotspot zones where the kinetic energy is concentrated along the canyon's main channel. The model also simulated the current pattern associated with turbidity currents and their sediment transport plume.

Experiments were carried out with the model set to reproduce the western boundary currents in the Campos Basin. The results showed hotspots of possible high rates of residual sediment transport in the portion of the continental slope at water depths of 1600–2000 m, where the Canyon São Tomé mouth is located. The action of the south-flowing NADW jet seems to explain the trajectory of the residual transport acting on sediment deposits carried downwards along the channel. As this work is concerned with the fate of sediments discharged from the Canyon São Tomé mouth, this was the only region of the continental slope that was investigated.

Further developments should include the dynamic coupling of sediment transport and hydrodynamics, and the effects of low-frequency western boundary currents along a complete Campos Basin cross-section from the shelf break to the lower continental slope.

The authors thank Petrobras for permission to publish this paper. Suggestions from I. Soares and an anonymous reviewer greatly helped to improve the manuscript.

References

BLUMBERG, A. F. & MELLOR, G. L. 1987. A description of a three-dimensional coastal ocean circulation model. *In*: HEAPS, N. (ed.) *Three-Dimensional Coastal Ocean Models*. American Geophysical Union, 1–16.

CAMPOS, E. J. D., GONÇALVES, J. E. & IKEDA, Y. 1995. Water mass characteristics and geostrophic circulation in South Brazil Bight: summer of 1991. *Journal of Geophysical Research*, **100**(C9), 18537–18550.

CASTRO, B. M & MIRANDA, L. B. 1998. Physical oceanography of the western Atlantic continental shelf located between 4°N and 34°S. *In*: ROBINSON, A. R. & BRINK, K. H. (eds) *The Sea, Volume 11*. Wiley, New York, 209–251.

EVANS, D. L., SIGNORINI, S. R. & MIRANDA, L. B. 1983. A note on the transport of the Brazil Current. *Journal of Physical Oceanography*, **13**(9), 1732–1738.

KIDD, G. D. 1999. Fundamentals of 3-D seismic volume visualization. *Leading Edge*, **18**(6), 702–712.

LE PROVOST, C., GENCO, M. L., LYARD, F., VINCENT, P. & CANCEIL, P. 1994. Spectroscopy of the world ocean tides from a finite element hydrodynamic model. *Journal of Geophysical Research*, **99**(C12), 24777–24797.

LIMA, J. A. M. 1997. *Oceanic circulation on the Brazilian shelf break and continental slope at 22°S*. PhD dissertation, University of New South Wales, Sydney.

MELLOR, G. L. & YAMADA, T. 1974. A hierarchy of turbulence closure models for planetary boundary layers. *Journal of Atmospheric Research*, **31**, 1791–1806.

PALMA, E. D. & MATANO, R. P. 1998. On the implementation of passive open boundary conditions for a general circulation model: the barotropic mode. *Journal of Geophysical Research*, **103**, 1319–1341.

PALMA, E. D. & MATANO, R. P. 2000. On the implementation of passive open boundary conditions for a general circulation model: the three-dimensional case. *Journal of Geophysical Research*, **105**, 8605–8627.

PETRUNCIO, E. T., ROSENFELD, L. K. & PADUAN, J. D. 1998. Observation of the internal tide in Monterey Canyon. *Journal of Physical Oceanography*, **28**, 1873–1903.

PETRUNCIO, E. T., PADUAN, J. D. & ROSENFELD, L. K. 2002. Numerical simulations of the internal tide in a submarine canyon. *Ocean Modelling*, **4**, 221–248.

SHEPARD, F. P. & MARSHALL, N. F. 1973. Currents along the floors of submarine canyons. *AAPG Bulletin*, **57**(2), 244–264.

SHEPARD, F. P. & MARSHALL, N. F. 1978. Currents in submarine canyons and other sea valleys. *In*: STANLEY, D. J. & KELLING, G. (eds) *Sedimentation in Submarine Canyons, Fans and Trenches*. Dowden, Hutchinson & Ross, Stroudsburg, PA, 3–14.

SHEPARD, F. P., MARSHALL, N. F. & MCLOUGHLIN, P. A. 1974. Currents in submarine canyons. *Deep-Sea Research*, **21**, 691–706.

SHEPARD, F. P., MARSHALL, N. F., MCLOUGHLIN, P. A. & SULLIVAN, G. G. (eds) 1979. *Currents in Submarine Canyons and other Sea Valleys*. AAPG Studies in Geology, **8**.

STOW, D. A. V., FAUGÈRES, J.-C., VIANA, A. R. & GONTHIER, E. 1998. Fossil contourites: a critical review. *Sedimentary Geology*, **115**(1–4), 3–32.

VAN RIJN, L. C. 1993. *Principles of Sediment Transport in Rivers, Estuaries and Coastal Seas*. Acqua Publications, Amsterdam.

VIANA, A. R. 1998. *Le rôle et l'enregistrement des courants océaniques dans les dépôts de marges continentales: la marge du bassin Sud-Est brésilien*. PhD thesis, Université de Bordeaux 1.

VIANA, A. R. & FAUGÈRES, J.-C. 1998. Upper slope sand deposits: the example of Campos Basin, a latest Pleistocene/Holocene record of the interaction between along-slope and downslope currents. *In*: STOKER, M. S., EVANS, D. & CRAMP, A. (eds) *Geological Processes on Continental Margins: Sedimentation, Mass-Wasting and Stability*. Geological Society, London, Special Publications, **29**, 287–316.

VIANA, A. R., FAUGÈRES, J.-C. & STOW, D. A. V. 1998. Bottom current controlled sand deposits—a review from shallow to deep water environments. *Sedimentary Geology*, **115**(1–4), 53–80.

Index

Page numbers in *italic* denote figures. Page numbers in **bold** denote table.